THE NATURAL HISTORY OF POLLINATION

THE NATURAL HISTORY
OF POLLINATION

Michael Proctor, Peter Yeo and Andrew Lack

Timber Press
Portland, Oregon

First published 1996

© Michael Proctor, Peter Yeo and Andrew Lack 1996

The authors assert their moral rights to be identified as the authors of this work
ISBN 0-88192-352-4 (Hardback)
ISBN 0-88192-353-2 (Paperback)

Designed and typeset by British Wildlife Publishing, Rotherwick, Hampshire
Printed and bound in Great Britain by The Bath Press

Timber Press, Inc.
The Haseltine Building
133 S.W. Second Avenue, Suite 450
Portland, Oregon 97204, U.S.A.
1-800-327-5680 (U.S.A. & Canada only)

Contents

List of Plates

Plate 1 Flowers and guide-mark patterns
a Alpine clover, *Trifolium alpinum*
b A sun-rose, *Halimium lasianthum*
c Half-section of flower of common dog violet, *Viola riviniana*
d *Androsace sarmentosa*: colour change as the flower ages
e Bittersweet, *Solanum dulcamara*, a buzz-pollinated flower
f Bastard-balm, *Melittis melissophyllum*
g A lousewort, *Pedicularis oederi*
h Half-section of flower of foxglove, *Digitalis purpurea*

Plate 2 Flower-visiting insects
a Sawfly, *Tenthredo* sp., on marsh marigold, *Caltha palustris*
b Solitary bee, *Colletes daviesanus*, on *Anthemis tinctoria*
c Solitary bee, *Andrena haemorrhoa* on pear blossom
d Solitary bee, *Anthophora plumipes*, on comfrey, *Symphytum* sp.
e Bumblebee, *Bombus pratorum*, on knapweed, *Centaurea nigra*
f Beefly, *Bombylius major*, visiting primrose, *Primula vulgaris*

Plate 3 Flower-visiting insects
a Hoverfly, *Rhingia campestris*, on red campion, *Silene dioica*
b Common blue butterfly, *Polyommatus icarus*, on fleabane,
       Pulicaria dysenterica
c Red admiral butterfly, *Vanessa atalanta*, on hemp agrimony,
       Eupatorium cannabinum
d Lulworth skipper butterfly, *Thymelicus acteon*, sucking nectar of
       red valerian, *Centranthus ruber*
e Broad-bordered bee hawkmoth, *Hemaris fuciformis*, hovering at
       field scabious, *Knautia* sp.

Plate 4 Wind pollination
a Scots pine, *Pinus sylvestris*: male cones shedding pollen
b Scots pine, young female cone at receptive stage
c Norway spruce, *Picea abies*: female cones at receptive stage
d The yellow male catkins and the red female flowers of hazel, *Corylus avellena*
e A spikelet of couch grass, *Elymus repens*, showing the large freely-exposed stamens
       and the feathery stigmas

Plate 5 Fly-pollinated flowers
a *Aristolochia elegans* (Aristolochiaceae), a liana, with a U-shaped tube and large entrance funnel
b *A. cretica*, cut open to show entrance to prison guarded like a lobster pot, receptive stigmas and undehisced anthers
c *Stapelia hirsuta* (Asclepiadaceae): a carrion-like flower borne near the ground, wit blowfly eggs
d *Arisaema propinqua* (Araceae), W. Himalayas: the translucent veins of the spath show up brightly behind the entrance
e 'Mouse plant', *Arisarum proboscideum*, (Araceae), with half of the spathe removed showing spadix appendage and flowers
f Umbels of hog-weed, *Heracleum sphondylium* (Apiaceae), with flies and an ichneumon wasp

Plate 6 Sexually-deceptive orchids
a Male solitary wasp, *Argogorytes mystaceus*, on fly orchid, *Ophrys insectifera*
b Mirror orchid, *Ophrys speculum*
c 'Yellow bee orchid', *Ophrys lutea*
d Self pollination of bee orchid, *Ophrys apifera*
e Ichneumon wasp, *Lissopimpla excelsa*, visiting Australian tongue-orchid, *Cryptostylis leptochila*
f Thynnine wasp visiting W. Australian hammer orchid, *Drakaea glyptodon*

Plate 7 Flower-visiting birds
a Bronzy hermit hummingbird, *Glaucis aenea*, feeding at passion flower, *Passiflora vitivfolia*, Costa Rica
b Broad-tailed hummingbird, *Selasphorus platycercus* (female), at scarlet gilia, *Ipomopsis aggregata*, USA
c Male purple sunbird, *Nectarinia asiatica*, at aloe, Oman
d Cape sugar bird, *Promerops cafer*, on *Protea*, South Africa
e Tawny-crowned honeyeater, *Phylidonyris melanops*, on *Melaleuca* sp., West Australia

Plate 8 Flower-visiting mammals
a Queensland blossom bat, *Syconycteris australis*, on *Mucuna macropoda*, Papua New Guinea
b Short-nosed fruit-bat, *Cynopterus sphinx*, visiting wild banana inflorescence
c Honey-possum, *Tarsipes rostratus*, on *Banksia coccinea*, West Australia
d Namaqua rock mouse, *Aethomys namaquensis*, feeding at *Protea* sp., South Africa

Editors' Preface

For a generation of pollination ecologists, the forerunner of this book was an inspiration and something of a bible. Its authors are professional biologists, and at the same time excellent naturalists with a love of books and with expertise in photography. By drawing together the hitherto rather separate botanical and zoological European literature of pollination ecology, and enriching it with their own wide experience and their excellent photographs, they drew many people into pollination ecology. This is now a fast-growing research area with many devotees throughout the world, and there is a risk that the rapid development of particular aspects will make its practitioners lose sight of the main field, and particularly of the basis in natural history that attracted many of us to the topic in the first place. The time has come for a fresh synthesis, particularly now that a decline in native pollinating communities threatens the viability of crops and wild flowers, so that the understanding and management of pollination systems is becoming a practical necessity.

Michael Proctor and Peter Yeo have been joined by a third author, Andrew Lack, to help them produce this new synthesis. It is a fresh book, with new content and a different title, but it retains the qualities that made its forerunner such an important stimulus to the development of interest in the subject of pollination ecology. Its scope has been widened; the British focus of the earlier book has been broadened so that its scope is now worldwide, including, for example, pollination by birds and bats and some remarkable interactions involving exotic orchids.

Since its inception, the New Naturalist series has been known for its illustrations. It is appropriate that this, with its focus on flowers and animals of such beauty and intricacy, should be among the most lavishly illustrated volumes in the series so far. In the New Naturalist tradition, it will appeal to the eyes as well as the minds of naturalists. Perhaps it will help to inspire the new generation of pollination ecologists who will be responsible for protecting and maintaining this delicate mutualism in the decades to come.

Authors' Foreword

Pollination has been a quintessential part of popular natural history for as long as any of us, or our parents or our grandparents, can remember, and the sight of insects busying from flower to flower holds a quite unabated fascination today. Awareness of the need for pollination of the date palm stretches back into classical antiquity, but general recognition of the significance of pollination, and the importance of flower-visiting insects, had to wait for the scientific reawakening which gathered pace from the closing decades of the seventeenth century onwards. As time went on, other agents of pollination – wind, water, birds and bats – were found to play a part. One of the main reasons for the enduring appeal of the flower–insect relationship must be that it is so fundamentally one in which both partners benefit. In our present go-getting age we can sometimes be more conscious of how the partners exploit one-another's services (and indeed there are some examples of very one-sided exploitation between flowers and their visitors). But in some ways this only adds to its appeal, and to the intellectual challenge of elucidating how the flower–pollinator relation has come into being, and how it is maintained in the competitive world of natural selection.

The New Naturalists already include *The Pollination of Flowers*, as No. 54 in the series. That book was published in 1973, near the beginning of a resurgence of interest in pollination biology. A period of sixty years or so, from the publication of Knuth's monumental compendium in the early years of our century until about 1965, can be seen with hindsight as a rather unproductive time for pollination research. Some notable contributions were indeed made, but the subject was something of a backwater from the mainstream of biological progress. Over the last thirty years all that has changed dramatically. Not only has pollination biology emerged as a dynamic research field, but it has taken on a wholly new complexion. The cause of this renaissance and the flood of new papers on pollination has been an explosion of interest in the ecology and evolution of pollination, and the implications of this for many general aspects of the ecology, evolution and genetics of populations. *The Pollination of Flowers* recorded the beginnings of this upsurge of new interest. Over the last three decades, pollination biology has contributed major insights into the way plant populations function and it has given us a much deeper understanding of how plants have evolved. It has also influenced our perceptions on the evolution of social insects, and has played a major part in developing current ecological and evolutionary theories. These wider implications have kept pollination studies at the forefront of ecology. This in turn has rekindled interest in the more 'classical' observational and descriptive aspects of pollination and these too have advanced significantly, so we see today a buoyant and active subject advancing on many fronts.

When the idea of updating *The Pollination of Flowers* was suggested in the late 1980s it soon became obvious that we should have to rewrite the book completely. This is the result. Taking the earlier book as a starting point, the first and largest change was to invite a third author, Andrew Lack, to join the team specifically to cover the work on ecology and evolution that had been done since 1973. The three of us together were responsible for the shape and character of the present book, which differs from its predecessor in two main ways. First, we have shifted the emphasis towards a more functional view of the benefits and costs of the pollination relationship to flowers and their visitors, and the ways in which these interact. Second, while we have written the book from a primarily 'British Isles' standpoint, we have taken a more liberal view of the New Naturalist brief ' ... to interest the general reader in the wildlife of Britain ...' than we did in the earlier book. The preface to *The Pollination of Flowers* called pollination 'an obstinately international subject', and it has become ever more so as years have passed. Much of the research on pollination biology since 1970 has been done in North America, with a significant quantity from tropical countries around the world and from the southern hemisphere. The results, along with those of continuing study in Europe, are often relevant to all pollination relationships, and to understanding the pollination of our native plants here in Britain. Added to that, travel has become easier and we have surely all become more international in outlook in the last quarter century. The countries of continental Europe, the Mediterranean coast, and increasingly North America, Africa and places even farther afield, are coming within the experience of holidaying (and sometimes working) Britons – part of our 'home range'. Perhaps even more significant is the much wider awareness of the world's flora and fauna generally brought about by the superb natural history films which have become so regular a feature of television since David Attenborough's *Life on Earth*.

Like its predecessor, this book is a selective distillation of a vast subject. We hope it gives a reasonably rounded view of pollination biology in the 1990s – and that it will give pleasure and interest to many readers. We realise that some people may not want to read the chapters in the order in which they appear here; we have tried to put as few difficulties in their way as possible. Some technical terms are inevitable, some are useful shorthand for otherwise cumbersome circumlocutions; we have tried to keep them to a minimum, and those that we use are explained in Chapter 2 or at their first appearance in the text. As in *The Pollination of Flowers* we have given numerous references to the scientific literature, for the benefit of readers who want to know more or to go to the source of statements in the text. We hope that will not detract from the readability of the book. Sources of unpublished information are acknowledged by name and initials with no date. For plant names, we have followed Stace's *New Flora of the British Isles* (1991) for British plants and *Flora Europaea* (Tutin *et al.*, 1964–1981) for other European plants; for non-European plants we have followed the published source quoted. Insect names follow the second edition of Kloet & Hincks's *A Check List of British Insects* (1964–1978).

In writing *The Natural History of Pollination*, M.C.F.P. was primarily responsible for Chapters 1, 2, 6, 7 and 9, P.F.Y. for Chapters 3, 4, 5, 8, 10, 11, and 13, and A.J.L. for Chapters 12, 14, 15 and 16. Throughout, we have all read and criticised each others chapters, and we have seen the book very much as a joint venture. Line drawings not otherwise acknowledged are original, mostly taken from *The Pollination of Flowers*. Unacknowledged photographs are by M.C.F.P.; other photographers are acknowledged in the captions.

Many other people have contributed directly or indirectly to the writing of this book, and we are grateful to all of the friends and colleagues, too many to name individually, with whom we have discussed pollination, or who have sent reprints and sometimes unpublished manuscripts and observations. We are particularly indebted to Prof. Christopher Cook, Dr Sally Corbet, Dr Paul Cox, Dr James Cresswell, Prof. Bertil Kullenberg, Dr Anders Nilsson, Dr Börge Petterson and Dr Jonathan Silvertown, all of whom read and commented on parts of the text in draft. Their help was of great value. M.C.F.P. remembers with appreciation the fine collection of living plants (some portrayed here) built up by the late Prof. John Caldwell in Exeter, and is grateful to Gavin

Wakley for technical help with the scanning elecgtron micrographs. A.J.L. acknowledges a special debt to Dr Quentin Kay and Dr Peter Gibbs for discussions on many subjects, and to Derek Whiteley for contributing such splendid line drawings to Chapters 12 and 14. We have greatly appreciated the patience and helpfulness of the HarperCollins natural history editors, and especially the cheerfulness, and efficiency of Isobel Smales through the long haul of turning a pile of evolving typescript and a large pool of assorted photographs into an almost finished book, and of Liz Bourne who coped with the final stages of bringing *The Natural History of Pollination* into the world. Finally, each of us owes a debt to his co-authors. Working together on this project has been a stimulating and enjoyable experience. We hope we can share that stimulus and enjoyment with our readers.

M.C.F.P.
P.F.Y.
A.J.L.

A NOTE ON THE INSECT PHOTOGRAPHS

Most of the photographs of insects visiting flowers were taken in Devon (a few in the Channel Islands and elsewhere) between 1964 and 1970. At that time small transistorised electronic flash units and 35 mm single-lens reflex camera with pentaprisms and instant-return mirrors were just becoming widely available, and had opened up new possibilities for insect photography in the field. The photographs reproduced here were taken on Ilford Pan F film, using Praktica and Pentax cameras, either with a 50 mm Tessar or 55 mm Takumar standard lens on extension tubes, or with the Pentax 55 mm macro lens (a rather longer lens is easier for work of this kind, but I did not have one at the time). For the earlier pictures a small standard flash unit (Mecablitz) was used, giving directional lighting; a number of the later pictures were taken with a Minicam ringflash. Ringflash lacks the 'modelling' given by a directional light source, and shiny convex surfaces (in which insects abound) tend to reflect distracting circular images of the flash tube, but for recording insect behaviour in the field these disadvantages are outweighed by the advantage of even and predictable lighting without strong shadows; what you see in the viewfinder is what you get on the film! The success-rate in field photography of this kind (as in many sporting activities) is heavily dependent on practice. Obviously some subjects are much easier than others. Over all, perhaps 50% of exposures yielded reasonably framed and acceptably sharp negatives, but the really worthwhile pictures probably averaged only two or three on a film. However, it is a rewarding activity, to be recommended. With modern improvements in fast films, much field insect photography can now be done without flash.

M.C.F.P.

1

The Study of Pollination: a Short History

The association of flowers and insects, and the need for pollination if flowers are to set seed, are very much a part of our everyday awareness. It is perhaps surprising to realise that the discovery of the pollination of flowers, and the part often played in this by bees and other insects, dates only from the late seventeenth and early eighteenth centuries. It was a product of that same great century of European science which saw the discovery of the circulation of the blood by Harvey, the establishment of the Royal Society of London and the French *Académie des Sciences*, and Newton's discoveries in mechanics, mathematics, optics and astronomy.

Early ideas

The ancient world was certainly familiar with the association of bees and flowers, and with the need for pollination of some plants if they were to set fruit. The Old Testament abounds with references to honey. Aristotle described the habits of bees in his *History of Animals* and Virgil devoted the fourth book of his *Georgics* to honey and the ways of bees. But the people of biblical and classical times seem generally to have been content to see the visits of bees to flowers and the production of honey simply as facts of nature, and not to seek any significance in it for the flowers. This seems all the more curious, since the Greek and Roman writers were familiar with the need for pollination of the date palm (see p.360). An excellent account of the way in which date palms were fertilised is given by Theophrastus (c. 373-287B.C.), who says 'With dates it is helpful to bring the male to the female; for it is the male which causes the fruit to persist and ripen ...The process is thus performed; when the male palm is in flower, they at once cut off the spathe on which the flower is, just as it is, and shake the bloom with the flower and the dust over the fruit of the female, and if this is done to it, it retains the fruit and does not shed it. In the case both of the fig and the date, it appears that the "male" renders aid to the "female" – for the fruit-bearing tree is called "female" – but while in the latter case there is a union of the two sexes, in the former the result is brought about somewhat differently.' Theophrastus repeatedly refers to plants as 'male' and 'female', and records that '...they say that in the citron those flowers which have a kind of distaff growing in the middle are fruitful, but those that have it not are sterile.' After this it is disappointing to read about the differences '...by which men distinguish the "male" and "female", the latter being fruit-bearing, the former barren in some kinds. In those kinds in which both forms are fruit-bearing the "female" has fairer and more abundant fruit; however, some call these the "male" trees – for there are those who actually thus invert the names. This difference is of the same character as that which distinguishes the cultivated from the wild tree ...' It seems that the sex of a plant meant no more to Theophrastus and his contemporaries than the possession of some characters associated with maleness or femaleness; he had little idea of a process in plants analogous to sexual union in animals. He seems to have rejected the idea of a real sexual union in the date palm because a similar state of affairs could not be seen in other plants. Theophrastus was certainly the greatest botanist of classical times, and his account of plants was not surpassed until the sixteenth century. The vague notion of sex in plants current in his day lingered on; a relic of it is preserved in the names of two of our common ferns, the male fern (*Dryopteris filix-mas*) and the lady fern (*Athyrium filix-femina*), whose only qualification for these names is that the lady fern is more delicate and ladylike than the other! It is hardly surprising that some botanists, like the Italian Caesalpino (1519-1603), rejected the idea of sex in plants altogether.

Discovery of the importance of pollination

Like many other great discoveries, the idea that a sexual fusion takes place in the reproduction of plants, and that the stamens are the male organs of the flower, seems to have developed independently in the minds of a number of botanists towards the end of the seventeenth century. In a paper on 'The Anatomy of Flowers' read before the Royal Society in 1676, and published in 1682 as part of his *Anatomy of Plants*, the English botanist Nehemiah Grew said that he had discussed the connection of the stamens with the formation of seeds with Thomas Millington (at that time Sedleian Professor of Natural Philosophy at Oxford), who had suggested to him that 'the attire [stamens] doth serve, as the male, for the generation of the seed ...', and that he, Grew, agreed with him.[1] In 1686 John Ray clarified and cautiously accepted Grew's opinion in his *Historia Plantarum*, adding 'This opinion of Grew, however, of the use of the pollen before mentioned wants yet more decided proofs; we can only admit the doctrine as extremely probable.' (Vol. 2, p.18).

The 'more decided proofs' were supplied a short time afterwards by Rudolph Jacob Camerarius (1665-1721), who was Professor of Physic at Tübingen in Germany. Camerarius carefully examined flowers, and carried out a number of experiments on pollination. He found, for instance, that when he removed the anthers from the male flowers of the castor oil plant, *Ricinus communis*, the female flowers failed to set seed, and maize failed to set seed when he removed the stigmas from the female flowers. Similarly he found that female plants of mulberry, mercury and spinach failed to produce viable seed in the absence of male plants. Many of the earlier botanists seem to have been worried by the occurrence of the two sexes together in plants. Camerarius, like Grew, quoted Swammerdam's discovery of hermaphroditism in snails, and he suggests that what is the exception in animals is the rule in plants. Camerarius set out his observations and his conclusions together with a long discussion of previous writings on the subject, in a dissertation entitled *Epistola de Sexu Plantarum* addressed to Michael Bernard Valentini (1657-1729), who was Professor of Physic in Giessen, on 25 August 1694.

Camerarius's conclusions were not everywhere accepted at once, or without contro-

---

[1] In 1671 Grew had considered 'The Use of the *Attire*... to be not only Ornament and Distinction to us, but also Food for a vast number of little Animals, who have their peculiar provisions stored up in these *Attires* of Flowers: each Flower becoming their Lodging and their Dining-room, both in one.' (Account of The Anatomy of Vegetables begun in *Phil. Trans.*, No. 78, p.3041 [1671]). Grew seems to have been thinking of insects: his 'little animals' evidently have no connection with the 'animalcules' Leeuwenhoeck was to describe from water and from the semen of animals a few years later.

versy. Some of his experiments had appeared inconclusive or contradictory, and in 1700 the great French botanist Tournefort, apparently in ignorance of Camerarius's work, still considered that the stamens served to excrete unwanted portions of the sap in the form of pollen, and he doubted the need for pollination of the date palm. For half a century little was added to Camerarius's experimental demonstration of the need for pollination, though sporadic experiments are recorded, of which the most interesting are those of Richard Bradley (Fellow of the Royal Society and Professor of Botany in Cambridge from 1724 until his death in 1732),[1] Philip Miller and James Logan. Bradley describes his experiments in his *New Improvements of Planting and Gardening*, published in 1717.

> 'I made my first experiment upon the *Tulip*, which I chose rather than any other Plant, because it seldom misses to produce *Seed*. Several years I had the Conveniency of a large Garden, wherein there was a considerable bed of *Tulips* in one Part, containing about 400 Roots; in another part of it, very remote from the former, were Twelve *Tulips* in perfect Health, at the first opening of the Twelve, which I was very careful to observe, I cautiously took out all of their *Apices*, before the *Farina Fecundens* was ripe or in any way appeared: these *Tulips* being thus *castrated*, bore no Seed that Summer, while on the other Hand, every one of the 400 Plants which I had let alone produced *Seed*.'

This experiment seems to be the first on hermaphrodite flowers. Bradley then commends to his reader the experiment of removing the young male catkins of an isolated hazel or filbert, which will then not bear unless the female flowers are dusted with pollen from 'Catkins of another Tree, gather'd fresh every Morning for three or four Days successively, and dusted lightly over it, without bruising its tender *Fibres*.' He goes on to describe the production of an artificial hybrid between a carnation and a sweet william by Thomas Fairchild (1667-1729), and looks forward to the use of artificial pollination for the selective breeding of plants.

Philip Miller (1691-1771) performed an experiment on tulips similar to Bradley's in 1721, apparently at the suggestion of Patrick Blair, a medical man and Fellow of the Royal Society who was sentenced to death (but reprieved) for acting as surgeon with the Jacobite forces during the rebellion of 1715. In a letter to Blair, dated 11 November 1721, Miller described how he had,

> '...experimented with twelve Tulips, which he set by themselves about six or seven Yards from any other, and as soon as they blew, he took out the *Stamina* so very carefully, that he scattered none of the Dust, and about two Days afterwards, he saw Bees working on Tulips, in a bed where he did not take out the *Stamina*, and when they came out, they were loaded with Dust on their Bodies and Legs; He saw them fly into the Tulips, where he had taken out the *Stamina*, and when they came out, he went and found they had left behind them sufficient to impregnate these Flowers, for they bore good ripe Seed: which per-

---

[1] Bradley was a prolific and popular writer, but in Cambridge he was evidently felt to be something of a charlatan, and his ignorance of Latin and Greek and his neglect of his teaching duties excited great scandal. In extenuation, it must be said that he seems to have contributed more to his subject than some of his more respectable contemporaries.

suades him that the *Farina* may be carried from Place to Place by Insects ...'
(Blair, 1721)

Miller included an account of his experiments in his *Gardener's and Florist's Dictionary* (1724) and in his *Gardener's Dictionary* (1731); the article on 'Generation' in which the account appears was omitted from the 3rd-5th editions of the *Gardener's Dictionary* but reinstated in the 6th edition (1752). James Logan (1674-1751), who was born in County Armagh, went as William Penn's secretary to Pennsylvania in 1699, and was Chief Justice and President of the Council of the province at the time of his experiments, described in a letter dated 20 November 1735 to his fellow-Quaker, Peter Collinson, FRS. In each corner of his garden in Philadelphia, Logan had planted a hill of 'Mayze or *Indian Corn*'.

'...from one of these Hills, I cut off whole Tassels, on others I carefully opened the Ends of the Ears, and from some of them I cut or pinched off all the silken Filaments: from others I took about half, and from others one fourth and three fourths &c. with some variety, noting the Heads and the Quantity taken from each; Other Heads again I tied up at their Ends, just before the Silk was putting out, with fine Muslin, but the Fuzziest or most Nappy I could find, to prevent the passage of the Farina: but that would obstruct neither the Sun, Air or Rain. I fastened it also very loosely, as not to give the least Check to Vegetation.'

He found that the plants in the group from which the male panicles had been removed produced no good grains – apart from a single large cob which had its stigmas fully exposed in the direction of one of the other groups of plants, and produced 20 or 21 out of a possible total of some 480 grains. Logan plausibly attributes this to carriage of pollen by wind. The cobs wrapped in muslin again produced no seed. On cobs from which he had removed some stigmas, Logan found seed in proportion to the stigmas he had left.

The early eighteenth century saw much speculation and argument over the way in which the pollen fertilised the ovules (Morton, 1981). Samuel Morland (1703) discussed whether the pollen grains passed down the 'tubes' of the styles to fertilise the ovules. He was unable to discover whether the ovules contained an embryo before pollination, but suspected they did not, and recommended 'the inquiry to those gentlemen who are masters of the best microscopes, and address in using them.' However, he observed that the 'seminal plant always lies in that part of the seed which is nearest to the insertion of this stylus, or some propagation of it into the seed vessel', and continues, 'I have discovered in beans and peas and phaseoli, just under the extremity of what is called the eye, a manifest perforation, discernible by the larger magnifying glasses, which leads directly to the seminal plant, and at which I suppose the seminal plant entered ...'.[1] At this period pollen was frequently referred to as the *Farina Fecundens*, and striking instances of the transmission of characters by the pollen were thought worthy of comment. Philip Miller described the motley progeny of

---

[1] Grew had described the micropyle of the seed in 1671, but it was not until 1830 that Amici was able to trace the path of the pollen-tube from the germinating pollen-grain to the ovule (Sachs, 1875, p.467; 1890).

a batch of seed saved from savoys which had grown close to red and white cabbages (Blair, 1721). The Hon. Paul Dudley described the transmission of seed colour between rows of maize plants in New England in 1724 – 'even at the distance of 4 or 5 rods: and particularly in one place where there was a broad ditch of water between them ...Mr D. is therefore of the opinion that the stamina, or principles of this wonderful copulation, or mixing of colours, are carried through the air by the wind ...' (*Phil. Trans.*, Vol. 35, p.194). Benjamin Cooke (1749) grew red and white maize together, and writes '...you may with pleasure observe the filament in the white plant, which has been struck with the red farina, discovering its alien commerce by a conscious blush, and by counting the threads thus stained, foretell how many corresponding seeds will appear red at the opening of the ear, when ripe.' Certainly by the time Gleditsch demonstrated the development of fruit of the palm *Chamaerops humilis* following artificial pollination in the Berlin botanical garden in 1749 (Sukopp, 1987), sexuality in plants was regarded as an established fact. The theory had champions whose influence carried great weight even though they added little new evidence, notably Sebastien Vaillant (1669-1722), whose *Discours sur la Structure des Fleurs* appeared in 1718, and that most influential of eighteenth-century botanists, Carl Linnaeus (1707-1778). In England, the account of the generation of plants in Patrick Blair's *Botanick Essays* (1720), which quotes Grew, Ray, Camerarius, Vaillant and Bradley, and which was reproduced in Miller's *Gardener's Dictionary*, was widely read.

In 1750, Arthur Dobbs observed bees around his home near Carrickfergus in County Antrim, confirming Miller's observation that flowers could be pollinated by insects, and Aristotle's brief comment on the flower-constancy of bees.[1] He communicated his observations to the Royal Society, and in the *Philosophical Transactions* we read:

' ...I have frequently follow'd a Bee loading the *Farina*, Bee-Bread or crude Wax, upon its Legs, through a Part of a great Field in Flower: and upon whatsoever Flower I saw it first alight and gather the *Farina*, it continued gathering from that Kind of Flower: and has passed over many other Species of Flowers, tho' very numerous in the Field, without alighting upon or loading from them: tho' the flower it chose was much scarcer in the field than the others; So that if it began to load from a Daisy, it continued loading from them, neglecting Clover, Honeysuckles, Violets &c.; and if it began with any of the others, it continued loading from the same Kind, passing over the Daisy. So in a garden upon my Wall-Trees, I have seen it load from a Peach, and pass over Apricots, Plums, Cherries &c. yet made no distinction betwixt a Peach and an Almond.'

Dobbs pointed out that this observation is confirmed by examining the pollen-loads carried by bees returning to the hive, and continues:

'Now if the Facts are so, and my Observations true, I think that Providence has appointed the Bee to be very instrumental in promoting the Increase of Vegetables ...

---

[1] 'On each expedition the bee does not fly from a flower of one kind to a flower of another, but flies from one violet, say, to another violet, and never meddles with another flower until it has got back to the hive...' (*History of Animals*, IX, 40, trans. D'Arcy Wentworth Thompson).

'Now if the Bee is appointed by Providence to go only, at each Loading, to Flowers of the same Species, as the abundant *Farina* often covers the whole Bee, as well as what it loads upon its Legs, it carries the *Farina* from Flower to Flower, and by its walking upon the *Pistillium* and Agitation of its Wings, it contributes greatly to the *Farina's* entering into the *Pistillium*, and at the same time prevents the heterogeneous Mixture of the *Farina* of different Flowers with it; which, if it stray'd from Flower to Flower at random, it would carry to flowers of a different species.'

The main credit for the demonstration of the significance of insects in flower pollination must go to Joseph Gottlieb Kölreuter (1733-1806), Professor of Natural History in the University of Karlsruhe. Kölreuter did experiments in hybridisation and made systematic observations on pollination, which he published between 1761 and 1766. His writings record a great deal of careful and critical observation, and some remarkable advances in floral biology. Kölreuter found that insect visits were necessary for the successful pollination of cucumbers and their relatives, irises and many Malvaceae, and he says, 'In flowers in which pollination is not produced by immediate contact in the ordinary way, insects are as a rule the agents employed to effect it, and consequently to bring about fertilisation also; and it is probable that they render this important service if not to the majority of plants at least to a very large part of them, for all of the flowers of which we are speaking here have something in them which is agreeable to insects, and it is not easy to find one such flower, which has not a number of insects busy about it.' He examined the nectar in many flowers, and concluded correctly that it was the source of the bees' honey, and that its significance to the flower lay in the attraction of insects. With remarkable patience he counted the numbers of pollen grains produced by various flowers, and found by experiment how many were needed to fertilise all the ovules in the flower; and he described the structure of the pollen grain with surprising accuracy, considering the crude microscopes of the time. Among other observations he described the sensitive stamens and stigmas which occur in a number of plants, and he noticed that the stamens of the willowherb and other plants ripen before the stigma – a fact whose importance in floral biology was soon to be realised.

The systematic study of pollination

The founder of the systematic study of the relations between flowers and insects was Christian Konrad Sprengel (1750-1816). The son of a clergyman, Sprengel was born in Brandenburg. He was not a botanist by training. He studied theology and philology, and for much of his working life was a teacher, first at the school of the Friedrichs-Hospital in Berlin, and then from 1780 to 1794 as Rector of the great Lutheran school at Spandau. According to his own account, Sprengel was drawn to the study of insect pollination of flowers in 1787 by his observation of hairs on the bases of the petals of the wood cranesbill (*Geranium sylvaticum*). 'Convinced that the wise Creator of nature has brought forth not even a single hair without some particular design, I considered what purpose these hairs might serve.' Sprengel came to the conclusion that, as the nectar was provided for the nourishment of insects, the hairs served to protect the nectar from being spoilt by rain. In the following year he examined the flowers of a forget-me-not, and recognised in the yellow ring surrounding the centre of the flower a 'honey-guide', leading insects to the nectar in the short tube at the centre of the sky-blue flower. From these and other observations in the next few years, Sprengel was led to distinguish four parts of the flower concerned with the secretion of nectar:

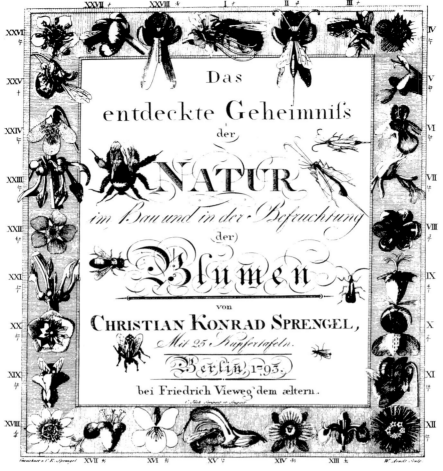

Fig. 1.1 Sprengel's title-page illustrating some of the floral mechanisms he observed. Notice the ichneumon on a twayblade flower (II), the bee on a *Salvia* (XV) and the wasp visiting a figwort (XXV).

the nectary itself, which prepares and secretes the nectar; the nectar reservoir, which receives and contains the nectar secreted by the nectary; the nectar cover, protecting the nectar from rain; and the parts that enable insects readily to find the nectar – corolla, odour and 'honey-guides'. In 1793 he published his classic book *Das entdeckte Geheimniss der Natur im Bau und in der Befruchtung der Blumen* – 'The revealed secret of Nature in the structure and fertilisation of flowers' – in which he described the floral adaptations of some 500 species of flowers, often with admirable lucidity, accuracy and detail. Sprengel pointed out the very wide occurrence of protandry (ripening of the anthers before the stigmas), and he was the first to describe the opposite condition of protogyny, which he found in the cypress spurge (*Euphorbia cyparissias*) in 1791.

Sprengel was an excellent observer, and he left little to add to his descriptions of the structural adaptations of many flowers to insect pollination. He also discussed wind-pollinated flowers, pointing out the much greater quantity of pollen produced by them than by insect-pollinated flowers and the significance of their exposed anthers and large, often feathery stigmas. From his observations he came to the conclusion that, 'Nature seems unwilling that any flower should be fertilised by its own pollen.' His near contemporary Thomas Knight (1758-1838), for many years President of the Horticultural Society of London, also concluded from his experiments on peas that cross-fertilisation was beneficial. Among the progeny of his hybridisations he found '...a numerous variety of new kinds produced, many of which were, in size, and in every other respect, much superior to the original white kind, and grew with excessive luxuriance ...' (Knight, 1799)[1].

Sprengel's work made little impact for over half a century, although his ideas seem to have been quite widely known and discussed, perhaps more by entomologists than by botanists. They are mentioned in all seven editions of Kirby & Spence's *Introduction to Entomology* between 1815 and 1867. Charles Darwin (1862b) wrote of Sprengel's book, 'This author's curious work, with its quaint title of "Das Entdeckte Geheimniss der Natur", until lately was often spoken lightly of. No doubt he was an enthusiast, and perhaps carried some of his ideas to an extreme length. But I feel sure, from my own observations, that his work contains an immense body of truth. Many years ago Robert Brown, to whose judgment all botanists defer, spoke highly of it to me, and remarked that only those who knew little of the subject would laugh at him.' (*Fertilisation of Orchids*, p.275, footnote).

The next important contributions to the study of flower pollination came from Charles Darwin (1809-1882) himself, who published many observations on the subject from 1857 onwards. In 1858, the year before the *Origin of Species*, Darwin showed that various papilionaceous flowers set seed less vigorously if they are covered with a net to prevent the visits of insects. In 1862 he published an account of the pollination mechanism of the primrose and other species of *Primula* that have flowers of two kinds, the first of several contributions on heteromorphic flowers (see pp.325-7), and his classic book on the fertilisation of orchids appeared in the same year. Like Sprengel and Knight, Darwin was drawn to the conclusion that 'Nature tells us in the most emphatic manner that she abhors perpetual self-fertilisation. ' The results of his experiments and observations on *The effects of Cross- and Self-fertilisation in the Vegetable Kingdom* appeared in 1876.

The *Origin of Species* and Darwin's work on pollination stimulated an upsurge of interest in the biology of pollination, and in the relations between plants and insects. The succeeding few decades are the classical period of floral biology, during which much of our knowledge of the pollination mechanisms and insect visitors of European and North American flowers was gathered. Flower pollination was a topic of lively current interest, to which the vigorous growth of science generally, expanding

---

[1] Knight's experiments on peas tantalisingly foreshadow those of Mendel. Knight observed that purple flower-colour was dominant to white, and tallness to dwarfness, and that reciprocal crosses produced the same results. He noticed segregation for flower and seed-colour in back-crosses between hybrid and white-flowered plants. It does not seem to have occurred to him to investigate further the inheritance of these striking characters, probably because that seemed irrelevant to the interest in practical plant breeding that led him to carry out his experiments.

popular interest in natural history and the countryside, and fascination with the profusion of exotic plants newly introduced into Europe, all contributed. Those who
made major additions to floral biology included some of the best-known botanists of
the period, and many lesser-known men besides. A few names stand out above the
others, as those who gave shape and direction to the study.

Asa Gray (1810-1888) in North America and Fritz Müller (1821-1897) in South
America both followed closely on Darwin, and each published many papers on pollination from the 1860s onwards. The mass of scattered information was rapidly growing, and Friedrich Hildebrand, Professor of Botany in Freiburg, published the first
comprehensive textbook on floral biology in 1867. Hildebrand classified all the floral
arrangements known to him, and a few years later Federico Delpino (1868, 1874) in
Italy produced a very much more elaborate classification. Perhaps the greatest of observers of relationships between insects and flowers was Hermann Müller (1829-
1883), brother of Fritz Müller mentioned above, who taught for most of his life at the
Realschule in Lippstadt. Hermann Müller was 37 when he became acquainted with
Darwin's *Origin of Species* and *Fertilisation of Orchids*, and from then on he devoted his
energies to the study of pollination. He observed and recorded a vast number of individual visits of insects to flowers, and published his results in three important works
between 1873 and 1881. Not only did his observations enable him to describe the
pollination mechanisms and insect visitors of many central European plants; he
showed too an awareness of ecological context that foreshadowed the rich developments in pollination ecology that were to come a century later. Müller seems to have
stimulated other botanists to follow his example to an even greater extent than Darwin twenty years before, and the literature of the remaining years of the nineteenth
century abounds in studies of pollination in particular districts, or in particular groups
of plants. No doubt many became interested in the pollination of flowers through the
account in Anton Kerner von Marilaun's *Natural History of Plants*. Ernst Loew (1895)
summarised the work on floral biology carried out in central and northern Europe in
the decade following the death of Hermann Müller, and the whole period fittingly
culminates in the publication of the monumental three-volume *Handbuch der Blütenbiologie* (1898-1905) by Paul Knuth (1854-1900), Professor in the Ober-Realschule at
Kiel, and himself the author of many papers on floral biology in the north German
islands and elsewhere. The first two volumes of this invaluable compendium were
translated into English as *Handbook of Flower Pollination*, published in 1906-1909.

The twentieth century

After the turn of the century, interest in classical floral biology waned. There were
probably several reasons for this. It was a period of ascendancy for experimental and
laboratory botany: palaeobotany and morphology, plant physiology, and the new sciences of genetics and cytology. Plant ecology was developing as a vigorous branch of
botany, demanding the attention of those interested in plants in their natural habitats.
It is probably also true to say that floral biology had reached the limit of its development in Europe and North America in the state of biology at the time; with the publication of Knuth's book, it must have seemed that few observations on pollination
remained to be made. There was still detail to be filled in, but major advances in the
understanding of flowers had to wait for the development of cytology and genetics,
ecology and the study of animal behaviour. For the time being floral biology passed
largely from the field of active research to the textbooks – a state of affairs exemplified
by A.H. Church's magnificent but uncompleted *Types of Floral Mechanism* (1908), in
which it is easy to feel that the beautiful and precise drawings of the details of the

flowers embody the finality of perfection – and overshadow thought of their functions. A professional biologist who bridged the gap between the post-Darwinian and modern periods was the Austrian Fritz Knoll, whose outstanding contribution was the elucidation of the insect-trapping mechanism of *Arum*. This was published in the same year as an account of the remarkably similar arrangements for trapping insects for pollination in an unrelated plant, *Ceropegia woodii*, by Leopoldine Müller (1926; see Chapter 10). But for the most part the tradition of Darwin and Hermann Müller lingered on in the hands of amateur naturalists, where it produced a notable twentieth-century addition to floral biology in the discovery of 'pseudocopulation' in the pollination of the orchids that mimic insects – and in so doing, solved a problem which had greatly puzzled Darwin and his successors (Chapter 7).

Many of the most significant of the newer observations in 'classical' floral biology have come from outside Europe. Pollination by both birds and bats is recorded in Knuth's *Handbuch*, but the importance of birds as pollinators almost everywhere in the world except Europe and northern Asia was only slowly recognised, and was not fully appreciated until well into the present century (Porsch, 1924). Recognition of the significance of bats as pollinators has come even more recently; thanks to the observations of Otto Porsch, Lennart van der Pijl, Herbert Baker, Stefan Vogel and others, we now know that bat pollination is widespread and important in tropical countries (Chapter 8).

If the centre of interest during the earlier part of the twentieth century passed from floral biology as such, it moved to subjects which illuminated many aspects of the functions of flowers and their relationships with their pollinating agents. When Darwin was writing about pollination and the setting of seed, little was known about the details of the way in which fertilisation was brought about once the pollen had reached the stigma. In the course of a few years around 1880, Eduard Strasburger (1844-1912) and Walther Flemming (1843-1915) independently elucidated the main features of the usual mode of division of cells and their nuclei (*mitosis*), and Strasburger observed the fusion of a nucleus from the pollen tube with the egg nucleus in the embryo-sac of the ovule. Discovery of the reduction division (*meiosis*), by which the number of chromosomes is halved in the formation of the pollen grains and the embryo-sac nuclei, followed a few years later. The main cytological details of the formation of the sexual parts and the process of fertilisation in flowers had been worked out by 1900, and chromosome cytology had become an established science, to remain an active field of research to the present day.

The breeding experiments on peas carried out by Gregor Mendel a few years after the publication of Darwin's *Origin of Species* were retrieved from obscurity in 1903, and provided the foundation on which a new science of genetics was built. The analogy between the behaviour of the various characters in Mendel's peas and the behaviour of chromosomes at meiosis suggested that the chromosomes of the cell nucleus are the bearers of the hereditary units, or *genes* as they were later called. Evidence soon accumulated to confirm this conclusion, which bound cytology and genetics inseparably together, and led to the 'neo-Darwinian synthesis' (Haldane, 1932; Dobzhansky, 1937; Huxley, 1942) which established the central role and mechanism of evolution in biology. Different developments were to lead to the discovery that deoxyribose nucleic acid (DNA) constitutes the essential genetical material of the chromosomes, and the elucidation of the structure of DNA by J.D. Watson and F.H.C. Crick (1953; Watson, 1968, 1970) opened the way to understanding of the molecular mechanism of inheritance.

Cytology and genetics provide a rational explanation of the inheritance not only of obvious features of colour and form, but also of many significant characteristics in plant reproductive biology. Conversely, the pollination relationships of flowers impinge on both genetics and ecology in the study of the genetical composition and microevolution of natural plant populations, and of gene flow within and between them. The kind of investigations pioneered by the Danish botanist Göte Turesson from 1922 onwards, often referred to as 'experimental taxonomy' or 'genecology', of which books such as G.L. Stebbins's *Variation and Evolution in Plants* (1950), Davis & Heywood's *Principles of Angiosperm Taxonomy* (1963) and Briggs & Walters's *Plant Variation and Evolution* (1969, 1984) have provided syntheses, has come to merge imperceptibly into the broader and very active field of plant population genetics (Chapter 16).

After following largely separate courses for the first half of the century, plant and animal ecology increasingly converged on fundamental areas of common interest. The development of population genetics, growing understanding of the flow of energy and matter through ecosystems, increased comprehension of the behaviour and interactions of populations, and increasing emphasis on theoretical concepts and models in ecology, all had a part in this. Plants, rooted to the ground, with often indeterminate modular growth, and nourished by photosynthesis and mineral nutrients from the soil, differ in their ecology from animals in fundamental respects. Yet there are important features in common, many of them emphasised by John Harper in his *Population Biology of Plants* (1977).

Recent decades

So much is history. The last 30 years or so are still too close to us, too interwoven with the present, for the detached view of a historical narrative. Developments in cytology, genetics, evolutionary studies and ecology have combined to bring pollination biology into the mainstream of biological research, and stimulated a resurgence of interest which has gathered pace from the 1960s onwards. It is probably not entirely coincidence that a period of less than 20 years saw the publication of H. Kugler's *Einführung in der Blütenökologie* (1955a), F. Knoll's *Die Biologie der Blüte* (1956), B.J.D. Meeuse's *The Story of Pollination* (1961), Mary Percival's *Floral Biology* (1965), Faegri & van der Pijl's *The Principles of Pollination Ecology* (1966), and our own *The Pollination of Flowers* (1973). Later, Meeuse & Morris's attractive and popular book *The Sex Life of Flowers* (1984) took full advantage of post-war advances in photography and colour reproduction in its beautiful illustrations. Pollination takes place in an environment populated by many species of plants and animals, competing, coexisting or interdependent. Flowering and seed production require resources, in some cases a large fraction of the food and mineral nutrients amassed by a plant in its lifetime; pollinators too need resources, met in varying proportions by the nectar, pollen and other 'rewards' provided by flowers, and by sources outside the pollination relationship (Heinrich & Raven, 1972; Heinrich, 1975a, 1979). And, of course, pollination systems are both the products of, and have profound effects upon, the genetic structure and evolution of plant populations. Real (1983), Jones & Little (1983), Lovett Doust & Lovett Doust (1988), Roubik (1989) and Wyatt (1992) provide syntheses of some of these fields of research. Modern methods for studying pollination are decribed by Dafni (1992) and Kearns & Inouye (1993).

Undoubtedly a further factor in the explosive development of studies in pollination ecology in recent decades has been a cultural shift in style and emphasis in scientific research (Mayr, 1982). Many of the Victorians saw 'science' largely in terms of adding to the edifice of 'knowledge'. The scientific method was seen as a process of extracting significant generalisations from a sufficiently large body of observed facts –

and progress as deriving from enlargement of that body of facts. Twentieth-century science has increasingly centred on how things are related and how they work, and increasingly recognised the role of intuition and imagination in formulating hypotheses that can be tested by observation and experiment. (Indeed, the greatest advances in science must always have happened in this way; some of the discoveries mentioned earlier in this chapter which we now take for granted needed major leaps of imagination from the thinking of their time.) This has been an immense stimulus to experiment, and to question and test much that we have been accustomed to take for granted in pollination relationships. Of course (as we·have seen), experiment is not new in the study of pollination. The experiments of Lubbock (1875) and Plateau (1885-1898) on the colour senses and responses of insects, and the researches of von Frisch (1914 onwards) and Clements & Long (1923) – and indeed the experiments of the early pioneers – were innovative in their day and a foretaste of an experimentally-minded era yet to come. Technical developments have also been important. Visible-light and ultraviolet photography, cinematography, electron microscopy, new and more sensitive methods of chemical analysis (especially liquid and gas chromatography), radioactive and other tracers, and biochemical and molecular-biological techniques, have all contributed to pollination biology and opened new fields of research. Perhaps no innovation has been more influential and all-pervading than cheap, fast computers, making possible the storage and analysis of data on a scale and in ways that would have been unthinkable 50 years ago. The last quarter-century has seen a very exciting broadening of our understanding of pollination ecology; the coming decades promise to be no less fruitful.

2

Flowers, Pollination and Fertilisation

Flowers are among the most complex and diversified objects in the plant kingdom; that is a great part of their fascination. But underlying this complexity and diversity there is much common ground – a large measure of unity and regularity in their structure. Some appreciation of this is needed to understand how flowers work, and in describing their structure some technical terms can hardly be avoided. This chapter, then, is background: an introduction to the essential features of flowers relevant to pollination, and in effect a discursive glossary. You may like to leave it aside, and refer to it only as needed to clarify matters considered in later chapters.

What is a flower?

It is generally much easier to recognise a flower as such than to give a definition of what a flower is. Typically, a flower is made up of four kinds of members: *sepals*, making up the *calyx*; *petals*, forming the *corolla*; *stamens*, sometimes collectively referred to as the *androecium*; and the *ovary* or *gynoecium*, made up of one or more *carpels*. These are borne on the *receptacle* – the conical or thickened end of the flower-stalk or *pedicel*. Thus in the flower of a buttercup (Fig. 2.1) there are five green sepals, which enclose and protect the developing bud, and five glossy yellow petals, each with a minute flap-like *nectary* at the base, which form the most conspicuous part of the flower. Next, there are a large number of stamens, each consisting of a *filament* bearing an *anther*

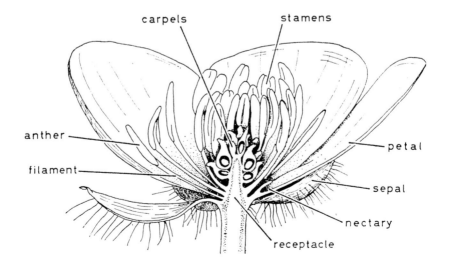

Fig. 2.1 Half section of a buttercup flower, *Ranunculus repens.*

which will open to release the powdery yellow *pollen*. Finally, in the centre of the flower there is a cluster of carpels, each one with a receptive *stigma* at the tip, and each containing an *ovule* which, after fertilisation, can develop into a *seed*.

A similar basic structure can be recognised in most flowers, but there is enormous variation in detail. First, there is variation in the number of parts. The buttercups usually have five sepals and five petals, a characteristic they share with a large propor-tion of dicotyledons, but other numbers are found, for instance, among the poppy family (Papaveraceae) and crucifers (Brassicaceae) which regularly have their parts in twos and fours, and among monocotyledonous plants (such as lilies and orchids) which typically have their parts in threes. Larger numbers are found, though less com-monly; the lesser celandine, unlike the related buttercups, has about 8-12 petals (but only three sepals), and in magnolias, water-lilies and cacti the petals may be very nu-merous. A buttercup has a large (and indefinite) number of stamens and carpels, but many plants have quite small and regular numbers of both. The stamens are often in rings (or 'whorls'), the members of each whorl usually equalling in number the sepals and petals, and the parts in successive whorls alternating in position. It is quite com-mon also to find as many carpels as corolla lobes. Thus in the cranesbill family (Ger-aniaceae), the flower parts are in fives throughout (with two whorls of stamens); in the lily and iris families (Liliaceae, Iridaceae), they are in threes. Often, however, there are fewer carpels than this. Flowers of many families have two carpels; flowers with single carpels are less usual, but the legumes (Fabaceae) are an important example. The number of ovules in a carpel varies greatly. In the buttercups and the grasses, among many other examples, there is only a single ovule in each carpel. At the other extreme, in an orchid of the genus *Maxillaria*, Fritz Müller estimated that a single capsule (com-prising three carpels) contained about one-and-three-quarter million seeds; each car-pel must have produced well over half-a-million ovules.

Flower-parts are often joined. This is brought about by the primordia of neighbour-ing flower-parts being carried up together on a complete ring of growing tissue early in the development of the flower. In this way, 'sepals' and 'petals' are often 'fused' into a tubular calyx and corolla, dividing into separate lobes some distance from the base. A calyx and corolla of this kind are called *gamosepalous* and *gamopetalous* (or *sympetalous*) to distinguish them from the *polysepalous* calyx and *polypetalous* corolla seen in flowers

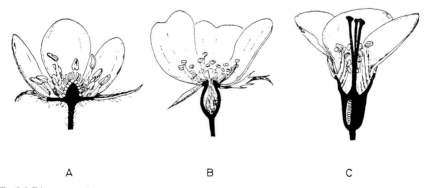

A   B   C

Fig. 2.2 Diagrammatic sections of flowers of **A**, strawberry (*Fragaria* × *ananassa*). **B**, dog rose (*Rosa canina*), and **C**, 'japonica' (*Chaenomeles speciosa*). Solid black indicates the extent of the receptacle and, in **C**, the carpel tissues fused to it.

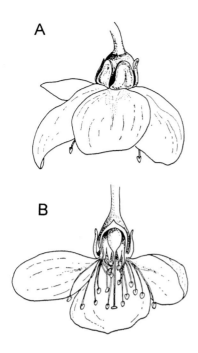

Fig. 2.3 Flower of cherry, *Prunus avium*.
A, side view. **B**, section showing sepals,
petals and stamens attached to the rim of
the cup-like receptacle, of which the inner
surface secretes nectar.

like the buttercup. In a plant with a
gamopetalous corolla, the stamens are almost always inserted on the inside of the
corolla-tube; they have become 'fused' to
it during development by exactly the
same process. Carpels are often fused into
a *syncarpous* ovary, contrasting with the
apocarpous ovary of the buttercup. The
stigma may be sessile on the carpel, as in
the buttercup, or it may be borne at the
tip of a more-or-less elongated *style*.
Where the ovary is fused, there may be a
single style and stigma, or a single style
branched at its tip bearing several stigmas, or there may be several styles. In the
last two cases, the number of styles or
stigmas usually indicates the number of
carpels.

The form of the ovary is closely bound
up with the growth of the receptacle.
Many flowers have a narrow, more-or-less
conical receptacle like that of the buttercup. In others the receptacle expands into
a disc, with the sepals, petals and stamens
inserted around the edge, and the carpels
in the centre. This is easily appreciated by
comparing a buttercup flower with a
strawberry (*Fragaria*) (Fig. 2.2a) or a saxifrage. In the flower of a plum or cherry
(Fig. 2.3) the receptacle forms a shallow
cup, secreting nectar, with a single carpel
at the centre. If a rose is cut in half, it will be seen that the receptacle forms a deep
flask enclosing the carpels, with a narrow opening at the top through which the styles
project in the centre of the flower (Fig. 2.2b). This suggests how it may have come
about that in many flowers the carpels are completely embedded in the tissues of the
receptacle, with the sepals, petals and stamens apparently borne on top of the ovary.
A flower like the buttercup is said to be *hypogynous* and to have a *superior* ovary. The
saxifrage, strawberry or rose are said to be *perigynous*. The flower of an apple or a
daffodil (Fig. 6.15, p.152) is said to be *epigynous*, and to have an *inferior* ovary.

So far we have assumed that a flower will contain all four kinds of floral members.
This need not necessarily be so. There may be only one whorl of members surrounding the stamens, or if there is more than one whorl they may be similar in colour and
texture, as in many lilies; in such flowers we speak of a *perianth* (made up of *tepals*),
rather than of a calyx and corolla. There may be clear evidence that particular
flower-parts are missing; thus there may be an obvious calyx, but no corolla, even
though a normal corolla is found in related plants. In some plants the perianth may
be missing altogether. Stamens are sometimes reduced to sterile *staminodes*, or they
may be lost without trace. The loss of a whorl of stamens may account for flowers in
which two adjacent whorls of floral members appear opposite one another (like the
stamens and petals of the primrose) instead of alternating in the normal way.

The buttercup has both stamens and carpels in the same flower; the flowers are hermaphrodite or bisexual, and this is the usual condition in the flowering plants. However, there are many plants which have stamens and carpels borne in separate *staminate* and *pistillate* (or *ovulate*) flowers – often loosely referred to as 'male' and 'female' flowers. In some families unisexual flowers are the rule, and seem to be a characteristic feature of long standing, as in many of our catkin-bearing trees. In other plants, for example the red campion (*Silene dioica*) (Fig. 12.10, p.341) and the shrubby cinquefoil (*Potentilla fruticosa*), the flowers resemble those of related hermaphrodite species; the pistillate flowers often have vestigial stamens and vice versa, and it is clear that they must be of relatively recent origin from hermaphrodite ancestors.

If male and female flowers are borne on the same individual, as in hazel (Fig. 9.6, p.275; Plate 4d) or marrow (*Cucurbita pepo*), the plant is said to be *monoecious*. If they are borne on separate individuals, as in red campion or the willows, the plant is *dioecious*. A species which bears hermaphrodite and female flowers on the same individual is *gynomonoecious* – and on different individuals *gynodioecious*. The corresponding terms *andromonoecious* and *androdioecious* are applied to plants that produce hermaphrodite and male flowers on the same individual and on different individuals respectively. In some species, the distribution of sex in the flowers is even more diverse, and there may be male, female and hermaphrodite either on the same or on different individuals, as in the salad burnet (*Sanguisorba minor*). Such species are described as *polygamous*. These

---

Box 2.1 The form and structure of inflorescences

Inflorescences are of two main kinds. In *racemose* inflorescences (A-D), the oldest flowers are at the base, and the youngest at the apex. If the individual flowers have no stalks the inflorescence is called a *spike* (A), if they are stalked it is a *raceme* (B). A branched raceme is called a *panicle* (C), and a raceme in which the pedicels of the lower flowers elongate to produce a flat-topped inflorescence with all the flowers at the same level is called a *corymb* (D).

In *cymose* inflorescences (E-F) the stem apex terminates in a flower, and subsequent growth is from side-branches lower down the stem. Two common types of cymose inflorescences are the *dichasial cyme* or *dichasium* (E), in which two branches are produced below each flower (as in many Caryophyllaceae and Gentianaceae), and the *scorpioid cyme* (F), in which the single branch below each successive flower is always produced on the same side, so that the upper part of the inflorescence curls like a scorpion's tail (for example, forget-me-not [*Myosotis*] and other Boraginaceae).

There are two types of inflorescence, the *umbel* (G) and the *head* or *capitulum* (H), in which it is often not obvious whether the inflorescence is fundamentally racemose or cymose. In an umbel, the individual flower stalks radiate from a point like the ribs of an umbrella; an umbel may be *compound*, consisting of an umbel of smaller umbels. In a head, the flowers are sessile (or nearly so), clustered tightly together at the tip of the stem, as in clovers (*Trifolium*) and composites (Asteraceae).

Inflorescences may be mixed in character. Thus many labiates (Lamiaceae) have racemose inflorescences of which the branches are dichasial cymes, and composites (Asteraceae) often have racemes or corymbs of capitula.

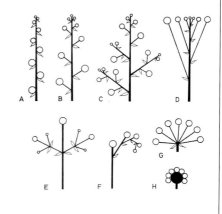

various states are discussed in more detail in Chapter 12.

Inflorescences

Individual flowers are often grouped together in inflorescences (Box 2.1). Some inflorescences are quite open, so that the individual flowers are borne wide apart; others are compact and the massing of flowers together can greatly enhance the floral display of a plant (both to our eyes and to visiting insects), as in many familiar species of the herbaceous borders and shrubberies in our gardens. In many dense inflorescences, most strikingly in the heads of composites (Asteraceae), it is the inflorescence rather than the individual flower that is the effective functional unit for pollinators; the non-botanist agrees with the insects in seeing a daisy or scabious head as 'a flower'. Faegri & van der Pijl (1979) have suggested the useful convention of using the word 'blossom' in a technical sense for this functional unit – most often a single flower (buttercup, mallow, bindweed, foxglove), sometimes a compact inflorescence (clover and composite heads, sallow catkins and many others), or occasionally part of a flower, as in *Iris*, where one flower forms three functional units (p.153).

Massing together of flowers in inflorescences enhances the floral display to attract pollinators, but it also increases the chance that a flower will receive pollen from the same plant; the form of inflorescence found in a particular species will generally reflect an evolutionary balance between conflicting demands (Wyatt, 1982).

The development and form of pollen

In the young bud, the stamen first appears as a projection on the developing receptacle, and the filament and anther are soon recognisable. The young anther becomes slightly four-lobed, and rows of cells, rather larger and with larger nuclei than their

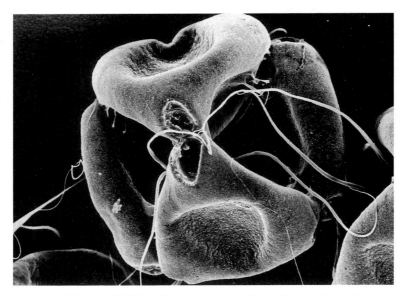

Fig. 2.4 In great willowherb (*Epilobium hirsutum*) the large pollen-grains remain together as tetrads at maturity. The 'viscin threads' streaming like ribbons from the grains help to attach the pollen to visiting insects. Scanning electron micrograph, × 500 (half the scale of Figs. 2.5-2.7).

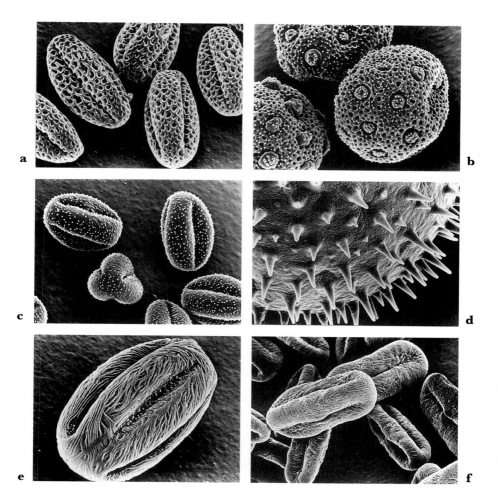

Fig. 2.5 Pollen grains. **a**, watercress (*Rorippa nasturtium-aquaticum*); three longitudinal furrows. **b**, red campion (*Silene dioica*) pores scattered over the surface of the grain. **c**, meadowsweet (*Filipendula ulmaria*). **d**, musk mallow (*Malva moschata*) the numerous pores on this very large grain are mostly hidden by the thick *pollenkitt*, but one can be seen top left. **e**, purple loosestrife (*Lythrum salicaria*); three long and three shorter longitudinal furrows. **f**, fennel (*Foeniculum vulgare*); three furrows, each with a pore in the middle.
Scanning electron micrographs of air-dry pollen, × 1000.

neighbours, become differentiated within each lobe. These cells divide by walls parallel with the surface of the anther; the outer cell-layers form the inner parts of the anther wall, while the inner layers divide a number of times to produce the *pollen mother-cells.* These then undergo reduction division (*meiosis*), each producing a *tetrad* of four *haploid* cells (with half the chromosome number of the *diploid* parent plant). In a few families of plants (e.g. the heather family, Ericaceae, and the rushes, Juncaceae) the four cells remain together so that the pollen grains occur in tetrads (Figs. 2.4, 2.6a). Usually the individual cells separate and round off before they develop the thick resistant wall characteristic of the mature pollen grain.

The outermost layer of the anther wall is a thin epidermis; the greater part of the thickness in a ripe anther is made up of the 'fibrous layer' immediately underneath. The innermost layer of the anther wall, the *tapetum*, is important because all the food material for the developing pollen mother-cells must pass through it. Towards the end of meiosis in the pollen mother-cells, the cells of the tapetum separate and begin to break down; in some families they lose their cell walls and form an amorphous mass of protoplasm around and between the tetrads. The development of the resistant, patterned outer layer of the pollen-grain wall, the *exine*, begins very early. The inner layer of the wall, the *intine* (made up of normal cell-wall material – cellulose and pectic substances), starts to develop soon after the break-up of the tetrads. Outside this, the *exine*, composed of *sporopollenin*, is initiated while the grain is still within the callose wall of the tetrad, but most of its thickness is added after the pollen-grains have been released from the tetrads, through the activity of the tapetum. The exine is extraordinarily resistant to decay or chemical attack, and is often elaborately sculptured into characteristic and sometimes strikingly beautiful patterns. Apart from providing nourishment for the developing pollen, the tapetum is the source of protein material which remains in the cavities in the exine wall and is important in 'sporophytic' self-incompatibility mechanisms (Chapter 12, p.324), and of the oily 'Pollenkitt' (from German, *Kitt* = cement) covering the surface of the mature grain (Heslop-Harrison, 1975a, b). These same pollen-surface proteins are also responsible for hay-fever in people who become allergic to them (Buisseret, 1982; Lichtenstein, 1993). Before the pollen grain is shed, its nucleus divides to form a *vegetative nucleus* (sometimes called the 'tube nucleus') and a *generative nucleus*.

Mature pollen grains are extraordinarily diverse in size and appearance (Figs. 2.5-2.7); around 30-40 µm is a common size, but pollen grains of some forget-me-not species (*Myosotis*) may be only 5 µm long (Fig. 2.6c), while those of some members of the cucumber family (Cucurbitaceae) may be 200 µm or more across. Pollen grains usually have one or more obvious pores or furrows in the wall, through which the pollen-tube emerges when the grain germinates (though a few, like the sedges and the poplars, lack obvious apertures). Many monocotyledons have a single germ-pore or furrow. Thus all grasses have a single pore (*monoporate*; Fig.9.3a); *Iris* and flowering-rush (*Butomus*) have a single furrow (*monocolpate*; Fig. 2.7e). Dicotyledons most often have three pores or longitudinal furrows round the equator of the grain, reflecting its relation to the three companion grains in the tetrad. Thus birch, hazel (Fig. 9.2b), stinging-nettle and genera in various other families have *triporate* grains, and *tricolpate* pollen grains with three longitudinal furrows are common in many families (e.g. Figs. 2.5a, 2.7a). Often there is a distinct pore in the middle of each furrow; and such *tricolporate* pollen grains also occur very widely (e.g. Fig. 2.7c-d). Many genera have

Fig. 2.6 (opposite) Pollen grains. **a**, strawberry tree (*Arbutus unedo*); pollen in tetrads, each grain has three longitudinal furrows with a central pore. **b**, hedge bindweed (*Calystegia sepium*); large grains with scattered pores. **c**, bittersweet (*Solanum dulcamara*); small rather smooth grains from a 'buzz-pollinated' flower (pp.125, 179). **d**, water forget-me-not (*Myosotis scorpioides*); tiny dumb-bell-shaped grains, each with six longitudinal furrows and three pores round the equator. **e**, honeysuckle (*Lonicera periclymenum*); a large grain with three short furrows. **f**, willow gentian (*Gentiana asclepiadea*). Scanning electron micrographs of air-dry pollen, × 1000.

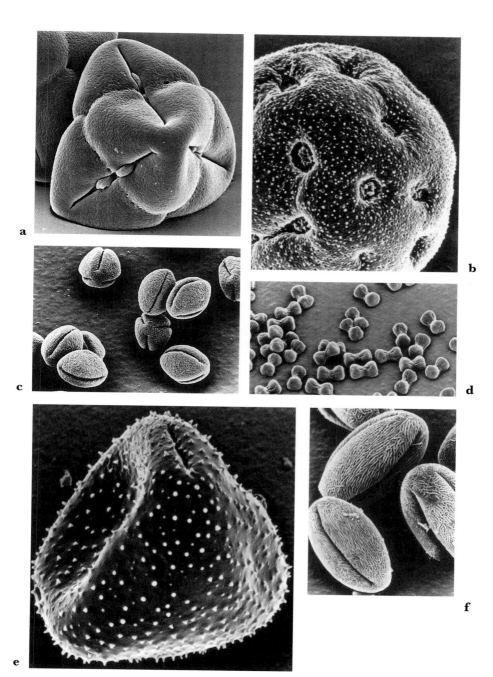

more than three pores or furrows, sometimes equatorial (as in the bedstraws [*Galium*] [Fig. 2.7b]), sometimes uniformly scattered (as in milkworts [*Polygala*], plantains [*Plantago*] [Fig. 9.3b] and mallows [*Malva*] [Fig. 2.5d]), or arranged in diverse other ways over the surface of the grain.

The resistance of the exines of pollen grains to destructive influences, and their remarkable diversity of form and surface ornamentation, together provide the basis of the technique of *pollen analysis*. This has been very widely used to reconstruct regional and local vegetation history from the pollen preserved in peats and sediments, for the dating of archaeological sites and artefacts, and in forensic investigations (Erdtman, 1969; Godwin, 1975; Faegri, Kaland & Krzywinski, 1989; Moore, Webb & Collinson, 1991). Pollen identification can help to establish the provenance of honey, and provide evidence for other forensic purposes, and analysis of the pollen loads of flower-visiting animals can provide information about their foraging and pollinating activities (e.g. Chapter 5, p.109ff.).

Pollination and fertilisation

The development of the ovule

At the time of flowering, each ovule consists of a roundish mass of tissue, the *nucellus*, closely surrounded by one or two *integuments* attached at the base of the ovule and leaving only a narrow open channel, the micropyle, at the apex. The opposite end of the ovule, the *chalaza*, is attached to the carpel by a short stalk, the *funicle*. Some time before the flower opens, the nucleus of a cell near the centre of the nucellus undergoes meiosis. The lowest of the resulting four cells enlarges, crushing the others out of shape, to form the embryo-sac.[1] Its nucleus divides into two, and each of the products divides twice to give two groups of four nuclei. Two nuclei, one from each quartet, come together and fuse to form a large single nucleus (the 'diploid fusion nucleus') in the centre of the embryo-sac. Of the three nuclei remaining at the micropylar end of the embryo-sac, one enlarges to form the egg. The other two, the 'synergids', and the three 'antipodals' at the chalazal end take no direct part in fertilisation. The embryo-sac enlarges at the expense of the neighbouring cells by a process of digestion, so that by the time it is ready for fertilisation it forms a large cavity in the nucellus, bounded by a very thin cell wall, and lined with dense, granular cytoplasm.

Subsequent events depend on the transfer of pollen grains to the stigma. This process of *pollination*, the means by which it is brought about, and its consequences, are the main subject of this book. Pollination is not an end in itself; it is merely a necessary prelude to *fertilisation* of the embryo-sac, and the development of the ovule into a *seed*.

Germination of pollen, the pollen-tube and fertilisation

On reaching the stigma, the pollen grain imbibes water and germinates, producing a pollen-tube. In some species, germination is easily observed under the microscope if pollen grains are placed in a sugar solution of suitable strength for a short time. With a few exceptions, such as the mallow, bellflower and cucumber families (Malvaceae,

---

[1] The type of embryo-sac described here is found in over 70% of the flowering plants that have been investigated. Other flowering plants differ considerably in details (but generally not in the essentials) of embryo-sac development and fertilisation.

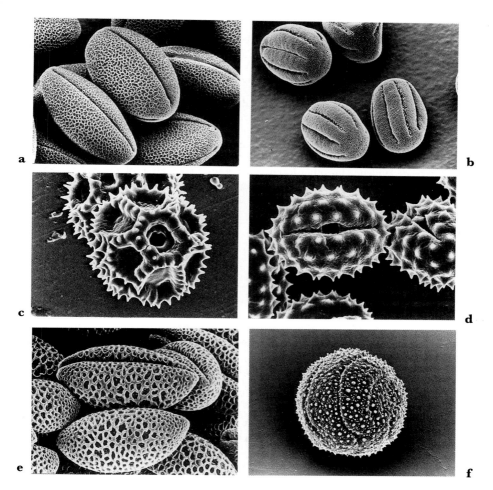

Fig. 2.7 Pollen grains. **a**, black horehound (*Ballota nigra*); three longitudinal furrows. **b**, marsh bedstraw (*Galium palustre*); grain with many longitudinal furrows. **c–d**, two composites (Asteraceae), both 'tricolporate' grains – with three furrows each with a pore in the middle. **c**, lesser hawkbit (*Leontodon saxatilis*). **d**, ragwort (*Senecio jacobaea*). **e**, flowering rush (*Butomus umbellatus*); grains with a single furrow. **f**, pipewort (*Eriocaulon aquaticum*); the furrow follows a spiral, like peeling an orange. Scanning electron micrographs of air-dry pollen, × 1000.

Campanulaceae, Cucurbitaceae) where a number of short pollen-tubes help to attach the grain to the stigma, only a single pollen-tube develops, whatever the number of germ-pores. The vegetative nucleus of the pollen grain passes into the growing pollen tube; it is followed by two *gametes* formed by division of the generative nucleus either before or after germination of the grain. The pollen-tube penetrates the (usually papillose) stigma surface, growing beween the cells of the tissues of the style, and so down through the style to the ovules.

The pollen-tube usually enters the ovule through the micropyle (more rarely through the chalaza), and then pierces the nucellus and the wall of the embryo-sac. The remaining living contents of the pollen tube are discharged into the embryo-sac, where one of the two male gametes fuses with the egg, and the other with the diploid fusion nucleus. The fertilised egg can now develop to form the embryo of the future seed. The nucleus formed by fusion of the second male gamete with the diploid fusion nucleus divides repeatedly in the course of formation of the *endosperm*, a tissue which provides for the nutrition of the young embryo, and often (as in the cereals) constitutes the food reserve of the ripe seed.

Box 2.2 The chemistry of flower pigments (illustration opposite)

(1) flavonoid pigments. These include the anthocyanins, responsible for the common blue, purple and pink colours of flowers. The core of an anthocyanin molecule is called an anthocyanidin. There are three common anthocyanidins – pelargonidin *(a)* giving scarlet colours, cyanidin *(b)* giving red and magenta, and delphinidin *(c)* giving mauve, purple and blue. In nature, the anthocyanidin is always compounded with one or more sugar molecules to form an anthocyanin, which is usually somewhat bluer than the corresponding anthocynanidin. The colour of anthocyanins can also be affected by complexing with metal ions (usually iron or aluminium) as for example in the blue flowers of cornflower (*Centaurea cyanus*) and *Hydrangea macrophylla*. The flavonoids also include the flavone, flavonol and anthochlor pigments, often collectively referred to as anthoxanthins. These are similar to anthocyanins, but the various molecules with a series of slightly different core structures produce a range of colours from pale ivory to deep yellow. The chalcone isoliquiritigenin *(d)* is one of the pigments of common gorse (*Ulex europaeus*); the flavone quercetagetin *(e)* is the principal pigment of the primrose (*Primula vulgaris*). The flavonoid pigments which appear ivory white to us absorb strongly in the ultra-violet region of the spectrum, so that they may appear 'bee-blue-green' to ultra-violet-sensitive insects (p.130ff.). Anthoxanthins and anthocyanins may jointly contribute to flower-colour, giving reddish and brownish shades, as in some garden snapdragons (*Antirrhinum majus*). In fact, full expression of the colours of all the common anthocyanidins requires the presence of flavone or flavonol 'co-pigments'. Co-pigmentation was formerly thought to be a special effect peculiar to plants with blue flowers,

but it now known that the increased blueness of these is due to a higher-than-usual ratio of flavone to anthocyanidin.

(2) betalains. These are related to the alkaloids, and their structure includes a 5- and a 6-membered heterocyclic ring, with a nitrogen atom in each. The betalain pigments are very restricted in their taxonomic distribution, occurring only in a small group of plant families in the order Centrospermae (Caryophyllales – but not in the Caryophyllaceae, which have flavonoid pigments), where they are responsible for some very vivid yellow, purple and magenta hues. Portulaxanthin *(f)* is a yellow pigment from flowers of *Portulaca grandiflora*; betanidin *(g)* gives the red-purple flower colours of cacti, Aizoaceae and *Bougainvillea*.

(3) carotenoids. Structurally, these important plant compounds are tetraterpenes in which the head-to-tail arrangement of the 5-carbon isoprene units from which they are built up is characteristically reversed in the centre of the molecule. The carotenes are tetraterpene hydrocarbons, yellow to orange or red in colour; β-carotene *(h)* is a common pigment, e.g. in daffodil flowers, lycopene *(i)* is the red pigment of tomatoes, also found in marigold flowers. The oxygen-containing xanthophylls, such as auroxanthin *(j)*, are very common in yellow to lemon-yellow flowers.

For more information on flower pigments see Goodwin (1976, 1988) and Harborne (1993). The distribution of anthocyanins among flowering plants is summarised by Harborne (1963).

Flowers and their pollinators: advertisement and reward

Insects and other animal pollinators obtain food from the flowers they visit, usually in the form of pollen or nectar. This is one side of a mutually beneficial relationship, the plants obtaining, in return, the services of the pollinators in carrying pollen from one flower to another. The interrelations between the adaptations of insect-pollinated flowers and those of the specialised flower-visiting insects are a classic instance of co-evolution (Chapter 14).

Although food is generally the tangible benefit pollinators get from flowers, they are

(a) pelargonidin

(b) cyanidin

(c) delphinidin

(d) isoliquiritigenin

(e) quercetagetin

(f) portulaxanthin

(g) betanidin

(h) β-carotene

(i) lycopene

(j) auroxanthin

usually attracted to the flowers in the first place by the flowers' colour or scent. There is thus an important distinction to be drawn between these *advertisements* (generally not of value in themselves to the visitor), and the all-important *rewards* of nectar and pollen (and sometimes other provisions) which for the visitors are often a crucial resource.

The colours of flowers: floral pigments

Coloured pigments occur in the floral parts of plants either dissolved in the cell sap, or in bodies called plastids in the cytoplasm. The commonest pigments found in the cell sap are *flavonoids*. They fall into two main classes – anthocyanins and anthoxanthins. All purple and blue shades and most red colours are due to anthocyanins, which occur commonly in leaves and stems as well as in flowers; their solubility in water will be familiar to anyone who has cooked red cabbage. Anthoxanthins range from yellow through ivory to white, and are the predominant pigments of many yellow, cream or white flowers. Another group of water-soluble cell-sap pigments, of much more restricted distribution, are the betalains. These are quite different chemically from the flavonoids, but fill a similar role in a limited group of families within the order Centrospermae. Betacyanins are red or purplish, and are responsible for the brilliant crimson and magenta colours of *Mesembryanthemum* and cactus flowers (Aizoaceae and Cactaceae) and *Bougainvillea* bracts (Nyctaginaceae) – and the red juice of beetroot (*Beta vulgaris*, Chenopodiaceae). The betaxanthins produce a range of yellow colours.

The pigments found in plastids are not water-soluble, but dissolve in oils and in fat-solvents. The best-known plastid pigment is chlorophyll, the leaf pigment responsible for absorbing light energy in photosynthesis. It also occurs regularly in green sepals and may sometimes influence the petal colour, or even, in the case of some green flowers, constitute the main floral pigment. The plastids which contain chlorophyll are known as chloroplasts, and they normally contain chlorophyll-a and chlorophyll-b, which are green, carotenes, which are orange, and xanthophylls, which are yellow. The last two also occur independently of chlorophyll as flower pigments; both belong to a class of pigments known as *carotenoids*, which include red and brownish colours as well as orange and yellow. Familiar examples of intense carotenoid pigmentation are carrot roots and tomato fruits; the pigmented plastids of a tomato are easily seen if a little of the flesh is teased out and examined under the microscope. (The fat-soluble carotenoids tend to accumulate in the drops of fat in a casserole, and in the grease round the washing-up bowl after the meal!) The plastid pigments occur in almost all yellow flowers and many others, and though not physically mixed with the cell-sap pigments, they may occur in the same petal. Such combinations of carotenoids and anthocyanins are responsible for the reddish and brown colours of wallflowers (*Erysimum cheiri*) and garden auriculas (*Primula* × *hortensis*).

A pigment is often confined to particular cell layers and two pigments in different layers can produce a difference in colour between the front and the back of a petal. Different distributions over the surface can produce patterning, as in heartsease (*Viola tricolor*), foxglove (*Digitalis purpurea*) and many other species which have *guide-marks* ('nectar-guides', 'honey-guides') on the corolla. More information on flower pigments can be found in Goodwin (1976, 1988). A useful summary of the distribution of anthocyanins among flowering plants is given by Harborne (1963).

The 'colour' of white petals is sometimes produced without the aid of any pigment, the effect of whiteness being a result of reflection and refraction at numerous cell surfaces and, in particular, the surfaces between the cells and air spaces within the tissue. The whiteness of snow has a similar cause. The glossy, matt or velvety textures of floral parts are determined mainly by the surface of the epidermal cells (Kay, Daoud & Stirton, 1981). An interesting case is the structure of the glossy yellow petals of buttercups (*Ranunculus* spp.). Except at the base (which is not glossy) the epidermal cells of the petals are thin and smooth-walled; they lack nuclei and are filled with an oily solution of a yellow carotenoid pigment. Beneath these outer cells is a layer of deep, thin-walled cells, densely packed with white starch grains, which ensure maximum reflection of light. In this way the brightness and intense yellow colour are achieved. The back of the petal is matt and coloured by a yellow plastid pigment (Parkin, 1928, 1931, 1935). *Ranunculus* species without yellow pigment, such as the water crowfoots and the alpine glacier crowfoot (*R. glacialis*), have flowers of a particularly brilliant and solid white. All these texture effects operate also in pigmented flowers, and provide the basis upon which the colours are displayed.

Attraction by scent: flower fragrances

The scents of flowers have always been a source of pleasure to the human race and have been the basis of a perfume industry since civilisation began – and the role of scent in attracting insects to flowers has been appreciated since the early days of pollination biology. Yet it is only in recent decades that it has become possible to study natural floral fragrances in a systematic and analytical way. This has come about through the development of sensitive gas-chromatographic techniques of chemical analysis, which make it possible to detect and identify minute quantities of volatile organic substances (Box 2.3).

Flower scents can be divided broadly into three categories. Commonest are the typical pleasant 'flowery' scents. These are usually due largely to terpenes and benzenoid compounds, though simple alcohols, ketones and esters and a wide range of other organic compounds may be present too (Box 2.3). The terpenes include such widespread fragrant substances as geraniol (in *Rosa* and other flowers), citronellol (rose petals) and limonene (*Citrus* and many other flowers), and many others. Scented compounds of the benzenoid type include eugenol (cloves, and flower scents), vanillin (various orchids including *Vanilla*: in many other flower fragrances too) and methyl salicylate ('oil of wintergreen'; many flower fragrances).

A more specialised group of fragrances are those that mimic insect pheromones, particularly the pheromones involved in sexual attraction of male bees by the females. Chemically, many of these are simple aliphatic hydrocarbons, alcohols, acids or esters, while some are terpenoid in nature. They thus embrace a similar range of structures to normal flower scents. Their effectiveness in manipulating insect behaviour seems often to depend not so much on single highly-specific substances as on broadly imitating a particular mix to which the insect responds, as in the orchids pollinated by scent-gathering euglossine bees (Chapter 7, p.216 – for these bees the scent is also a 'reward', not just an advertisement). Thus the volatile substances of the fly orchid (*Ophrys insectifera*) are rich in aliphatic hydrocarbons, and those of *Ophrys fusca* and *O. lutea* are rich in aliphatic alcohols and terpenoids, in both cases resembling secretions of the females of the species that pollinate them (Bergström 1978, Borg-Karlson 1990). The part played by fragrances in the pollination of orchids is discussed further in Chapter 7.

Flowers which rely for pollination on attracting dung or carrion-breeding flies have unpleasant dung or carrion-like smells. These are largely due to amines, ammonia and indoles, all substances produced in the normal course of the decay of proteins (Fig. 10.13). Examples of flowers producing foul smells of this kind are lords-and-ladies (*Arum maculatum*) and the African desert succulent *Stapelia* (Chapter 10, p.296).

Box 2.3 The chemistry of flower scents (illustration opposite)

Most flower scents are due to complex mixtures of volatile organic substances, and a very wide range of compounds have been detected in them (Knudsen, Tollsten & Bergström, 1993). These are of three main kinds: aliphatic substances (hydrocarbons, alcohols, ketones, esters, etc.), terpenoids, and benzenoid ('aromatic') substances.

(1) Aliphatic compounds (a-d opposite). These are often chemically simple substances. Pentadecane (*a*) is a paraffin hydrocarbon, in the scent of, e.g. *Magnolia, Rosa* (rose), *Actinidia* (kiwi-fruit) and *Ophrys* species; the alcohol hexanol *(b)* has a similarly wide distribution; ethyl acetate (*c*) is one of many fruity-scented esters commonly found in flower scents; jasmone (*d*) is a cyclic ketone, occurring in jasmine, honeysuckles and other flowers.

(2) Benzenoid compunds (e-h). These contain an 'aromatic' benzene ring. Vanillin (*e*) occurs in *Vanilla* and other orchids, and in many other fragrances; methyl salicylate (*f*) ('oil of wintergreen') occurs in many flower fragrances;

eugenol (*g*) is in cloves and various flower scents; methyl cinnamate (*h*) is an important ingredient of the orchid scents gathered by South American euglossine bees (p. 216).

(3) Terpenoids (isoprenoids) (i-o). These substances can be thought of as being built up of 5-carbon (isoprene) units (Vickery & Vickery, 1981). The 10-carbon monoterpenes (*i-l*) include geraniol (*i*), in *Rosa* and many other flowers, limonene (*j*), a major ingredient of the scent of *Citrus* flowers, but very widespread in other flowers too, linalool (*k*) and α-pinene (*l*), both common and widespread constituents of flower scent, and many others.

Sesquiterpenes with (normally) 15 carbon atoms (*m-o*) include α-farnesene (*m*) and caryophyllene (*n*), both very common and widespread, and as an irregular member with only 13 carbon atoms, the intensely-scented β-ionone (*o*) of sweet violets (*Viola odorata*), present also in *Freesia* and various orchids.

Flower scents are generally secreted rather diffusely over broad areas of flower parts, particularly the corolla. But scent-production is often in varying degrees localised within the flower. Thus the corona is responsible for the scent of fragrant *Narcissus* flowers. In the roses *Rosa rugosa* and *R. canina*, the scent of the flower as a whole is dominated by terpenoid and benzenoid alcohols produced by the petals; sepal odours include a high representation of sesquiterpenes, and anthers and pollen have a diversity of compounds, few of which are shared with the perianth (Dobson *et al.*, 1987, Bergström, 1991). Many flowers show areas of more concentrated scent-production, frequently coinciding with visible or ultraviolet guide-marks, which help to guide visiting insects to the source of nectar, as in field buttercup (*Ranunculus acris*, Bergström *et al.*, 1995). In some flowers there are 'scent-marks' of this kind without corresponding optical guide marks; the small scales at the base of the limb of the petals in white campion (*Silene latifolia*) are an example. The foul smells of deceptive fly-pollinated flowers are often diffused from specialised *osmophores* ('scent-bearers', Chapter 10).

Rewards to pollinators: pollen and nectar

Probably the first visitors to flowers, back in the Mesozoic, ate pollen, sucked up stigmatic secretions and chewed at the softer parts of the flower – all, to some extent, damaging activities to weigh against the benefit to the flower in more efficient transport of pollen. As a reward for pollinators, pollen is 'expensive' to the plant, because it is rich in nitrogen and phosphorus, two elements often in growth-limiting short supply in natural habitats. The provision of carbohydrate in the form of nectar makes lighter demands on the plant's resources; water, carbon dioxide and sunlight are in comparatively abundant supply – though profuse nectar production can still consume a substantial fraction of the plant's total photosynthetic production during the flower-

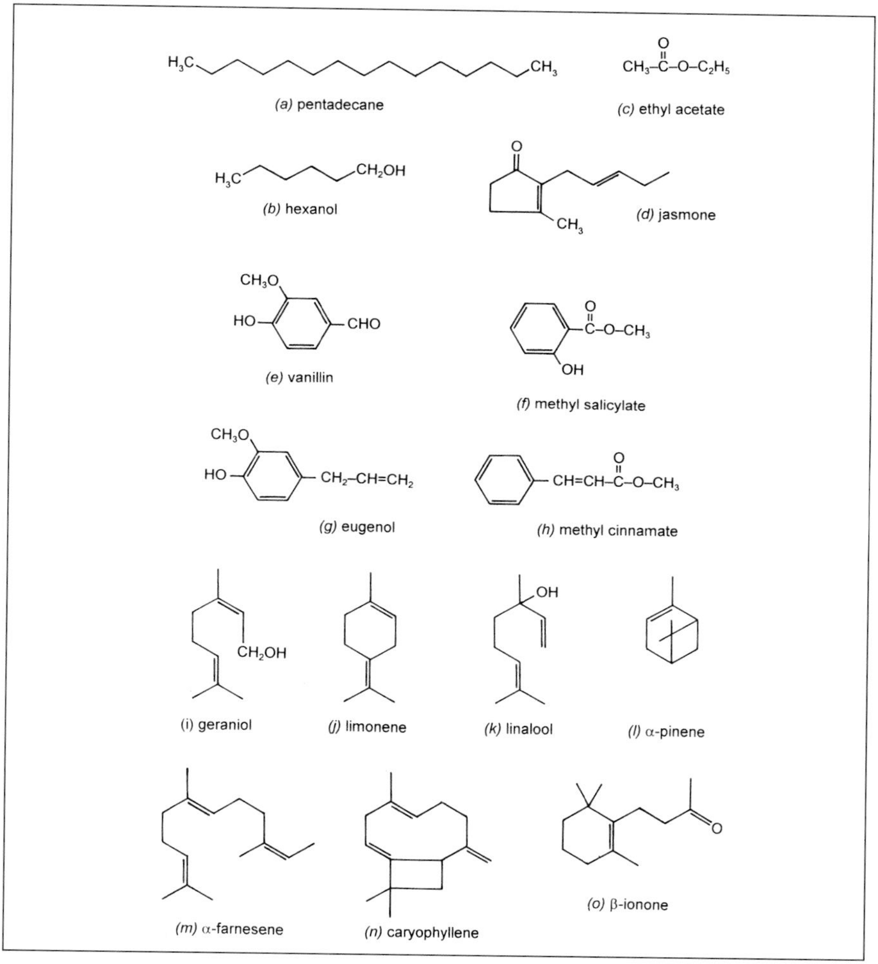
H3C
CH3
(a) pentadecane
CH3–C–O–C2H5
(c) ethyl acetate
H3C
CH2OH
(b) hexanol
O
CH3
(d) jasmone
CH3O
HO
CHO
(e) vanillin
C–O–CH3
OH
(f) methyl salicylate
CH3O
HO
CH2–CH=CH2
(g) eugenol
CH=CH–C–O–CH3
(h) methyl cinnamate
CH2OH
(i) geraniol
(j) limonene
OH
(k) linalool
(l) α-pinene
(m) α-farnesene
(n) caryophyllene
O
(o) β-ionone

ing period (Southwick, 1984; Pyke, 1991). Carbohydrate, as a source of energy,[1] is the principal food requirement of adult winged insects, whose larvae mostly get the protein necessary for growth from sources other than flowers. Bees, however, differ from other insects in that their larvae feed on flower foods collected by the adults. Since the larvae, in addition to nectar, require much protein for growth, this creates a large demand for the protein-rich pollen. In order to feed their young as well as themselves the bees make far more visits to flowers than do other insects, enhancing their effectiveness as pollinators. Their evolution must have brought about selection for in-

[1] 1 g of sugar or other carbohydrate yields about 17 kJ of energy.

creased pollen production in many insect-pollinated flowers, in spite of the fact that nectar is more economical to produce.

Pollen and pollen-flowers

Pollen is a very nutritious material. It contains 16-30% or more of protein, 1-7% starch, 0-15% free sugar and 3-10% of fat by dry-weight (Harborne, 1993), as well as significant amounts of phosphates and all the other essential ingredients of living cells. The exine of the pollen grain is completely indigestible, but is no barrier to the digestive enzymes of the insects that feed upon it, and assimilate the cell contents; the pollen grains generally burst open in the insect's gut. In one sense, from the plant's point of view, pollen eaten by insects is pollen wasted. On the other hand, even the most specialised pollen-gathering insect cannot groom its body completely clean of pollen grains, and this pollen on the body surface can bring about effective pollination.

Some bee-pollinated flowers have become specialised pollen-flowers, offering visitors abundant pollen but little or no nectar. Examples are the poppies (*Papaver* spp.), peonies (*Paeonia* spp.) and the sunroses and rockroses (*Cistus* and *Helianthemum*; Cistaceae), whose big colourful flowers with masses of stamens yield profuse quantities of pollen, and the conspicuous feathery inflorescences of the bee-pollinated meadow-rue species *Thalictrum flavum* and *T. aquilegifolium*, with their numerous cream or lilac-coloured stamens. A specialised group of pollen flowers, including bittersweet (*Solanum dulcamara*) and kiwi-fruit (*Actinidia deliciosa*) are adapted to 'buzz pollination' by bumblebees (Buchmann, 1983; Harder & Barclay, 1994; Chapter 6, p.179).

Nectar and nectaries

Nectar is essentially an aqueous solution of sugars, ranging in sugar content from about 15% to 75% by weight. Only three sugars occur in quantity, sucrose, fructose and glucose. Sucrose (cane sugar) is a *disaccharide* which can be broken down into equal parts of the two *monosaccharides*, glucose and fructose (Fig. 2.8), by the action of the enzyme invertase. Nectars may contain sucrose only, or mixtures in various proportions of all three or any two of the sugars; a complete range can be found between sucrose-dominated nectars and nectars in which hexoses make up 90% or more of the sugars present; glucose and fructose are not necessarily present in similar amounts in nectars (Percival, 1961; Baker & Baker, 1983a, b). There is some relationship between the nectar type and the form of the nectary and type of visitor (Table 2.1), though many exceptions occur. Generally, sucrose-rich nectars tend to be associated with

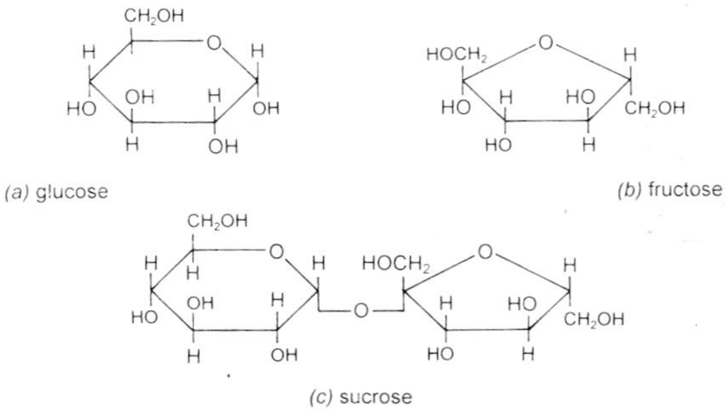

Fig. 2.8 The common sugars in nectar. *a*, glucose; *b*, fructose; *c*, sucrose. Sucrose is split by the enzyme invertase into equal parts of glucose and fructose, both *hexose* sugars with 6 carbon atoms. (Glucose is shown in the α form, and fructose with carbon atom 2 at the left of the ring.)

Table 2.1 Pollinators and the sucrose/hexose composition of nectar.

The figures in the table are percentages of the total number of species of flowers examined that fell into the four categories of sugar composition, for each of the major categories of pollinators. Of the less common pollination-types, wasp flowers (18 spp.) appear to show a similar range of nectar composition to those pollinated by long-tongued bees; beetle flowers (9 spp.) probably roughly match those mainly visited by Lepidoptera, but more species would need to be examined to confirm this. (Simplified and recalculated from Baker & Baker, 1983a, b.)

Principal pollinators	Sugar ratio [sucrose/(glucose + fructose)]				Number of species
	<0.10	0.10-0.50	0.50-1.00	>1.00	
Hummingbirds	0	13	32	55	140
Sunbirds and honeyeaters	74	23	3	0	57
Hawkmoths	3	13	31	53	61
Other moths	7	32	26	35	43
Butterflies	7	23	32	38	75
Short-tongued bees	44	39	11	6	263
Long-tongued bees	6	37	24	33	203
Bats	29	62	6	3	34
Flies	40	38	10	12	72

deep (or 'concealed') nectaries and with long-tongued bees, butterflies and moths, or birds. Hexose-rich nectars are often associated with freely-exposed nectaries and visits by short-tongued bees and flies; the nectar of tropical bat-pollinated flowers (Chapter 8) is also typically hexose-rich. Bird, bat and butterfly-pollinated flowers tend to produce rather dilute nectars (15-25%), whereas the nectar of bee flowers is often more than 50% sugar. However, the concentration of nectar can vary widely through evaporation, or uptake of moisture from the air.

Much of the nectar collected by the social bees is stored as honey. The average sugar content of the nectar collected by bees is probably about 40%. In the processing of nectar to honey, much of the sucrose is broken down to glucose and fructose by the enzyme invertase, added to the nectar by the bees, and water is lost by evaporation. The sugar content of honey is typically about 80%. Converting nectar to honey requires the expenditure of a good deal of energy by the bees; honeybees seldom collect nectar with a sugar content below 20%, and they may not make a net energy gain for the colony until the sugar content is at least 30% (Butler, 1954). Heinrich (1975a) has

Table 2.2 Pollinators and the amino-acid content of nectar (from Baker & Baker, 1983b).

Principal pollinators	Mean amino-acid concentration (μmol/ml)	Number of determinations
Carrion and dung flies	12.50	9
Butterflies	1.50	118
Non-hovering moths	1.06	78
Bees, butterflies	1.02	257
Wasps	0.91	44
Bees	0.62	715
Flies (general)	0.56	97
Hawkmoths	0.54	65
Hummingbirds	0.45	150
Bats	0.31	23
Passerine birds	0.26	21

Box 2.4 The chemistry of floral oils

The floral oils that have been analysed are made up mainly of free fatty acids with carbon-chain lengths from C_{14} to C_{20} (usually acetoxy-substituted in the 3-position of the carbon chain) such as 3-acetoxy-octodecanoic acid (*a*, below, one of several acetoxy-substituted free fatty acids typically found in the floral oils of *Calceolaria* and other genera),

or their mono and diglycerides; the 1,2-diglyceride of 3-acetoxy-*trans*-11,12-octadecanoic acid (*b*) is a major component of the floral oils of *Calceolaria* and *Lysimachia*. Floral oils may also contain smaller amounts of long-chain hydrocarbons, aldehydes and esters. (See Buchmann, 1987.)

(a)

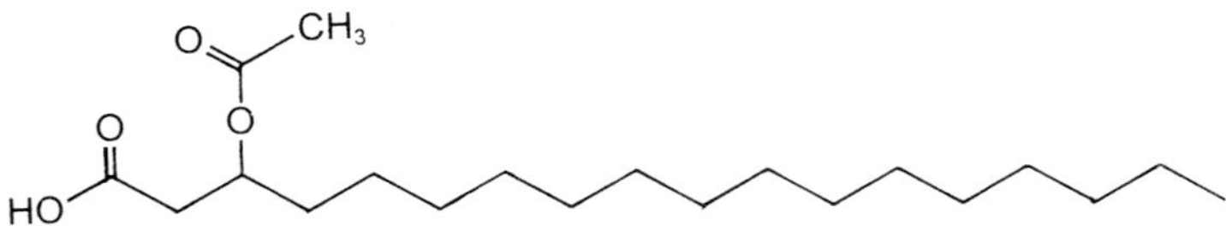

(b)

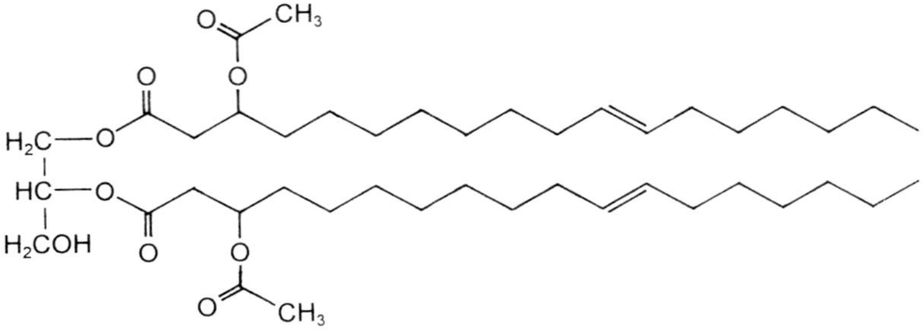

calculated that, 'One pound of white clover honey represents approximately 17,330 foraging trips. Since the bees visit about 500 flowers during an average foraging trip of 25 min., each pound of honey represents the food rewards from approximately 8.7 million flowers, and 7,221 hr of bee labor.' White clover has very small flowers, each yielding only about a twentieth of a milligram of sugar, but they are numerous and well placed for efficient foraging. Other flowers may provide several milligrams of sugar at a visit, but more time and energy may be required to find and fly between them. Honey is remarkably cheap, considering the labour involved!

In fact nectar is *not* simply a sugar solution. Almost all nectars contain (amongst other things) measurable quantities (~ 1 mMol/L) of amino-acids (Baker & Baker, 1973; 1986), and these may be nutritionally significant to many flower visitors, especially those that do not also feed on pollen or other protein-rich materials. Even if its main requirement is for energy, a long-lived adult insect still has a need of nitrogenous material for body maintenance. There are correlations between the amino-acid content of nectar and the type of visitor (Table 2.2). The nectar of butterfly flowers averages over twice the amino-acid content of that of bee flowers. The nectar of specialised bird-pollinated flowers, and flowers pollinated mainly by nectar-seeking flies, seems generally to be low in amino-acids, averaging about half the content found in bee flowers. The nectars richest in amino-acids are found in flowers that specifically attract carrion or dung flies.

In different species, nectaries may involve almost any part of the flower. Nectar secretion is often associated with the bases of the petals and stamens; in perigynous flowers nectar is commonly secreted from the disc of the receptacle. Whole floral members may be specialised as nectaries (as in the horn-shaped nectaries of the hellebores, *Helleborus*), or nectar secretion may be confined to a small part of such a member, as in the tiny pockets at the bases of the petals of buttercups, or the 'tails' at the bases of the lower stamens in pansy and violet (*Viola*) flowers. *Viola* species are an example of the not uncommon situation where nectar is secreted by one floral member, but accumulates in a structure formed from a neighbouring but different part of the flower, in this case the spur at the base of the lowermost petal.

Oil-providing flowers

Nectar and pollen are not the only floral rewards. In 1969 Vogel described 'flowers offering fatty oil instead of nectar', and flowers in genera of five families were confirmed as having specialised oil-secreting 'elaiophores' – glandular hairs or specialised regions of the floral epidermis. Since that time, many other flowers have been added to the list, so that by 1987 it was possible to write, 'Flowers offering fatty oils instead of, or in addition to, nectar and/or pollen are found in ... 10 families, 79 genera, and 2,402 species of flowering plants worldwide.' (Buchmann, 1987). Oil flowers are found most abundantly, but by no means exclusively, in tropical American savannas and forests. They occur in families of very diverse taxonomic affinities. About 7.5 % of the iris family (Iridaceae), 3.7% of orchids (Orchidaceae), 97% of Malpighiaceae, 7.3% of Scrophulariaceae and 6% of Gesneriaceae are known to possess floral elaiophores. Some genera made up largely or wholly of oil-providing flowers occur within families in which the usual reward is nectar or pollen. Thus in orchids, the genus *Disperis* and its allies are all oil flowers, as are almost all the species of the large genus *Calceolaria* in the figwort family (Scrophulariaceae), and some 40% of the species of the widespread north-temperate genus *Lysimachia* (yellow loosestrife and related flowers) in the primrose family (Primulaceae) (Box 2.4; see also pp.112, 115).

Some bees use the floral oils with or in place of nectar in the pollen provisions for their larvae. Others also use the oils for water-resistant cell-linings, and perhaps also for their own nutrition. The floral oils are a rich source of energy; on a dry-weight basis their calorific value is about twice that of carbohydrates such as the sugars in nectar.

Flowers of the tropical genera *Dalechampia* (Euphorbiaceae) and *Clusia* (Clusiaceae [Guttiferae]) produce resins (apparently mixtures of triterpenes [C_{30}]) which are collected by visiting bees and used in nest construction (Armbruster, 1984).

Flowers and insects: some general considerations, and examples

Flower form and pattern: guide marks

The previous few pages have given a matter-of-fact description of the structure of flowers, part by part. But, of course, we see flowers as integrated wholes, and the overall effect of the form and pattern of the flower is important also to insect visitors. Petals and corolla-lobes often bear *guide marks* (Plate 1) – so-called 'nectar-guides' or 'honey guides'. Guide marks often show up more strongly in the UV, and many flowers show strong UV guide marks but none in visible light (Kay, 1987; Harborne, 1993). These marks highlight the form and architecture of the flower as the visitor approaches. Heinrich (1975) stressed their importance as 'close-in signals', especially in flowers with large petals, or massed into inflorescences, saving the visitor time and energy in its foraging. Scent and touch are probably also major cues in the actual location of nectar. In some plants the petals or the guide marks change colour when the flower has been pollinated and the ovules begin to develop; the flowers then cease to engage the attention of visiting insects, but the petals may remain unwithered for some time and still contribute to the general floral display of the plant (Weiss, 1995). The ways in which visiting insects respond to visual and other features of flowers are considered in Chapters 3-5.

Alighting places for pollinators, and the size and position of flower parts

Flying insects need a landing-place, and this is usually provided by the perianth; zygomorphic flowers such as deadnettles (Chapter 6) typically have a prominent lower lip which serves this purpose. In some flowers, the styles and stamens form a more-or-less well-defined column in the centre, and this is often grasped and used as a foothold by visiting insects, especially bees, as they manoeuvre themselves in the flower. Compared with the flowers of wind-pollinated species, with their flexible, pendulous catkins, or long filaments, large hinged anthers, and feathery stigmas, insect-pollinated flowers typically show greater rigidity in their floral parts. This gives the insects something firm to grip, while stamens and styles need to be rigid enough to make firm contact with visitors' bodies – and to withstand a degree of rough handling. Flowers borne singly are generally rather larger, but not enormously larger, than the insects that pollinate them; in particular, it is obviously essential that the size and disposition of the stamens and stigmas should be appropriately matched to the flower's pollinators. This size relationship tends to set an upper limit to the size of individual flowers, and hence to how conspicuous they can be from a distance.

Where flowers are massed in dense inflorescences, it is the surface of the inflorescence rather than the individual flower that provides the alighting platform, and the dimensions of the flowers can be matched to the mouth-parts rather than the bodies of their visitors. These features are very obvious in species which bear their flowers in corymbs or umbels; an inflorescence of ragwort or hogweed spreads out a broad table over which visitors can wander unhindered as they feed on nectar or pollen from the small individual flowers.

Cross- and self-pollination, dichogamy, self-incompatibility

Darwin's belief that 'Nature ... abhors perpetual self-fertilisation,' has been quoted already; it was reasonable enough in the light of what was known at the time. But for the following half-century or more it was commonplace to regard the structures of flowers rather simplistically in terms of mechanisms favouring cross-pollination – in a way that now often looks to us unthinking and uncritical. The breeding systems of plants (Chapter 12) are very diverse, the balance between outcrossing and selfing varying widely depending on the life-history and ecology of the particular species. Many plants are, indeed, largely or even exclusively cross-fertilised. This is often brought about by self-incompatibility to their own pollen, rather than because of their pollination mechanism. But pollination is important in regulating breeding systems – in determining the balance beween cross- and self-fertilisation – and all gradations exist between flowers in which self-pollination is mechanically almost impossible and those (such as the 'cleistogamous' flowers of violets, self-pollinated before the flower opens) in which it is inevitable and regularly results in self-fertilised seed.

In flowers which rely on insects for cross-pollination, the anthers and stigma are usually separated by at least a small gap, and sometimes quite widely as in a number of the flowers described in Chapter 6. Transfer of pollen from the stamens to the stigma of another flower depends on the pattern of movement of the visitor as it enters and leaves the flower. There is also often a separation in time – *dichogamy* – between dehiscence of the anthers and the shedding of pollen, and maturation of the stigma. Flowers which shed their pollen before the stigmas are ready for pollination are *protandrous*, and those in which the stigma matures first are *protogynous*. Neither dichogamy nor spatial separation of anthers and stigmas within the individual flower will prevent pollination of the flower by pollen from another flower on the same plant. Such *geitonogamy* (from Greek, 'neighbour-marriage') is genetically no different from pollination of a flower by its own pollen (*autogamy* – 'self-marriage'); how likely it is depends on how many flowers are open on a plant at the same time, on pollinator behaviour, and on the level of pollen carryover (p.419). Often less than 20% of an insect's pollen load is deposited on each flower it visits, so a mass floral display may be less than it seems at first sight (Robertson, 1992).

The examples in Figs. 2.9-2.16 illustrate illustrate some of the structural and other features of flowers and other topics that have been discussed in this chapter.

Fig. 2.9 **a, b** Creeping buttercup (*Ranunculus repens*, Ranunculaceae). Buttercups are visited by a very wide range of insects for nectar and pollen; the nectar secreted beneath a small scale at the base of the petals is accessible even to short-tongued visitors like the small hoverfly *Neoascia podagrica* in **a**; the hoverfly (*Melanostoma* sp.) in **b** is feeding on pollen. The flowers are generally self-incompatible (Harper, 1957; Coles, 1971). Buttercups are an almost uniform yellow to us, but the centre absorbs UV strongly, so appears a contrasting colour to visiting insects (compare marsh marigold, *Caltha palustris*, Fig. 5.20, p.142). In the Norwegian mountains, Totland (1993, 1994) found the field buttercup, *R. acris*, visited mainly by muscid and anthomyiid flies. During periods of observation at three sites, a flower was visited on average once every 40 minutes to 2 hours; a single fly visit in July produced an average of 5.3 seeds -*c.* 18% of the total potential seed set.

Fig. 2.10 Lent rose (*Helleborus orientalis*, Ranunculaceae). The large pale green to purple-red sepals are the most conspicuous part of the flower; the tubular nectaries in a ring beween the sepals and the stamens correspond to the petals of a buttercup flower. *H. orientalis* is a native of the east Mediterranean. In cultivation in north-west Europe it produces little or no nectar, but the flowers are visited by pollen-collecting bees on sunny days.

Fig. 2.11a Spring pasque flower (*Pulsatilla vernalis*, Ranunculaceae), a characteristic plant of the thin turf on high exposed ridges in the Alps. The flowers are visited and pollinated by a variety of bees. In pasque flowers (and anemones) the conspicuous brightly coloured 'petals' are in fact the sepals, as in the hellebores. Nectar is secreted at the bases of the outermost stamens.

Fig. 2.11b Wood anemone (*Anemone nemorosa*, Ranunculaceae). The wood anemone produces no nectar, but the flowers are visited for pollen by small beetles and a wide variety of flies and bees. Self-fertilisation is prevented by the self-incompatibility of the flowers to the plant's own pollen.

Fig. 2.12 Corn poppy (*Papaver rhoeas*, Papaver- aceae). The anthers and stigmas mature together; the self-incompatible flowers last only a day. There is no nectar. Bumblebees and honeybees visit the flowers for pollen, scrambling round the flower on their sides, as they work the pollen tinto their pollen-baskets. Bees show a high degree of constancy to particular poppy species, probably based on flower shape and subtle colour differences (McNaughton & Harper, 1960).

Fig. 2.13 Bramble (*Rubus fruticosus* Rosaceae). Bramble flowers produce abundant nectar and pollen, and are visited by a very wide range of insects, including beetles, butterflies, honeybees and bumblebees, and flies – especially hoverflies, like *Volucella bombylans* seen here. Most brambles produce seed by agamospermy (Chapter 12, p.348), but are *pseudogamous*, requiring pollination to initiate endosperm and seed development.

Fig. 2.14 Greater stitchwort (*Stellaria holostea*, Caryophyllaceae). The flowers are protandrous, the two whorls of stamens maturing successively and shedding much of their pollen before the stigmas become receptive. The flowers are visited for pollen and nectar by a wide range of flies (Parmenter, 1952a), bees (here a species of *Nomada*) and other insects, and may also be self-pollinated.

Fig. 2.16 (above) Wood spurge (*Euphorbia amygdaloides*). The 'flowers' of spurges are compound structures, made up of one reduced female flower and several reduced male flowers surrounded by a cup, called the cyathium, resembling a perianth but derived from bracts. They are typically yellowish green; the rim of the cup carries large horizontal 'glands' (here crescent-shaped) which secrete nectar over the whole of their upper surface. The styles of the stalked female flower emerge first; after pollination the stalk elongates and the ovary hangs down over the side of the cup (visible here just left of the insect). After this, the male flowers emerge and shed their pollen. The main visitors are small flies of various families, like the hoverfly *Baccha elongata*, seen here, but sawflies, ichneumons, ants and occasionally beetles, wasps and bees also visit the flowers (Figs. 5.2, 5.6).

Fig. 2.15 (above) Meadow cranesbill (*Geranium pratense*, Geraniaceae). This flower behaves in a broadly similar way to greater stitchwort, five stamens at a time becoming erect, shedding their pollen and then returning again to lie close to the petals. However, the stigmas become receptive so late in relation to the pollen-shedding of the inner anthers that self-pollination cannot take place. Bees are the chief visitors to these showy blue flowers, usually foraging for nectar. The bees usually alight on the central organs of the flower, so the stamens need to be erect in order to be touched by them. See also Fig. 12.5, p.331.

3

The Insect Visitors
I: Beetles, Flies and Some Others

To understand the structure and behaviour of flowers in their interactions with potential insect pollinators, we need to know how the insects are built, and something about
their sensory abilities and behaviour. These are the subjects of this chapter and the
two that follow it.

Differences in the mouth-parts and sensory capacities of insects provide the background against which the form, colours and scents of flowers can be interpreted
(Chapters 2 and 6). The insect sense of touch depends on hair-like organs scattered
over the body with varying density. The perception of chemicals in the vapour state
(sense of smell) and in solution (sense of taste) comes through thin-walled discs in the
cuticle, or thin-walled projecting or sunken hairs. These occur on the feet, the antennae and the mouth-parts. Nerve cells at the base of chemo-sensory hairs have processes running up inside each hair; these are exposed to the environment through a
pore at the tip of the hair in the case of a taste-organ, but through hundreds of much
smaller pores in the hair if the function is olfactory. The insect compound eye is a
highly developed organ which can produce a detailed retinal image. Its facetted structure results in good ability to detect movement. Movement and length of contour
seem to be important in resolving patterns. Most insects are sensitive to ultra-violet
radiation but have little or no sensitivity to red, so that their visible spectrum covers
the wavelengths 300 nm to 650 nm, compared with 400 nm to 750 nm in man (Burk-
hardt, 1964; Menzel, 1990). Many insects possess colour vision, and much work on
this faculty has made use of the insects' visits to flowers. Social bees have some ability
to respond to sound (vibrations in the substratum or the atmosphere). (For further
information, see the general account of insect senses in Imms [1947, in this series] and
the textbook by Mordue *et al.* [1980] or the account in non-technical German or Eng-

Fig. 3.1 Wild cockroach
(*Ectobius pallidus*,
Dictyoptera), on sea carrot
(*Daucus carota* ssp. *gummifer*);
Alderney.

Table 3.1 Orders of insects recorded as flower visitors.

APTERYGOTA

Collembola, springtails
Generalised mouth-parts. Feeding on pollen of *Chrysosplenium* and resting.

EXOPTERYGOTA

Dermaptera, earwigs
Mandibles like those of beetles. Hiding in and eating flowers of many species.

Dictyoptera, cockroaches
Ectobius licking nectar of *Spiraea*, *Filipendula* and Apiaceae (Fig. 3.1).

Psocoptera, booklice
Passing visitors.

Odonata, dragonflies, damselflies
Passing visitors.

Orthoptera, grasshoppers, crickets
Passing visitors.

Plecoptera, stoneflies
Mandibles like those of beetles. Feeding on pollen of *Caltha*, *Helianthemum*, *Alchemilla*, *Rosa*. Feeding on nectar of *Listera ovata*.

Hemiptera, bugs
Piercing and sucking mouth-parts. Aphididae: feeding on plant sap, sometimes in the flowers. Miridae (Capsidae): regular visitors for pollen from open flowers in Asteraceae, Apiaceae. Anthocoridae (Cimicidae): feeding on nectar from *Salix* etc. Lygaeidae: *Lygaeus* sucks juices, perhaps a minor pollinator of the pollen flower *Adonis vernalis*.

Neuroptera, lacewings
Biting mouth-parts. Recorded on Apiaceae, possibly feeding on nectar.

Mecoptera, scorpion-flies
Biting mouth-parts on long snout. Feeding on nectar of Apiaceae and Asteraceae and commonly on *Polemonium caeruleum*.

Trichoptera, caddis-flies
Mouth-parts adapted for liquid feeding. Feeding on nectar of *Hedera helix*, *Listera ovata*, Apiaceae, *Valeriana officinalis*, *Nuphar* spp.

Thysanoptera, thrips (note: thrips is both singular and plural) (Fig. 3.2)
Some species are pollinators, see text.

ENDOPTERYGOTA

| Coleoptera | Lepidoptera | ⎤ of definite importance for pollination and |
| Diptera | Hymenoptera | ⎦ dealt with separately in this book |

(Burkill, 1897; Müller, 1883; Knuth, 1906-1909; Willis & Burkill, 1895-1908; Hagerup, 1950a; Pigott, 1958; Porsch, 1957; Popham, 1961; Corbet, 1970; Borge Pettersson; MCFP)

lish by Barth [1982, 1985] which is related to pollination.)

The insects may be classified into three major groups. The most primitive of these comprises four orders with no trace of wings (Apterygota) and includes the Collembola, or springtails. The other insects are winged as adults and are divided into those in which the developing insect becomes gradually more like the adult at each moult (Exopterygota), and those in which the larva, usually very unlike the adult, enters a more or less dormant pupal phase during which the transformation to adult takes place (Endopterygota) (Table 3.1).

Apterygota and Exopterygota are incidental visitors to flowers or occasional con-
sumers of nectar or pollen. They may sometimes transfer pollen, but their importance
as pollinators must be negligible; exceptions to this are the stone-flies (Plecoptera),
bugs (Hemiptera) and scorpion-flies (Mecoptera) as shown in Table 3.1, and the
thrips, described separately below. Regular specialised flower visiting (other than by
thrips) is recorded only in the four orders of the Endopterygota. Of these, the beetles
(Coleoptera) and the flies (Diptera) are described in this chapter.

Thysanoptera: thrips

These minute insects have a pupal phase but are classed as Exopterygota because the
wing buds are visible in the larval stages. They are usually 2 mm long, or less, and
narrow-bodied (Fig. 3.2). Their four wings are each composed of a central shaft and
a row of hairs to the front and back. Together with staphylinid beetles of similar size,
they are popularly known as thunder-flies, as they enter houses in hot weather (and get
behind picture-glass!). They have specialised asymmetric piercing and sucking mouth-
parts and are sometimes pests of crops, either through feeding on them or by transmitting virus diseases. However, some species inhabit flowers and are specialised
feeders on pollen, sucking out the contents of the grains after piercing them with their
mouth-parts (Kirk, 1984b). Most thrips caught in flowers carry pollen grains, and the
number of bristles on the body (which varies among species) affects the number of
grains likely to be carried (Lewis, 1973). In a few cases, pollen-eating thrips may be
significant as pollinators. For example, Hagerup (1950a) found that in the Faroes,
where bees and butterflies are rare, thrips were pollinators of ling (*Calluna vulgaris*),
cross-leaved heath (*Erica tetralix*) and common catsear (*Hypochaeris radicata*) (Hagerup &
Hargerup, 1953). Some species go through their whole life-cycle in flowers. One example of this is *Ceratothrips* (*Taeniothrips*) *ericae*, which lives in *Calluna* flowers, where it
often causes self- and cross-pollination. Strangely enough, species of these minute
insects are regular pollinators of huge forest trees of the family Dipterocarpaceae in
South-east Asia (pp.311, 392-3). In addition, crop plants in which some degree of
pollen transfer by thrips has been recorded are onion (*Allium cepa*), plum (*Prunus domestica*), French bean (*Phaseolus vulgaris*), flax
(*Linum usitatissimum*), sugar beet (*Beta vulgaris*), pyrethrum (*Tanacetum cinerariifolium*)
and cacao (*Theobroma cacao*) (Lewis, 1973).

The flower-visiting thrips show colour
'preferences'. They respond more to the
colours white (without ultraviolet, i.e. 'bee
blue-green', p.130-1), blue and yellow
than to green, red, black or white-with-ul-
traviolet ('bee-white') (Kirk, 1984a). They
also respond to the vapour of anisalde-
hyde, a flower scent (Kirk, 1985).

Coleoptera: beetles

This is the largest order of insects, though
not the largest in the British Isles. The
beetles are not very important in flower
pollination in cool-temperate climates but
are more so in arid areas and in the moist

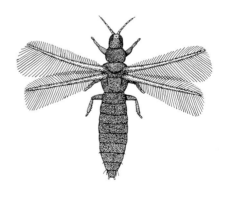

Fig. 3.2 A typical thrips (Thysanoptera):
insect about 2mm long. (Derek Whiteley).

Fig. 3.3 Garden chafer, *Phyllopertha horticola* (Coleoptera: Scarabaeidae, subfam. Rutelinae), on hogweed, *Heracleum sphondylium.*

Fig.3.4 Soldier beetles, *Rhagonycha fulva* (Coleoptera: Cantharidae), on hogweed, *Heracleum sphondylium.*

Fig. 3.5 Oedemera nobilis (male) (Coleoptera: Oedemeridae) on mouse-ear hawkweed (*Pilosella officinarum*).

tropics. In southern Africa different groups of the family Scarabaeidae, feeding on either pollen or nectar, are important in pollinating the rich annual flora of Namaqualand (including many Asteraceae) and also some species of *Protea*; other groups are probably pollinators too (Whitehead *et al.*, 1987). In the moist tropics, they are of especial interest as pollinators of some primitive flowering plants and even some gymnosperms (Chapter 14), as well as particular groups such as palms, waterlilies and aroids (Chapters 10 & 11). The beetles are classified on structural and anatomical characters into two main sub-orders: Adephaga and Polyphaga. The first of these is mainly predatory and its members do not visit flowers. The second is much the larger group and includes the flower-devourers and the pollinators. Even so, beetles have primitive mouth-parts (Box 3.1).

The European beetles listed as flower visitors by Kugler (1984) represents 18

Box 3.1 Primitive insect mouth-parts

Diagrams A-E show the mouth-parts of a cockroach, *Blatta* (Dictyoptera), illustrating the primitive condition in insects. The capital letters in the labels indicate abbreviations used in other figures; for 'compass-points' see Box 3.2. A, labrum-epipharynx; B, one mandible of pair; C, one maxilla of pair; D, hypopharynx; E, labium, with one each of paired appendages. Diagrams F and G show a beetle: F, side view of head, with parts shown in A to E attached in same sequence from 'a' to 'b' (distance 'a' to 'b' exaggerated); G, head from below, with paired parts shown singly (parts shown in A, B, C and E are visible, that shown in D is concealed) (G after Crowson, 1956).

The mandibles of beetles are the biting and chewing parts, while the maxillae and labium taste and manipulate the food; the labium also helps, in conjunction with the labrum, to form a chamber in which the food can be confined during mastication. Thus a beetle bites and chews the food while it is still outside the mouth, and its mouthparts perform roughly the functions which in the mammal are performed by mouth, lips, jaws, tongue and, in some cases, fore-limbs.

Modifications towards flower feeding in beetles consist of a tilting upwards of the head which brings the mouth-parts forward, the development of the head behind the eyes to form an additional, neck-like region, transformation of the usually short and broad first segment of the thorax into a long and narrow segment, and a lengthening of the hairs on the lobes of the maxillae. These modifications increase the insects' ability to reach sunken nectaries and lick up the nectar efficiently. The greatest depth to which European beetles can reach with their foreparts for nectar is about 6 mm (in *Strangalia*). In South America, however, the genus *Nemognatha* is adapted to reach deep-seated nectar, having its maxillae modified to form a slender tube 12 mm long (Müller, 1883).

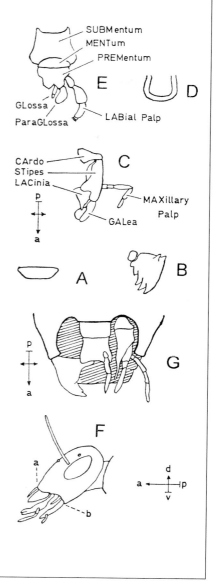

genera. Some of these, and the families to which they belong, are listed in Table 3.2. In some instances, the nearest relatives of flower-frequenting (anthophilous) species are not known to visit flowers; on the other hand, in some whole genera or even families the adult beetles feed exclusively on flower foods (Müller, 1983). This habit has

Table 3.2 Beetles recorded visiting flowers in Europe.

Mordellidae (all) Oedemeridae (all) (Fig. 3.5) Melyridae (formerly Malachiidae) (all)	} ?destructive pollinators

Chrysomelidae
   Hydrothassa marginella
   Cryptocephalus aureolus (Fig. 3.8)
	} eating pollen of
	Ranunculus spp. and
	Asteraceae-Lactuceae

Donacia spp.	visiting Nymphaeaceae and other waterside plants

Cantharidae
   Rhagonycha fulva (Fig. 3.4)
   Cantharis spp.
	Apiaceae, Asteraceae
	Asteraceae; *Coeloglossum viride*
	(frog orchid, Finland)

Nitidulidae *Meligethes* spp. (flower-beetles, Fig. 3.6)	*Ranunculus ficaria*, Brassicaceae, etc.

Staphylinidae
   Anthobium spp.
   Omalium spp.
	} many species

Scarabaeidae *Cetonia aurea* (rose chafer)	*Cornus* (*Thelycrania*)*sanguine*a (dogwood)
Phyllopertha horticola (garden chafer, Fig. 3.3) *Melolontha melolontha* (cockchafer)	*Heracleum sphondylium* many species, destructive

Elateridae
   Athous haemorrhoidalis (Fig. 7.8)
   other species
	Listera ovata (twayblade)
	many species, destructive

Cerambycidae several genera (Fig. 3.7)	plants ranging from those with exposed to moderately deep- seated nectar, depending on the level of adaptation of the beetle genus

(Hobby, 1933; Kugler, 1984; Müller, 1883; Parmenter, 1956; MCFP)

in some evolutionary lines of beetles than in others. The fossil records show that bee-tles were abundant during the Mesozoic; at least some must have been flower visitors in the times of the earliest angiosperms (p.378) and probably earlier (p.370). The pre-sent-day association of beetle pollination with primitive woody angiosperms such as *Magnolia* (Thien, 1974) and *Calycanthus* (Grant, 1950a) is probably ancient, going back to their evolutionary origins. Some families of flower-visiting beetles are found as fos-sils back into the Mesozoic (e.g. Elateridae, Nitidulidae, Cerambycidae). Others (e.g. Scarabaeidae, Oedemeridae, Cantharidae) probably arose in the burst of evolution-ary radiation of flowers and insects in the Tertiary which saw the origin of bees, but-terflies, and other specialised flower-visitors that now make up the majority of pollinators (pp.381-2; Willemstein, 1987). In *Calycanthus* and some beetle-pollinated flowers described in Chapters 10 & 11 beetles eat nutritive tissues which, apparently, are provided especially for them.

Fig. 3.6 Flower beetles (*Meligethes* sp., Coleoptera: Nitidulidae), on wall rocket (*Diplotaxis muralis*).

Fig. 3.7 Longhorn beetle (*Judolia cerambyciformis*, Coleoptera: Cerambycidae) on bramble (*Rubus fruticosus* agg.).

Fig. 3.8 Leaf beetle, *Cryptocephalus sp.* (Coleoptera: Chrysomelidae) on flower of buttercup, *Ranunculus* sp.

Though the flower-visiting beetles are more active than many beetles with different habits they are usually less active than insects such as flies and Hymenoptera, and therefore less useful to the plants they visit except for those that are highly specialised to them. Beetles tend to protect themselves by their horny exterior and their repellent secretions rather than by flight, and may linger in the same flower or inflorescence for hours. In an assessment of the interactions between *Oedemera* species (Fig. 3.5) and flowers, Kugler (1984) found that they readily moved from flower to flower and that the beetles were well adapted to living on pollen and nectar and were effective pollinators of the flowers that they visited; however, they accounted for a tiny minority of insect visits to these flowers.

The small pollen beetles in the family Nitidulidae (Fig. 3.6) and small rove beetles (Staphylinidae), which may be abundant in many flowers, mainly creep about eating pollen and nectar, only sometimes causing pollination. However, Marsden-Jones (1935) proved that the pollen beetle *Meligethes* (Nitidulidae) moves between the flowers of lesser celandine (*Ranunculus ficaria*); he removed the beetles from all the flowers of certain plants and found that the flowers were again occupied by *Meligethes* a few hours later.

Adaptations against damage to the ovules would be expected in flowers specialised for beetle-pollination. Grant (1950b) suggested that this protection might be provided by the sinking of the ovules into the receptacle (by epigyny or perigyny) or by the close massing together of the flowers so that the ovaries cannot be reached, as in some of the more robustly constructed Asteraceae (see also pp.174-8).

In beetles, the sense of smell plays a large part in feeding and in finding a place for egg-laying. Fritz Knoll, in his study of the plant *Arum nigrum* (1926; see also p.302), obtained evidence that dung-frequenting beetles found their way to its inflorescences by scent. The smell of this plant is produced by its spadix, and as the beetles flew past they suddenly turned in flight towards the spadices on coming within about 30 cm of them. They then either flew straight towards them, with repeated brief reversals of direction, or flew round, gradually getting nearer. These flight paths are indicative of the insects' finding their way by scent; this becomes specially apparent when they are compared with the flight paths of insects finding their way to flowers by sight (p.63).

Scents specifically associated with beetle-pollination (in preference to pollination by short-tongued insects in general) are found mainly in plants from outside Europe; examples are the spicy scent of certain crab-apples (*Malus* species), of *Paeonia delavayi* (a tree peony) and wintersweet (*Chimonanthus praecox*), and the smell of fermenting fruit produced by *Calycanthus* (Grant, 1950a).

Colour vision was demonstrated in some beetles by Schlegtendal (1934), using a method of investigation independent of feeding reactions. She found that Chrysomelid beetles could clearly distinguish yellow and orange from blue, and violet from green; dung-beetles (*Geotrupes*) could distinguish yellow, orange and violet from blue, and also yellow-green and light green from other colours. Species of *Amphicoma* (Scarabaeidae subfamily Glaphyrinae) in the eastern Mediterranean region are strongly associated with red bowl-shaped pollen-flowers (p.179) and have been found to respond preferentially to red plastic cups, so these beetles are presumably sensitive to red (Dafni *et al.*, 1990).

Diptera: two-winged flies

This is another major order of insects, and in the British Isles it substantially outnumbers the Coleoptera in species, with about 5,200 as against 3,700. Structurally the true flies are easily distinguished by their possession of only one pair of wings, the second pair being reduced to stalked knobs, the halteres or balancers. In the adult state the vast majority of species have well-developed wings and fly readily. Very many flies are flower visitors and the order is therefore an important one in flower pollination.

Nearly all flies have suctorial mouth-parts, and there are two main types of these; one type is adapted for penetrating the tissues of animals in order to imbibe their internal fluids, while the other lacks penetrating organs and is adapted for mopping up exposed liquids. The second is much the commoner and the piercing type has probably arisen repeatedly from the mopping-up type. Many of the flies that feed on exposed fluids can also eat small solid particles, including pollen grains. The ability of flies to take in solid particles through their tubular proboscides is dependent on the suspension of the particles in a liquid, and this is always available because saliva is conveyed nearly to the apex of the proboscis. In some groups of flies the proboscis has become elongated in adaptation to feeding at the longer-tubed flowers. The main organ of the mouth-parts of Diptera is the labium, with its labella which may correspond to the paraglossae of other orders (see Box 3.1). In the mopping-up type, the labella collect the food and the labium conveys it to the mouth. Details of fly mouth-parts are given in Boxes 3.2-3.6.

In Diptera the taste organs are chiefly in the region of the mouth, but they often also occur in the tarsi of the legs. When they taste food with their legs insects automatically begin to lower the proboscis in order to feed; it is, therefore, easy to investigate the sensitivity of the taste organs in the legs. In this way it has been found that the legs of blow-flies (*Calliphora vomitoria* or *C. vicina*) are 100 to 200 times more sensitive to the taste of cane sugar (sucrose) than the human tongue.

The scent organs of Diptera are found on the antennae. The thin-walled olfactory hairs were found by Liebermann (1925) to be either solitary on the surface, or sunk, singly or in groups, into pits. The pits, when large, may be partially subdivided and they are always oblique, the openings facing towards the apex of the antenna. The olfactory hairs occur only on the terminal joint of the three-jointed antennae. In the species investigated, the total number of olfactory hairs in the two antennae ranges from 300 in the male of *Psila rosae* to 9,260 in the male of *Calliphora vomitoria*. On average, there were more olfactory hairs on the males of dung- and carrion-frequenting species than on those of flower-frequenting species. The females were left out of this comparison because some species feed on flowers as adults but lay eggs in dead organic matter. In these, the females have more olfactory hairs than the males. Lieber-

Fig. 3.9 St Mark's fly (*Bibio marci*, Diptera: Bibionidae), and a smaller species of *Bibio* on alexanders (*Smyrnium olusatrum*).

mann thought that the explanation of these differences lay in the fact that vision cannot play such a big part in finding decaying matter as it can in finding flowers.

Most Diptera that visit flowers have large eyes with many ommatidia (facets), and it
seems probable that at least all the higher Diptera have colour vision. The lesser house
fly (*Fannia canicularis*) was included in Schlegtendal's investigation (see p.56) and was
found to have similar colour vision to that of the chrysomelid beetles.

The following diagram shows how the Diptera are classified and the sequence (from
left to right) in which they will now be described.

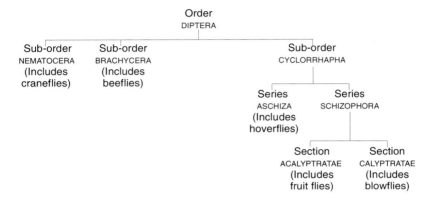

Diptera, suborder Nematocera

The flower visitors of the sub-order Nematocera comprise about eleven families. In
general they are not very important pollinators, for most are very small and all except
Culicidae have very short mouth-parts, even the relatively large *Bibio* (Fig. 3.9) and
Tipulidae having a proboscis only a millimetre or two in length. Table 3.3 shows that
most of the flowers visited by them are either flat or bowl-shaped and have well-exposed nectar, or are tubular but so small that the nectar is easily reached. (These small
tubular flowers are frequently assembled into dense clusters to form 'brush-blossoms'
[see p.178], or into a well-defined head or 'capitulum' [see p.174].) Nectarless flowers
are presumably visited for their pollen. Nematocera appear, from the records of Willis
& Burkill (1895-1908), to visit flowers chiefly for nectar, but these authors record that
Bibionidae, Mycetophilidae and Scatopsidae eat pollen as well. Not unexpectedly,
mosquitoes visit flowers at night (Parmenter, 1958; Corbet, 1970). Brantjes & Leemans (1976), in their study of *Silene otites* in the Netherlands, found that mosquitoes of
both sexes were flower-visitors, drinking the nectar and causing pollination.

The very small insects of the remaining families of Nematocera visit a range of the
flowers already mentioned and, in addition, those of some dwarf herbaceous plants
of damp shady places that are particularly suited to them: *Chrysosplenium* species (gold-
en saxifrage) and *Adoxa moschatellina* (moschatel) (Grensted, 1946; Hagerup, 1951;
Kimmins, 1939; Knuth, 1906-1909; Parmenter, 1952b; Smart, 1943; Willis & Burkill,
1895-1908). Such flies are often involved in pollination systems that deceive them, for
example, the relationship between lords and ladies (*Arum maculatum*) and the owl
midges (family Psychodidae). Outside the British Isles, Nematocera are special pollinators of various plants. For example, mosquitoes pollinate a small North American

Box 3.2 Mouth-parts of Diptera-Nematocera: craneflies and mosquitos

Both the piercing and the mopping-up types of mouth-parts are found in this sub-order. In *Tipula* (cranefly or daddylonglegs, family Tipulidae), shown here (after Hammond, 1874), the proboscis is quite short. The mouth-parts are borne on a tubular prolongation of the head (Diagrams A-C; note the compass-points shown: anterior, posterior, dorsal, ventral, left and right). The mandibles are missing and there is little of the maxillae apart from their very long, five-jointed palps (C); the remaining structures are the labium, with no palps, and the labrum. The labrum is a simple structure attached above the mouth (C). The labium, attached below the mouth (A), is more complex; it is partly covered with soft cuticle, but its short and stout basal portion is supported at the back by a sclerite (a hard plate) corresponding to the mentum of beetles. On the front is a furrow which leads out from the mouth, beneath the labrum, to the gap between the labella, where the salivary duct opens into the furrow (B). The bladdery labella (A,B,C) contain in their surface a number of sclerites between which they can be folded up; they are presumably kept in a distended condition by blood-pressure as in the blowfly (*Calliphora*), described in Box 3.6. They are partly covered with a system of microscopic furrows called pseudotracheae (because of their resemblance to the tracheae, or breathing tubes, of insects) (A,D,E). The furrows are kept in shape by transverse, incomplete rings of hard chitin (E); there are

four major pseudotracheae in *Tipula* (A,D) and numerous minor ones that lead into them. The pseudotracheal system acts as a filter by which liquids can be mopped up from a wet surface and solid particles left behind. The fluid is presumably drawn into the pseudotracheae by capillary action; the main pseudotracheae lead into the groove on the front of the labium (D), and the liquid is drawn into the mouth by the sucking action of the pharynx exerted through a tube formed by the labrum and the groove on the front of the labium. *Tipula* differs from most Diptera in the lack of a free tongue-like or lance-like hypopharynx with the salivary duct inside it and opening at its tip.

The piercing Nematocera are mosquitoes and gnats, black-flies and biting-midges. The largest of these are the mosquitoes, such as *Culex* and *Anopheles*. They have a long narrow labium with small, equally narrow labella; the labrum and hypopharynx are long and sword-like and are placed together to form a feeding tube which lies in the groove on the front of the labium. There are also two needle-like maxillae and, in the female, two mandibles of the same form. All these long slender organs are used by the female to pierce the skin of the food-animal. The male mouth-parts are not used for blood-sucking, the mandibles being absent. Thus the mosquitoes (and, similarly, the black-flies and biting-midges) lack the mopping-up type of labella of the cranefly, but the proboscis can be dipped into drops of nectar.

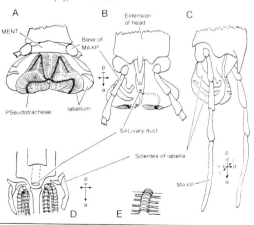

Table 3.3 Diptera-Nematocera and the flowers they visit in Britain.

A. Flowers visited by Tipulidae and Culicidae
(craneflies and mosquitoes)

Silene otites	Spanish catchfly	small tubular flowers
Parnassia palustris	grass of Parnassus	nectar well exposed
Saxifraga hypnoides	mossy saxifrage	nectar well exposed
Filipendula ulmaria	meadowsweet	nectarless flowers
Potentilla palustris	marsh cinquefoil	nectar partly concealed
Euphorbia	spurge	nectar well exposed
Frangula alnus	alder buckthorn	nectar well exposed
Hedera helix	ivy	nectar well exposed
Apiaceae	umbellifers	nectar well exposed
Mentha aquatica	water mint	small tubular flowers
Valeriana dioica	marsh valerian	small tubular flowers
Valeriana officinalis	common valerian	small tubular flowers
Listera cordata	lesser twayblade	nectar well exposed

B. Flowers visited by Bibionidae
(St Mark's flies – *Bibio* spp., and fever flies – *Dilophus* spp.)

Euphorbia	spurge	nectar well exposed
Acer pseudoplatanus	sycamore	nectar well exposed
Crataegus spp.	hawthorn	cup-shaped flowers
Sorbus aucuparia	rowan	cup-shaped flowers
Malus	apple	cup-shaped flowers
Apiaceae	umbellifers	nectar well exposed
Euonymus europaeus	spindle-tree	nectar well exposed

(Records in section A from: Brantjes & Leemans, 1976; Corbet, 1970; Knuth, 1906-1909; Müller, 1883; Parmenter, 1952b; Willis & Burkill, 1895-1908; in section B from Drabble & Drabble, 1927; Knuth, 1906-1909; Parmenter, 1952b; Willis & Burkill, 1895-1908.)

orchid, *Habenaria obtusata* (Dexter, 1913; Thien, 1969a & b), and these and other Nematocera are the pollinators of the orchid genus *Pterostylis* (p.212) and several of the other exotic insect-trapping or insect-deceiving plants mentioned in Chapter 10, as well as some members of the family Sterculiaceae, including the cocoa plant (p.311) (Schumann, 1890-1893; Young *et al.*, 1984).

Diptera, suborder Brachycera

Flower visitors in this suborder are spread over several families (Table 3.4). Distinctive types of flower-visitor are covered in five families described here.

Stratiomyiidae are medium-sized flies, with flattened and often brightly coloured bodies suggestive of hoverflies. However, they are rather slow in flight and usually frequent damp places and waterside habitats (Fig. 3.10). The proboscis is well-developed but rather short; it has large labella with numerous pseudotracheae, and thus functions like the proboscis of the cranefly. Knuth (1906-1909) records that Stratiomyiidae have been seen at the flowers of 22 different families. British records, however, are rather scarce. They visit Apiaceae freely and the small tubular flowers of mint (*Mentha*), but are also recorded at a number of Asteraceae, which they perhaps visit for pollen, since either the depth or the narrowness of the floral tubes in this family would probably prevent the flies from reaching the nectar.

The Empididae is a large family of small flies (Fig. 6.2) in which only exceptionally

Fig. 3.10 Soldier fly
(*Chloromyia formosa*, Diptera:
Stratiomyiidae), on
hogweed (*Heracleum
sphondylium*).

large species reach a length of 10 mm. They are predatory on other insects, chiefly
Diptera, and have a rigid, piercing proboscis (Box 3.3) which is also suitable for taking
nectar from small tubular flowers, including the longer-tubed Asteraceae, such as
knapweed (*Centaurea*) and *Cirsium* spp. (thistles). They are presumably effective pollinators of such flowers. The records of Willis & Burkill (1895-1908) indicate *Empis* as
being much the most important genus of flower visitors in this family in Britain, having been seen at the flowers of about 50 species of plants, while Hobby & Smith
(1961) list 20 species of flowers visited by the large species *Empis tessellata* (Fig. 3.11).

The Bombyliidae include some of the most highly specialised flower-feeders in the
Diptera. The genera *Bombylius* and *Phthiria* both have a long, slender, rigid proboscis
normally held pointing forwards horizontally but having some flexibility in its joint
with the head (Box 3.3). *Bombylius* species (bee-flies, Plate 2f) are rather large and visit
chiefly large, long-tubed flowers, but also some flowers with more exposed nectar (Table 3.5; see also Table 3.7).

Bombylius and *Phthiria* each have a proboscis similar to that of the Empididae but use
it exclusively for feeding at flowers. *Phthiria* is a small insect, so cannot reach deep-

Table 3.4 Alphabetical list of families of Diptera-Brachycera for which records of flower-visiting are
available.

Acroceridae	small-headed flies
Apioceridae	flower-loving flies
Asiliidae	robber-flies
Bombyliidae	bee-flies
Dolichopodidae	longheaded-flies
Empididae	empids
Lonchopteridae*	pointed wing-flies
Nemestrinidae	(South African)
Phoridae*	scuttle-flies
Rhagionidae	snipe-flies
Stratiomyiidae	soldier-flies
Tabanidae	horse-flies
Therevidae	stiletto-flies

*alternatively, this is placed in sub-order Cyclorrhapha

(Disney, 1980; Knuth, 1906-1909; Whitehead *et al.*, 1987; Willis & Burkill, 1895-1908)

Box 3.3 Mouth-parts of Diptera: empids and beeflies

Diagram A shows the complete mouth-parts of *Empis* and B the labium and 3-toothed tip of the labrum. Diagrams C-F show *Bombylius*; as in *Empis* the feeding tube is formed by the labrum and hypopharynx (C-F), supported by the labium (seen from beneath in D). The folds at the base of the proboscis (C, E) allow some extension and retraction of the labium. The labella are long and narrow (D), forming an extension of the tube, and each carries only a few pseudotracheae,

joined at the base (Becher, 1882; Gouin, 1949). The margins of the pseudotracheae are raised above the general surface of the labella (Peterson, 1916). The maxillae are fine bristles (C). Müller (1883) considered that the hypopharynx and labella were used for boring into soft tissues, as he had often seen these flies putting their tongues into the nectarless flowers of common St John's wort. (Diagrams A-C after Peterson, 1916; F, D based on Kugler, 1955a; E after Gouin,

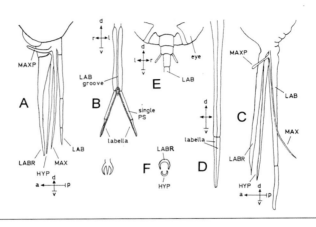

seated nectar. *Bombylius* species, however, look like small bumblebees, and the larger ones have a proboscis 10-12 mm long.

Bombylius usually hover when feeding but may sometimes hold the flower with the legs, thereby tilting the body if necessary so that the proboscis can enter the tube (Knoll, 1921; Simes, 1946). These flies move rapidly from flower to flower and are clearly highly developed nectar-feeders, though some are known to eat pollen as well (Knoll, 1921; Beattie, 1972). They are on the wing early in the year and may be quite important as pollinators of some spring flowers, although they are more or less ineffective as pollinators of wide-throated flowers such as ground ivy (*Glechoma hederacea*), even though they may visit them freely.

The long-tongued hovering mode has reached a remarkable level in the South African Tabanidae and Nemestrinidae. The proboscis is up to 47 mm and 70 mm long in these groups respectively, and it may be four times the length of the body. These insects have perhaps evolved to fill the niche of butterflies where these are absent (Vogel, 1954) (see p.390). The flowers they visit have their own distinctive characteristics (Whitehead *et al.*, 1987).

The remaining families of the Brachycera, such as Dolichopodidae (Fig. 3.12), are infrequently seen at flowers and they mainly visit those with exposed or slightly concealed nectar and small tubular flowers. The robber-flies (Asiliidae), large insects that are similar in their method of catching prey to Empididae, have been recorded at flowers of field scabious (*Knautia arvensis*) and bilberry (*Vaccinium myrtillus*), with relatively deep-seated nectar.

Knoll (1921), working in Dalmatia, tested the flower colour discrimination and preferences of *Bombylius fuliginosus*. By presenting a board with 16 contiguous squares of various shades of grey and one of blue-violet, like the flowers of the grape hyacinth (*Muscari neglectum*), he demonstrated that the flies could distinguish its violet colour from grey, for they paused only over the violet square. Flies that were visiting *Cerastium litigiosum*, a white flower similar to greater stitchwort (Table 3.5), approached flowers of all colours except red (poppies, which were ignored), but yellow only rarely.

When foraging, bee-flies clearly used sight to find flowers, since they flew quickly and straight from flower to flower in the characteristic manner of visual flower seekers. Where flowers were scarce they would make wide sweeps over the

Fig. 3.11 Empid fly (*Empis tessellata*, Diptera: Empididae) on flowers of hawthorn (*Crataegus monogyna*).

ground, slowing down near attractive flowers and resuming their sweeps if the flower was unsuitable. They usually passed from one plant to another in a constant direction, even though at each plant they had gone all round the inflorescence (Fig. 3.12a), which suggests that they orientate by the sun. The flight-line was independent of wind direction, so scent was not involved. If the inflorescences of grape hyacinth were covered by inverted glass tubes, the flies pushed against the glass where it covered the

Table 3.5 Flowers visited by bee-flies, *Bombylius*.

Long-tubed flowers		Flowers with more accessible nectar	
Cardamine pratensis	lady's smock	*Stellaria holostea*	greater stitchwort (Fig. 2.14)
Aubrieta deltoidea	aubrietia	*Malus sylvestris*	apple
Viola riviniana	dog violet	*Limonium* spp.	sea lavender
Oxalis rubra	pink oxalis	*Tussilago farfara*	coltsfoot
Pulmonaria spp.	lungwort	*Myosotis* spp.	forget-me-not
Ajuga reptans	bugle	*Salix* spp.	sallows
Glechoma hederacea	ground ivy	*Hypericum perforatum*	St John's wort
Vinca minor	lesser periwinkle		
Primula elatior	oxlip		
Primula vulgaris	primrose (Plate 2f)		
Primula veris	cowslip		

(Christy, 1922; Knight, 1968; Knuth, 1906-1909; Müller, 1883; Parmenter, 1952a; Scott, 1953; Verrall, 1909; PFY, including observations from gardens)

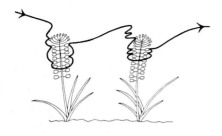

Fig. 3.12a Flight path of the beefly *Bombylius medius* (Diptera Bombyliidae) visiting *Muscari comosum*. After Knoll (1921).

flowers and not, even at this close range, at the source of the scent emanating from the end of the tube a little way below the inflorescence. When presented with scented and old scentless flowers, the flies sucked nectar from the the scented ones but only hastily alighted on scentless ones, suggesting that scent decided whether a proper feeding visit was made.

The normal behaviour of the bee-flies in relation to *Muscari* was to approach very closely the light violet, sterile, nectarless flowers that top the inflorescence, and then to descend and feed at the lower, dark violet, fertile flowers (Fig. 3.12a). When inflorescences of the tassel hyacinth, *Muscari comosum*, which has violet sterile flowers and brown scentless fertile flowers, were placed in a habitat of *M. neglectum*, *B. fuliginosus* went to the sterile flowers repeatedly but did not find the fertile flowers. In its natural habitat, *M. comosum* was ignored by *B. fuliginosus* but it was a favourite flower of *B. medius*, which fed from it in the same way as *B. fuliginosus* did at *M. neglectum*.

Diptera, suborder Cyclorrhapha: Aschiza, including hoverflies

This last sub-order of the Diptera is by far the largest and is itself subdivided into series (p.58), the first of which to be dealt with is the Aschiza. This includes, apart from a few small families, the Syrphidae, the most important family of flower visitors among the Diptera.

The Syrphidae, or hoverflies (Figs. 2.16, 3.13, 3.14, 3.16 & 3.17), comprise nearly 250 species in the British Isles. They get their popular name from their habit of remaining stationary in the air, a habit shared with Bombyliidae. Many of them are brightly coloured, while the darker species are often highly polished. The bright colouring often consists of a pattern of yellow and black which gives them a resemblance to wasps, or red and black, which again makes them resemble certain kinds of wasp,

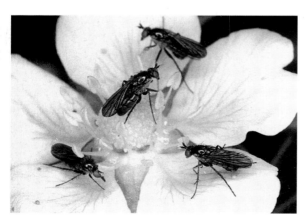

Fig. 3.12 'Long-headed flies' *(Dolichopus* sp., Diptera: Dolichopodidae) on flower of creeping cinquefoil *(Potentilla reptans)*.

Box 3.4 Mouth-parts of Diptera: hoverflies

The lower part of the head of *Eristalis* is produced into a conical snout (rostrum or fulcrum); this is partly membranous and partly hardened for support. Two of the supporting structures embedded in the front of the rostrum seem to be the basal parts of the maxillae. At the apex of the rostrum are the small laciniae and the larger palps of the maxillae (Diagram A). The rostrum is continued downwards by the labium (A, B), which is again partly membranous and partly hardened, with a contractile basal part and a main channelled part supported underneath by a sclerite (C). Emerging from the rostrum, and normally lying in the labial channel, are the usual labrum and hypopharynx, the latter housing the salivary duct (A, B) (Becher, 1882). The labella are well developed and have abundant pseudotracheae (B), similar to those of the blowfly *Calliphora* (Dimmock, 1881). (Diagrams are based on Müller, 1883; Peterson, 1916; and Gouin, 1949.)

Food suspended or dissolved in the saliva is sucked back into the mouth, often having first been taken up by the pseudotracheae. Gilbert (1981) noted that bristles are present on the labella, their length and number varying with the species. The food is drawn first into a canal formed by the labrum and hypopharynx when muscles in the labral part contract and widen the tube. When these muscles relax, this tube closes by its own elasticity. As this happens, muscles in the rostrum contract and dilate the next part of the food canal, so that the liquid is drawn into this. At the inner end of this region is the true mouth, hitherto kept shut by muscles. Relaxation of the rostral muscles closes the rostral passage while (contrarily) relaxation of the mouth muscles opens the mouth and so the food continues its journey (Schiemenz, 1957).

During feeding on nectar, the labella can either be spread out if the liquid surface is wide enough or pressed together in narrowly tubular flowers. Müller (1883) reported that in this case the labium is retracted sufficiently for the labrum and hypopharynx to protrude and suck the nectar directly. Gilbert (1981) noted that the nectar-specialists are all large species, possibly because these have heavier energy requirements than the small ones. The density of the pseudotracheae is correlated with the percentage of pollen taken in the diet, but the role of the pseudotracheae in pollen-feeding is not fully understood. In long-tongued hoverflies both the rostrum and the labium are lengthened, in contrast to *Bombylius* where only the labium is elongated and the proboscis cannot be folded.

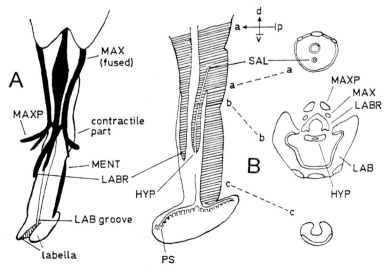

bee or ichneumon. Some species are stout and furry, and in their colouring closely mimic various species of bumblebees. Many species are to be found in sunny places but some are frequent in and on the edges of woods.

One of the first people to study the proboscis of hoverflies was Müller (1883) and one of the most recent was Schiemenz (1957). Both worked on species of Eristalis, and details are given in Box 3.4.

Hoverfly feeding behaviour has been the subject of much confusion and contradiction in the literature (Gilbert, 1981). Gilbert himself studied eight common hoverfly species at two dissimilar sites in Cambridge. He found considerable differences between species in the relative numbers seen feeding on pollen and nectar. Three were principally pollen feeders (*Syrphus ribesii*, *Episyrphus balteatus* and *Melanostoma scalare*), two fed more on nectar than pollen (*Eristalis arbustorum* and *E. tenax*) and three divided their time about equally between the two types of food (*Metasyrphus corollae*, *Platycheirus albimanus* and *Syritta pipiens*). These differences in habits are reflected in the form of the proboscis. The pollen feeders (those mentioned above, together with *Meliscaeva auricollis*) have a short thick proboscis with broad labella, the nectar feeders (which include also *Sphaerophoria scripta* and *Platycheirus peltatus*) have a long proboscis with relatively

Fig. 3.13 Drone fly (*Eristalis* cf. *arbustorum*, Diptera: Syrphidae) on hogweed (*Heracleum sphondylium*).

Fig. 3.14 Long-tongued hoverfly (*Rhingia campestris*, Diptera: Syrphidae), on herb robert (*Geranium robertianum*).

quite small labella with only short bristles, while the mixed feeders are intermediate or like the nectar feeders. Gilbert also found that, with one exception, male hoverflies took pollen less frequently than females of the same species, though newly emerged males may need pollen to provide the nutrients required for the production of sperm. Males of the mixed feeders spend a lot of time hovering and their feeding at sources of sugar presumably meets their energy needs for this activity.

Pollen is usually taken directly from the anthers of the flower by rubbing them between the labella. Gilbert saw rapid movements in which the labial gutter seems to move up and down relative to the labral sucking tube; these are perhaps actuated by the contractile zone at the base of the labium (see Box 3.4). However, species of *Melanostoma* feeding on wind-pollinated plants with pivoted anthers grasp these with their forelegs, insert their labella into the anther cells and systematically clean them out (Holloway, 1976, in New Zealand; Stelleman & Meeuse, 1976, in the Netherlands). During feeding, hoverflies frequently pause to clean the face and proboscis and they consume the pollen thus collected. The forelegs clean the head, the middle legs and one another; the hind tibiae clean the abdomen. Intermittently during grooming, *Eristalis tenax* (Fig. 6.50) holds either the forelegs or the hind legs out beyond the body and scrapes the tip of the tibia and the tarsus of one leg against those of the other. This causes transfer of pollen to spirally grooved, pollen-retaining hairs on the tarsus.

Table 3.6 Flowers visited by long-tongued hoverflies (Syrphidae).

A. *Narrow-tubed flowers visited by* Volucella *(Kugler, 1955a)*

Melilotus	melilot	Fabaceae-Faboideae
Trifolium spp.	clover	Fabaceae-Faboideae
Stachys	woundwort	Lamiaceae
Armeria	thrift	Plumbaginaceae
Knautia	scabious	Dipsacaceae
Centaurea	knapweed	Asteraceae
Cirsium	thistle	Asteraceae

B. *Narrow-tubed flowers visited in Britain by* Rhingia campestris

Viola spp.	violets	Violaceae
Silene dioica	red campion	Caryophyllaceae
Geranium robertianum	herb robert	Geraniaceae
Primula vulgaris	primrose	Primulaceae
Glechoma hederacea	ground ivy	Lamiaceae
Ajuga reptans	bugle	Lamiaceae
Hyacinthoides non-scripta	bluebell	Liliaceae

Table 3.7 Flowers visited by some Conopidae as well as by *Bombylius, Rhingia* and bees.

Species	Common name	Flower characteristics
Succisa pratensis	devil's-bit scabious	flowers small, tubular
Scabiosa columbaria	small scabious	flowers small, tubular
Trifolium pratense	red clover	flowers large, tubular
Echium vulgare	viper's bugloss	flowers large, tubular

Sometimes the fly hovers and scrapes all its legs against one another as they dangle below the body; this brings pollen from the hind legs to the forelegs, whence it can be eaten (Holloway, 1976). *E. tenax* was found by Holloway to obtain all or nearly all the pollen it consumed indirectly by grooming in this way after visits to nectar-producing flowers. Its dense covering of hairs, including branched, plumose and curly-tipped types, is apparently related to this habit. This fly thus presents an interesting parallel with the honeybee. The grooming movements are the same as those used purely for cleaning in *Melanostoma fasciatum*.

Various views have been put forward as to the treatment of pollen gathered by hoverflies but, as undamaged grains have been found in the gut by Gilbert and others, it seems that they are not physically crushed or broken. Nutrients are probably extracted by enzymatic penetration of the pollen grains.

The proboscis length of six of the eight hoverflies in Gilbert's (1981) study fell between 2 mm and 3.5 mm, while that of the other two, *Eristalis arbustorum* (Fig. 3.13) and *E. tenax*, was 5.36 mm and 7.85 mm respectively. *Volucella* (Fig. 2.13) has a similarly long proboscis, while that of *Rhingia campestris* (Fig. 3.14) measures 12 mm (Gilbert, 1981). All the Syrphidae can fold up the proboscis into a cavity under the beaked head; the size of this beak corresponds with the size of the proboscis, and it is therefore a conspicuous feature of flies such as *Volucella* and *Rhingia* (Plate 3a).

Gilbert found a very strong correlation between the proboscis-lengths that he measured and the depth of the flowers that the insects visited for nectar; the flower/insect relations at the upper limits of tube-length are as shown in Table 3.6. However, there are also records of visits by *Rhingia* to many small tubular flowers (Table 3.7) and some with exposed or only slightly concealed nectar. The flowers in these last groups belong to the families that appear to be most favoured by hoverflies in general: Ranunculaceae, Brassicaceae, Caryophyllaceae, Rosaceae, Apiaceae and Asteraceae (for the importance of the grasses, see later). Gilbert found that the mixed feeders *Episyrphus balteatus* (Fig. 6.11B) and *Syrphus ribesii* relied for their pollen requirements particularly on the nectar-producing flowers of Apiaceae and Asteraceae.

Two English naturalists, who recorded visits by 123 species of flies, nearly all hoverflies, to flowers of 35 species of plants (Drabble & Drabble, 1917 and 1927), concluded that Asteraceae were especially favoured by hoverflies. However, they found inexplicable differences in attractiveness; thus autumn hawkbit (*Leontodon autumnalis*) was very attractive, and dandelion (*Taraxacum officinale*) and a hawkweed (*Hieracium boreale*) were well visited, but the similar flowers of nipplewort (*Lapsana communis*), common catsear (*Hypochaeris radicata*) and smooth hawksbeard (*Crepis capillaris*) were visited by only one or two species each. Only nipplewort, however, showed this lack of visits

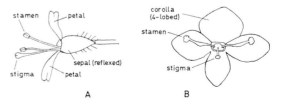

Fig. 3.15 **A**, flower of enchanter's nightshade (*Circaea lutetiana*) from the side. **B**, flower of germander speedwell (*Veronica chamaedrys*) from the front. After Kugler, 1938.

a

b

Fig. 3.16 **a–b**, hoverflies (Diptera: Syrphidae) on flowers of *Veronica chamaedrys*. **a**, *Baccha elongata*. **b**, *Melanostoma* sp.

in the observations of Willis & Burkill (1895-1908).

In some plants, there appears to be a degree of specialisation to hoverflies; examples are germander speedwell (*Veronica chamaedrys*) and enchanter's nightshade (*Circaea lutetiana*) (Kugler, 1938). These are members of different families but have a similar arrangement of stamens and stigma (Fig. 3.15). In the speedwell, the corolla is bright blue and the two stamens are weak near the base. When an insect clings to one or both stamens they droop so that the underside of the insect's body comes into contact with the stigma. The anthers also touch the underside of the insect so that repeated visits to different flowers of this species are likely to lead to cross-pollination (Fig. 3.16). Although hoverflies frequently alight on the stamens, they sometimes cling to the surface of the corolla, in which case they may avoid coming into contact with anthers and stigma. Small bees, however, which frequently visit the flowers, almost always alight on the stamens and are therefore more reliable pollinators than the hoverflies. The stamens of the white-petalled enchanter's nightshade droop in the same way when an insect alights upon them, and pollination also takes place in the same way. In a natural woodland habitat Kugler found that Diptera were the only visitors to this flower, but in parkland where there was open ground near at hand small bees also visited it. The hoverflies best suited to these two flowers are small species which inhabit shady situations (those shown in Fig. 3.16 and *Neoscia* [Fig. 2.9] and *Syritta* species).

In England, the Drabbles found that germander speedwell and brooklime (*Veronica beccabunga*) received few visits from hoverflies but that these insects did operate the pollination mechanism.

The common British bindweeds (family Convolvulaceae) (Fig. 6.11) have been repeatedly noted as favourites of hoverflies, and especially of *Rhingia*, which has been recorded feeding on the pollen (Bennett, 1883; Drabble & Drabble, 1927; Verdcourt, 1948; Parmenter, 1948; Baker, 1957). A plant that is visited almost exclusively by the

Fig. 3.17 Hoverfly, *Sphaerophoria ruepellii* male (Diptera: Syrphidae), on flower of tormentil (*Potentilla erecta*).

smaller hoverflies is small balsam (*Impatiens parviflora*), which has pale yellowish flowers and provides the insects with both nectar and pollen (Coombe, 1956).

The many reports of hoverflies feeding on the pollen of wind-pollinated plants have been listed by Stelleman & Meeuse (1976). This is an important habit of certain genera, the principal ones being *Melanostoma* and *Platycheirus* (Gilbert, 1981; Holloway, 1976; Stelleman & Meeuse, 1976). The grasses (Poaceae) are particularly important pollen sources for such hoverflies. For example, Drabble & Drabble (1927) found that *Melanostoma mellinum* ate the pollen of timothy grass (*Phleum pratense*) and cocksfoot (*Dactylis glomerata*) and apparently caused pollination. *Melanostoma* has also been seen feeding on the pollen of a sedge (*Carex binervis*) (MCFP). Hoverflies can effectively pollinate the apparently wind-pollinated ribwort plantain (*Plantago lanceolata*) (Stelleman & Meeuse, 1976). Another instance of pollen-feeding by a hoverfly involves the use of thoracic vibration to obtain pollen from flowers of *Solanum* ('vibratory pollen-collection', see p.125). The fly concerned, *Volucella mexicana*, is a mimic of a bee of the genus *Xylocopa* that exploits the flower in the same way (Buchmann, 1983).

Syrphidae that visit zygomorphic flowers such as those of Scrophulariaceae and Lamiaceae behave, apparently instinctively, in a manner suitable to the form of the flower, unlike most members of the fly families Muscidae and Calliphoridae (see later), which tend to walk all over the flowers at random.

Flower-constancy is another attribute that is highly developed in Syrphidae, as demonstrated by Kugler (1950) in a habitat with abundant hoary alison (*Berteroa incana*, Brassicaceae) and scentless mayweed (*Tripleurospermum inodorum*, Asteraceae, named for its scentless leaves). One individual of *Eristalis tenax* visited 47 mayweed heads and two unopened yellow composite heads; it made nine approaches to other flowers, but none to *Berteroa*. Another *E. tenax* showed a similar constancy to *Berteroa* and did not approach *Tripleurospermum*.

Differential visitation by hoverflies was reported by Parmenter (1958). Yarrow (*Achillea millefolium*, flowers white), common catsear (*Hypochaeris radicata*, yellow), and knapweed (*Centaurea nigra*, pinkish purple) grew in neighbouring patches of approximately equal size, and *Eristalis arbustorum* made 132 visits to yarrow, 34 to catsear and one to knapweed, whereas *Helophilus parallelus* made three visits to yarrow, none to catsear and 24 to knapweed.

The reaction of *Eristalis tenax* to colour was tested with artificial flowers containing sugar-water (Ilse, 1949). Yellow 'flowers' were presented among others with a range of shades of grey; although there were five grey flowers to each yellow one, many more visits were made to the yellow ones. This and other experiments showed that *Eristalis* had a general preference for yellow, in contrast to *Bombylius*, but that periods of training on several colours modified the choice (Kugler, 1950). Blue always had a low stimulatory effect and deep red was not distinguished from grey. Tests with models of a fixed area but varied length of outline showed that the flies could distinguish the models but had no inborn preference (Kugler, 1950).

The response of *Eristalis tenax* to scent was tested by adding artificial carnation scent to various flowers that the flies were constantly visiting in a natural habitat (Kugler, 1950). Only 2% of approaches to these resulted in normal visits, while 46% resulted in short visits (a normal approach followed by alighting and flying away) and 52% resulted in no visit. When yellow model flowers were used, treated either with sugar-water and flower scent or salt solution, they were equally attractive. However, with blue flowers with sugar-water or salt and scent, 83% of visits were to flowers with sugar; here, the increased (negative) effect of scent was probably caused by the low stimulatory value of the colour blue. It is interesting that scent can inhibit visits.

Diptera, suborder Cyclorrhapha: Schizophora (Acalyptratae)

The Acalyptratae (p.58) comprise about 280 genera in numerous families, one of which, Conopidae, will be dealt with on its own and the rest as a group. The Conopidae is a small family, some of whose members have a long proboscis (see Box 3.5). These Conopidae sometimes visit tubular flowers of the type favoured by other long-tongued insects (Table 3.7) but most of them show a partiality for flowers with exposed or easily reached nectar such as Asteraceae, Apiaceae and Rosaceae. The rare *Leopoldius signatus* is recorded in Britain only at ivy which has exposed nectar (Harper & Wood, 1957; Smith, 1959, 1961).

Of the remaining 39 families of the Acalyptratae, species belonging to 23 were recorded at flowers (Knuth, 1906-1909; Willis & Burkill, 1895-1908). These are mainly rather small flies with a short proboscis of the mopping up type (see Box 3.5), and many have the habit of waving their wings, often alternately, as they walk about. In some, the wings are banded or marbled (giving rise to the name 'picture-wing flies') (Fig. 3.18). Members of one of these families, the Trypetidae, lay eggs in the capitula of Asteraceae, so that although they may transfer pollen, the capitula are later damaged by the larvae. Three genera of Trypetidae are recorded as specialised flower

---

Box 3.5 Mouth-parts of Diptera: Conopidae and Acalyptrate flies

Sicus (in Conopidae) has a slender proboscis about 6 mm long, including the rostrum; the labium, 2.5 mm long, is directed forwards, while the terminal part, 3 mm long, made up of the greatly elongated labella fused at the base, can be directed downwards or, when out of use, folded right back underneath the labium. In *Conops quadrifasciatus* the proboscis is 4 mm long, with shorter labella (Diagrams A, B). *Conops* and *Sicus* feed on nectar, but are said not to eat pollen; the nectar flows between the labella, into the groove on the labium and then up a tube formed by the labrum and hypopharynx. Diagram C shows the mouth-parts of *Sepsis* (Sepsidae), which are more like those of *Eristalis*.

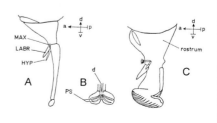

Fig. 3.18 'Picture-winged flies'
(*Herina frondescentiae*, Diptera:
Otididae), on shrubby
cinquefoil (*Potentilla fruticosa*).

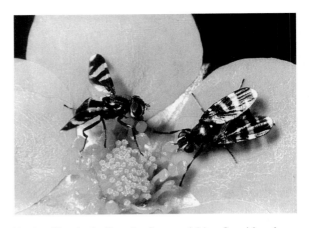

visitors (Knuth, 1906-1909). Six families, including the flower-visiting Sepsidae, have
the wing-waving habit but clear wings. Species of *Sepsis*, small dark brown flies with
the first abdominal segment narrow, and resembling ants, are frequently seen on the
umbels of Apiaceae, and this and some other genera have been recorded from a
number of flowers. They have the mopping-up type of mouth-parts.

The Lauxaniidae have been recorded at several rather diverse flowers by Willis &
Burkill (1895-1908), while Chloropidae visit Apiaceae, Asteraceae and forget-me-not
(*Myosotis*). Chloropid flower-visitors include the frit fly (*Oscinella frit*), the larvae of
which are a pest of cereals, and *Chlorops*, a small yellowish fly with longitudinal black
and yellow stripes on the thorax. Chloropidae (together with Chalcid wasps) are pollinators of the tiny flowers of the musk orchid (*Herminium monorchis*) which they visit for
nectar (see pp.204-5; Nilsson, 1979b). The flower-visiting genus *Siphonella* has a long
proboscis similar to that of the hoverfly *Eristalis*. Members of several acalypterate
families are attracted to the trap-flowers lords-and-ladies (*Arum maculatum*), large
cuckoo pint (*A. italicum* subsp. *neglectum*) and *Aristolochia* (see Chapter 10). This seems to
be especially the case with Drosophilidae, which breed in decaying fruit, carrion and
excrement (among other things), and with Sphaeroceridae, which are often dung-frequenting (Knuth, 1906-1909; Grensted, 1947; Prime, 1954).

Diptera, suborder Cyclorrhapha: Schizophora (Calyptratae)

There are about 280 genera in the Calyptratae; the common flower-visiting members
are grouped into seven families. These flies are generally larger than the Acalyptratae
and are more important as flower visitors and pollinators, although they are nearly all
unspecialised in feeding habits. In many genera the proboscis has 'prestomal teeth'
between the labella which can be used for scraping. In all four families, feeding on
both nectar and pollen is recorded.

Tachinidae (=Larvaevoridae) is a family of rather bristly flies, often greyish, whose
larvae are internal parasites of insects and other arthropods. Much the commonest
member of this family in Willis & Burkill's (1895-1908) lists is *Siphona geniculata* which
is only about 6 mm long, but has a slender proboscis twice as long as the head (Box
3.6). It visits plants with small tubular flowers and avoids certain flowers with exposed
nectar favoured by less specialised Diptera. *Prosena (Calirhoe) siberita*, a fly about 10 mm
long, is also long-tongued and visits various narrowly tubular flowers (Box 3.6). *Eri-*

othrix rufomaculatus, a small black and red fly, and *Tachina* (=*Larvaevora*=*Echinomya*) *fera*, a large coarsely bristly black and orange fly (Fig. 3.19), are also recorded on Asteraceae, Apiaceae, *Mentha* etc. *T. fera* has a proboscis 5-6 mm long and is found at tubular flowers more often than the shorter-tongued members of the family. Pollinating visits to the burnt orchid (*Orchis ustulata*) have been recorded for *Tachina magnicornis* (Vöth, 1984) and to *Veratrum album* (Liliaceae) for unspecified Tachinidae (Daumann, 1967). (Records of flower-visits by other members of the family are given by Colyer & Hammond, 1951; Andrews, 1953; Parmenter, 1941, 1952a & b; Grensted, 1946; Harper & Wood, 1957; and Willis & Burkill, 1895-1908.)

Eight of the 27 genera of Sarcophagidae and Calliphoridae were recorded at flowers by Willis & Burkill (1895-1908). *Pollenia*, *Lucilia* (green-bottles), *Melinda* (*Onesia*) and *Sarcophaga* (e.g. *S. carnaria*, the common flesh fly, a familiar large black and grey fly with red eyes) are predominantly flower-feeders, while *Calliphora* (blue-bottle or blowfly) (Fig. 3.20) and *Cynomya* feed chiefly on carrion and excrement but also visit flowers; *Pollenia vespillo* feeds at flowers and sap only (Kugler, 1955b). The Calliphoridae have a short proboscis (2-4 mm) and they visit flowers with well exposed nectar such as Apiaceae and yellow mountain saxifrage (*Saxifraga aizoides*) or small tubular flowers

Fig. 3.19 Tachina fera (Diptera: Tachinidae) on ragwort (*Senecio jacobaea*).

Fig. 3.20 Blow-fly (*Calliphora vomitoria*, Diptera: Calliphoridae) feeding at ivy (*Hedera helix*).

Box 3.6 Mouth-parts of Diptera: blowflies and other calyptrate flies
(illustration opposite)

Flies specialised for flower-visiting may have long and very slender tongues but attain this condition in different ways, as in *Siphona* (Diagrams A and B) and *Prosena* (C).

The proboscis of one calyptrate, the blowfly, *Calliphora*, is known in detail from the work of Graham-Smith (1930). The basic structure is the same as in the hoverflies and *Sepsis* (Boxes 3.4, 3.5), but the maxillae are further reduced and only the palps remain (D, E). The way the feeding tube is formed is shown in sections a-d of Diagram F, corresponding to levels with the same letters in Diagram D. The edges of the labrum are grooved to receive the edges of the hypopharynx (Fb), but the tube ends just short of the end of the labium; here the labial groove is roofed over by pairs of interlocking folds arising from its edges so that the tube is continued forwards (Fc). The pseudotracheae are completely closed over when the labella are used for filtering, and the liquid passes into each pseudotrachea through short sloping passages at right angles to its length (J, K). The diameter of the pseudotracheae in *Calliphora vicina* ranges from .01 to .02 mm and the sloping lateral passages are .004 to .006 mm in diameter; supporting sclerites for these are seen in J (right) and K. Between the labella there are four rows of prestomal teeth (Fd, G, H).

Several different feeding positions of the labella are described by Graham-Smith. He fed films of drying milk and drying solutions of sugar containing Indian ink to the flies, and it was from the impressions left by their proboscides in these and other foods that he made his discoveries about these feeding positions. In the filtering position, the labella are inflated and spread horizontally on the food surface; they are also pressed together so that the labial tube has no connection with the food other than through the pseudotracheae (G). Other feeding positions allow the prestomal teeth to reach the food and are presumably not used for feeding at flowers. Filtering is entirely abandoned when the direct feeding position is adopted (H), as it is for viscid materials. The movements of the labella are produced by muscles acting on the furca (G, H) but the labella are inflated by blood pressure. When not in use, the proboscis is folded up and retracted into a recess in the head (E). The pseudotracheae and the tubes leading to the mouth can all be opened along one side; this may be of importance in allowing cleaning if the passages become blocked.

---

such as Asteraceae and alpine bistort (*Persicaria vivipara*) and water mint (*Mentha aquatica*). They also feed on the pollen of certain large tubular flowers, the nectar of which is inaccessible to them. Kugler (1956) found that *Lucilia* was sometimes constant in its visits to grass of Parnassus (*Parnassia palustris*) and other species in nature.

In the Scathophagidae only one genus (out of 22) appears to have been recorded at flowers in Britain; this is *Scathophaga* (=*Scopeuma*), which as larvae live in dung and as adults prey on insects, piercing them with their prestomal teeth. *S. stercorarium* (Fig. 3.21) is the common yellow dung fly, the males of which are golden-yellow furry insects. *Scathophaga* visits flowers to seek prey and to feed on pollen and nectar. Colyer & Hammond (1951) record it at blackthorn (*Prunus spinosa*), hawthorn and bramble, while Knuth (1906-1909) gave a long list, including many Ranunculaceae, Brassicaceae, Apiaceae and Asteraceae.

The very large families Anthomyiidae and Muscidae, and the family Fanniidae, together comprising about 80 genera and about 450 species in Britain, are the most important flower-visiting families among the Diptera apart from the hoverflies. Willis & Burkill's (1895-1908) records cover about 28 genera. Examples are: *Musca autumnalis* (Fig. 3.22), a close relative of the house fly (*M. domestica*) with orange-yellow markings

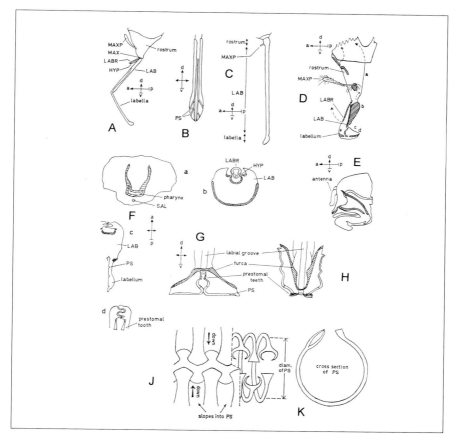

on the abdomen; *Mesembrina meridiana*, a large black fly; *Dasyphora cyanella*, *Orthellia viridis* and *O. cornicina*, green-bottles similar to *Lucilia* of the Calliphoridae; *Polietes lardaria* and *Graphomya maculata* (Fig. 3.23), two flies resembling the common flesh fly in their colouring, though this applies only to the female in *Graphomya*; and *Fannia canicularis*, the lesser house fly.

The anthomyiid, muscid and fannid flies do not appear to be specialised flower-visitors, but many of them are occasional or regular ones. Willis & Burkill (1895-1908) found that they visited species of thistle beside *Cirsium arvense*, which has the most accessible nectar, and recorded them feeding on pollen at flowers with inaccessible nectar. Totland (1993) found Muscidae and Anthomyiidae the commonest flower visitors at his site in the Norwegian mountains. The Drabbles recorded one species feeding on the pollen of grasses (cf. Syrphidae, p.70).

These families of the Calyptratae comprise the great majority of the short-tongued Diptera that visit flowers. These flies are an important part of the insect fauna generally and they provide a greater proportion of the flower-visitors in north Europe (as in Britain and Norway) than in Germany, where bees and beetles are more abundant. At Scarborough (coast of north-east England), in summer, 61% of 1,800 individual visits

Fig. 3.21 Yellow dung fly (*Scathophaga stercoraria*), on buttercup.

Fig. 3.22 Musca autumnalis (Diptera: Muscidae), a close relative of the housefly, on flower of coltsfoot (*Tussilago farfara*).

Fig. 3.23 Graphomya maculata (Diptera: Muscidae) on hogweed (*Heracleum sphondylium*).

to six species of plant were by short-tongued Diptera, as were 48% of visits to 27 species in spring (Willis & Burkill, 1895-1908). Willis & Burkill played down their importance as pollinators, in spite of their abundance, whereas Drabble & Drabble (1927) were convinced that they are effective pollinators, at least for Asteraceae. The less specialised flowers that are most visited by short-tongued Diptera are also those most visited by insects in general.

Many of the flowers visited by short-tongued Diptera are sweet-scented, for example, meadowsweet (*Filipendula ulmaria*) and rowan (*Sorbus aucuparia*) with heavy scents, crab apple (*Malus sylvestris*) with a lighter scent, and sallows (*Salix* spp.), lady's bedstraw (*Galium verum*), thrift (*Armeria maritima*) and white mignonette (*Reseda alba*), also with a light scent differing from that of crab apple but all much like each other (PFY). Some have a tang of stale dung or urine with their sweet scent, as in many umbellifers such as cow parsley (*Anthriscus sylvestris*), alexanders (*Smyrnium olusatrum*); the crucifers treacle mustard (*Erysimum cheiranthoides*), sweet alison (*Lobularia maritima*) and swede or rape (*Brassica napus*); some composites such as goldilocks (*Aster linosyris*); and, to some people, hawthorn (*Crataegus monogyna*). These scents, even to human noses, therefore show a correlation with the prevailing habits of these unspecialised flower visitors.

The responses of Calyptratae to the sensory stimuli offered by flowers are broadly similar to those of Syrphidae. Liebermann (1925) covered an isolated plant of cowbane (*Cicuta virosa*, family Apiaceae) with large leaves and showed that the flies could find the flowers by scent alone. Flies of the genera *Lucilia, Dexia, Pyrellia, Sarcophaga* and others, when passing within 2 m downwind, turned towards it and flew to and fro, then alighted and crawled under the leaves to the flowers; on the windward side, none was deflected towards it. In another experiment, a plant of *Cicuta* was placed in a room behind a curtained window, which was then opened 5 cm. On the wall of the house near the window many flies that were resting in the sun became restless, and after some flying to and fro before the opening, flew through it and on to the flowers. Flies were also attracted from a distance, first settling on the wall and then entering the room. This shows that insects which are not actively seeking food may respond to stimuli associated with it.

Knoll (1926) found that flies that were attracted to *Arum nigrum* were attracted by the scent, as were the beetles; they were species normally associated with human excrement.

Kugler (1956) carried out experiments on calypterate flies hatched in captivity (naive flies). *Lucilia, Calliphora* and *Sarcophaga* more or less ignored flower-models (coloured discs) until scent was brought near them, whereupon visits and proboscis-reactions took place. When scent was placed on some of the models these were preferred. Both fragrant and excremental scents were used, but *Calliphora* and *Sarcophaga* preferred yellow models over brown-purple in the presence of sweet scents, and brown-purple over yellow and white in the presence of excremental scents.

Other experiments did not use naive insects. Captive *Lucilia*, presented with artificial flowers of the type used for *Eristalis* (Syrphidae, p.71), made 62% of their visits to hawthorn-scented models and 38% to scentless models. If the smell of ammonia, associated with egg-laying, was substituted, the flies visited both models equally, showing no instinctive association of ammonia smell with food. Rewardless models smelling of ammonia were avoided, showing that the flies perceived the scent and learnt to associate it with the absence of food.

In experiments involving real flowers of scentless mayweed (*Tripleurospermum inodorum*) and artificial models, Kugler (1951) found that *Lucilia* responded in a similar way to *Eristalis* (Syrphidae, p.71): visual stimuli were used to find the flowers from a

Fig. 3.24 **a, b** Hoverfly (*Neoascia podagrica*) on grass of Parnassus (*Parnassia palustris*); **a**, probing one of the false nectaries; **b**, sucking nectar from the disc of the receptacle.

distance, but olfactory cues triggered alighting. The reactions of the flies to shape, and to yellow, blue and grey colours were also the same as those of *Eristalis* (Kugler, 1955b, 1956), and they preferred the form of a funnel to a disc shape.

Calliphoridae and Syrphidae are visitors to the curious flowers of grass of Parnassus (*Parnassia palustris*). These have five green nectaries that produce scent, and five clusters of glistening knobs on stalks. Daumann (1932, 1935) found that the approach of the hoverflies to the flowers was visual but that alighting was induced by scent; the flies also located the individual nectaries by scent. Removal of the glistening knobs made no difference to the flies' behaviour, and the knobs therefore appeared to be functionless. However, Kugler (1956), using naive *Lucilia* and hoverflies, showed that they were initially much more attracted to the knobs (see Fig. 3.24), but in successive tests were increasingly likely to touch the nectaries first. Experiments showed that the flies were attracted to small glistening objects placed on an artificial flower. Thus the knobs may, indeed, be 'false nectaries', as earlier supposed.

Conclusions on the behaviour of Diptera

With naive or resting flies, scent is mainly excitatory, inducing the insects to seek and alight on objects of appropriate colours, either yellow and white, as the colours of sweetly scented flowers most commonly suited to them, or brown and dark purple, as the colours of excrement and carrion, sometimes imitated by flowers (see Chapter 10). Sometimes scent may guide insects all the way to its source. However, with foraging flies the stimulus that elicits approach from a distance does not usually enable the insect to tell whether the flower is suitable for it. When it is close to the flower, different sensory perceptions come into play which determine whether a visit takes place. The reaction to a flower may be divided into three phases: response to a distant signal leading to approach; response to a short-range signal, leading either to alighting or flying away; and response to internal flower-signals, including the presence of food, leading to uptake of food. Sometimes alighting takes place before the response to the short-range signal. The first two stages, with a visual distant signal and an olfactory short-range signal, have been repeatedly noted in this chapter.

Diptera can distinguish colours when feeding, and in general tend to visit white, pink, yellow and green flowers most readily. Red-blindness is proved to exist in the blowfly (*Calliphora*) and the drone-fly (*Eristalis tenax*), and it probably occurs in the bee-fly as well. Visits to purple and blue flowers are commonly made only by the longer-tongued genera of the Brachycera and Cyclorrhapha, and all of these also visit flowers of other colours quite readily. The blue and purple flowers usually have more deeply seated nectar than flowers of other colours. These colour differences therefore help the insects to find the flowers most suited to them. The preferences of the insects may be inherited in some flies, for example the aversion of *Bombylius fuliginosus* to yellow, but learned in others (see remarks on trainability in the corresponding section of Chapter 4). Innate preferences for certain colours might be expected to have evolved once the association between particular colours and deep-seated or exposed nectar had become established.

Diptera with long tongues are scattered through the Brachycera and Cyclorrhapha; these flies can reach nectar not available to shorter-tongued insects and they show some diminution of attention to flowers with well exposed nectar. This is particularly true of *Bombylius*, *Rhingia* and *Siphona* and is a sign of their specialisation which has evolved independently.

4

The Insect Visitors
II: Butterflies and Moths

The butterflies and moths (Lepidoptera) are represented by about 2,000 species in the British Isles. Their caterpillars usually feed on plants, and while most of them eat leaves, some feed in wood, in flower-heads or as leaf-miners. With few exceptions the adults feed only on liquids, and the range of these is much the same as in Diptera, namely nectar, fruit juice, exudates from plants and fluids occurring on excrement and carrion. Foods other than nectar may be consumed optionally as, for example, by the peacock butterfly (*Inachis io*), or they are consumed either predominantly or exclusively, as by the Camberwell beauty butterfly (*Nymphalis antiopa*) and the death's head hawkmoth (*Acherontia atropos*).

Four sub-orders of Lepidoptera are recognised, and one (the Ditrysia) contains all the butterflies and nearly all the moths (Richards & Davies, 1977). Any generalisations which follow apply to this sub-order; the other three are small and will be dealt with individually. The so-called Microlepidoptera are distributed among all sub-orders in this classification. Butterflies are classified as two superfamilies in the midst of a long series of superfamilies of moths. The popular appeal of the Lepidoptera to the general public is attested by the two volumes of the 'New Naturalist' series devoted to them (Ford, 1945 and 1955).

Structure and habits of the main sub-order of Lepidoptera

The long slender proboscis typical of butterflies and moths (Plate 3b-e) is very uniform compared with that of the Diptera and very differently constructed (Box 4.1). It is clearly adapted for reaching nectar at the base of narrowly tubular flowers, though it can be used for sucking up exposed liquids. Because of its length it is coiled when at

Fig. 4.1 Small skipper butterfly (*Thymelicus sylvestris*, Lepidoptera: Hesperiidae), feeding on nectar from water mint (*Mentha aquatica*); note the 'knee bend' in the butterfly's proboscis.

Table 4.1 Proboscis lengths of some European Lepidoptera (from Knuth, 1906-1909).

Species	English name	Family	Length in mm
Pyrausta		Pyralidae	4-9
Autographa gamma	silver-Y moth	Noctuidae	15-16
Macroglossum stellatarum	hummingbird hawkmoth	Sphingidae	25-28
Sphinx ligustri	privet hawkmoth	Sphingidae	37-42
Agrius convolvuli	convolvulus hawkmoth	Sphingidae	65-80
Papilio machaon	swallowtail butterfly	Papilionidae	18-20
Parnassius apollo	apollo butterfly	Papilionidae	12-13
Coenonympha pamphilus	small heath butterfly	Satyridae	7
Aglais urticae	small tortoiseshell butterfly	Nymphalidae	14-15
Inachis io	peacock butterfly	Nymphalidae	17
Boloria pales	shepherd's fritillary	Nymphalidae	9-10
Pieris brassicae	large white butterfly	Pieridae	16

rest (Box 4.1, Diagram C), and when in use it has a 'knee-bend' that facilitates entry to slender flower-tubes (Fig. 4.1). Pollen of the flowers visited is conveyed involuntarily by the insect on the proboscis or head. Different species vary greatly in the length of their proboscis (Table 4.1); some are therefore confined to shallower flowers (Fig. 4.2), while others can reach the deep-seated nectar of the very long-tubed ones which must have evolved with them in mutual adaptation. Such insects and the flowers they pollinate are clearly highly specialised. The longest proboscis of any European lepidopteran is that of the convolvulus hawkmoth (Table 4.1), but hawkmoths with tongues several times longer occur in Madagascar (see p.214 and p.382).

The tip of the proboscis is armed with numerous fine spines which are thought to be used for breaking open the tissues of nectarless flowers to release the sap, as butterflies are occasionally seen apparently sucking at these flowers (for example common centaury [*Centaurium erythraea*]). It has been found that the proboscis spines are used by moths to release a fluid that is held in copious supply by thin-walled cells lining the nectary-cavity of an orchid growing in Florida. The plant, *Epidendrum anceps*, is pollinated by these moths (Adams & Goss, 1976). In addition, Lepidoptera pierce the intact skin of fruits and may thereby become pests (F. Darwin, 1875; Knoll, 1922).

A special method of feeding is adopted by the New World tropical butterfly genus

Fig. 4.2 Green hairstreak butterfly (*Callophrys rubi*, Lepidoptera: Lycaenidae), sucking nectar from flower of creeping buttercup (*Ranunculus repens*).

Box 4.1 Mouth-parts of butterflies (illustration opposite)

Whereas the sucking tube of the Diptera is formed by the great development of the labrum, hypopharynx and labium, that of the Lepidoptera is formed by the terminal lobes (galeae) of the two maxillae. These are enormously elongated and lie touching one another, their contiguous surfaces being hollowed out so that a tube, circular in cross-section, is formed between them. This is the food canal (Diagrams D, E), of which the base connects with the mouth and the apex is open for taking in the food. The galeae are themselves hollow internally (D, E), and their cavities are continuous with the general body cavity which contains the blood. The proboscis is rolled up under the head when not in use (C) and is kept in this condition by a longitudinal elastic bar in each galea which is coiled when at rest. The outer surface of the galea is constructed to facilitate this rolling, since it consists of alternate transverse bands of thin membrane and thickened rings (F); the wall of the food tube is comparatively rigid, but it has a laminated structure which permits coiling in the vertical plane while preventing movement from side to side. During uncoiling, small oblique muscles inside each galea (D, E, stippled areas) contract and cause an alteration in the cross-sectional shape of the proboscis, such that the upper surface changes from flat (D) to convex (E). Convexity of the upper surface is incompatible with coiling, and the proboscis is therefore forced to unroll. The effect is analogous to the effect of the convexity of a

coiled steel rule when released from its casing. The operation of the mechanism is also dependent on the rigidity of the walls of the food canal, the presence of longitudinal partitions inside the galeae, and the maintenance of blood pressure in them by the closure of a valve at the base of each. The 'knee-bend', some distance from the base of the proboscis, is kept in being by a set of muscles which counteract at this point the uncoiling effect of the main set of muscles. The angle of the bend can be varied at will. The galeae are held together on their lower side by a series of closely interlocking teeth, and on their upper side by a series of larger overlapping plates which slide over one another during coiling and uncoiling. A glandular secretion provides lubrication and helps to seal the joint which is formed by the plates. The other mouth-parts (A, B, C) are a pair of structures which may be rudimentary mandibles, the labrum and the labium, the last of which bears two palps of moderate size that may act as 'feelers' and are often furry so that they conceal the proboscis when it is rolled up (Eastham & Eassa, 1955, based on the large white butterfly [*Pieris brassicae*]).

Further analysis of the way the Lepidopteran proboscis is engineered has been provided by Hepburn (1971), who found that the elastic bar in each galea is made of resilin, a structural protein with rubber-like qualities, first discovered in 1960. It also emerged that considerable differences of detail occur in different genera.

Heliconius. The insect gathers a ball of dry pollen under its head and then squirts a drop of clear liquid into it; it kneads this pollen for some hours by movement of the proboscis. The liquid, now enriched with amino-acids, is then sucked back into the gut. The protein-building material acquired in this way sustains the butterfly in a much longer life than is usual and is also put into egg production (Gilbert, 1972). The exuded fluid does not contain digestive enzymes, but it is known that some pollens moistened with sugar solutions release their protein and amino-acids.

The hummingbird hawkmoth (*Macroglossum stellatarum*) can only drink solutions that are quite fluid; it strives in vain to suck up very thick syrupy solutions of sugar (Knoll, 1922). The silver-Y moth (*Autographa gamma*) (Fig. 6.61), on the other hand, secretes saliva copiously and this enables it to dilute and suck up such materials.

Some flowers that seem to be adapted to pollination by butterflies are described on p.182 and others are listed in Table 4.2. These flowers are chiefly blue or deep pink.

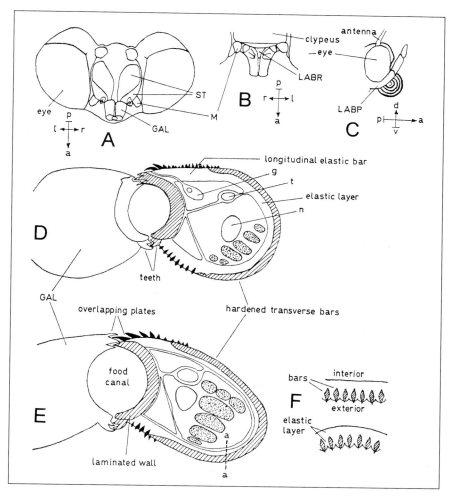

However, a campion from the Mediterranean region (*Lychnis chalcedonica*) and many tropical and subtropical plants pollinated by butterflies have scarlet flowers. As the tubular part of the flower into which the thread-like proboscis has to pass is usually very slender, the proboscis does not easily avoid touching the stamens and stigmas.

The scents of flowers adapted to pollination by Lepidoptera are usually sweet and sometimes heavy, as in honeysuckle (*Lonicera periclymenum*), hyacinth (*Hyacinthus orientalis*), lilac (*Syringa vulgaris*), wallflower (*Erysimum cheiri*) and carnation (*Dianthus caryophyllus*). One of the best-known butterfly flowers of gardens, *Buddleia davidii*, produces a scent containing a balsamic element, which is perhaps related to its power of attraction for the butterflies of intermediate feeding habits (p.80).

Some of the flowers visited by butterflies are also visited at night by moths, *Anacamptis* (Chapter 7) being an example of these. The different species of night-flying moths have their characteristic emerging times. The flowers that are specially adapted to

Table 4.2 The British flowers apparently most adapted to butterflies.

Certain Caryophyllaceae, such as *Lychnis flos-cuculi* (Fig. 4.3) and pinks, *Dianthus* species (p.182)
Forget-me-not, *Myosotis* (Boraginaceae)
Pyramidal and fragrant orchids, *Anacamptis* and *Gymnadenia* (Figs. 7.16 & 7.14)
The valerians, Valerianaceae (Plate 3d)
Some Rubiaceae
Some Asteraceae, such as hemp agrimony, *Eupatorium cannabinum* (Plate 3c), and common fleabane,
    Pulicaria dysenterica (visited by Lycaenidae – 'blues') (Plate 3b)
Some Lamiaceae in which the corolla lacks an upper lip such as bugle, *Ajuga reptans*, and
    wood-sage, *Teucrium scorodonia*
Mentha, Lamiaceae, with small tubular flowers

pollination by them are again represented in Caryophyllaceae (p.182) and Orchidaceae (p.200). Although Britain has a good range of flowers adapted to Lepidoptera (Table 4.2), the most visited are not limited to these specialised flowers, as is shown by Table 4.3, which includes an important element of small tubular flowers, namely *Valeriana*, Asteraceae and *Thymus*. These are visited by butterflies, but not by the white admiral (*Ladoga camilla*), which feeds on tree sap and excrement as well as certain other flowers (in England especially brambles, *Rubus fruticosus* agg.).

Some of the most frequently occurring moth visitors in the rather scanty British records of Willis & Burkill (1895-1908) are listed in Table 4.4. These mostly belong to the families Noctuidae (Fig. 4.4) and Geometridae, the two largest families of moths in Britain. Certain families of moths are quite absent from Willis & Burkill's lists, and the reason for this in some cases is that the adults never require food and have a short and non-functional proboscis. Hawkmoths (family Sphingidae) (Plate 3e) were scarce, no doubt because members of this family are rare in Scotland, where most of the observations were made.

In fact, members of this last family appear to be particularly effective as pollinators. They fly rapidly from flower to flower, and usually take the nectar without settling; their long tongues are no doubt helpful in enabling them to do this, and the longer-tongued species can reach deeply-seated nectar. Exceptions to the rule of feeding in flight have been noted both in California, where hawkmoths have been seen to crawl

Fig. 4.3 Green-veined white butterfly (*Pieris napi*, Lepidoptera: Pieridae) sucking nectar from ragged robin (*Lychnis flos-cuculi*).

Table 4.3 Flowers most visited by British Lepidoptera (Willis & Burkill, 1895-1908; Bennett, 1883; Scorer, 1913; and others).

Scientific name	English name	Family
Silene	campion and catchfly	Caryophyllaceae
Lychnis	campion or ragged robin	Caryophyllaceae
Lotus	birdsfoot trefoil*	Fabaceae-Faboideae
Prunus spinosa	blackthorn	Rosaceae
Rubus fruticosus	bramble	Rosaceae
Rubus idaeus	raspberry	Rosaceae
Hedera helix	ivy	Araliaceae
Lonicera periclymenum	honeysuckle	Caprifoliaceae
Valeriana officinalis	valerian	Valerianaceae
Centranthus ruber	red valerian	Valerianaceae
Dipsacus fullonum	teasel	Dipsacaceae
Succisa pratensis	devil's-bit scabious	Dipsacaceae
Eupatorium cannabinum	hemp agrimony	Asteraceae
Pulicaria dysenterica	common fleabane	Asteraceae
Solidago	golden rod	Asteraceae
Achillea	yarrow, etc.,	Asteraceae
Senecio jacobaea	ragwort	Asteraceae
Centaurea	knapweed	Asteraceae
Cirsium	thistle	Asteraceae
Calluna vulgaris	ling	Ericaceae
Stachys	woundwort	Lamiaceae
Thymus	wild thyme	Lamiaceae
Salix	sallow	Salicaceae

*bees pollinate it, Lepidoptera thieve the nectar

into large trumpet-shaped flowers (Baker, 1961; compare also report of visits to 'large flowered *Convolvulus*', p.149), and in Brazil, where they alighted on the corolla-lobes of a narrow-tubed flower (Silberbauer-Gottsberger, 1972). Most hawkmoths fly in the evening, but the hummingbird hawkmoth (*Macroglossum stellatarum*), which migrates to Britain and is sometimes common, flies by day and visits a great variety of flowers of the types favoured by butterflies. In the Alps, Müller (1881, quoted by Knuth, 1906-1909) observed individuals of this species differing in their choice of flowers at the

Table 4.4 The moths most commonly seen at flowers by Willis & Burkill (1895-1908), working mainly in Scotland.

Scientific name	Family	English name
Pyrausta spp.	Pyralidae	
Celaena haworthii	Noctuidae	Haworth's minor moth
Cerapteryx graminis	Noctuidae	Antler moth
Amphipoea oculea	Noctuidae	Ear moth
Autographa gamma	Noctuidae	Silver-Y moth
Psodos coracina	Geometridae	Black mountain moth
Chloroclysta citrata	Geometridae	Common marbled carpet moth
Coenotephria salicata	Geometridae	Striped twin-spot carpet moth
Glyphipterix fuscoviridella	Glyphipterigidae	a 'micro-moth'

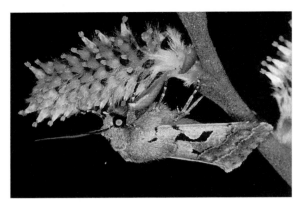

Fig. 4.4 Hebrew character moth (*Orthosia gothica*, Lepidoptera: Noctuidae subfamily Hadeninae), a common nocturnal visitor to sallow (*Salix cinerea*) sucking nectar from female catkin.

same locality. Thus one visited *Primula integrifolia*, another visited three species of gentian and *Viola calcarata*, while two others visited only *V. calcarata*. The rapidity of the visits is shown by the fact that one insect visited 106 flowers of the *Viola* in four minutes.

Structure and habits of the minor sub-orders of Lepidoptera

Mouth-parts specially modified for unusual functions are found in the yucca moths, which belong to the Monotrysia, one of the small sub-orders mentioned on p.80 (see also Chapter 11). In another sub-order, Dacnonypha, are the small moths of its only family, the Eriocraniidae, several of which occur in Britain. They have small mouth-parts which are believed not to be reduced but to represent an early stage in the evo-lution of the typical lepidopteran proboscis (Box 4.2).

The most primitive Lepidoptera, the Zeugloptera, again comprise only one family, the adults of which feed on pollen and other powdery vegetable matter (see Box 4.2). *Micropterix calthella* is a common British representative of this family and is one of the characteristic visitors to meadow buttercup (*Ranunculus acris*) (Fig. 4.5). Willis & Burkill (1895-1908) found it also at bog asphodel (*Narthecium ossifragum*) and both feeding on pollen and sucking nectar at bramble (*Rubus fruticosus*), tormentil (*Potentilla erecta*) (Fig. 3.17) and lady's bedstraw (*Galium verum*), while MCFP has seen it feeding on the pollen of a wind-pollinated plant (*Carex*).

A species of *Sabatinca* (of the same family) has a strong association with small trees

Fig. 4.5 Micropterix calthella (Lepidoptera: Micropterygidae), a small primitive moth with chewing mouth-parts, feeding on pollen of creeping buttercup (*Ranunculus repens*).

Box 4.2 Mouth-parts of primitive moths

Details of the mouth-parts of yucca moths are given in Chapter 11. In the Eriocraniidae (Diagram A) each galea, though considerably elongated, is nevertheless shorter than the five-jointed maxillary palps. The two galeae are opposed to each other and channelled on their opposing faces as in Ditrysia; they are softly chitinised and delicately ribbed transversely. They end at the base in a heavily chitinised segment. Above their bases is a wide soft labrum, and below them a narrower hypopharynx, above which opens the mouth. These two organs close over the bases of the galeae, and nectar sucked up through the tube passes between them and into the mouth. The mandibles are more distinct than in Ditrysia but cannot be moved.

In the Micropterigidae, the well-developed maxillae (B) and mandibles recall those of allied orders with biting mouth-parts. However, the two mandibles are dissimilar and have a complicated arrangement of interlocking teeth. The muscles that work them also cause the compression of the highly specialised pouch-like hypopharynx,

which is lined inside with a rasping surface. The epipharynx (i.e. under-surface of the labrum) has specialised brush-like developments, and the whole system is adapted to pushing the food between the mandibles and into the hypopharyngeal pouch, grinding it thoroughly in the process (Tillyard, 1923).

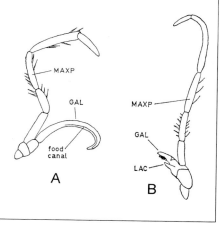

---

of the primitive genus *Zygogonum* (family Winteraceae) in New Caledonia. The moths of both sexes assemble on and near the flowers to find mating partners, but they also eat the pollen and cause pollination. They are attracted to the flowers by strong scents (Thien *et al.*, 1985).

The senses and behaviour of butterflies and moths

As with the Diptera, the reactions of Lepidoptera to colours and scents have also been given much attention. Müller observed in the Alps instances of butterflies preferring flowers similarly coloured to themselves; he thought that the colour preference in choosing a mate must have become transferred to the choice of flowers (Knuth, 1906-1909). Many Lepidoptera produce scent which attracts the opposite sex; these scents are often like those of flowers, and Müller thought that the scents of lepidopterid flowers might also have been evolved as an adaptation making use of the already existing attractiveness of these scents. A similar suggestion has been made to account for the scents of bat-pollinated flowers (p.255).

The following main characteristics of vision have been listed for butterflies by Silberglied (1984a,b):

1. the spectral range is wide, from ultraviolet to red (300-700 nm);
2. the resolving power of the eye is low, perhaps tens of thousands of times lower than that in man; and
3. temporal resolution is high, flicker-fusion frequency being 150 images per second in the silver-washed fritillary (*Argynnis paphia*) (this is the flicker frequency above which the illumination appears continuous; in man it is 40 images per second).

Noctuid moths

One of the commonest British flower-feeding moths, the silver-Y (*Autographa gamma*, family Noctuidae) (Fig. 6.61), was studied in Austria by Schremmer (1941a). In natural conditions the moths sometimes fly and feed during the day, but their main period of activity is in the evening twilight, continuing until after dark in moonlight or on clear starlit nights. If the moths are disturbed when at rest during the day they alight on the herbage and settle with their heads downwards and the eyes shaded from the sun. In feeding flights the moths fly against the wind, if any; otherwise they fly about irregularly. The moths fly low when seeking food, so that they often approach flowers from below and readily discover those hidden in the grass. Though the wings continue in motion during feeding, only occasionally stopping momentarily, the flowers are gripped with the legs, and in fact a silver-Y moth lacking forelegs has great difficulty in feeding from a flower. When the nectar in a flower is nearly exhausted, or when the flower contains no nectar, the proboscis is repeatedly almost withdrawn and pushed in again. This probing is accompanied by a noticeable activity of the antennae, and is no doubt an attempt to ensure that no nectar has been overlooked. Schremmer noticed that the antennae were used for feeling around the entrances to the flowers of red clover (*Trifolium pratense*) while the proboscis was seeking entry; after the proboscis had entered, the antennae were raised.

On one occasion, Schremmer found that the types of flower visited early in the evening were later neglected in favour of others; those visited first were a pink (*Dianthus carthusianorum*) and a yellow-flowered scabious (*Scabiosa ochroleuca*), while those visited later, when it was almost dark, were white (white campion [*Silene latifolia*], and bladder campion [*Silene vulgaris*]).

Varying degrees of constancy to a particular kind of flower were recorded; for example, one moth visited 68 pinks and four heads of the mauve field scabious, meanwhile making many approaches to the yellow S*cabiosa ochroleuca*, none of which led to visits; in roadside habitats with a varied flora, however, the insects commonly visited, one after another, several different flowers with different structure and colour.

The feeding flights of the silver-Y moth against the wind suggest that it seeks flowers by scent, but Schremmer pointed out that to travel downwind at high speed would make it difficult for the moths to alight on a chosen flower. However, on one occasion he saw a moth apparently finding bladder campion after dark by scent. On the other hand, the use of sight to the exclusion of scent was also demonstrated, using the inverted glass tube test (see pp.63-64) on *Dianthus* flowers. In laboratory experiments, using boxes large enough to allow the moths to fly about, it was found, as with Diptera, that scent is the activator of food-seeking behaviour and that moths could locate a hidden food source entirely by scent when within 10-15 cm of it. Inexperienced moths used scent alone; thus it was both activator and guide. On the other hand, after a little experience they began to use vision for the approach to flowers.

Colour discrimination was demonstrated using techniques similar to those applied to Diptera. Tests with scentless flower models and a scent source slightly apart from them showed that when the moth smells the flower it follows the concentration gradient of the scent; if it then meets a visually conspicuous object it will usually approach it and sometimes visit it, but then continues to follow the scent until it reaches the source.

To test for trainability to scent, Schremmer caught moths and fed them from flowers of soapwort (*Saponaria officinalis*) concealed in black containers inside the flight box. Next evening six soapwort flowers and four phlox flowers were presented close together intermixed on a board. All the soapwort flowers were visited and the phlox flowers were repeatedly approached, once with unrolling of the proboscis, but never alighted upon or fed from. Both types of flower were white and were very similar in size and outline, so recognition was presumably by scent. Training to the scent of soapwort for three days induced a constancy to that flower, coupled with a neglect of white campion, when the two types of flower were presented visibly and intermixed, or invisibly in black containers. A converse training to white campion gave the converse result. But when an insect trained to white campion was given a mixture of this flower without food and soapwort with food, the original training was broken, though with some difficulty, and finally almost replaced by constancy to the new flower.

The possibility of training the moths to visit flowers with only one particular colour and one particular scent was also investigated, and Schremmer succeeded in training a moth to feed chiefly at transparent blue discs masking flowers of soapwort, which were offered together with yellow-masked soapwort flowers, blue discs and yellow discs.

Investigation of another noctuid species, the shark moth *Cucullia umbratica*, by Brantjes (1976) showed that the sense organs in its antennae could detect dampness of a piece of filter-paper held near it. After being caged for a day, the moths would drink water. If the insects were satiated with water, unrolling of the proboscis could still be elicited by contact with a paper soaked in sugar solution or with the petals of *Silene latifolia*; even a single detached petal had this effect. The sense organs involved with this response are situated both in the antennae and the legs. In addition, contact of the proboscis tip with the coronal scales of the petals of this flower led to insertion of the proboscis into the flower tube. The nature of these contact stimuli is unknown; the responses are independent of the sense of smell, which was also demonstrated in the moths and enabled them to find hidden flowers.

Hawkmoths

Important contributions on the behaviour of Lepidoptera were made by Knoll, who experimented with four species of hawkmoth and a butterfly. His first object of study was the hummingbird hawkmoth (*Macroglossum stellatarum*), which flies by day. This work (Knoll, 1922), like that on the bee-flies, was carried out in southern Dalmatia.

The hummingbird hawkmoth feeds in flight, without the use of the legs. The food-seeking reaction can be recognised by the unrolling of the proboscis which occurs whether the stimulus is accompanied by food or not. When the insect is among flowers it flies directly, but fairly slowly, from one to another; when away from flowers it flies much faster, making wide sweeps over the ground as if searching, as *Bombylius* does (p.63).

In experiments with insects taken into captivity and with newly emerged moths, Knoll showed that the hummingbird hawkmoth can distinguish certain colours and variations in intensity of colour. It can develop constancy to a particular colour which in nature may be induced by the productivity of certain kinds of flower, but is also maintained by a tendency of the insect to recognise its accustomed flowers by the shape of the entry to them, which it perceives by tactile sense organs in the proboscis. In feeding, it apparently does not use its sense of smell either at very close range or at short distances away from the flowers. This applies to newly hatched as well as experienced individuals. The existence of the hummingbird hawkmoth's sense of smell, however, was shown in its egg-laying behaviour. Females in egg-laying condition were attracted to green, yellow and orange-yellow objects; only if these were made to smell of the larval food-plant, bedstraw (*Galium*), by the addition of its juice, did they lay eggs on them.

The flowers of common toadflax (*Linaria vulgaris*) (Fig. 6.46) were much visited by the hummingbird hawkmoth late in the season in Dalmatia, and Knoll used these yellow snapdragon-like flowers (p.169) to test this moth's reaction to guide-marks.[1] Normally the proboscis drummed on the palate until it found the entrance, where-upon the insect lunged forward and pushed the tip of the proboscis right down to the tip of the spur. A variant form of the flower, which was plain yellow without orange on the palate, was offered to the moth, and it fed from it in a normal way. Another abnormal form had a wide open mouth but normal colouring. With this the moths drummed on the palate but rarely found the entrance which was now some distance from the palate. Thus the deep orange guide-mark, if present, guides the insect but its presence is not essential to the finding of the nectar, at least for experienced insects.

This subject was further investigated with the aid of the 'proboscis-trace method'. In this, an object to be investigated is placed between glass plates: the insect is allowed to feed on sugar-water from an adjacent flower or model and, when it attempts to feed at the object behind the glass, its proboscis leaves a trace of syrup on the glass. After the experiment the traces are dusted with lead oxide, the surplus powder is knocked off, and the glass is heated until the sugar is dry. The traces are then quite hard, and the plate can be used to make a photographic contact print. The method shows ex-actly where the proboscis touched the glass, and it can be used in light too dim for direct observation. Some of the objects placed between the glass plates by Knoll are illustrated in Fig. 4.6. They were all light violet in colour with dark violet markings. With A and B the traces were over the rings, and with C they were over the conver-gence of the lines; with D the traces were all over the ellipse, and with E they were concentrated over the group of dots. Thus nearly all these artificial guide-marks had a directive effect, and circles were of major importance. The effectiveness of converg-ing lines in the absence of a ring (object C) varied with their thickness.

The striped hawkmoth (*Hyles lineata* subsp. *livornica*) and the convolvulus hawk-moth (*Agrius convolvuli*) were also investi-gated by Knoll (1925 and 1927). The proboscis of the first is 26 mm long (simi-lar to that of the hummingbird hawk-moth) and of the second is 65-80 mm. In the experimental 'flight box', 50 cm cubed, both moths flew and continued to feed in twilight until it was almost com-pletely dark.

The proboscis-trace method was used to investigate how the striped hawkmoth finds the entrance to the flower. On *Lo-nicera implexa* (a honeysuckle similar to *L. periclymenum*, Fig. 4.7A), all the marks were

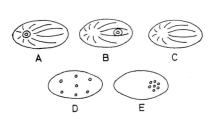

Fig 4.6 Designs used by Knoll (1922) to test the effect of guide marks on the hummingbird hawkmoth (*Macroglossum stallatarum*).

---

[1] This term is here used to replace the traditional 'nectar guide', which is inappropriate when such markings occur in nectarless flowers (see also pp.36, 43).

on or near the roughly rectangular upper lip. On models of *Nicotiana tabacum* (Fig. 4.7B), the traces were all over the star-shaped surface, with no concentration at holes backed by grey paper that had been provided to imitate the entrance. The moth, therefore, apparently does not find the entrance to these flowers visually, and in fact it frequently fails to find the entrance at all, being more attracted to the bright parts of the flower. It seems as though funnel-shaped flowers are better adapted to pollination by this insect, but the flowers of *Lonicera implexa* are probably also well adapted to it, the moths finding the entrance as described for *L. periclymenum* on p.92. In nature the striped hawkmoth has been recorded at flowers of all these shapes.

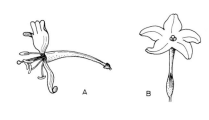

Fig 4.7 **A**, flower of honeysuckle (*Lonicera periclymenum*); **B**, *Nicotiana* flower of the type used in Knoll's experiments. (B, after Knoll, 1925).

In training experiments using the proboscis-trace method it was established that both moths had a well-developed colour sense even in extremely dim light, that optical stimuli were entirely sufficient for finding the entrance to a flower and that light colours were not necessary. Both species could be trained to constancy to either light colours or violet. A preference was shown by the convolvulus hawkmoth for coloured circles 35 mm in diameter over circles 14 mm or less in diameter.

Knoll's efforts to demonstrate a reaction to scent in feeding were unsuccessful with both moths. The same is true for the elephant hawkmoth (*Deilephila elpenor*) (Knoll, 1927) though it is common at collectors' 'sugar', which is an aromatic preparation (Knoll, 1925). This apparent lack of reaction to scent by hawkmoths was caused by their unexpectedly good vision at very low light intensities. However, Brantjes (1973), in experiments with hawkmoths in 60 cm flight boxes, showed several effects of scent.

As light fades in the evening hawkmoths prepare for flight in two stages. The first is called activation: the insect rises on its legs and brings the antennae forward from their resting position. The second is the warming-up stage, in which the wings are raised and vibrated; this takes place after a further sharp drop in light-intensity and it leads to flight. Brantjes found the following effects of introducing concealed fragrant flowers of honeysuckle (*Lonicera periclymenum*) into a cage with *Deilephila elpenor*: (1) scent supplied during activation led to warming-up without further reduction in light-intensity; (2) the first flight of the evening lasted consistently longer if scent was present; (3) the apparently aimless flight in the absence of scent changed to 'seeking flight' within three seconds of the introduction of the scent (this was lost again three minutes after the removal of scent). Honeysuckle flowers were placed in a box half way up the cage and with its entrance away from the light. Moths made sweeping flights and then approached the hole from below, swinging from side to side. They then rose up and alighted on the rim of the hole, unrolling the proboscis. Visible honeysuckle flowers were visited within ten seconds; the moths usually approached with the light behind them, except sometimes on the first visit, when they were perhaps guided by scent. The first approach was also from below, whereas later ones were from the same level as the object or above. After some visits to flowers, scentless objects were also visited by the moths. Training to scentless artificial flowers was quickly lost, scented objects being preferred. Moths previously trained would only visit the training object if scent was present.

The feeding procedure at *Lonicera periclymenum* is that the proboscis is repeatedly pressed against the upper lip of the flower until the entrance is found; the moth then moves forward and holds the flower with its legs while still hovering. When the nectar is finished the head is slightly withdrawn and the proboscis pushed in again several times, in the manner of the silver-Y moth, as if the insect is testing for more nectar. The legs, head and body become smeared with pollen.

Out of a number of natural and artificial scents tested, only two had any effect on the behaviour of *D. elpenor* and only one, that of *L. periclymenum*, induced feeding. Moths that had not fed would feed from artificial flowers at the end of the evening in the absence of scent.

Another hawkmoth, *Manduca sexta*, differed only in that food sources continued to be visited for a long time after the scent-source was removed, and that three artificial scents and that of *Petunia × hybrida* were effective as a stimulus. Colour preferences were learned and retained next day, but only displayed in the presence of scent. In this case, it was concluded that vision sufficed to find the flowers for the first time and that after some experience it was the only method used. Scent promoted the location of flowers and helped in the discrimination of visually equal flowers.

Even with the hummingbird hawkmoth Brantjes (1973) demonstrated an effect of scent. Wild-caught moths with artificial flowers of four colours visited none of them during the first day. Next morning the fragrant, nectarless flowers of Spanish broom (*Spartium junceum*) were supplied: within ten seconds two approaches with extended proboscis were made to the *Spartium* and then the artificial flowers were visited.

Brantjes concluded that the various hawkmoths have an innate releasing mechanism to initiate feeding and that the sign-stimulus corresponding to it is a particular scent or scents. The relative importance of smell and sight in attraction varies: at close range *Deilephila elpenor* is fully dependent on smell for its first flower visits, whereas for *Manduca sexta*, *Hyles livornica* and *Macroglossum stellatarum* vision is sufficient or is the dominating factor. Scent can allow discrimination between flowers, but vision is always adequate for the final approach. It is this that Knoll was studying. Attraction from a distance by scent is still not ruled out.

Much of the hawkmoth behaviour reported by Brantjes is similar to that of the silver-Y moth. Brantjes was able to amplify his results by studying electrical responses in the antennae by inserting micro-electrodes (see p.96) into the nerves. The oscillograph display in which the responses are seen is called an electro-antennogram (EAG). These showed that both *Deilephila elpenor* and *Manduca sexta* responded neurally to all 22 scents tested, despite behavioural response to only two by *Deilephila* and five by *Manduca*.

The general importance of scent that has now been demonstrated fits in with the attributes of flowers believed to be adapted to hawkmoth pollination. The flowers are very pale in colour and have a heavy scent produced mainly in the evening and at night. Nevertheless, the experiments show that nocturnal and crepuscular hawkmoths can exploit flowers lacking one or more of these attributes.

Vogel (1954) has pointed out that moth-pollinated flowers tend to have narrower perianth lobes than related plants pollinated by day-Lepidoptera. The resulting floral shape should help the moths to find the entry, as successive probings of the bright perianth will tend to be aligned in a radial direction, and not in all directions as they might be on a more disc-like flower. Flowers of this star-like shape can combine a large diameter with a restriction of the probed area, and do not suffer the loss of brightness caused by the shadow inside a funnel.

Hawkmoth weights range from 0.1 g to 6 g. An oxygen consumption of 60 ml per gram of body-weight per hour was cited by Heinrich (1975a) for hawkmoths in flight (compared with 77 for bees). If the oxygen is used to oxidise sugar (not lipids), then in a hawkmoth weighing 1 g, 1.25 kilojoules of energy are produced per hour; converting this to joules per second gives a power consumption of 350 milliwatts. The sugar equivalent to this energy production, based on 1 mg of sugar yielding 16.7 joules, is 75 mg or 1.25 mg per minute (a figure similar to that reported on p.232). At the upper limit of hawkmoth size, consumption per gram is less: a moth weighing 6 g was found to consume 5.7 mg sugar per minute (Heinrich, 1983). Such a large insect needs flowers with larger nectar supplies than are needed by bees (Heinrich, 1983). However, hovering, though expensive in energy, permits much higher rates of flower-visiting than other methods of foraging. Warming up is relatively inexpensive of energy for animals of this size; one example (Heinrich, 1975a) is of a moth which used 30 calories (125 joules) to warm up, equivalent to the cost of 3.7 minutes in flight (compare bees, p.139). Similar power consumption figures can be calculated for other hawkmoths, based on a summary by Casey (1989, table 1), in which oxygen consumption per gram of body-weight per hour ranges from 43 to 60 ml, but little is known about the energetics of butterflies.

Butterflies

The way butterflies find food varies according to species even more clearly than it does with moths. At one extreme is *Charaxes jasius*, a butterfly that ignores the flowers visited by other butterflies and feeds on tree sap and ripe or overripe fruit. It always arrives upwind at such a food source or, if the wind is changeable or the air is still, by zigzag flight (Knoll, 1922). Short-range searching after the insect has alighted evidently excludes visual observation. The same was true for Knoll's captive insects presented with coloured artificial unscented flowers containing sugar-water. When odour was introduced by adding plum juice, unrolling of the proboscis occurred, but location of food was apparently entirely by means of the antennae.

An intermediate condition was found by Lederer (1951) in some other butterflies that feed exclusively or characteristically on food other than flower-foods. He found that tree sap and excrement were always found by smell by the white admiral (*Limenitis camilla*), the purple emperor (*Apatura iris*) and allied species. The newly hatched white admiral was also directed by scent in its first visits to flowers, but in later visits it was often guided visually; its behaviour was therefore like that of the silver-Y moth. The peacock butterfly (*Inachis io*) found tree sap by scent and flowers by sight. The scents associated with food had an activating effect in all three butterflies. Lederer found that the distance at which the butterflies first reacted to scent ranged from 20 cm to 30 m (or, with a favourable wind, to 60 m).

Butterflies were extensively investigated by Ilse. Several species were kept in a greenhouse and were offered artificial flower-models containing sugar-water (Ilse, 1928). In six species, newly emerged insects showed a statistically significant inherited positive colour reaction when tested with coloured models among a series of greys. Colour preference data obtained from chequerboard tests with up to 16 colours, covering the whole visible spectrum, were presented for each species as a graph showing the number of visits to each colour, the colours being arranged in spectral order. Training experiments were also carried out. The results of Ilse's work are summarised in Table 4.5. The butterflies that preferred the colours blue, violet and purple, to-

Table 4.5 Food-seeking responses of butterflies (Ilse, 1928).

| Butterfly | Natural colour preference | Trainable to other colours | Approach to food | | Type of food |
			Visual	Olfactory	
Small tortoiseshell *Aglais urticae*	yellow and blue	yes	yes	no	flowers
Large tortoiseshell *Nymphalis polychloros*	yellow and blue	yes	more or less	yes	flowers and sap
Peacock *Inachis io*	yellow and blue	yes	more or less	yes	flowers chiefly
Large white *Pieris brassicae*	blue to purple	no	yes	scarcely	flowers
Brimstone *Gonepteryx rhamni*	blue to purple	no	yes	scarcely	flowers
Swallowtail *Papilio machaon*	blue to purple	no	yes	scarcely	flowers
Camberwell beauty *Nymphalis antiopa*	none	yes	when trained	yes	flowers and sap
Purple emperor *Apatura iris*	none	no	no	yes	dung
Red admiral *Vanessa atalanta*	?	?	yes	yes	flowers, dung & sap

gether with the small tortoiseshell, gave the full feeding reaction to coloured objects covered by glass – that is, in the absence of any flower scent. This was called the *Pieris*-type of approach, and it represents the opposite extreme to the *Charaxes*-type described earlier. Some butterflies behave consistently according to one or other of the extreme types, whereas others are flexible according to circumstances. The butterflies that could not be trained were, firstly, the three that feed exclusively on flowers and find food visually and, secondly, the one that feeds on dung, which it finds by scent. On the other hand, flexibility in choice of food was correlated with flexibility in the sense employed to find it and with trainability. As with moths, non-feeding butterflies could be induced to search for food by scent.

Later, the colour-reactions of the large white butterfly were studied in all phases of its activity (Ilse, 1941), and it was found that it could distinguish at least three groups of colours, as follows: (a) red-to-yellow, (b) green-to-blue-green, (c) blue-to-violet. These results applied to the families Pieridae and Papilionidae in general.

The effects of the size and form of coloured objects which elicit feeding reactions in butterflies were earlier investigated by Ilse (1932). Using the silver-washed fritillary (*Argynnis paphia*), the peacock and the small tortoiseshell, she found that the larger the coloured surface, the more attractive it was – a contrast to the results with the hummingbird hawkmoth; the largest surface tried was 30 × 50 cm and the colours used were blue and yellow, each presented on a black background. Using the peacock and the small tortoiseshell, the effects of form were studied with blue objects on a black or white ground; a ring of external diameter 5 cm was greatly preferred to a disc of diameter 3.5 cm but with the same area of colour; similarly, a blue/white or blue/black chequerboard was more favoured than the same area of colour presented as a single square. The preferred figures had the greater extension and longer outline. When the figures to be compared were made equal in extension, the preference, though somewhat reduced, was still for those with a longer outline, although these now had a smaller area of colour.

These results did not hold when yellow objects (on a black background) were used instead of blue. To a yellow ring and disc of equal coloured area (dimensions as before) visits were about equal; when the ring and disc were equal in extension, the disc received two-thirds of the visits (compared with one-third when the colour was blue). When offered a choice of yellow and blue, the insects usually showed a strong preference for yellow, which perhaps explains why, with yellow, area is more important than the length of outline. However, chequerboards showed that the longer outline was preferred if the difference in the length of outline of two figures of identical yellow area was made sufficiently great. Nevertheless, a limit to the increase in attraction caused by subdivision into chequerboards was reached when the small squares were reduced to 8 mm square. From further experiments Ilse concluded that no particular form was preferred.

Bennet's field observations (1883) relate to some of the species used in Ilse's experiments. Species of *Pieris* visited many purple, violet and white flowers but no yellow ones, which accords with Ilse's results. The small tortoiseshell, on the one occasion on which it was observed, visited only yellow flowers of a single species, while the allied painted lady (*Vanessa cardui*) invariably visited only violet and purple flowers. Members of two other genera of butterflies were also observed; the meadow brown (*Maniola jurtina*) visited white and yellow flowers, and the common blue (*Polyommatus icarus*) went to yellow, pink and purple flowers.

Bennett many times observed single visits, or three or four successive visits, to one species of flower, but instances of marked constancy were also seen, with seven to 23 and more successive visits to the same flower on the part of various butterflies. He also described numerous painted ladies spending their time visiting only the closely related *Centaurea nigra* and *C. scabiosa* and often flying a considerable distance between visits.

Experimental studies of New World tropical butterflies, using artificial flowers in large insectaries, have been carried out more recently. Spontaneous preference, effects of training and colour discrimination were looked at. *Papilio troilus* showed a strong spontaneous preference for blue and a weak one for orange (and so was similar to *Aglais urticae*). Tests of choice were always carried out at the beginning of the day after training the previous day, so all positive results represent retention in the memory overnight. After training to models reflecting longer wavelengths (green-to-red), *P. troilus* showed an increased response to orange (which is within the green-to-red range). *Heliconius charitonius* also showed peaks of preference for orange-to-red and blue-to-blue-green when offered a wide range of feeding choices. When tested with three shades of grey, white without and with ultraviolet, black and three shades of orange, 70% of visits were to orange, 20% to white and 10% to grey, indicating true colour discrimination. Tests with a wide range of colours after feeding only on yellow for two days (training) showed 49% of visits to yellow, compared with 9% before training. Training to green resulted in 55% of visits to green (only 2% before training).

Discrimination was possible between blue-to-blue-green, green, yellow, orange-to-red and magenta (Swihart & Swihart, 1970). Tests designed to show discrimination between closely similar colours in the yellow region were positive with two yellows reflecting at 540 nm and 590 nm. In the more preferred blue region of the spectrum, the response was less precise (C.A. Swihart, 1971). Training of *Heliconius charitonius* and *H. erato* to a bicoloured model followed by testing with a range of one-coloured models did not result in a preference for the principal colour of the training model. It was suggested, therefore, that the insects had developed a 'bicoloured' search-image which was not satisfied by the one-coloured models (C.A. Swihart, 1971; S.L. Swihart, 1972). In the butterflies studied by the Swiharts, there was no suggestion of response to ultraviolet.

The Swiharts have also made neuro-physiological studies of butterfly vision, using micro-electrodes inserted into the retina of the eye or into the nervous system. This is a great technological achievement, as probes have to be a thousandth of a millimetre thick, or less, and be hollow (Barth, 1985). Response to colour depends on the existence in the retina of photochemically different cells, reacting to light of different wavelengths. Skipper butterflies (family Hesperiidae) have two such types of retinal cells (vision bichromatic) (S.L. Swihart, 1969); other butterflies have three (vision trichromatic) (S.L. Swihart, 1970, for family Papilionidae) and some moths may have four (Silberglied, 1984a,b).

The bichromatic vision of skipper butterflies is associated with a flower-visiting preference for the range blue-to-magenta and for yellow. For this type of eye there is a neutral point between the two perceived colours at which the light is indistinguishable from grey. In fact white, grey, green and blue-green were found to be unattractive (S.L. Swihart, 1969). The Papilionidae, with trichromatic vision, are considered to be more advanced. The receptor cells respond to blue, green-and-white, and red. Each type sends messages to the brain along different nerve-fibres (neurones). In addition, 'higher-order' colour neurones were detected in the brain, processing the basic data, and showing response over narrower bands of wavelength at the extremities of the visual spectrum and also between the peaks shown by the receptors of the eye (S.L. Swihart, 1970). This will be mentioned again when we deal with bee vision. The butterflies with the most advanced colour vision belong to the family Heliconiidae (S.L. Swihart, 1972). The variety of types of neurones in the brain concerned with vision is greater than in Papilionidae and the insect is considered to be more advanced in its use of colour. In particular, it responds to red, which is present in the wings in a patch that is significant in courtship.

A problem that affects the interpretation of experiments on colour vision is caused by 'subjective brightness'. To man, some colours seem more conspicuous than others. To investigate this in insects the animal is presented with light sources of varying wavelengths but equal energy. The strength of neural or retinal response is then recorded over the range of wavelengths. This produces a curve of subjective brightness. In *Heliconius charitonius* it was found that this showed little or no correlation with attractiveness to the insects.

Role of the senses in Lepidoptera

Use of the senses in feeding by Lepidoptera is broadly similar to that in Diptera. In experiments described in this and the previous chapter, insects were tested with flowers covered with glass vessels or plates, so that the effective scent source was separated from the flower, and in all cases the insects flew direct to the flowers. It is important to make it clear that these experiments merely show that once the insect has brought the flower within its range of vision, it uses vision for its final approach. When one considers that the insects concerned are all strong fliers, highly dependent on vision for guidance, and that any moderately developed sense of sight must always be able to provide more precision than scent, the results of this kind of experiment are hardly surprising: vision would be expected to take precedence over any perceived scent. This precedence of vision is well shown by the case of the silver-Y moth that followed a scent gradient and was temporarily diverted by a bright object on the gradient but not at the scent source. To demonstrate sense of smell the flowers must be out of sight, as in some other experiments with this moth. However, the seemingly improbable alternative of failing to employ the sense of sight in food location is seen in *Charaxes*. Restriction of an animal's responses to only one of a number of available signals is a common feature of instinctive behaviour, often leading to absurd results in abnormal situations such as those set up by experimenters.

Apparent anomalies in the behaviour of the nocturnal hawkmoths in Knoll's experiments (Knoll, 1925, 1927) can perhaps be explained with the aid of Schremmer's (1941a) observations on the silver-Y moth. Knoll found that insects allowed to drink, with little or no prior training, from scented flowers made proboscis traces over adjoining flowers sandwiched between glass plates. Thus they were not deflected to the edges of the glass plates where the scent could escape (which, as we have just seen, is hardly to be expected), nor were they put off by an absence of scent at the place where the flower was. This is like the behaviour of the newly hatched silver-Y moth which was presented with phlox flowers and unscented white models and, being untrained, visited first the flowers and then the models. On the other hand, a hawkmoth which refused to visit a white flower, despite its strong scent, had been trained to visit scentless yellow models for the previous three days. Its behaviour was perhaps, therefore, conversely analogous to that of another silver-Y moth which was trained to phlox and then refused to visit unscented models. The hawkmoth's apparent disregard of scent may, in fact, have been an all too tenacious regard for it! All that Knoll deliberately tested in the way of response to scent was the possibility of guidance at short range and this was ruled out by the unexpectedly good vision in dim light which his experiments revealed.

As in Diptera, the commonest effect of scent is to alert the insects and cause them to start searching. Frequently, it acts as a recognition signal for food sources that have been used previously. It can also attract from a distance and, if the scent-source is concealed, can lead the insect all the way to it. Indeed, we have seen that scent is the means of guidance from a distance and is the sole means of finding food in some non-flower-visiting butterflies.

Researches have shown that colour vision is widespread in the Lepidoptera. Red-blindness is proved to exist in the hummingbird hawkmoth, but most butterflies can see red, though some of them may not use it in their feeding behaviour. It has been proved that guide-marks have some significance for *Macroglossum* and for the striped hawkmoth (*Celerio livornica*), but in an unmarked flower the striped hawkmoth finds the entrance by more or less random probing. Trainability – which implies ability to develop constancy – to both colour and scent occurs in many moths and in some, but not all, butterflies. Some butterflies will visit yellow and blue but prefer yellow; others prefer blue. Yet others respond only to blue and violet in feeding; they are untrainable and do not use scent in finding food. As the bee-fly *Bombylius fuliginosus* is likewise known to refuse to visit yellow in nature and does not use scent for guidance (pp. 63, 64), it seems similar to these butterflies and it would be interesting to know whether it lacks trainability. Butterflies of the genera *Aglais*, *Inachis* and *Argynnis* prefer a greater extension of objects and length of outline to a lesser. When the food object shows the less favoured colour, which Kugler characterises as having a low stimulatory value, it is found that outline becomes much more important for these butterflies.

5

The Insect Visitors
III: Bees and their Relatives

The Hymenoptera, which includes the bee, are a very diverse group of extreme biological interest, with about 6,200 species in Britain. These insects have four wings, those on each side being held together by a system of easily disengaged hooks. In the female the abdomen terminates in either an ovipositor or a sting. Also characteristic are the well-developed mandibles and the labio-maxillary complex. This is formed by the partial fusion of the maxillae with the labium (Box 5.1); it has a central tongue-like glossa (sometimes bilobed) but we shall use the word 'tongue' for the whole complex. The homology of the mouth-parts with those of primitive insects (Box 3.1) is clearer than in the Diptera or Lepidoptera. The order is divided into two sub-orders, the Apocrita, comprising gall-wasps, ichneumon-wasps, bees, wasps and ants, and the Symphyta, a more primitive and less numerous group, the sawflies.

Sawflies

The Symphyta are distinguished from the other Hymenoptera by their lack of a narrow 'waist' (Fig. 5.1, Plate 2a). They are somewhat wasp-like in appearance but have proportionately larger wings, usually soft clumsy bodies, and comparatively slow movements. The flower-visiting species range in length from about 5 mm to 22 mm. The larvae of most sawflies are like lepidopteran caterpillars in appearance and live externally on plants. The adults use the mandibles to chew solid food and the tongue to lick up solids. There are three main types of food: (1) insects, (2) moisture (rain, dew,

Fig. 5.1 Sawfly (*Tenthredo* sp., Hymenoptera Symphyta: Tenthredinidae), on buttercup (*Ranunculus* sp.).

Table 5.1 Flowers visited by sawflies, Hymenoptera-Symphyta ('L' indicates larval food-plant of the flower-visiting species).

Scientific name	English name
Some Brassicaceae	Crucifer family
Reseda lutea	Wild mignonette
Geranium dissectum	Cut-leaved cranesbill
Geranium sanguineum	Bloody cranesbill
Acer campestre	Field maple
Acer pseudoplatanus	Sycamore
Rubus spp	Bramble, raspberry, etc. (L)
Sedum telephium	Orpine
Saxifraga hypnoides	Mossy saxifrage
Apiaceae	Umbellifer family
Euphorbia cyparissias	Cypress spurge
Salix spp.	Sallows and willows (L)
Scrophularia spp.	Figwort, etc. (L)
Verbascum spp.	Mullein
Ajuga spp.	Bugle, etc. (L)
Polygonum bistorta	Bistort
Galium spp.	Bedstraw
Symphoricarpos rivularis	Snowberry (L)
Valeriana officinalis	Common valerian
Scabiosa spp.	Scabious (L)
Cephalanthera longifolia	Narrow-leaved helleborine

(Jones, 1945; Knuth, 1906-1909; Kugler, 1955a; Poulton, 1932; Willis & Burkill, 1895-1908; PFY)

'cuckoo-spit', honey-dew and damaged ripe fruit), and (3) flowers and leaves. The floral materials include nectar, pollen, stamens and petals, and are taken chiefly by females (Benson, 1950). Nectar and pollen have to be well exposed, as the mouthparts are usually 2 mm long at most. In fact, the flowers most favoured by sawflies are Apiaceae, Rosaceae with large inflorescences, yellow Asteraceae and Ranunculus (Benson, 1950; Willis & Burkill, 1895-1908). Sawflies must certainly be effective as pollinators, and they are often common enough to form a significant fraction of the pollinator fauna, but their benefit to the plants is sometimes offset by the injuries which they do to the flowers. The larvae of sawflies are often closely restricted to certain food-plants, and there is a marked tendency of the adults to visit mainly the flowers of the larval food-plant (Table 5.1). Such attachments are particularly strong with species feeding on willows (Salix), some of them confining themselves entirely to the flowers of the species on which the eggs are laid, even when other Salix species are available (Benson, 1950, 1959).

One of the more striking flower-visiting sawflies is Abia sericea (family Cimbicidae), a large blackish insect with a green metallic sheen and a wrinkled abdomen; it has been recorded as frequent at small scabious (Scabiosa columbaria) (Chambers, 1947). Many members of the large family Tenthredinidae visit flowers; an example is Tenthredo arcuata, a black insect with yellow bands, which was the sawfly most commonly recorded at flowers by Willis & Burkill (1895-1908) in their observations in the Cairngorms; it is a regular visitor to buttercups (Ranunculus spp.) (Harper, 1957). Another member of the family that visits buttercups is Athalia bicolor, a medium-sized species which is black except for the orange abdomen – a very common colour-pattern among sawflies. Sawflies (together with ichneumon-wasps, see later) have been found

Box 5.1 Mouth-parts of short-tongued Hymenoptera

Diagrams A and B (after Bischoff, 1927) show the relatively primitive condition of the labio-maxillary complex in a sawfly (*Cimbex*). The social wasp *Vespula* is not very different (C, D – with the right maxilla removed; after Duncan, 1939), but the glossa is bilobed and, like the paraglossae, covered with flattened hairs among which liquids can be imbibed (E). This hairy part of the tongue can be brought upwards into a pouch formed by the hypopharynx, maxillae and epipharynx, and suction can then be applied to draw the imbibed liquid into the mouth. A filter is formed by the fitting together of comb-like rows of hairs on the hypopharynx and on the galeae of the maxillae. Another filter is formed by a similar row of hairs across the narrow slit-like mouth, and this prevents all but microscopic particles from entering the mouth. Material filtered out by this second comb is temporarily stored in the pouch. Key to lettering: GALea, GLossa, HYPopharynx, LABial Palp, LABRum, LACinia, MAXillary Palp, MENTum, ParaGLossa, SALivary duct, STipes, SUB-Mentum.

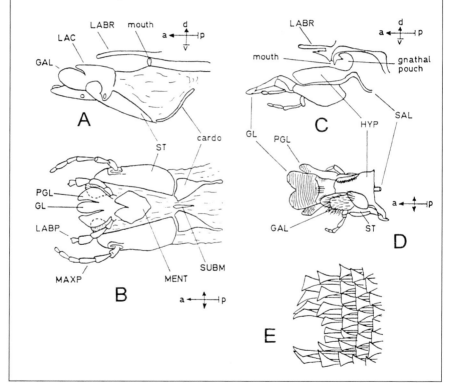

to be important pollinators of the twayblade orchid (*Listera ovata*) in Sweden (Nilsson, 1981). In both groups, more males than females were found visiting this and other flowers, even though in general the populations of these insects show a female-biased sex-ratio. One sawfly species in Australia effects the pollination of the orchid *Caleana major* by the process of pseudocopulation, in which the males act as if mating with the flowers, which in form and scent resemble the female insect (pp.205 & 212). The precise positioning and the movements involved in mating make it a suitable process for exploitation by orchids, since their method of pollination, involving the transfer of pollen masses to the stigma (Chapter 7), also requires accurate positioning of the insect.

The larvae of the stem-sawflies (family Cephidae) live in the stems of plants, including those of cereal grasses; the adults are small or medium-sized with long and very slender bodies, and in colour they are black with narrow yellow bands. The adults visit flowers, especially *Ranunculus* spp.

Hymenoptera-Parasitica

The great majority of the Hymenoptera belong to the sub-order Apocrita and may be divided into those with stings (Hymenoptera-Aculeata) and those without (sometimes referred to as Hymenoptera-Parasitica, since most of them are parasitic). The stingless Hymenoptera, which can be referred to as wasps of various kinds, are very numerous in species but are not of great importance in pollination. The flower-visiting species are distributed across about ten families.

The most noticeable of the Parasitica are the ichneumon-wasps (family Ichneumonidae); these are parasitic in the larval stage, usually on the larvae of other insects. Some of the adults are very small, but a good proportion of them are of moderate or large size (with a body length of 1-2 cm). They are particularly slender insects with long antennae and long legs (Fig. 5.2), and they are usually very active. They consume sap, honey-dew, nectar and pollen; the nectar includes that obtained from extra-floral nectaries (p.108) and the pollen that has been trapped by dew-drops (Leius, 1960). The mouth-parts, even of quite large species, are usually under 1 mm long; they are rather similar to those of sawflies but the glossa is larger than the paraglossae.

The Ichneumonidae visit much the same flowers as the short-tongued Diptera (Table 3.3). Laboratory experiments carried out in Canada with three species of ichneumon-wasp showed that one of them would visit only Apiaceae, among plants of several families offered, while the other two visited various flowers but made most visits to Apiaceae. Such investigations are carried out mainly with a view to finding out the food requirements of Ichneumonidae used for biological control of insect pests, since a lack of suitable food for the adult wasps after their release might lead to failure of projects. In general, any destruction of wild flowers tends to reduce the natural populations of these useful parasites of crop pests (Leius, 1960; van Emden,

Fig. 5.2 Ichneumon (Hymenoptera Parasitica: Ichneumonidae) feeding at wood spurge (*Euphorbia amygdaloides*). (See also Fig. 7.6.)

1963).

Among the more distinctive of the ichneumon-wasps is the genus *Pimpla*, the species of which range from about 6 mm to 20 mm long, and are black with brown legs and a transversely wrinkled abdomen. The species of the genera *Ichneumon* and *Amblyteles* are also conspicuous, being over 10 mm in length and decorated with bands and spots of cream, yellow and light red in various patterns on the black body. All these insects are common at Apiaceae, especially hogweed (*Heracleum sphondylium*) and parsnip (*Pastinaca sativa*), while *Amblyteles uniguttatus* is one of the ichneumons recorded at tway-blade (next paragraph). Willis & Burkill (1895-1908) designated the Hymenoptera-Parasitica as merely injurious but this is erroneous on the evidence of the Ichneumonidae alone.

The twayblade orchid (pp.192-3 and Fig. 7.6) is often given as an example of a rare class of flowers called ichneumon-wasp flowers, and Knuth gives records of six genera of these insects visiting its flowers; it is, however, also visited to some extent by small Diptera and Coleoptera, as well as sawflies (see above). An Australian species of ichneumon, *Lissopimpla excelsa* (*L. semipunctata*), is involved in pollination by pseudo-copulation of orchids of the genus *Cryptostylis* (Plate 6e; see also p.210).

Much less conspicuous among the flower-visiting Hymenoptera-Parasitica are several families constituting the chalcid wasps and gall-wasps, insects that are mostly only 1-4.5 mm in length. One of the largest chalcid wasps, *Brachymeria minuta*, is easily recognised, having a short, stout shining black body with a pointed abdomen and swollen hind legs. However, most chalcids are moderately slender in form and are usually black with a green or blue metallic sheen (Fig. 7.19). Some chalcids have vegetarian larvae that live in seeds. Included here is the family Agaonidae, which are specialised for living in the tissues of figs which they cross-pollinate in the course of their egg-laying activities (see Chapter 11). The gall-wasps (Cynipidae) are so called because most of them form galls on plants (the rest are parasitic on insects). The eggs of the gall-formers are laid inside the plants and the presence of the larvae stimulates the development of abnormal growths which they then consume. These small insects visit much the same flowers as Ichneumonidae (Harper, 1957; Knuth, 1906-1909; Willis & Burkill, 1895-1908), including three tiny orchids, *Hammarbya paludosa* (bog orchid), *Listera cordata* (lesser twayblade – Ackerman & Mesler, 1979) and *Herminium monorchis* (musk orchid – Nilsson, 1979b) (see also Chapter 7). Kevan (1973) has surveyed the occurrence of parasitic Hymenoptera at flowers in the High Arctic of Canada; he found that flowers were important to them as food sources, and that some of them carried pollen, but they were probably insignificant as pollinators compared with other groups of insects.

Our third group of Hymenoptera-Parasitica is the Chrysididae, which are closely related to the true wasps and have been placed in this position only for convenience of description in this book. These wasps have brilliant metallic colouring: the prevailing pattern is blue or green on the head and thorax and red on the abdomen, giving rise to their popular name of rubytail-wasps. They range in length from 3 mm to 10 mm, are rather scarce, and in habits they are parasitic on the larvae of solitary bees and wasps. The British species, which have short mouth-parts, have been recorded in Britain almost exclusively at the flowers of Apiaceae. However, the members of the continental chrysid genus *Panorpes* are specialised flower-visitors with a proboscis 6-7 mm long. In this genus the glossa is elongated and rolled at the sides to form a tube, while the galeae of the maxillae are also elongated and cover the glossa.

True wasps

The Hymenoptera that have stings are classed as Hymenoptera-Aculeata. They mostly build nests in which food for the larvae is stored by the adults, and they comprise the true wasps, the ants and the bees. In many families, however, there are species or whole genera that have the habits of the cuckoo, the adults laying eggs in the nests of other Hymenoptera, and the resulting larvae feeding on the food supply of the rightful owner; such insects are called inquilines. Apart from the inquilines, there is a group of wasp families regarded as the most primitive of the aculeates, which are entirely parasitic on other larvae. This group, the superfamily Scolioidea, is represented in Britain by the large velvet-ant (*Mutilla*) and several smaller species that visit the same sorts of flowers as the Hymenoptera already described. In most of them, the females are wingless. Also in Scolioidea are large, usually hairy wasps, such as *Campsoscolia* (Fig. 7.20), noted as a pollinator of the mirror orchid (*Ophrys speculum*) by pseudocopulation (p.206), and a variety of Australian species that also pollinate orchids by pseudocopulation (Jones & Gray, 1974; Stoutamire, 1974) (see pp.210-13 for details, including plant and insect names and further references).

The great majority of the true wasps capture insects or spiders as food for their larvae. Usually the prey is stung or mutilated and stored in cells in the nest, one egg being laid on the food supply in each cell. Sometimes these predatory wasps feed on the juices exuded by their victims, but otherwise they take the same liquid foods as the Hymenoptera-Parasitica, namely sap, honey-dew and nectar. Their nesting requirements are the same as those of solitary bees (p.109).

Table 5.2 A selection of the flowers visited by non-social short-tongued wasps of the superfamilies Pompiloidea, Sphecoidea and Vespoidea.

Scientific name	English name	Flower characteristics
Apiaceae	Umbellifers	
Parnassia palustris	Grass of Parnassus	Nectar well exposed
Listera ovata	Twayblade	
Ranunculus spp.	Buttercups	
Cakile maritima	Sea rocket	
Reseda lutea	Wild mignonette	
Geranium spp.	Cranesbills	
Rubus fruticosus	Bramble	Nectar partly concealed
Potentilla erecta	Tormentil	
Bryonia dioica	White bryony	
Symphoricarpos rivularis	Snowberry	
Epipactis palustris	Marsh helleborine	
Leucanthemum vulgare	Ox-eye daisy	
Cirsium arvense	Creeping thistle	
Senecio jacobaea	Ragwort	
Solidago spp.	Golden rod	
Jasione montana	Sheepsbit	Small tubular flowers
Calluna vulgaris	Ling	
Thymus spp.	Thyme	
Mentha arvensis	Corn mint	

(Chambers, 1949; Knuth, 1906-1909; Kullenberg, 1956a; Richards & Hamm, 1939; Spooner, 1930, 1941; PFY)

Fig. 5.3 Solitary wasp (*Mellinus arvensis*, Hymenoptera Aculeata: Sphecidae), on sea carrot (*Daucus carota* ssp. *gummifer*).

There are three superfamilies of predatory wasps: the Pompiloidea, the Sphecoidea and the Vespoidea. The family Pompilidae, or spiderhunting-wasps, comprises about 40 species in Britain. With their long legs and slender bodies, 5–14 mm long, they are somewhat ichneumon-like. In colour, they are mostly black with red on the fore-part of the abdomen, but some are entirely black or black with white spots or bands. They pursue their prey largely on foot, skimming the ground in short flights from time to time. Spiderhunting-wasps visit flowers (Table 5.2) but are absent from Willis & Burkill's (1895-1908) records from the Cairngorms, probably because, like most aculeate Hymenoptera, the Pompilidae require warm conditions. At the beginning of their adult lives, they seem to spend a lot of time feeding on nectar and may briefly be a significant component of the pollinator fauna for Apiaceae. Pseudocopulation is known in Pompilidae (see p.205).

The family Sphecidae contains most of the non-social wasps and has about 100 species in Britain. The length of the body ranges from 3-25 mm, and its shape is very variable, especially that of the abdomen, whose narrowed fore-end can have many different conformations. The colour may be black (or black with red on the fore-part of the abdomen) but often it is black and yellow, the yellow being either confined to small spots or occurring in large spots or bands so that the insect conforms to the layman's idea of a wasp (Fig. 5.3).

The flowers they visit are listed in Table 5.2 but most visits seem to be to Apiaceae (see also p.358). The tongue length of two fairly large species measured by Kugler was only 1.5 mm; however, in the sand-wasp, *Ammophila sabulosa*, it is 3 mm and the sides of the glossa are rolled back to form a tube. This wasp visits a variety of flowers, including bramble, thistles and snowberry. Species of *Argogorytes* pollinate the fly or- chid (*Ophrys insectifera*) by pseudocopulation, as described on p.207. Like *Campsoscolia*, but unlike *Lissopimpla*, *Argogorytes* carry the pollinia on their heads (Plate 6a; Fig. 7.20).

The Sphecidae, like Pompilidae, are probably significant as pollinators only occa- sionally. No doubt chiefly on account of their need for warmth they are completely absent from the records of Willis & Burkill (1895-1908), but possibly their ability to avoid capture also played its part. As in the Chrysididae, there is a continental repre- sentative of the Sphecidae with an elongated proboscis 7 mm long; this is *Bembex rostrata*, which is able to obtain nectar from and pollinate the explosive pea-type flow- ers of lucerne (*Medicago sativa*).

Fig. 5.4 Mason wasp (*Ancistrocerus* cf. *trifasciatus*, Hymenoptera Aculeata: Eumenidae), on hogweed (*Heracleum sphondylium*).

The members of the third superfamily of predatory wasps, the Vespoidea, are distinguished by having the fore-wings folded lengthwise when at rest. They comprise the non-social families Eumenidae and Masaridae, and the social Vespidae. The non-social groups have the same life-history as the Sphecidae. The family Eumenidae includes the potter wasp, *Eumenes*, with one species in Britain, and the mason-wasps, about 20 species in Britain, belonging to *Odynerus* and closely related genera. These are all yellow and black (or rarely white and black) wasps, about 8-12 mm in length (Fig. 5.4). Flowers visited are listed in Table 5.2. *Eumenes coarctata* is found on heaths in Britain, and it probably feeds from the flowers of ling (*Calluna vulgaris*). Its tongue is 4 mm long, and the glossa is rolled back at the edges to form a tube, as in *Ammophila* (Kugler, 1955a). This species (males only) has also been found as a visitor to marsh helleborine (*Epipactis palustris*) (Fig. 7.3, 7.4), whose dimensions its body fits exactly (Nilsson, 1978b). The Masaridae are almost all flower-feeders, occurring mostly in warm-temperate areas. They collect pollen on hairs on the face and transfer it to the mouth with the forelegs; it is then mixed with nectar in the crop, as in the bee *Hylaeus*. The one Central European species (*Celonites abbreviatus*) has a proboscis nearly as long as the body. It visits especially flowers with a lipped corolla, such as *Teucrium* (Mauss & Treiber, 1994). In California the species visit bee-adapted flowers, such as certain *Penstemon* species.

In the Vespidae, each colony is founded by a queen, who lays eggs and feeds the offspring as they grow. These offspring are the workers and after they become adult the queen remains in the nest and lays eggs, while the workers bring in the food and extend the nest. In temperate countries, the colony usually dies out in autumn and only the new generation of queens survives the winter. These social wasps are represented in Britain by the hornet (*Vespa crabro*) and the insects normally recognised by the layman as wasps. In Britain, these comprise several species each of the genera *Vespula* and *Dolichovespula*. Elsewhere, there are many slender-bodied members of the family that make nests without an envelope. They constitute the subfamilies Polistinae and Stenogastrinae; a common European example is *Polistes*.

In *Vespula* and *Dolichovespula* the tongue is slightly longer than in most Sphecidae (for description see Box 5.1). Although the adults of *Vespula* feed the larvae chiefly on insects, they also feed both themselves and the larvae on liquid foods, often obtaining these from flowers. It used to be thought that the larvae were fed only on animal food,

Fig. 5.5 Common wasp (*Vespula germanica*, Hymenoptera Aculeata: Vespidae), feeding on exposed nectar of ivy (*Hedera helix*).

but the transport of large quantities of a solution of sugar and honey to the nest was observed by Verlaine (1932b), and it was later shown that sugary fluids are essential to the diet of the larvae of *Dolichovespula sylvestris* (Brian & Brian, 1952). On the other hand, pollen is not a wasp food (Duncan, 1939).

Vespidae visit some of the flowers in each of the groups listed in Table 5.2. Of the flowers with well-exposed nectar, the Apiaceae are the most visited, especially by the social species. In addition, ivy (*Hedera helix*) is visited abundantly (Fig. 5.5), and both the hornet and the common wasps can be seen on it in October. They spend a lot of time taking nectar, though they may also be attracted by the presence of flies for prey.

Individuals of *Vespula* and *Dolichovespula*, when visiting heath and cross-leaved heath (*Erica cinerea* and *E. tetralix*), take the nectar through borings, the entrances to the flowers being much too narrow for them to get their heads in (Willis & Burkill, 1895-1908). Such borings may be made by the wasps themselves (Block, 1962) but they are more often made by the shorter-tongued species of bumblebee and, of course, they by-pass the pollination mechanism.

In addition to the flowers already named, Vespidae (and to some extent Eumenidae)

Table 5.3 'Wasp-flowers' (favoured by Vespidae) and similar forms.

Scientific name	English name
Berberis spp.	Barberry
Ribes uva-crispa	Gooseberry
Cotoneaster spp. (in gardens)	Cotoneaster (Fig. 6.59)
Frangula alnus	Alder buckthorn
Vaccinium myrtillus	Bilberry (Fig. 6.7)
Vaccinium vitis-idaea	Cowberry
Scrophularia nodosa	Common figwort (Fig. 6.58)
Scrophularia aquatica	Water figwort
Symphoricarpos rivularis	Snowberry
Epipactis helleborine	Common helleborine orchid (Fig. 7.4)
Epipactis purpurata	Violet helleborine orchid

(Brian & Brian, 1952; Chambers, 1949; Müller, 1883; Shaw, 1962; Spooner, 1930; Trelease, 1881; Willis & Burkill, 1895-1908; Yarrow, 1945; PFY)

visit a range of small, more or less globular flowers that Müller (1883) informally referred to as 'wasp-flowers'. These are described on pp.180-82 and listed in Table 5.3. De Vos (1983) found that, on average, one visit by *Dolichovespula sylvestris* transferred seven times as many pollen grains to the stigma of *Scrophularia nodosa* as a bumblebee visit, though the total of visits by honeybees and bumblebees was much greater. The genus *Cotoneaster* is divided on flower form into two parts, one with dingy red nodding globular wasp-flowers (Fig. 6.58C), and one with upright flowers with spreading, pure white petals that are apparently not visited by wasps. The barberry, gooseberry, alder buckthorn, bilberry and cowberry, listed in Table 5.3, are all mainly visited by bees and have never been classed as 'wasp-flowers', but their points of similarity to them suggest that the wasps are attracted by the same means. The fact that pendent flowers discourage visits by short-tongued flies and encourage visits by bees was noted by Willis & Burkill (1895-1908). As some of these flowers are more or less nodding, it appears that wasps share with bees a readiness to cling underneath a flower to get nectar.

Ants

The ants, classed in the superfamily Formicoidea, form a great group of social insects which often form perennial colonies. They are great lovers of nectar and regularly collect it from flowers. Since the worker-ants are wingless and have to reach the flowers by crawling up the stems, they are very unlikely to cause cross-pollination between different plants, and their method of entry to the flower will in many cases permit them to take the nectar without effecting pollination at all. However, there is a rare category of plants which have become adapted to ant-pollination. Typically these are prostrate or low-growing plants and in any case they have small inconspicuous flowers close to the stem. Different individuals intertwine and plants are self-incompatible. A British example is rupture-wort (*Herniaria ciliolata*, family Caryophyllaceae), which in Cornwall is pollinated by *Lasius niger* and *Formica fusca* (L.C. Frost). A similar but unrelated North American plant, *Polygonum cascadense* (family Polygonaceae), has been dealt with in detail by Hickman (1974), who listed other examples and set out the above-mentioned features of the ant-pollination syndrome, to which is added a hot dry habitat in which ants are abundant (for the syndrome concept, see Chapter 6). The characters of the plant *Diamorpha smallii* (family Crassulaceae), found to be mainly ant-pollinated by Wyatt (1981), proved to fit this syndrome very well. The plant grows on hot dry granite outcrops in the south-east United States. In the Mediterranean region *Paronychia* species, identical in habit to *Herniaria*, belonging to the same family and growing in soil pockets on coastal rocks, receive flower-visits from ants (PFY; see also Fig. 5.6). A few cases of orchid-pollination by ants are known, one of which involves winged male ants in a pseudocopulatory relationship (Peakall, 1989; see p.212). Further examples of ant pollination are listed by Peakall *et al.* (1991).

Except on such plants as these, ants are liable to be harmful visitors and the adaptations of plants to exclude them were treated in detail in a little book by Kerner (1878). The two main ways by which plants exclude ants are the formation of impassable barriers between the ground and the flowers, and the provision of additional nectaries away from the flowers to act as decoys. Impassable barriers are found in teasel (*Dipsacus*) in the form of pools of dew and rainwater around the stem, held by the united bases of each pair of leaves, and in some species of catchfly (*Silene*) in the form of sticky zones on the upper parts of the stem, though they may have other functions as well. Decoy nectaries ('extra-floral nectaries') occur on the stipules at the bases of the leaves of some vetches (*Vicia* spp.) and near the base of the leaf-blades in

Fig. 5.6 Ant (*Formica fusca*), feeding at inflorescence of Portland spurge (*Euphorbia portlandica*).

the cherry laurel (*Prunus laurocerasus*) which is commonly grown in gardens.

In many tropical plants nectar is specially provided outside the flowers to attract ants. The ants, being powerfully equipped for biting and stinging, then protect the plant from various kinds of attack, including nectar-robbery by corolla-piercing. Among the tropical plants which secrete nectar on the leaves, bracts or calyces, are some that are pollinated by the large and powerful *Xylocopa* bees. Since the floral mechanism makes it difficult to obtain entry to the internal nectar, these bees are often tempted to pierce the corollas from the outside. The 'ant-guard', however, effectively deters them from doing so. The flowers seem to be provided with a chemical means of keeping the ants away from the inside of the corolla (van der Pijl, 1954). In Trinidad, the nectar of the tropical weed *Hippobroma longiflora*, which is consumed by hawk-moths, elicits a strong negative reaction by ants (and is also distasteful to humans) (Feinsinger & Swarm, 1978).

Bees

General biology and flower preferences

The Hymenoptera dealt with so far are of slight significance as pollinators when compared with the one remaining group of this order, the bees. The different groups of bees are sometimes treated as subfamilies of a single family, the Apidae, or sometimes as separate families (O'Toole & Raw, 1991), which is more convenient for us. Unlike the wasps, they are nearly all complete vegetarians. The adults of both sexes feed on nectar and, sometimes, pollen, and the larvae on both nectar (after its conversion to honey) and pollen. As with true wasps, the larval food is collected by the female adults. Nectar is a solution of one or more of the three sugars, sucrose, glucose and fructose (see Chapter 2). After collection, the nectar is carried in the bee's crop and any sucrose in it is enzymically converted to glucose and fructose (one molecule of sucrose makes one each of glucose and fructose). The product is honey. The vast importance of the bees as pollinators arises from the fact that the larvae, unlike all other insect larvae, have large quantities of flower-food brought to them or stored up for them in the nest by the female adults. Many flowers are specially adapted to pollination by bees, and many others, less specialised, benefit from their activities.

There are about 240 species of bees in Britain. The bumblebees (*Bombus*), belonging to family Apidae, have exactly the same life-cycle as the social wasps, with fertilised queens founding colonies in spring, and being later helped by their worker offspring. They number about 20 species. The honeybee (*Apis mellifera*) is different from all other British social bees and wasps in that the colony remains in being all the year round, and new colonies are founded by queens that are accompanied by part of the worker population of the parent colony. The life-history of the non-social majority is much like that of the non-social wasps but a great many bees appear in March or early April, whereas in Britain very few non-social wasps emerge before about the end of May. Varying levels of social organisation occur in the mainly non-social Halictidae and some species of *Halictus* and *Lasioglossum* have a similar system to that of *Bombus*, but two or three females may combine to start a nest in spring, and the colonies are much smaller. A considerable proportion of bee species are inquilines, laying their eggs in the nests of other species, as described for wasps.

Numbers of honeybees and their distribution in the habitable world are unnatural, being governed by the provision and movement of hives (see Chapter 13). In addition, they have a longer season of activity than any other kind of bee. When left to themselves they nest chiefly in hollow trees and cavities in buildings. Bumblebees also have a rather long season, but their numbers are disproportionately small early in the year; they nest in grass tussocks or underground (usually in ready-made holes) according to species. The non-social bees, together with the social species of *Halictus* and *Lasioglossum*, burrow in the ground or make nests in hollow stems, in rotten wood, in beetle holes in wood, in snail shells, in holes and crevices in masonry, or under stones. The situation chosen depends on the kind of bee concerned, and the occurrence of non-social bees is thus somewhat restricted by availability of nesting sites. The ground-nesting species may also be restricted in occurrence by their soil preferences for nesting purposes. The season of activity of many species is short but many genera contain both early and late species, while some species can be seen over almost as long a period as the honeybee, this being achieved by having two or three broods during the year so that there are periods of absence or scarcity between broods (see Box 5.2).

Specialisation by bees in foraging

Owing to the bees' need for pollen, which is usually easily accessible, general compilations of flowers visited by the various species of bees do not show much correlation between the tongue-length of the insect and the accessibility of the nectar (see Knuth,

Fig. 5.7 Bumblebee, *Bombus terrestris* (female; Hymenoptera Aculeata: Apoidea), on ramsons (*Allium ursinum*).

1906-1909; Westrich, 1990). Kugler (1940) reported, for instance, that species of *Lasioglossum* readily collect pollen from flowers whose nectar they cannot reach. However, tongue-length (Table 5.4) must limit a bee's choice of flowers for nectar-gathering except where holes are bitten in the corolla, as they often are by bumblebees, some of which appear to be more resourceful than the solitary species in getting food in unconventional ways. The particular preferences of bees for certain flowers already mentioned often concern flowers that provide both nectar and pollen; most of the family Asteraceae have flowers which provide good supplies of both, and the same applies to many Fabaceae-Faboideae. These two families are particularly suited to the bees with abdominal pollen brushes as, with few exceptions, they present their pollen from below and the bees can scrape it directly into the brush. Sometimes bees will take pollen from some flowers and nectar from others; this is commonly the case with the social species, but it occurs also with solitary bees; for example, *Anthophora plumipes* visits peonies (*Paeonia* spp.) in gardens for pollen only (PFY).

However, it seems that some long-tongued bees never visit certain flowers with

Fig. 5.8 Non-social short-tongued bee (*Andrena* sp., female; Hymenoptera Aculeata: Apoidea), on dandelion (*Taraxacum officinale* agg.).

Fig. 5.9 Non-social bee: a small species of *Lasioglossum* (female), on common daisy (*Bellis perennis*).

Box 5.2 The seasonal succession of British bees

The first bees to appear in spring are honey-bees (Figs. 5.16, 6.47), and these are followed by the first queen bumblebees (Figs. 5.7, 6.8, 6.32) and, among the non-social bees, the early andrenas (Andrenidae, Figs. 5.8, 5.14) and two species of *Anthophora* (Anthophoridae) (Plate 2d). Honeybees have medium-length tongues and visit practically any available flower. Bumblebees have long tongues but their choice of flowers in early spring is very limited, the chief natural flower for them at this time being sallow catkins (*Salix* spp.). *Andrena* is the largest genus of bees in Britain, with over 60 species, all of which nest in the ground. Most of them are very short-tongued. It is the larger species of the genus which emerge first, and these may be as large as or larger than the honeybee. The thorax of these larger species, and sometimes the abdomen also, is densely hairy; the females of some could be mistaken for the honeybee but others are beautifully patterned and coloured. The andrenas active in March and April visit chiefly sallows and the early yellow composites (Asteraceae) (Fig. 9.9a). As the sallows are dioecious, bees collecting pollen only will not cause pollination. The early species of *Anthophora* (*A. retusa* and *A. plumipes*) are rounded furry bees rather like small bumblebees but they frequently hover; the males are mainly brown while the females are entirely black except for their rust-coloured pollen brushes. Their movements are extremely quick and they are more easily frightened by the human presence than the bumblebees. Their tongues are very long and they visit mainly large tubular flowers.

In March or April the first *Lasioglossum* species appear; these early ones are among the smallest of their genus – tiny blackish bees with rapid oscillating flight (Fig. 5.9). In gardens in April these small bees are often seen round Spanish bluebell (*Hyacinthoides hispanica*) which has large, bell-shaped flowers which they can crawl right into. The smaller andrenas now emerge and these are similar to the small *Lasioglossum* though a little larger and with more flattened abdomens. Both these groups of small bees visit yellow composites and birdseye speedwell (*Veronica chamaedrys*; see p.69). A common bee in gardens in April is the reddish-brown ma-

son-bee *Osmia rufa* (Megachilidae, Fig. 5.10), with a furry body about 12 mm long. The pollen brush is on the undersurface of the abdomen instead of on the hind legs as in most bees. It has an extremely long tongue but visits a wide variety of flowers, including the wide open fruit blossom flowers with partly concealed nectar, which also attract the larger andrenas at this time.

In late April and during May further species of *Andrena* (Plate 2c), *Lasioglossum* (Fig. 5.12) and *Osmia* (Fig 5.10) appear, as well as representatives of some other genera. These include *Sphecodes* (Halictidae), which are black with a red band on the fore-part of the abdomen, and *Nomada* (Anthophoridae, Fig. 2.14), which are banded with black and yellow, or brown and yellow, and look like wasps. In both genera the hair clothing is sparse, and both are inquilines, *Nomada* usually on *Andrena*, *Sphecodes* on *Halictus* and *Lasioglossum*. The blossoms of hawthorn (*Crataegus* spp.) are much visited by *Andrena*, *Halictus* and *Lasioglossum* during May.

In the course of June, worker bumblebees become numerous and representatives of most of the remaining genera also appear. Among these are *Hylaeus* (Colletidae), small shiny black bees, called in Germany 'mask-bees' because of their pale yellow facial markings. *Hylaeus* have few body hairs and no pollen brush (Fig. 5.11); the pollen required for the larvae is carried in the crop with the honey. The proboscis is very short. A wide variety of flowers is visited, but the species are particularly attracted by wild mignonette and weld (*Reseda* spp.). *Colletes* (Colletidae) also appear in June; these resemble medium-sized andrenas and the common garden species visits chiefly yarrow (*Achillea millefolium*), cultivated *Achillea* species, tansy (*Tanacetum vulgare*), and related Asteraceae, all with heads of closely-packed short tubular florets (Plate 2b). Friese (1923) noted that species of *Colletes* restrict themselves to a smaller selection of plants than *Hylaeus* and *Lasioglossum* and that their emergence is closely related to the flowering of their favourite plants. *Panurgus* (Andrenidae) may also appear in June; the two British species are blackish-brown with orange pollen brushes on the hind legs. They nest in sandy soil and are particularly fond of the yellow ligulate-flowered composites. The three remaining genera of Megachilidae, bees that collect pollen on the underside of the abdomen, also appear in June. These are *Megachile* (the leafcutter-bees), which favour Asteraceae, Campanulaceae and Fabaceae-Faboideae; *Chelostoma*, which favour Campanulaceae; and *Anthidium*, of which the only British species, the carder bee (*A. manicatum*), visits chiefly the large tubular flowers of Scrophulariaceae and Lamiaceae, as well as frequenting Fabaceae-Faboideae. *Anthidium* is unusual in that the males are larger than the females instead of smaller and hold territories based on plants at which the females forage; they are large brown bees, capable of hovering, with yellow on the face and yellow spots on the abdomen. Other bees that emerge in June are *Lasioglossum leucozonium*, one of the larger species of the genus, with similar flower-preferences to *Panurgus*, and *Anthophora quadrimaculata*, with similar preferences to *Anthidium*.

In July the species of *Colletes* that visit ling (*Calluna vulgaris*) emerge, as well as various species of *Andrena* found on sandy heaths, some of which favour ling and bell heather (*Erica cinerea*). The longer-tongued *Andrena marginata* also emerges in July, when its favourite flowers, scabious (*Knautia arvensis*, *Scabiosa columbaria* and *Succisa pratensis*) and knapweed (*Centaurea*) are out. *Macropis europaea* (Melittidae) also emerges in July and visits yellow loosestrife (*Lysimachia vulgaris*) almost exclusively for pollen and nutritive oil (Vogel, 1976b, 1986) (see pp. 43 & 115).

Fig. 5.10 Long-tongued solitary bee, *Osmia rufa* (mason bee, probably a female), sucking nectar from wallflower (*Cheiranthus cheiri*).

Fig. 5.11 Short-tongued solitary bee (*Hylaeus* sp.), on wallflower (*Cheiranthus cheiri*); the bee can reach the pollen but not the nectar.

Fig. 5.12 Non-social bee: a
medium-sized species of
Lasioglossum foraging for
pollen on white clover
(*Trifolium repens*).

rather easily accessible nectar, for there are no records of visits of *Megachile* or *Anthidium* to ragwort (*Senecio jacobaea*) according to Harper & Wood (1957). Even more restricted in their habits are *Macropis europaea* and *Andrena marginata* (already mentioned), and *Andrena praecox*, a very early bee which takes pollen exclusively from sallows (Chambers, 1946). *Andrena bicolor* visits a great variety of flowers in its first brood, while in certain localities the second brood rarely visits anything but *Campanula* and *Malva*, and in some other localities nothing but dandelion (*Taraxacum officinale*) and bramble (*Rubus fruticosus*) even when *Malva* is available. Such situations are rather frequent in *Andrena* and they are paralleled in their inquilines, *Nomada*, which may actually specialise in the same flowers as their hosts (Friese, 1923). Bees that visit only one or a few species of flowers for food are described as oligotropic, while those showing a similar restriction for pollen supplies are called oligolectic. Oligolecty appears to be rather rare in British bees but a good example is provided by *Melitta* (Melittidae); its four species specialise respectively on certain Fabaceae-Faboideae, *Onobrychis viciifolia* (also in Fabaceae), *Campanula rotundifolia* (Campanulaceae) and *Odontites verna* (Scrophulariaceae). However, oligolecty is a common and striking phenomenon in some parts of the world and examples from America are quoted in Chapter 13. In some of these cases, the flowers concerned are the meeting place of the sexes; this is so with the British bee *Macropis europaea*, in which the females leave a special scent on the *Lysimachia* flowers which makes these attractive to other females (even when all pollen has been removed) and especially to males (Kullenberg, 1956b). In a particular experiment concerning four species of bumblebee, with varying lengths of tongue, Brian (1957) found that when gathering nectar in competition with each other these species tended to restrict themselves to the flowers most appropriate to their tongue length, although at least the longest-tongued species was not so selective when there was little competition.

Chambers (1945, 1946) noticed that andrenas collect pollen from certain wind-pollinated trees, and honeybees are particularly noted for visits to a variety of wind-pollinated plants, including even a gymnosperm, yew (*Taxus baccata*) (Hodges, 1952). Wind-pollinated flowers are inconspicuous but they usually produce very large quantities of pollen which, however, is not sticky. This does not seem to make it difficult for *Andrena* to collect it, although these species do not moisten pollen with honey as do some bees, including the honeybee. Since many wind-pollinated flowers are unisexual

Table 5.4 Lengths (in mm) of proboscis and body of some British bees (both sexes, unless otherwise stated).

Species	Proboscis length (mainly from Knuth, 1906-1909)	Body length
Colletes daviesanus	2.5-3	8
Hylaeus communis	1.25	6.5
Halictus rubicundus	4-4.5	10
Lasioglossum morio	2.5	5.5
Lasioglossum leucozonium	4	7-9
Sphecodes reticulatus	2	8.5
Andrena argentata	2.5	8
Andrena bicolor	2.5	7.5-9.5
Andrena marginata	3.5-4	8-9.5
Andrena pubescens	3.5	9.5-13.5
Anthophora plumipes	13	14
Anthophora quadrimaculata	8	9-10.5
Melecta luctuosa	11	14
Nomada goodeniana	4	9-11.5
Megachile centuncularis	6-7	9-11
Coelioxys elongata	4.5	12
Osmia caerulescens	8	6.5-9
Osmia rufa	7-9	9-11
Bombus pascuorum (queen)	11-14	15-17
Bombus pratorum (queen)	10-12	16
Bombus sylvarum (queen)	14	16
Bombus hortorum (queen)	18-19	20
Bombus lapidarius (queen)	14	22
Bombus terrestris (queen)	10	24
Apis mellifera (worker)	6	13

the female flowers are not visited by the pollen-collecting insects, which thus fail to pollinate the plants.

The scents of bee-pollinated flowers

The scents of bee-pollinated flowers are sometimes indistinct to us, but those which we can clearly perceive are rather varied. The honeybee and short-tongued bumble-bees freely visit many of the flowers visited by Diptera, some of the scents of which are described on p.77. The flowers that are more specially adapted to bees have generally a sweet scent which, even when strong, is more delicate to the human nose than the heavy scents common in Lepidoptera-pollinated flowers (p.83), but as with that group, the same scent may be found in many unrelated flowers. For example, the scent of sweet violet (*Viola odorata*) is like that of *Iris reticulata* (a commonly cultivated Middle Eastern species which, incidentally, flowers at about the same time as this violet and is similar to it in colour). Very similar to these, but more strongly scented, is the cultivated mignonette, *Reseda odorata*, from North Africa. Honey-like scents are common in Fabaceae-Faboideae, examples being white clover (*Trifolium repens*), tree lupin (*Lupinus arboreus*, a Californian plant introduced into Britain), and spanish broom (*Spartium junceum*, from the west Mediterranean region). In the last two, the scent is very powerful; both species have pollen flowers, like gorse (*Ulex europaeus*), which smells of coconut. Other recurring scents of bee-pollinated flowers are the plummy scent of the grape hyacinth (*Muscari neglectum*) and oxlip (*Primula elatior*), the disagreeable smell of

Helleborus foetidus, and the pleasant scent of the garden pansy (*Viola* × *wittrockiana*) which is to be found in species of michaelmas daisy from both Europe and North America (for example, *Aster sedifolius* and *A. puniceus*, which are visited mainly by bees and hoverflies) and in some other flowers (PFY).

Unusual bee-flower relationships

Many species of bees are involved in the relationship of pseudocopulation with orchids (*Ophrys* spp.) in the Mediterranean region; the bee genera usually concerned are *Andrena* (Andrenidae) and *Eucera* (Anthophoridae) (see pp.208-9). Males of some Apidae (subfamily Bombinae, tribe Euglossini) are involved in the strange pollination processes of certain tropical American orchids, during which they collect fragrant compounds for later deposition at sites where females may appear (see pp.216-20). The fragrance is collected by brushes of hairs on the forelegs and then stored in the hind tibiae which are specially enlarged and chambered to take the liquid (Vogel, 1966). These bees also visit members of several other plant families that provide similar fragrances. One such plant is *Dalechampia* in the Euphorbiaceae. This genus is particularly interesting in that most of its species, instead of supplying fragrances to male euglossine bees, offer resin for building or lining nests to female euglossines (and other bees) (Armbruster & Webster, 1979; Armbruster *et al.*, 1989). In addition, bees specialised to collect oil in place of nectar from certain flowers (as *Macropis* does from *Lysimachia* [p.112]) occur in various parts of the world, for example *Rediviva* (Melittidae), which gathers oil by inserting its enlarged forelegs into the twin-spurred flowers of *Diascia* (family Scrophulariaceae) in southern Africa (Vogel, 1974, 1984; Steiner & Whitehead, 1988). This is a striking case of co-evolution. For transport, the bee adds the oil to its pollen load (see also Vogel, 1990).

The mouth-parts of bees

These are adapted for both nectar-collecting and nest-building. The tongue ranges in length from shorter than the insect's head to longer than its body. The bristly glossa is greatly developed compared with that of other Hymenoptera and, except in the shortest-tongued forms, its edges are rolled back so that it looks like a tube with a slit along the back. Most probably this is not a feeding tube, and liquid uptake is by the hairs on its outside. The galeae of the maxillae have evolved to form a sucking tube, though in a very different manner from those of the Lepidoptera (Chapter 4). The mouth-parts of various bees are described in Boxes 5.3-5.5 (fullest details are available for the honeybee, Box 5.4).

Evolution has undoubtedly progressed in the direction of greater proboscis length, but this process is better thought of as specialisation than as improvement. The gaining of access to deep-seated nectar is accompanied by decreased ease of collecting fully-exposed nectar, as observed by Kugler (1943) in bumblebees.

Pollen-collecting arrangements

The collection of pollen as food for the larvae is carried out in various ways by different bees. In *Hylaeus* pollen is carried entirely in the crop mixed with the nectar. Curiously, *Hylaeus* resembles all other bees in having some of the hairs branched, which is a feature generally regarded as an adaptation to pollen collection.

The abdominal pollen-collectors (Megachilidae) all have the underside of the abdomen thickly clothed with hairs which curve slightly towards the tail. Saunders (1878) found that the pollen brushes in this family consist of different kinds of hairs; thus in mason-bees (*Osmia*) and leafcutter-bees (*Megachile*) the hairs (unlike those elsewhere on the body) are unbranched, those of *Megachile* being spirally grooved, whereas in *Chelostoma* the hairs are waved and branched. The legs are used to gather up the pollen and transfer it to the abdominal brush. They are adapted to this by bearing bristles which are developed into a stiff brush on the inside of each metatarsus (compare Fig. 5.13A) and into a comb on the inner edge of each of the fore- and mid-tibiae towards their tips. These stiff hairs may be used both to collect pollen from flowers and to groom the body of the insect. While leafcutter-bees are flying from flower to flower their legs hang down and are scraped together, the pollen being passed back to the hind legs. These are then raised and the pollen transferred to the abdominal brush. However, pollen can sometimes be transferred to this brush straight from the flower (see p.110), with or without the assistance of the hind legs.

The bees that carry home the pollen on their legs resemble the abdominal collectors in having stiff pollen-gathering hairs on the insides of the metatarsi (present even in *Hylaeus* but there used only for cleaning the body), but these hairs sometimes extend to other joints of the tarsi. Andrenas carry home the pollen on the main joints of the hind legs and also on parts of the thorax (Figs. 5.13 & 5.14). The hind tibiae are densely clothed with branched and unbranched hairs, which form a large pollen-carrying brush. On the other hand, the hind femora carry part of their pollen in a brush on the front surface and most of it on the lower surface in a basket-like structure

Fig. 5.13 Pollen-collecting apparatus of *Andrena denticulata*. **A**, the insect seen from above (front legs, other parts of the body, and certain details omitted); **B**, right hind leg (coxa and details of tarsus and metatarsus omitted): f, femur; mt, metatarsus; t, tibia; ta, tarsus; tr, trochanter. Length of body parts shown is 11.5 mm. Individual hairs shown at greater magnification. For 'compass-points' see Fig. 3.10.

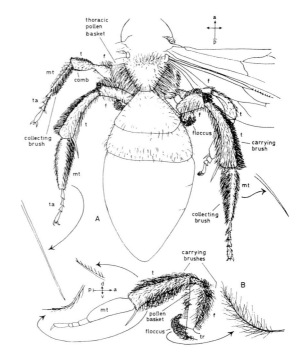

Fig. 5.14 Non-social bee (*Andrena* sp.), foraging at common melilot (*Melilotus officinalis*), and carrying a load of pollen.

formed by fringes of branched hairs. The trochanter, one of the two small joints between the femur and the body, carries a group of beautiful long plumose hairs which descend and then curve rearwards, constituting the floccus. These hairs themselves enclose pollen and also help to close in the baskets of the femora.

Like the leafcutter-bees, andrenas commonly carry out pollen-packing when in flight from flower to flower, but sometimes they do it before leaving the flower. In addition, some species make the pollen easier to transport by moistening it with nectar regurgitated from the crop.

The pollen-collecting apparatus in *Colletes* and *Lasioglossum* is closely similar to that of *Andrena*. In both genera, however, the hind femora carry all their pollen in a basket on the underside, and the branched hairs corresponding to the floccus of *Andrena* arise at the base of the femur. The branches of these hairs diverge widely and overlap to form a mesh; similar hairs clothe much of the thorax of *Colletes*. *Lasioglossum* has no thoracic pollen baskets, but it collects some pollen in the dense hairs beneath the front of the abdomen. In both these genera, the curved hairs on the lower side of the hind tibiae show a fan-like development of branches near the tip.

Both *Lasioglossum* and *Colletes* pack their pollen before leaving the flower, brushing each foreleg several times very rapidly on the middle leg, and each middle leg on the hind leg. These movements take place only on one side at a time, but both hind legs can be employed simultaneously to clean the under surface of the abdomen.

A simpler system for carrying pollen on the legs is found in Melittidae and Anthophoridae. The hind metatarsus is more enlarged, and this joint and the hind tibia have large brushes of backwardly directed pollen-carrying hairs, usually confined to their outer surfaces. *Eucera* and *Melitta* moisten their pollen with nectar. *Dasypoda* differs in having some pollen-carrying hairs on the hind femur and and a dense brush of very long feathery hairs over the entire surface of the tibia and the large metatarsus, the last-mentioned joint having no stiff combing bristles on the inner surface. On account of its enormous pollen-carrying capacity, *Dasypoda* is illustrated in almost every book on pollination, though the more complex arrangements in *Andrena* are more interesting. *Panurgus*, in the Andrenidae, has a similar distribution of pollen-carrying hairs to *Dasypoda*, with some branched hairs on the femur, but their capacity is slight compared with those of the tibia, which is unusually well clothed with pollen-carrying hairs on the inner surface. These hairs are waved and pinnate with numerous short branches.[1]

---

[1] This account of pollen-collecting apparatus has been based so far on the work of Müller (1883), Braue (1913) and Kugler (1955a), supplemented with observations by PFY.

Box 5.3 The mouth-parts of short-tongued bees (illustration opposite)

The glossa is short in the genera *Hylaeus* (Diagrams A, B) and *Colletes* (Colletidae) (C, D), *Andrena* (Andrenidae) (F, G), *Halictus* (J) and *Lasioglossum* and also in *Sphecodes* (all Halictidae). The first two of these are the shortest-tongued of all bees and are peculiar in having the glossa bilobed as in Vespidae. The detailed structure of the glossa and paraglossae in these two genera is adapted to the job of lining the nest with a fluid secretion (Demoll, 1908) but the hair-fringe of the glossa (C) presumably absorbs nectar. By drawing the glossa back into the space above the mentum and beneath the galeae of the maxillae, pressure and suction can probably be applied to it so that nectar imbibed by the hairs can pass into the mouth. Demoll (1908) believed that the nectar first passes through a very narrow passage between the glossa and the paraglossae. In any case, the nectar then travels backwards through a tube formed by the mentum and the overlapping maxillae (Saunders, 1890). These structures are linked towards the base by a membranous bag, the 'throat-membrane' (E shows this in transverse section at the level of the mouth [after Demoll, 1908]; the epipharynx is hanging down in front of the mouth). In the upper side of this are two folds which connect the edges of the maxillae to the labrum (H, J) forming a covered passage which links the mouth with the basal ends of the maxillae, so that fluid passing back from here is unable to escape. The folds are probably kept together by tension when suction is taking place, and narrow rods present in their edges may help to keep the gap closed. The underside of the membranous bag continues back under the head where it lines the cavity into which the mouth-parts are retracted when not in use. Embedded in it are the sclerites that join the mouth-parts to the hard parts of the head and project and retract them (H, Box 5.4 Diagram A, and Box 5.5 Diagram E); the cardines move the entire apparatus to and fro while the lorum can move the mentum in relation to the maxillae. In order to project the mouth-parts, the cardines have to swing forwards and downwards and this great movement is made possible by the skin of the throat-membrane loosely linking the hardened joints together. The lowered position of the mouth-parts when nearly fully projected can be seen in B, H and J and their retracted position in Box 5.4 Diagram L.

In *Andrena* the proboscis is again short and constructed much as in *Colletes* and *Hylaeus*, but the galeae are large, horny and opaque, instead of partly translucent. The paraglossae are well-developed and the undivided tip of the glossa forms a sort of scoop (F). *Andrena marginata* is one of a few species that have a longer proboscis (G,H). All the parts are somewhat elongated in comparison with those of other species and the glossa is more tongue-like, and covered with long hairs towards the tip. In *Halictus* and *Lasioglossum* the ratio of the proboscis-length to the body-size is about the same as in *Andrena marginata*, but the proportions of the parts are different, the stipes (the basal part of the maxilla) being relatively longer (J). The mentum is also longer as it has to match the length of the stipes, while the hairy convex glossa is short. *Sphecodes* species (mostly inquilines of *Halictus* and *Lasioglossum*) have similar but slightly shorter mouth-parts.

A third and final group of bees that collect pollen on their legs consists of the social bees, *Apis* (Fig. 5.15) and *Bombus*. In these, the collection of pollen is similar to that of other bees in its first stages but quite different in its final stages.

In the honeybee (*Apis mellifera*) (Fig. 5.16) pollen is scraped off the head and fore-part of the thorax by the antenna-cleaners and combs of the forelegs. The middle legs clear pollen from the hind part of the thorax and the hind legs clear it from the abdomen. All this pollen is worked into the brushes on the inner surface of the metatarsi of the middle legs, and is moistened by regurgitated honey. After this each middle leg in turn is placed between the two hind legs and then drawn forward; this action scrapes the pollen into the metatarsal brushes of the hind legs. When sufficient pollen

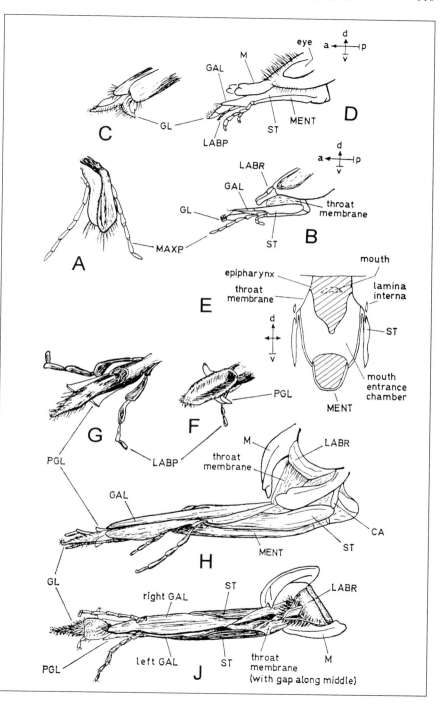

Box 5.4 Mouth-parts of the honeybee

The separated mouth-parts of *Apis* (family Apidae) are shown in Diagram A (after Snodgrass, 1956) (for mouth-parts of a long-tongued bee in their natural extended position see Box 5.5). The glossa of *Apis* is moderately long (A, D). In all bees it is a flexible structure with a springy rod running along it (E); muscles attached at the base of the rod can induce a great deal of movement in it, and there is in addition always a mechanism for drawing the base of the glossa some way back into the gutter-shaped mentum so that the whole is able to move to and fro in the front part of the food channel. The secretion of saliva takes place on the mentum near the attachment of the glossa (J).

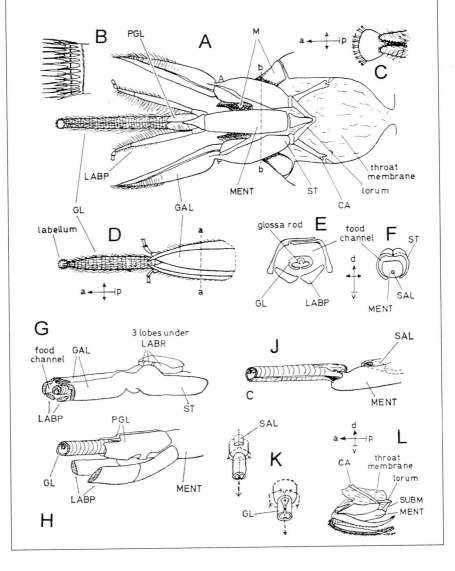

Box 5.4 (continued)

At least in the honeybee the mouth-parts also contain the taste organs. As with the bees already described, the stipites of the maxillae, together with the labium, form a food channel (F), while food is licked up by the glossa; however, in the honeybee the galeae of the maxillae and the labial palps are greatly developed, so continuing the food channel forward around the longer glossa (D, E).

The slit along the underside of the glossa is closed by dense fringes of hairs; its surface is transversely ringed and each ring is the seat of a row of stiff hairs (B). The shape and position of the glossa rod is seen in E; towards the base it has the possibility of some movement to and fro within the glossa, and as a result it is able to tighten or slacken the skin of the glossa so that the hairs of the glossa stand up or lie down. When the glossa is fully extended, the hairs spread and can thus absorb liquid between them, and as it is drawn back they close up and release the liquid into the food channel. The glossa rod is apparently able to produce bending movements at the tip of the glossa which assist in licking up nectar. These movements can be seen in wide flowers, such as a small *Crocus* visited by *Anthophora* (PFY), or in captive bumblebees drinking from artificial containers (Friese, 1932), when the galeae and palps are seen to be firmly united around the glossa as it moves to and fro, sweeping the

fluid into the tube. The proboscis may be extended fully if this is necessary to reach the nectar (Müller, 1883), but when the food has been imbibed among the hairs of the glossa, the proboscis has to be slightly retracted; this brings the raised part on the base of each maxilla back into contact with the lobes on the underside of the labrum. The parts fit neatly together and close over the food channel right back to the mouth, just as the throat membrane does in *Halictus*, etc. (G).

So far we have seen that the food channel lies between two concentric tubes in its distal part (E, section corresponding to a-a of D) and in a tube formed by the maxillae over-arching the mentum in its proximal part (F, section corresponding to b-b of A). Another part of the system is formed by the large flattened paraglossae, which ensheath the apex of the mentum and the base of the glossa (H, corresponding to G but with maxillae removed); their shape is such that the saliva, produced from the salivary pouch near the tip of the upper surface of the mentum (J, corresponding to H but with the paraglossae and labial palps removed), is carried downwards on either side and passes into the tube formed by the glossa, entering through an opening on the underside (J, K). The saliva emerges at the extreme tip of the glossa where there is a little flattened disc called the labellum (C, D). The folded mouth-parts of the honeybee are shown in L.

has accumulated on the hind metatarsi, it is transferred to the pollen baskets (or corbiculae) on the outside of the hind tibiae. This is made possible by a structure known as the pollen press, which is found only in the social Apidae and is, perhaps, the most striking adaptation to pollen collection found among all bees. As can be seen from Fig. 5.16C, D, the hind tibiae and metatarsi are flattened and greatly widened compared with those of the other legs. The metatarsi, however, are attached to the tibiae only at the front by a narrow joint (Fig. 5.16C); this allows the press to be opened by the bending down of the metatarsus and closed by its bending up. The proximal end of the metatarsus is produced (except at the actual joint) into an outwardly directed flange which slightly overlaps the apex of the tibia; this flange is called the auricle (Fig. 5.16E). Two actions are required to transfer pollen from the inner surface of the metatarsus of, say, the right leg to the pollen basket of the left leg. In the first, the rake of spines on the apex of the left tibia is pushed downwards through the hairs of the right metatarsus. This scrapes pollen into the space between the rake and the left auricle.

Box 5.5 Mouth-parts of other long-tongued bees (illustration opposite)

Other families in which a long proboscis is found are Andrenidae, Melittidae, Anthophoridae and Megachilidae. In some of these bees the mouth-parts are of a slightly more primitive kind than in the honeybee. In *Panurgus* (Andrenidae), for example, the glossa is only moderately long and the labial palps are very slender, apparently playing no part in the formation of the food channel, while the maxillary palps resemble those of the bees with a short glossa (Diagram A; after Saunders, 1890). The galeae are long and strongly tapered, so they appear well-adapted to sucking the slender tubular florets of Asteraceae, which are the favourite flowers of *Panurgus*. Similar mouth-parts are found in *Dasypoda* and *Melitta* (Melittidae), the latter having a specialised arrangement of long curved hairs at the tip of the glossa, making it look rather like a bottle brush. A closer resemblance to the honeybee is found in the mouth-parts of *Nomada* and especially *Epeolus* (Anthophoridae), two genera of inquilines.

In *Anthophora* the tongue is much longer than in the honeybee, and the proportions are rather different (B). When the galeae are folded back out of use, they project well back under the thorax of the bees. In *Anthophora* and the closely similar *Eucera* the hairs of the glossa have oar-like flattenings which increase their surface area for absorption (C). The bumblebees (*Bombus*) are rather similar in their mouth-parts to the bees just described.

The remaining bees (Megachilidae) all have rather similar mouth-parts (D, showing *Osmia coerulescens*, a bee 8 mm long, with tongue extended to 6.5 mm). The glossa is long and is sheathed by the galeae for a greater proportion of its length than in other bees with a long proboscis (Demoll, 1908). The labrum is very large and curved down at the sides, and the galeae are slightly curved. This group consists of the bees which carry pollen on the underside of the abdomen (see pp.116-17) and the related inquilines, *Stelis* and *Coelioxys*. Demoll suggested that the extensive sheathing of the glossa in these bees fits them to feeding from the flowers of Fabaceae-Faboideae, in which the opening is so small that it might squeeze nectar from an unprotected glossa on withdrawal from the flower.

The second movement is a bending upwards of the metatarsus which closes the gap between the auricle and the rake (Fig. 5.16E). The pollen is thereby forced upwards and outwards into the pollen basket (Fig. 5.16F) as a compact and sticky mass. Pressure applied by the springy auricle and its bristles causes successive masses of pollen to be plastered one upon the other. A solitary bristle is present on the surface of the

Fig. 5.15 Honeybee foraging for pollen on flower of white rockrose (*Helianthemum apenninum*). A mass of pollen can be seen in the pollen basket on the hind leg of the bee.

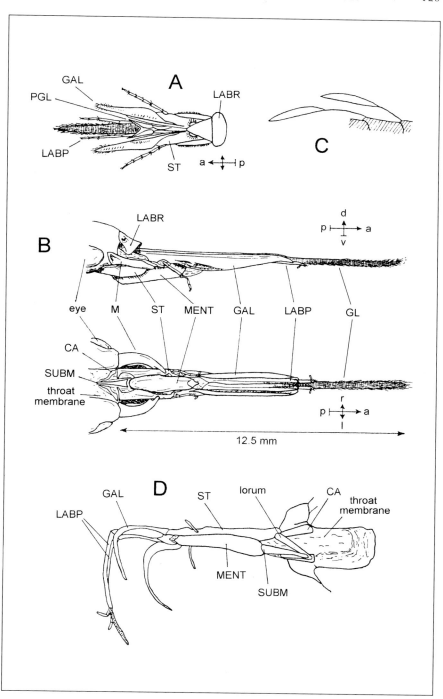

Fig. 5.16 Pollen-collecting apparatus of the honeybee. **A**, left fore-leg, front view; **B**, right mid-leg, back view; **C**, right hind leg, turned forward, back view; **D**, both hind legs, back view (tibial rake of left leg is about to be pushed through metatarsal brush of right leg); **E**, two views of left pollen press, similar to those seen in 'D' (hairs of pollen basket omitted in left-hand drawing); **F**, loaded pollen-basket. c, coxa; f, femur; mt, metatarsus; t, tibia; ta, tarsus; tr, trochanter. For 'compass-points' see Fig. 3.10. A–C, E after Snodgrass (1956), D, F after Hodges (1952).

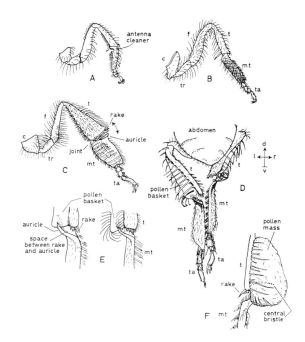

Fig. 5.17 Honeybee (*Apis mellifera*), foraging for nectar on garden sage (*Salvia officinalis*). The empty pollen basket shows as a clear shining area, fringed with bristles, on the bee's hind leg.

tibia which forms a pin through the pollen mass and is apparently important in holding it in position. The tarsi of the middle legs shape the pollen mass, which is eventually kidney-shaped. The efficiency of the pollen press is probably greatly dependent on the moistening of the pollen, which also makes it possible for the rather sparse unbranched hairs of the pollen basket to carry a large quantity of compacted pollen. Honeybees which collect pollen accidentally when concentrating on nectar-gathering (Fig. 5.17) sometimes retain it but may discard it. In discarding pollen, they make movements similar to those made when packing pollen, but the position of the legs is different and probably the press is kept closed, so that the pollen just drops away from the rake. Both processes are normally carried out while the insect is in flight. The pollen-packing apparatus of the bumblebee

Fig. 5.18 Bumblebee (*Bombus lucorum* worker) buzz-foraging for pollen on bittersweet (*Solanum dulcamara*).

(*Bombus*) differs in small details.[1]

Bumblebees, but not honeybees, are among the wide range of bees that practise vibratory pollen-collection from certain flowers with dry pollen (p.179). The vibration comes from activity in the indirect flight muscles, the main function of which is to cause deformations of the thorax that impart movement to the wings in flight. For pollen-collection, however (as also when warming-up [p.139], and for communication by sound-production [p.133]), the wings are uncoupled from the flight mechanism so that they vibrate at low amplitude. This sets up a resonance either in the individual anthers or in a space that they enclose, energising the pollen grains, which stream out of the flower. The flowers are usually pendent (Fig. 5.18) and it would seem that the greatest proportion of the ejected pollen will reach the insect in this case. As a rule, these flowers give pollen as the only reward, but a few also give nectar or oil (pp.115, 179). The pollen-carrying brushes of the bees are dense, with interstices that match the rather small size of the pollen grains (p.179). This process, often (inaccurately) called buzz-pollination, was reviewed by Buchmann (1983) and discussed by Corbet, Chapman & Saville (1988).

The senses and behaviour of Hymenoptera

Studies of the senses and behaviour of Hymenoptera that are relevant to pollination have mostly been carried out on social species, and it is mainly these that are covered here. Progress in research on the senses of social bees can be traced in books by von Frisch (1993), Ribbands (1953), Butler (1954), Lindauer (1961), Barth (1982, 1985), O'Toole & Raw (1991), Goodman & Fisher (1991) and many others.

Wasps and sawflies

The use by social wasps of their sense of smell in finding food is common knowledge. Verlaine's (1932a) experiments with scented sugar-water showed that certain scents (heliotrope, violet, jasmine, bergamot and aniseed) attracted wasps while others (cin-

---

[1] The discovery of the method of pollen-packing in *Bombus* and *Apis* is due to Sladen (1911a, b, 1912a-c) and descriptions of it in *Apis* are given by Hodges (1952) and Snodgrass (1956).

namon, lily-of-the-valley, creosote and turpentine) repelled them. A constancy to a particular scent was observed when a choice of attractive scents was offered, and this persisted when the strength of the preferred scent was greatly reduced. This behaviour is similar to that of Diptera and Lepidoptera, described in Chapters 3 and 4.

Verlaine (1932b) found that when workers have discovered a good food source, they display some excitement on returning to the nest and that when they leave again other workers follow them and may be led to the food directly; others also appear to be alerted and, though not able to follow the direct route to the food, they begin to search for it, probably knowing its scent from encountering it in the nest. An experiment was carried out with unscented sugar-water 25 m from the nest. A wasp that was deliberately introduced to the liquid returned once to four times every ten minutes throughout the day, but no other wasp from the nest came to this food, presumably because it was unscented and rather far from the nest. Communication between wasps relating to food sources is thus of a rudimentary nature.

An investigation of the red-blindness of *Vespula rufa* was carried out by Schremmer (1941b), who found a wasps' nest with a conveniently situated entrance – a knot-hole 2 cm in diameter in the side of a white-painted hut. He screened the hole with a white card, so that the wasps could emerge but, on returning, could not see the black entrance hole to the nest. Discs of various colours and 2 cm in diameter were placed on the surface of the hut near the nest hole, and the behaviour of the returning wasps was observed. The wasps mistook the red and purple discs for their black hole; yellow, green and blue discs were ignored, however, evidently being distinguished from black. These results showed that to these wasps red appears as black. Similar methods have been used for some of the Sphecidae, and red-blindness has been found to occur among them also (Molitor, 1937). Spectral sensitivity curves for *Vespula vulgaris, V. germanica* and a sawfly have been published by Menzel (1990). The first two are red-blind, and the third red-sensitive but blind to ultra-violet.

Sense of smell in bees

The olfactory organs of bees are similar to those of other insects and are on the antennae, where they are mixed with tactile hairs. In the worker honeybee, the organs of smell are absent from the first four joints of each antenna and present in the remaining eight. These bees can detect concentrations of scents ten to 100 times weaker than those just perceptible to man, and they are also very good at discriminating between slightly different mixtures of scents (Ribbands, 1955). It was found by Lex (1954) that the different parts of a flower often smell differently to human beings. She then used flowers cut up into these parts for experiments with bees and found that the honeybee could easily distinguish between the scents of the parts. The presence of the organs of smell in the mobile antennae enables bees to explore the exact distribution of the smell of an object. They could thus easily use the scented guide-marks of a flower, which often coincide with the visible guide-marks, to assist them in finding the nectar. Bolwig (1954), however, found that honeybees did not follow linear scent traces made on coloured models, although coloured objects with scent at one end only were visited chiefly at the scented end.

The role of scent in the approach of honeybees to food was investigated by von Frisch (1954), who trained bees to feed from a blue cardboard box containing jasmine scent, two similar but empty and uncoloured boxes being presented at the same time. The bees were then shown one plain empty box, one plain jasmine-scented box and one blue empty box. They approached the blue box directly from a distance, but on reaching the entrance hole they appeared startled and roamed around outside instead of going in. If they chanced to come within a few inches of the jasmine-scented hole they went in there. Observations show again and again that, just as with Diptera, vision is important in guiding the insects to a food source from a distance, but that scent is taken account of at close range and exerts a powerful influence on their behaviour. Kugler (1940) found that *Lasioglossum* could distinguish accurately between flower-heads of rough hawksbeard (*Crepis biennis*) and greater hawkbit (*Leontodon hispidus*), although they are very similar visually; discrimination took place only at very short range and was doubtless dependent on scent.

It was found by Butler (1951) that honeybee scouts (see p.136) were attracted to dishes of sugar-water scented with extracts of hawthorn (*Crataegus*) or white clover (*Trifolium repens*), but the bees were hardly attracted at all if the dishes were unscented or if they were scented with *Spiraea arguta*. The experiments were done before these plants had come into flower so that the young bees could not be familiar with the scents, and their reactions were presumably inborn. However, in experiments carried out by Free (1970a), bees trained to yellow and a particular scent, and therefore experienced bees, made more visits to scentless models than to models with a strange scent. This is the reverse of a result obtained with the fly *Lucilia* by Kugler (p.77).

Bees, like flies, can be prevented from visiting a flower if an unaccustomed scent is present. For example, both bumblebees and honeybees, when trained to visit rose-scented artificial flowers, refused to alight on models scented with a mixture of rose and lavender, although they approached them closely (Manning, 1957). Similar examples are given for bumblebees by Oettli (1972) and Manning (1956b). Even placing strongly smelling oil-of-thyme among three types of flower that honeybees had been visiting was enough to put a stop to foraging (Butler, 1951). Some solitary bees (*Halictus* or *Lasioglossum*) observed by Kugler (1940) made their first approaches to flowers visually and were put off from alighting on field bindweed (*Convolvulus arvensis*) (Plate 21a) by the application of clove-oil to the flowers.

The majority of flowers visited by bees do not have a very strong scent, and in fact the scents to which the bees pay attention at short range may be very faint to our noses. The houndstongue flowers used by Manning and some of the flowers which Lex found to have internal scent differences are not normally thought of as being fragrant. The use of such faint scents by bumblebees was clearly demonstrated by Kugler (1932a,b). The method was first tested using the strongly scented sweet pea (*Lathyrus odoratus*). Then three types of weakly scented flower (*Lycium halimifolium*, *Echium vulgare*, *Linaria vulgaris*) were tested in separate experiments. The result was that all visits to real flowers were accompanied by proboscis reactions; visually very imperfect unscented models received few visits and no proboscis reactions; similar but scented models received an intermediate number of visits, accompanied by proboscis reactions in some cases. Kugler showed in further experiments that the bees were responding to the specific flower scents, and not merely to the smell of vegetable matter or flower scent in general.

Among the louseworts (*Pedicularis* spp.) of North America, two species with little or no scent to humans have been shown, even when concealed, to be attractive to their bumblebee pollinators. Although the petal colours could not be seen, the bees unerringly alighted on that portion of the small muslin cage closest to freshly matured flowers. Moreover, individual bees went only to the cages containing the species they had been visiting previously, showing that they could distinguish the louseworts entirely by their smell. A third species of lousewort was found to have a much a stronger scent, and it was suggested that this was related to its shady habitat, in which scent might be more important than in the open (Sprague, 1962).

In locating a food source by scent, insects turn into the wind, and can recognise small increases in intensity, so that they reach its neighbourhood. Near the scent source bees can perceive differences in the intensity of stimulation received by each antenna and they then orientate themselves so as to equalise the stimuli. A bee deprived of one antenna waves the other from side to side to test for the same inequality (Barth, 1982, 1985).

Among honeybees, scent plays an important part in the communication of information about sources of food. Returning foragers bring into the hive the scent of the flowers from which food has been obtained, and this scent is used by other bees to find the same kind of flower. Von Frisch (1950) demonstrated this in the following way: a bowl of fragrant cyclamen flowers was put out near the hive, the flowers having been filled with sugar-water. Not far off, another bowl of cyclamen flowers was put out side by side with one containing phlox flowers which are also fragrant, but none of these flowers had food added. Some bees, presumably alerted by the finders of the original bowl of cyclamen, found the other two bowls, whereupon they ignored the phlox and persevered in searching for food in the cyclamen flowers. If the original cyclamens were replaced by phlox flowers containing sugar-water, bees interested in phlox began to appear at the site of the other two bowls of flowers. Von Frisch also discovered that when a rich source of food lacking scent is found the alerted bees visit scentless flowers. Furthermore, he found that the bees in the hive could learn a flower-scent either from another bee's body or from the nectar she had collected, which is normally passed round among the bees in the hive. Sometimes, however, the scent on the body is lost during the flight back to the hive, so that the second method is the more reliable. Honeybees are also able to produce a scent themselves, and they sometimes use this when they are on a good food source to attract other bees in the neighbourhood towards them.

Sense of sight in bees

The flicker-fusion frequency of the honeybee (see p.87) is ten times that of man. The reactions of honeybees to two-dimensional shapes were closely investigated by Hertz (1935). Honeybees are very poor at distinguishing shapes but, like butterflies (Chapter 4), they are quite sensitive to differences in length of outlines. Tests with differently shaped coloured figures of equal area showed that there was an innate and persistent preference for figures with longer outlines. A further property for which the bees had an inherited preference was unsteadiness of outline. Thus a group of small crosses was more attractive than a group of small circles, even though their total outline was of identical length. The results were the same for various colours. In addition, if leaves of various shapes were put out on a white background and covered with glass, the honeybees were attracted and alighted over the leaves, preferring the more compound shapes and any teeth or lobes.

Manning (1956a) offered bumblebees coloured paper shapes with a diameter of about 12 cm – large enough for him to follow the bees' reactions to different parts of the pattern. In a test with uniformly coloured models in the form of a circle, a six-pointed figure and a six-lobed figure, nearly all the visits were to the edges of the models, where the bees hovered, dipped down to the surface, and sometimes alighted and walked round the edge. Having approached the centre of the 'flower' along the side of one 'petal', they followed the margin of the next petal and so receded from the centre again. Manning also noticed that some natural flowers are large enough to reveal this preference for colour boundaries; for example, bumblebees and honeybees visiting *Magnolia* repeatedly reacted to the edges of the petals, and some of the bees failed to find the stamens and flew off.

Having found out the reaction of bumblebees to large plain models, Manning now introduced models of similar size, but with guide-marks. He used blue or yellow models, with a thin line of the other colour along the middle of each 'petal', reaching nearly to the centre. He now found that the bees made $1\frac{3}{4}$ times as many dips down to the centre as to the edge of a model, and often the bees reached the centre by flying along one of the lines. A bee visiting a model would make several dips over it, and it was found that the first dip was much more often at the edge than at the centre. On the other hand, there were many more subsequent dips to the centre than to the edge. What was apparently happening was that the bees, on sighting a model from a distance (usually not more than 50 cm), went straight to its edge without perceiving the guide-marks. The effect of the guides was therefore exerted only at very short range.

An extensive study by Free (1970a), in which the honeybee was tested for the effects of size, shape, colour and presence and form of guide-marks, gave broadly similar results but showed occasional differences from bumblebees. It was clear that bees see and respond to some shapes and some patterns of guide-marks; the response may be in the frequency of visits or, in the case of guide-marks, the frequency of alighting at the centre of the flower rather than elsewhere. Some guide-marks had little effect in attracting the bees spontaneously to the centre, but after training, which took effect rapidly, the effect was considerable. Correlations occurred between the effect of guide-marks and the outline shape of the flower-model. Size alone was not used by the bees as a distinguishing feature. Bees trained to a particular scent associated with a particular colour were tested with the training colour plus another scent, and the training scent on another colour; the tests showed that scent was more important than colour.

The effect of three-dimensional models on honeybees was studied by Hertz (1931), and it was found that white structures against a white background were attractive in proportion to the depth of shade in their darkest parts. The reactions of bumblebees to similarly coloured hollow cones and flat discs were compared by Manning (1956a). It was found that when the bees made clear choices of target from a distance of 50 cm or more there was no discrimination, but that dipping in flight after arrival was much more frequently to the centre than to the edge of the cones (which had their vertices downwards and were tilted at 30° to the vertical). Bees visiting the cones often flew right into them and quite frequently alighted near their centres. In addition, a flat disc that was coloured more intensely towards the centre was found to receive more dips at the centre than a uniformly coloured one. Kugler (1943) found that hollow cones received more visits from bumblebees than convex structures. For this experiment the bees were in small flight-boxes, so that they must have made their choices at short range, and this doubtless enabled them to discriminate between the different types of model.

In Chapters 3 and 4 it was seen that most Diptera and Lepidoptera can distinguish colours of the blue group from those of the yellow group, and that many of them have a preference for one group or the other. It was also shown that some butterflies react to red as a colour of the yellow group, whereas others are red-blind. Among bees, red-blindness has been found in *Hylaeus* (Knoll, 1935, using the method that was used for *Vespula* by Schremmer) and *Anthophora plumipes* (Menzel, 1990), as well as the honeybee and bumblebees. But, like other red-blind insects, the honeybee and bumblebee can perceive blue-green and ultra-violet as distinct colours (von Frisch, 1950, 1954; Kugler, 1955a). As most of the flowers that we see as white do not reflect ultra-violet, the bees see them as coloured because they do not reflect the complete bee-visible spectrum (see below). 'Bee-white' is rare in flowers and it is difficult to train bees to it. Ultra-violet perception has also been confirmed in several solitary bees in America (Grant, 1950c).

A thorough investigation of the colour vision of honeybees was carried out by Daumer (1956) using an elaborate specially-built 'spectral colour-mixing apparatus'. The equipment permitted the comparison of 'bee-white' light with various coloured lights of varying intensity. Bees were trained to a particular colour and then tested for ability to distinguish it from other colours spread through the spectrum. It was found that for bees there were three main colour groups: yellow, blue and ultra-violet. These groups were almost totally distinguished from one another at all intensities used. Within each group some power of discrimination was also found, but it was not very accurate in the yellow and blue groups, and in both these groups it fell off with decreasing intensity. In the yellow group the colours distinguished were orange, yellow and green; in the blue group they were blue and blue-violet. The ultra-violet group was different in that two light-wavelengths quite near together were distinguished at all intensities. Between the yellow and blue groups, the bees saw a colour (blue-green) which they could distinguish accurately from the main groups and, similarly, they could distinguish a colour (violet) between the blue and ultra-violet groups (Fig. 5.19). The situation in man is exactly comparable: there are three main ranges: red, green and blue; and two intermediate regions: yellow and blue-green. Man can see a further colour (purple) which is produced by mixing light from the opposite ends of the spectrum (red and blue). In bees, likewise, Daumer found that by mixing light from the opposite ends of the bee-visible spectrum (yellow and ultra-violet) another colour can be produced which is distinct to the bees ('bee-purple').

For man, when red is removed from white light the remaining light is a mixture of blue and green, and is seen as blue-green. If pure blue-green is added to red, the effect is white, and the colours are therefore described as complementary. Similarly, for bees, white light with ultra-violet removed appears as blue-green, and if blue-green is added to ultra-violet the effect is 'bee-white'. Blue-green and ultra-violet are therefore complementary for bees. The complementary colour to yellow is violet for bees, and the complementary colour to blue is 'bee-purple' (Fig. 5.19).

Blue-green for man, besides being a spectral colour, can also be produced by mixing blue and green, while yellow can be produced by a mixture of red and green spectral light. The colours of the two transitional zones of the bee spectrum can similarly be produced by mixing the spectral lights on either side of them. Daumer found that when these transitional colours were produced by mixing, it was possible to make different mixtures which were distinguishable by bees. Thus, three easily distinguishable violet colours were produced by mixing blue and ultra-violet in different proportions. Similarly, with 'bee-purple', which has no pure spectral counterpart, different mixtures of yellow and ultra-violet produced two easily distinguishable colours. The high sensitivity of bees to ultra-violet is brought out by the fact that a mixture of 2% ultra-violet and 98% yellow was distinguished by the bees from yellow, whereas 50% yellow had to be added to ultra-violet to make it different from pure ultra-violet to the bees. Mixtures of colours within the main colour groups also give intermediate colours; for example, a mixture of orange and green is confused by bees with yellow.

It has been shown that the eye of the honeybee, like those of the blowfly and most Lepidoptera, is trichromatic, with three types of light-sensitive cells, though their spectral sensitivities are different, there being a numerous type sensitive to ultra-violet, and two less numerous types sensitive respectively to blue and yellow. A hypothesis to

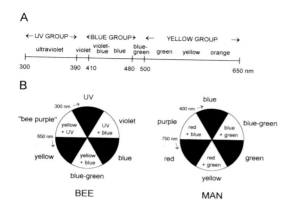

Fig. 5.19 Spectrum of bee-visible light (**A**) and colour circles for man and bee (**B**). In B, opposite segments are complementary and when combined produce 'white'. Main colour groups shaded. nm=nanometres. After Daumer (1956).

explain how these give rise to the colour circle shown in Fig. 5.19B is given by Burk-hardt (1964).

The good discrimination shown by bees among tints produced by mixing the primary colours is matched in the intermediate wavebands of the spectrum (as reported also for other insects). Helversen (1972) found that the minimum wavelength-separation for colour-discrimination was in the blue-green region, where honeybees could make 70% correct choices when the separation was as little as 4.5 nm. Helversen's results, however, did not show any better discrimination in the ultra-violet region than in the other primary regions, blue and yellow. Bearing in mind that each of the three types of colour receptor in the eye responds best to a particular wavelength, and decreasingly on either side of this, it is not to be expected that minor differences of wavelength near these peaks can elicit much qualitative difference in the response. On the other hand, as the wavelength nears the limit at which one type of visual pigment can respond, a response sets in in the pigment whose sensitivity is adjacent in the spectrum, for the sensitivities of the pigments overlap. Thus borderline wavelengths stimulate, albeit presumably weakly, two kinds of cells and their corresponding neurones. The brain, therefore, can make comparisons. This seems to be the explanation for the enhanced power of discrimination in these spectral regions shown by insects and, indeed, other classes of animal (Barth, 1982, 1985).

Daumer followed his remarkable investigation of the honeybee's colour discrimination by an equally outstanding study of the colours of flowers and the reactions of bees to them (Daumer, 1958). Two hundred kinds of flowers were photographed, each through three interference filters: one ultra-violet, one blue and one yellow. This showed that creeping cinquefoil (*Potentilla reptans*), for example, which looks yellow to us, reflects 7% of 'bee-white' light, while the remaining 93% is made up of yellow and ultra-violet in the proportion 94.5 to 5.5. To bees, this flower therefore appears 'bee-purple', very slightly diluted with white. The constitution of the flower colours found in this way is shown in Table 5.5. Daumer also found that the bees could distinguish 'bee-yellow' flowers from green foliage more easily than might have been expected since, on average, the 'bee-white' element in the colour of the leaves is six times as great, making them appear greyish. He confirmed part of his analysis of flower colour by experiments with bees. When three similar flowers (one 'bee-yellow' and the others each a different shade of 'bee-purple') were presented under a cover of glass opaque to ultra-violet, the bees confused them completely, although they had previously been able to distinguish them accurately. Similar results were obtained with a 'bee-blue' and a 'bee-violet' flower (two species of *Scilla*). Many flowers which have some ultra-violet reflection show patterns caused by the absence of ultra-violet from certain regions. These patterns, invisible to man, act as guide-marks to bees in the same way as the guide-marks we can see (Fig. 5.20). An extensive survey of such patterns was carried out by Kugler (1963, 1966). He found that, whereas only 30% of the flowers investigated had patterns visible to the human eye, a further 26% could be added by including ultra-violet patterns. These 26% thus have patterns visible only to an eye sensitive to ultra-violet. However, many flowers with patterns visible to man also have patterns formed by ultra-violet absorption. No types of ultra-violet pattern were found that were not also found in the 'visible' range. The frequency of patterning in general increases with the complications of exploiting the flower, and it is thus higher among zygomorphic than actinomorphic flowers and, correspondingly, among bee-visited flowers. However, among these the frequency with which the pattern is formed *only* by ultra-violet absorption is much below average. This blending of ultra-violet absorption with absorption in the 'visible' range may perhaps lead to more varied colourings than are found in the simpler types of flower. A similar study by Tanaka (1982) gave comparable results. When the presence of guide-marks was considered in relation to presumed pollinators, it was found that 'bee' or 'bee-plus-fly' flowers usually had guide-marks, whereas purely 'fly' flowers rarely did. Tests on the significance of these patterns were carried out by Daumer (1958) using petals with an ultra-violet-free spot at the base. The petals were arranged in the form of a flower, either the natural way, with the spots central, or with the spots at the periphery. The position of the spots made little difference to the approaches of honeybees from a distance, but after alighting the bees made proboscis movements over the spots, wher-

Table 5.5 Flower colours, as seen by bees and man (Daumer, 1958).

Reflected light	Colour to bees	Colour to man	Example
yellow + ultra-violet	bee-purple	yellow	creeping cinquefoil (*Potentilla reptans*)
yellow	yellow	yellow	cowslip (*Primula veris*)
green	yellow	green	stinking hellebore (*Helleborus foetidus*)
green + blue + red	blue-green	white	wild cherry (*Prunus avium*)
blue	blue	blue	wood forget-me-not (*Myosotis sylvatica*)
blue + red	blue	purple	ling (*Calluna vulgaris*)
blue + ultra-violet	violet	blue	bird's-eye speedwell (*Veronica chamaedrys*)
blue + ultra-violet + red	violet	purple	purple loosestrife (*Lythrum salicaria*)
ultra-violet + red (rare)	ultra-violet	red	corn poppy (*Papaver rhoeas*)
red (very rare)	black	red	(non-British) (*Nonnea pulla*)

ever they happened to be. A similar result was obtained with honeybees from an experimental colony which had never seen flowers or coloured flower models, and this showed that the attraction of the spots was inborn. Further surveys of UV-reflection continue to be made (for example, Biedinger & Barthlott, 1993; Burr & Barthlott, 1993; Burr, Ross & Barthlott, 1995).

Kevan (1978, 1979) has stressed the importance of considering flower-reflectance as a whole, rather than as reflectance in a particular waveband, and in relation to the light-environment and the backgrounds against which flowers are seen. He emphasised the fact that there is not much ultra-violet in the daylight spectrum, which means that the high sensitivity of insects to ultra-violet is a useful attribute. Both soil and vegetation are poor reflectors of ultra-violet and dull overall, so that most flowers will stand out well for insects, as they do for man.

Foraging honeybees navigate by the sun and they can tell the direction of the sun even if the sky is completely overcast, provided there is no obstruction other than the clouds. It has also been found by von Frisch (1950) that they can make use of the polarisation of light from a blue sky to orientate themselves.

Powers of communication in bees

One of the most remarkable powers of the honeybee is its ability to inform members of its colony of the direction, distance, abundance and nature of a valuable source of food, the direction being indicated only if the food source is more than about 20 m from the hive. We have already mentioned that information about the nature of the food is conveyed by its scent. The rest of the information is conveyed by dances executed by the bees that have found the food. This behaviour is well-known and has been described by its discoverer, von Frisch (1950, 1993), and also by Butler (1954)[1].

The bee dance is carried out in the darkness of the hive on the vertical combs. The movements of the dancing bee are followed by other bees who jostle her. Esch (1967) found that the dancing bee normally also makes a sound by vibrating her wings; sometimes the sound is not made, and in this case the workers who have followed her dance fail to leave the hive. In addition, the sound plays a part, along with the tempo of the dance, in indicating the distance of the food-source from the hive.

Among the small stingless bees and sweat bees of the tropics, which are also social, there is no dance, but returning foragers do give information about food, not only by the transfer of the food and its scent, but by making sounds with their wings. In *Melipona* the sound is pulsed, and the length of the pulse indicates distance (Esch, 1967). Direction is communicated by a rather complex leadership ritual outside the nest. Among the species of *Trigona*, the sound conveys no information about distance and merely causes other bees to go searching. However, one species leaves scent-marks at intervals as it returns from foraging and then guides its fellows along the trail to the feeding station (Lindauer, 1961; Esch, Esch & Kerr, 1965). Experiments in providing sound without scent induced the bees to go to the nest entrance, but not to go out foraging.

[1] In the late 1960s, new reaearch by P. Wells and A.M. Wenner raised doubts about von Frisch's interpretation of his results. There ensued a long controversy which stimulated many more experiments, leading ultimately to a confirmation of von Frisch's main conclusions (Gould, 1976; Gould & Gould, 1988).

Sense of time in bees

Honeybees have a highly developed time sense. If they are fed at certain times of the day in a certain place or places, they will appear at the feeding places at the appropriate times, and if, as an experiment, food is withheld, they will remain searching the accustomed area during the period when food is normally provided. In addition, if honeybees are fed on one day at a site in a particular direction from the hive, they will set off in the same compass direction the following morning, even though the hive has been moved overnight to a new and entirely unfamiliar place and faces another way. Although this does not bring the bees to the food, it shows their skill in navigating by the sun and allowing for its position according to the time of day (von Frisch, 1954).

Foraging behaviour of bees

The constancy of the individual bee to a particular species of flower is an important factor in cross-pollination. Little is known about foraging habits among solitary bees, but Chambers (1946) investigated the pollen loads of some species of *Andrena* that visit fruit blossom and found that *A. varians*, *A. haemorrhoa* and *A. armata* were very constant in collecting fruit tree pollen (cultivated plum, cherry, pear and apple, and wild blackthorn [*Prunus spinosa*]), *A. varians* particularly so. *A. armata* later changed its constancy to sycamore (*Acer pseudoplatanus*) when this came into flower. Evidence of appreciable and often substantial constancy has been found from pollen analysis of the loads of *Andrena*, *Lasioglossum*, *Anthophora* and *Megachile* in America (Grant, 1950c). Direct observations on constancy in *Lasioglossum* have been given by Kugler (1940). In one instance, one of these bees consistently visited the yellow flowers of rough hawkbit (*Leontodon hispidus*) without paying the slightest attention to various violet and purple flowers growing in the same place. Another *Lasioglossum* repeatedly visiting dandelion (*Taraxacum officinale*) occasionally alighted on flowers of meadow buttercup (*Ranunculus acris*), but immediately flew off and continued to collect food from the dandelions; both kinds of flower are 'bee-purple' according to Daumer, though yellow to our eyes. At a place where greater hawkbit was growing with a similar yellow composite (*Crepis biennis*), one *Lasioglossum* fed only at the hawkbit and another only at the *Crepis*. In addition to behaviour of this type, Kugler saw *Lasioglossum* visiting alternately two species of flower on the same flight, but here also there were generally a few successive visits to the same type of flower.

Some non-social bees of the seasonally dry tropics that feed on flowering trees forage in groups and move rather frequently from one tree to another. The reason is not understood but the behaviour is beneficial to the plants in promoting cross-pollination. A cloud of up to 300 group-foraging bees may also disturb other pollinating insects to the extent that they move to another tree, and the same effect may arise from the aggressive behaviour of territory-holding bees (Frankie & Baker, 1974; Frankie, 1976). In another tropical study, movements by solitary bees between trees 1.2 km apart were detected (Frankie, Opler & Bawa, 1976).

Another foraging strategy involving long-distance movement was described by Janzen (1971) for long-lived 'quasi-social' (nest-sharing) bees of the tribe Euglossini (family Apidae) in the New World tropics. Firstly, Janzen found by transportation experiments that these bees could return home to their nests from distances up to 23 km, and one bee covered 20 km in 65 minutes. The speed of return suggested that the bees knew the district up to the limits from which they returned. Secondly, it was found that the plants at which they foraged occurred at low densities in undisturbed forest and that these plants produced few flowers but did so over a long season. A series of observations and experiments indicated that each bee specialised on a few species of plant, and visited each individual plant by flying a set route each day. A pollen-collecting bee might take up to 50 minutes to collect a load, and be away from the nest for two hours, leaving 70 minutes for getting to and from a foraging area, during which it might be flying at 20 km per hour. The bees clearly knew where 'their' plants were, approaching with direct flight, and they regularly visited plants that had been artificially deflorated, as has also been found for bumblebees (see p.137); in fact, bumblebees and some other solitary bees (as well as some vertebrate flower-visitors) exhibit the same behaviour. This foraging strategy became known as trap-lining, based 'on the analogy of a trapper following a line of traps from home and back in the temperate mammal hunting season,' (H.G. Baker, letter to PFY, 1991. The idea was Janzen's but the expression was not used in his 1971 paper; it was left to Baker to promote its use in publications, beginning with Baker, 1970, p.103).

As an example of an investigation on the foraging behaviour of the honeybee, we may quote that of Ribbands (1949). He studied the foraging of individual bees on a plantation consisting of five species of flower, planted together, three of them in long rows contiguously, and all five nearby, each in large or small rectangular beds, some contiguous, and some separated by distances of a metre or so. The bees were marked and watched continuously for long periods. A bee that worked numerous flowers of Californian poppy (*Eschscholzia californica*) ranged over a large part of the available crop, and generally progressed steadily in one direction throughout each foraging trip; she often made successive visits to newly opened flowers or returned to them after visiting one or two other flowers. Evidently, the young flowers were supplying more pollen than those that had been open longer. Another bee, collecting pollen from Shirley poppy (*Papaver rhoeas* cultivars), showed a remarkable constancy to a single, very productive bloom, visiting it on each of 19 successive trips, usually visiting it first on each trip, and on many trips collecting from it all, or nearly all, her load. This showed a very exact awareness of the position of this flower. (The Shirley poppies were particularly variable in their pollen production, several of them being attractive, though less so than the one just mentioned.) These two bees provided examples of, respectively, a very extensive and an extremely restricted foraging area. Cases of moderate restriction were provided by bees collecting pollen from another of these specially planted flowers, the nasturtium (*Tropaeolum majus*); the bees all confined themselves to a square metre or so of the available crop, though each had a different foraging area.

The three flowers mentioned so far supplied the bees only with pollen, whereas nectar (and a little pollen) could be obtained from the remaining two kinds of flower in the experiment. A bee that had been working poached egg flower (*Limnanthes douglasii*) for nectar began work one day at 6.30 a.m., collecting pollen from Shirley poppy. Then, at about 9.00 a.m., it gradually transferred its attention to the *Limnanthes*, which were by that time producing nectar, and abandoned the Shirley poppy for the day. The bee's time sense, combined with its previous experience, was presumably responsible for its not attempting to visit *Limnanthes* early in the morning, though these factors are not infallible guides because the time of greatest discharge of pollen or nectar is affected by the weather. The value of this use of time sense is increased when alternative crops are far apart, as it reduces the number of unsuccessful journeys to the second crop.

Another bee was found on two successive days to be collecting both pollen and nectar from *Limnanthes*, and pollen from Californian poppy. During periods when the latter was not productive, the bee visited it occasionally until pollen production began, when it increased its visits to this flower.

Ribbands observed that bees working a crop that was rapidly becoming unproductive became increasingly restless, and began to move hastily over the whole foraging area or even beyond it, instead of moving short distances from flower to flower. (Later studies of bumblebees put this into measurable terms: as bees find themselves on less rewarding ground, so the flight distance between visited flowers increases and the angle between the directions of successive flights decreases; conversely, in a rich habitat flight distances are short and the turning angle is large, resulting in a more intensive investigation of the area [Heinrich, 1983].) It is rather curious that, although some bees will change to an alternative crop when one is exhausted for the day, others return to the hive until it is time for their favoured crop to be productive again (von Frisch, 1954; Butler, 1954; Free, 1963). Bees may work an adequately productive crop for a long time before changing over to a better one that has been available all the time. Apart from scout bees (see below), they do not appear to spend time sampling all accessible crops and choosing the best, so that excessive competition on the best crop is avoided. This situation has also been noted for bumblebees (Heinrich, 1976d).

There were fairly frequent changes in the crops worked by the bees observed by Ribbands, but he noted an exclusive attachment of 12 days by a single bee to a single pollen crop, and one of 21 days to a single nectar crop. Bees that worked only one crop at a time invariably changed from a less fruitful to a more fruitful pollen crop, or from a pollen to a nectar (or nectar plus pollen) crop. Thus it seems that nectar crops are more attractive to the older bees – bees having a foraging life of only a few weeks – in spite of the fact that a bee has to do more work to collect a load of nectar than a load of pollen (in certain other studies, however, no evidence for such a foraging sequence has been found [Free, 1963]). The number of flower-visits needed to make up a load of pollen was found by Ribbands to be one to 27 for Shirley poppy, 66 to 178 for nasturtium, and intermediate for Californian poppy. The time required ranged from a minimum of three minutes for Shirley poppy to a maximum of 18 minutes for nasturtium. Only two loads consisting purely of nectar were fully observed; each required over 1,000 flower-visits and a time of more than $1\frac{3}{4}$ hours. Bees collecting both pollen and nectar from *Limnanthes* took a minimum of 250 flower-visits and 27 minutes to complete a load.

By contrast, the foraging statistics for the solitary bee *Andrena complexa* visiting *Ranunculus* for pollen show a time of $1\frac{1}{2}$ hours to complete a load, and a pollen-foraging rate of three loads per day. However, four or five loads suffice to supply the pollen needed for one cell (Linsley & MacSwain, 1959).

Various factors influence the number of visits required for a particular type of load. For example, some individual honeybees are more thorough than others in collecting pollen, and therefore complete a load with fewer flower visits and in a shorter time. Furthermore, an increase in the number of visits required to complete a load of pollen or nectar may be caused either by a decrease in production (which often takes place late in the day) or by competition from other bees. The time required to collect a load is also increased by a drop in temperature, which lowers the rate at which the bees can work on the flowers.

Though most honeybees show strong crop-constancy, there are always some bees in a colony that actively explore – at least for part of their lives – coloured and scented objects, and in this way discover new sources of food. These bees are known as scout bees. If they find a rich source of food, they remain constant to it for a short time and communicate their knowledge to other members of the hive. Scout bees may be recognised from their behaviour, and it has been found that they are proportionately more numerous early and late in the season when there are fewer types of flower available (Butler, 1954). Crop-constancy of honeybees is greater when they are collecting pollen than when they are collecting nectar, and it has been shown that this is because the bees have difficulty in packing pollen from more than one source in the same load (Zahavi *et al.*, 1984).

The foraging behaviour of bumblebees was studied by Manning (1956b) with illuminating results. The special interest of his investigation arises from his choice of extreme flower types: houndstongue (*Cynoglossum officinale*), with its inconspicuous flowers (see p.147), and foxglove (*Digitalis purpurea*) (Plate 1h & Fig. 6.43), with very showy ones. The reaction of the bees to each of these flowers (mainly nectar-producing) was investigated separately, but in each case the plants were in a natural habitat, some in a group and others widely scattered in the neighbourhood, being 3 m or more away from the group and each other. Some of the plants were growing naturally, but others were grown in pots and put out in positions which completed the required arrangement. The potted plants made it possible to alter the arrangement and then to restore it to its former state.

The houndstongue flowers have a short tube and a concave five-lobed limb; they are dull brownish-red when young and purplish later, but bumblebees tend to avoid visiting the purple flowers. On each flowering branch there are usually two flowers in the red stage and two in the purple; normally the flowers are nodding, but the first few to open face upwards. In both the seasons when houndstongue plants were observed, there was an interval of 11 to 12 days between the opening of the first flower and the visit of the first bumblebee. During this time the flowers held copious uncollected nectar and, unlike the later flowers that received visits from the bees, they set no seed. The bumblebees began to visit the houndstongue when one of them, passing within 60 cm of a flower, alighted. Visits were at first rather tentative, but were soon repeated and became more purposeful. After a few visits, the bees became conditioned to the form of the houndstongue plant as a whole; this was shown by the fact that they visited houndstongue plants without any flowers, as well as other plants with a similar growth habit, sometimes searching in the leaf axils, where in the houndstongue the flowering branches are produced. Manning found that a houndstongue plant with no flowers was reacted to at a distance of 2 m by these conditioned bees, whereas a plant with flowers but no leaves was reacted to at not more than 60 cm. The strong mousy smell of the plant did not seem to help the bees to find it. Bees moving from one plant to another in the main group used a slow, apparently exploratory flight, and kept reacting to other similar species, until they gradually learnt the approximate positions of the houndstongue plants. Each bee, having visited all the plants in the group, flew off fairly slowly, making wide sweeps over the ground 40-60 cm above it in an exploration flight. The bees now reacted as before to houndstongue plants and similar species, ignoring plants with conspicuous flowers. In this way, all the houndstongue plants were found, the farthest away being 24 m from the central group and some being so hidden in bracken that they could be seen only at distances of less than 60 cm. Bees leaving each of the isolated plants made orientation flights, and after a few visits had a very exact knowledge of their positions. Orientation flights are important in the lives of all nest-building Hymenoptera, and are used when foraging. The insect flies round the object in which it is interested in gradually increasing circles, and may fix its position by quite distant objects in the landscape (Butler, 1954). The bumblebees re-visiting isolated houndstongue plants usually followed a particular course, and flew fast and directly. In other words, they were trap-lining (p.135). One bee, which had visited a houndstongue only once before, returned to the spot on three occasions, although the plant was taken away after the first visit.

One difference between the behaviour of bees at the central group and that of those at the distant plants was observed by removing a plant in each situation. In the central group, some bees did not appear to miss the absent plant, while others visited its site diminishingly in the course of half an hour. The removal of a distant plant, however, did not reduce the number of bees approaching its site, and they searched the site for much longer. This behaviour reflected the bees' approximate knowledge of plants in the central area and their exact knowledge of the positions of the distant plants.

In the foxglove experiment, the bees started to visit the plants as soon as the flowers showed colour, sometimes forcing their way in before they were fully open and before any nectar had been secreted. The bees quickly found all the distant plants, showing only slight signs of sweeping exploration flights, and flying directly to plants up to 4 m away. Very few orientation flights were seen round the distant plants, and only one of these was a perfect performance. The bees could presumably see the foxgloves some way off, so that an awareness of their approximate positions was all they needed to be able to find them again. It thus seems that bees must estimate the conspicuousness of a plant and have the power to determine how much orientation is required. The removal of a plant from the central group reduced the number of visits to its site, as in the houndstongue experiment; bees visiting the sites of distant plants that had been removed, however, searched for them over a wider area than in the houndstongue experiment, being less sure of their positions. Bees visiting flower-less foxglove plants spent a negligible time searching them, and other plants with a similar growth habit were not searched at all, although the bees would visit foxglove flowers lying on the ground. It is to be expected that plants intermediate in conspicuousness between houndstongue and foxglove will elicit intermediate behaviour. Such a case is provided by *Campanula barbata*; from observations in Central Europe it was concluded that the bumblebees foraging on this plant made approaches from a distance to a patch of colour which did not have to be the exact colour or shape of the *Campanula* inflorescence (Oettli, 1972). It was found that the several species of bumblebee observed in these experiments, together with a few honeybees that worked houndstongue, all behaved in the same way.

Many insects, and particularly bumblebees, when confronted with an inflorescence of columnar form, such as that of the foxglove, visit first the lowest available flowers and then work upwards. Bee-pollinated plants with this type of inflorescence show a corresponding behaviour. The lowest flowers open first, but there is an overlap of flowering between adjoining flowers, and the flowers are protandrous. This means that the bee first visits the oldest flowers which are in the female stage, and any pollen deposited in them is likely to come from another plant. As the bee moves upwards it acquires pollen from the younger flowers in the male stage; after visiting the highest available flower the insect must fly off to another inflorescence (which usually means going to another plant) in order to avoid revisiting flowers that have just been depleted. The protandry of flowers in such an inflorescence thus constitutes a highly efficient outbreeding mechanism. In rosebay willowherb (*Epilobium angustifolium*), which has an inflorescence of this type, the visitors cling to the projecting and slightly drooping stamens and style, and because of their weight assume an almost upright posture. Benham (1969) has suggested that this could be an arrangement to ensure that the insects move in the upward direction and that the occurrence of such arrangements may have led to upward movement becoming instinctive. A further feature of this plant which might reinforce the tendency to start at the bottom is the fact that nectar secretion is greatest during the female stage.

Nectar gathered by foraging bees must cover the cost of collection and provide a surplus, as the bee, or the colony of which it is a part, has to survive around the clock and provide for offspring. To find the energy consumption of a bee in flight, its oxygen consumption is determined experimentally. The amount of sugar that is equivalent to this is known, so we can establish that a honeybee, which weighs about 0.1 g, uses about 10 mg of sugar per hour for flight (Heinrich, 1975a). Many insect-pollinated flowers contain quite small amounts of nectar-sugar, usually under 1 mg. For example, 0.05 mg of sugar per floret has been found in red clover (*Trifolium pratense*), while a bee foraging on golden-rod (*Solidago canadensis*) may be able to harvest only 0.01 mg per minute (Heinrich, 1975a, 1983). However, some of the flowers that are difficult to enter, and often visited by bumblebees, may have immensely larger rewards. Thus blooms of the turtle flower (*Chelone glabra*, Lamiaceae) contain up to 3.3 mg of sugar. The 10 mg of sugar per hour used for flight by a honeybee represents 385 calories per gram body-weight per hour, or 1610 joules, and thus 161 joules in a bee of 0.1 g. The power consumption (joules/second) of the insect is then 45 milliwatts. A bumblebee has about the same energy requirement per gram body weight as a honeybee, so if it weighs 0.5 g it consumes 5 times as much energy in flight (225 mW). Another bumble-bee study gave slightly lower energy consumptions of 175-200 mW for insects weighing 0.5 g (Ellington *et al.*, 1990).

The foraging of the honeybee in cool weather is aided by the fact that the hive is warmer than the 'outdoors', so that the bees are warm enough to fly and may then keep up their temperature by activity in foraging. Bumblebees have considerable control over their temperature. Like hawkmoths, they are insulated by a dense coat of hair and can warm up by 'shivering' their thoracic muscles; again, their work in foraging may keep the temperature up: an American species of bumblebee foraging in air at only 2°C was found to have a thoracic temperature of 37°C (Heinrich, cited by Heinrich & Raven, 1972). However, on some plants with flowers massed together (for example, North American species of *Solidago canadensis* [golden-rod] and *Spiraea*), where they can forage for long periods without flying, they may get too cold to fly. When they are ready to leave the plant, they then have to spend time warming up (Heinrich, 1972). In the same study it was found that if they were setting out in the morning to forage for pollen, the bumblebees left the nest with a full crop. The energy requirement for warming up is modest in small animals: for a bumblebee weighing 0.5 g, a rise in temperature from 13.5° to 38°C needs 31.5 joules, or enough for three minutes' flying, and the temperature may rise at a rate of about 12°C per minute (Heinrich, 1975a).

Foraging bumblebees can often be seen rejecting certain flowers of a type on which they are working. This may be partly based on age-changes in the flower, such as a looseness among the petals of Fabaceae, withered petals in general, or colour changes in older flowers (which are quite pronounced in the horse chestnut [*Aesculus hippocastanum*], for example). Such signals indicate flowers that have become unrewarding through age. Sometimes, however, bumblebees and honeybees have been seen rejecting apparently suitable flowers, which suggests that they may be able to recognise recently-visited flowers (which might be empty), possibly by means of scent-marking (van der Pijl, 1954; Yeo, 1972; Mackworth-Praed, 1973). However, in some circumstances recent visitation can increase the number of visits that honeybees and bumblebees make to a flower-model (for example: Free, 1970; Cameron, 1981; Schmitt & Bertsch, 1990). Cameron's work showed that a pheromone (scent signal) was left on the flower by the bee. Corbet *et al.* (1984) consider the possibility that two pheromones of contrary effect may be involved, with the response being sometimes predominantly to one, sometimes to the other, or that there may be one compound which elicits a response that switches from negative to positive, or vice versa, as the scent fades. Schmitt & Bertsch's experiments also suggested that there are two pheromones, one enduring from one day to the next and the other fading quickly. Their work involved flower-models that were either rewarding only on the first visit or were replenished ten seconds after every visit. The bumblebees being tested apparently marked the latter; their foraging efficiency was substantially reduced when the discs of the flower models were cleaned to remove the scent-marks. However, nature's productive flowers must normally take much longer than ten seconds to replenish their nectar, so the experimental conditions seem unnatural in this respect.

Kugler (1943) pointed out that many of the flowers most clearly adapted to bumblebees (mainly long-tubed or funnel-shaped) have features which make the nectar difficult to find. These often take the form of physical barriers that have to be pushed aside; such barriers are found in toadflax (*Linaria* spp.), louseworts (*Pedicularis* spp.), antirrhinum, delphinium, comfrey (*Symphytum officinale*) and Fabaceae-Faboideae (see Chapter 6). To us, the bees may not seem particularly clever at finding their way into flowers, but their good memory enables them to repeat with ease their first successful visit to a flower. Furthermore, even after a bumblebee has learnt to handle the flower correctly, its speed in doing so continues to increase for some time (Heinrich, 1983). Add that these 'difficult' flowers usually offer a high reward (perhaps ten or even 100 times that of a simple flower) and it becomes clear that it makes sense to spend time learning to operate them and then to remain constant to them. Bumblebees, however, not having the services of scout bees, regularly sample flowers other than the one on which they are 'majoring' (Heinrich, 1979).

When all the flowers regularly visited by bumblebees are listed, it is found that those that have a reputation for being especially adapted as bumblebee flowers represent a comparatively small proportion. This is because we get our idea of the bumblebee flowers mainly from those preferred by the long-tongued species. In a study of four species of bumblebee in a limited area, Brian (1957) reported marked differences in their behaviour. The short-tongued species *Bombus lucorum* visits a wide range of short-tubed and fully open flowers. In addition, it collects honey-dew and bites holes in flowers as a short cut to the nectar (Fig. 6.49). In the type of flower it visits and in its opportunistic behaviour it resembles the honeybee. At the opposite extreme is the very long-tongued *B. hortorum*: in most of the flowers it visits the nectar is out of the reach of other species; it does not collect honey-dew, nor bite flowers, although Brian found that its jaws are strong enough for it to do so. The two other species studied by Brian were intermediate in tongue-length and behaviour, *B. pratorum* being more like *lucorum*, and *B. pascuorum* (*B. agrorum*) (Fig. 6.8) more like *hortorum*. *B. terrestris* is closely similar in habits to *lucorum* (Leclercq, 1960). Leppik (1953) found that bumblebees have a strong preference for bilaterally symmetric (zygomorphic) flowers over radially symmetrical ones, while honeybees show the reverse preference.

Brian (1957) compared the tongue-lengths of the bumblebees with the tube-lengths of the flowers they visited, and found that on average the tubes were a few millimetres shorter than the tongues that drained them. Each species, therefore, had different preferences, but no bee normally had to extend its proboscis fully. A bee cannot constantly forage at the limit of its reach because the nectar level will quickly drop out of reach in every flower probed. (See also Prys-Jones & Corbet, 1991.)

Like the honeybee, bumblebees will sometimes work two or more kinds of flower simultaneously (Hulkkonen, 1928), and they can be simultaneously trained to visit models with either of two scents, which they then prefer to models containing scents to which they have not been trained (Kugler, 1932a,b). Foraging at two kinds of flower is not at all the same as promiscuous visiting; it is probably induced by insufficient abundance of any one species of plant. When a crop of foxglove came into flower, the bumblebees which began working it did not at first confine themselves to it as there were not enough flowers open to keep them busy; later, when there were more fox-glove flowers, some of the bees restricted themselves to them, while some others occasionally visited red campion (*Silene dioica*) (Manning, 1956b). Kugler (1943) saw a bumblebee working in an area where there were three yellow-flowered species of different genera growing together; the bee was working only two of them, and though it was attracted by the third, it never collected any food from it. In addition, both honeybees and bumblebees are well known to visit different colour forms of the same species growing together, as they often do in gardens (Grant, 1950c). It therefore seems that bees continue to investigate flowers coloured differently from the one on which they first found food, and that the discovery of the right scent induces them to alight.

Bumblebees were believed by Müller (1881) to show a strong preference for blue and purple over white and yellow flower colours. Circumstantial evidence obtained by Kugler (1943) and Brian (1957) suggests that there is no inherited preference in this respect, and the preference noted by Müller probably therefore reflects the prevailing colours of the flowers adapted to pollination by the longer-tongued species. The colour preference of honeybee scouts, which is probably inherited, was investigated by Butler (1951), using white, green, pink, blue and yellow papers; blue and yellow were almost equally attractive, and much more so than the other colours.

Bumblebees, like honeybees, may restrict their foraging to a small area in a large crop (Free & Butler, 1959) and, also like honeybees, they may persist in working one crop without giving attention to others, with resulting neglect of superior crops (Kugler, 1943, p.302). A very important difference from honeybees is shown by bumblebees in their almost complete lack of communication of information about food sources. The only trace of communication arises when a bee sees others on flowers; then it may be induced to alight on the same flower, or one nearby (Brian, 1957).

According to a review article by Brian (1954) bumblebees in general work more quickly than honeybees, but the different species have different rates of working which increase in proportion with tongue-length.

Although bees are, on the whole, extremely reliable pollinators, their efficiency, especially that of the social species, appears to have gone beyond what is best for flower-pollination. The biting of holes in flowers and other kinds of illegitimate visiting are obviously bad for the plants (Figs. 6.48 & 6.49). In addition, the frequent concentration of social bees on very limited foraging areas means that, considering the number of flower-visits made, they contribute rather meagrely to outbreeding among plants. The organisation of the honeybee colony makes for the ruthless exploitation of every food source. Social life makes possible communication, co-operative effort and division of labour. The honeybee seems more or less to have taken charge of its floral environment, and there is perhaps here a faint parallel with the control of the environment achieved by man. The evolution of an almost total dependence on flower food by bees means that flowers in general must provide entirely for the nourishment of these insects. Of this nourishment, moreover, each species of bee-pollinated plant must provide a supply worth exploiting, if it is to retain its pollinators. The floral ecosystem can, however, bear a limited amount of cheating by plants (see Chapters 7 & 10) as it does cheating by bees.

If the proboscis lengths of the Hymenoptera as a whole are compared with those of Diptera, a very similar picture is found in each order. In both, there is a large number of quite unspecialised short-tongued visitors to flowers and other food sources, some of these insects being occasionally induced by special devices to visit highly specialised flowers. Then in both orders there are groups which feed exclusively on flowers, and they mostly have tongues of moderate length, a few having really long ones. However, long-tongued insects, it should be noted, are much commoner among the bees than among the flies.

Fig. 5.20 Flowers of marsh marigold, *Caltha palustris*. These flowers show no sign of a guide-mark pattern when photographed by visible light, as in the top photograph. The lower photograph, taken through a filter which blocks the visible spectrum but passes ultraviolet, shows that the centre of the flower is almost black to UV, and will appear yellow to a bee while the edges appear 'bee-purple'. Thus the flowers which are uniformly yellow to our eyes appear in brightly contrasting colours to visiting insects able to see in the UV region of the spectrum. The buttercups (*Ranunculus* spp., Fig. 2.9) which belong to the same family as *Caltha* (*Ranunculus*) show a similar pattern. Many other flowers have guide-mark patterns which either appear only in UV, or are much stronger in the UV part of the spectrum. Photographs by Adrian Davies.

6

The Diversity of
Insect-Pollinated Flowers

Introduction

A glance at a range of the flowers to be found in any region of the world quickly shows two things. First, many plant families themselves embrace a wide range of flower form. Examples are the buttercup family, Ranunculaceae, including flowers as different as the buttercups (*Ranunculus*), the meadow rues (*Thalictrum*) and the monkshoods (*Aconitum*), the figwort family (Scrophulariaceae) which includes the mulleins (*Verbascum*), foxgloves (*Digitalis*) and louseworts (*Pedicularis*), and that most diverse in flower form of all plant families, the orchids (Orchidaceae). Much of this diversity is obviously related to adaptation to different pollinators within a particular family or genus; it reflects what has been called *adaptive radiation*. Second, there are certain broad patterns of form and structure that appear repeatedly in different families of flowering plants. There are simple, superficially buttercup-like flowers in various different families. Similarly, a number of families produce large, flat heads of small white flowers, the best known being the carrot family, (Apiaceae; umbellifers). Many families include genera with large, bilaterally-symmetrical flowers pollinated by bumblebees. Such *syndromes* of form and adaptation suggest that flower-insect relationships offer particular possibilities which different families of flowering plants have evolved independently to exploit (van der Pijl, 1961). In terms of pollination mechanisms, they offer some striking instances of *evolutionary convergence*. Flowers are a prime reminder that we are almost always looking not at a tidily-branching evolutionary tree, but at an evolutionary bush or thicket, with the 'shoots' sometimes diverging, sometimes parallel from the same or different origins, sometimes tangled and intertwined.

Floral tubes and the 'concealment of nectar'

Several broad trends of adaptation to specialised pollinators have evidently oper-

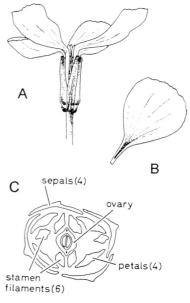

Fig. 6.1 Wallflower (*Erysimum cheiri*). **A**, whole flower. **B**, petal. **C**, transverse section of flower tube, showing that there are two nectar passages.

Fig. 6.2 **a, b** Male and female
empid flies (*Empis* cf. *pennipes*),
on flower of herb robert
(*Geranium robertianum*). Notice
the pollen grains on the insects.

a

b

ated very widely among the families of flowering plants. One of the most important
of these was described by the early-twentieth-century floral biologists in terms of
'concealment of nectar'. Leppik (1957) described another facet of essentially the
same evolutionary process in terms of the development from two-dimensional 'actinomorphic' and 'pleomorphic' flowers to three-dimensional 'stereomorphic' flowers.
In these, the nectar is so placed that it can be reached only by insects with a tongue of
some length, and the flower can only be worked quickly and effectively by an insect
with considerable powers of perception of three-dimensional form.

Leppik used the term 'actinomorphic' in a narrow sense for flowers with stamens,
pistil and nectar at the same level, with radial symmetry, and with relatively large and
variable numbers of floral members, and 'pleomorphic' for flowers with floral parts
(especially the petals) in definite numbers, a feature which he considered important in
the recognition of these flowers by visiting insects (Leppik, 1956). But the development of regular numbers of floral members undoubtedly opened up possibilities of
more precise and integrated further evolution of flower form, and this in itself may
have been an important selective pressure in its favour. In general, flowers with calyx-
tubes or corolla-tubes, or showing notably precise adaptation to their pollinators, have

small and regular numbers of calyx or corolla lobes.

Although union of perianth parts is especially characteristic of flowers pollinated by long-tongued insects, adaptation to these visitors has been achieved by a wide variety of plants in which the sepals and petals are free from one another. In the wallflower (*Erysimum cheiri*) the sepals are elongated and pressed firmly together, forming a long narrow tube (Figs. 6.1, 5.10). The petals are modified accordingly, having an expanded blade surmounting a slender claw. The flowers are freely visited and pollinated by long-tongued bees. Many other large-flowered members of the same family (Brassicaceae; crucifers), such as cuckoo flower (*Cardamine pratensis*) and the cultivated brassicas, have similar flowers, pollinated mainly by bees or sometimes by Lepidoptera. A similar floral arrangement in a different family, with flower parts in fives instead of fours, is seen in herb robert (*Geranium robertianum*) (Fig. 6.2) and related species. This is an interesting case, because the nearly related cranesbills which make up most of the genus *Geranium*, including the large-flowered bee-pollinated species such as meadow cranesbill (*G. pratense*) (Fig. 2.15), have open bowl-shaped flowers with the petals hardly clawed at the base. In the buttercup family (Ranunculaceae), the larkspurs (*Delphinium* spp.) have petaloid sepals, one of which is produced into a long hollow spur, while in the columbines (*Aquilegia* spp.) a long spur is formed by each of the five petals.

A fairly common way in which the flower tube is lengthened without union

Fig. 6.3 Honeybee collecting nectar from flower of raspberry (*Rubus idaeus*).

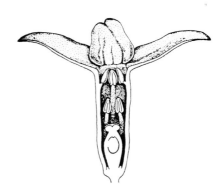

Fig. 6.4 Mezereon (*Daphne mezereum*). After A.H. Church.

of the sepals or petals is deepening of the receptacle so that it forms a tube itself. In the wild cherry (*Prunus avium*) (Fig. 2.3) the numerous stamens add to the depth of the receptacle, which is only 3-4 mm long. The receptacles of the blackcurrant (*Ribes nigrum*) and the gooseberry (*R. uva-crispa*) are even shorter, and their effective depth is slightly augmented by the small erect petals of the flowers. In raspberry (*Rubus idaeus*) the petals and stamens set round the edge of the nearly flat receptacle combine to achieve the effect of a short tube (Fig. 6.3). A very deep receptacle is found in purple loosestrife (*Lythrum salicaria*) (Fig. 12.3). The receptacle is proportionately deeper still in species of *Daphne* and other members of its family (Thymeleaceae); there are no

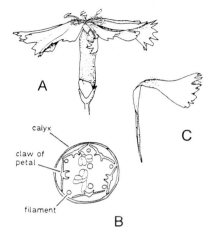

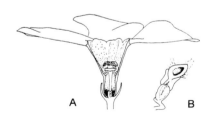

Fig. 6.5 Cheddar pink (*Dianthus gratianopolitanus*). **A**, side view of flower. **B**, single petal. **C**, transverse section of flower-tube about halfway up the calyx. The channel on the claw of each petal is probably a proboscis guide; it fades out below. The stamens elongate in succession from the bottom of the tube; five cut filaments and two anthers just below the cut are shown in **C**, but three even shorter stamens are omitted. The nectaries are at the bases of the stamens.

Fig. 6.6 Lesser periwinkle (*Vinca minor*). **A**, side view of flower with two corolla-lobes and the corresponding part of the tube removed. **B**, stamen. The stamens and style are very specialised in structure. The pollen is shed in coherent masses on the non-receptive top surface of the style. It will adhere to the tongue of a visiting insect only after this has been made sticky by the secretion from the receptive zone which encircles the widest part of the head of the style.

petals and the whole of the receptacluar tube and sepals are often petaloid, as in mezereon, *D. mezereum* (Fig. 6.4). But undoubtedly the most spectacular of all receptacular tubes are those of the big, gaudy flowers of the cactus family (Cactaceae).

There are two important groups of 'long-tubed' flowers in which the sepals are united into a tube while the petals remain separate. One of these is that part of the family Caryophyllaceae comprising the pinks, campions and their allies. An example is the Cheddar pink (*Dianthus gratianopolitanus*) (Fig. 6.5), which is similar to the wallflower apart from its united sepals and greater number of floral parts. The pinks, with their slender-tubed flowers, are adapted to pollination by Lepidoptera (Erhardt, 1990), whereas the red campion (*Silene dioica*) (Fig. 12.10, Plate 3a), with a shorter and wider tube, is adapted to bees and long-tongued flies. The second group comprises the vetches, peas and their many allies in the pea family (Fabaceae). They are like the pinks and campions in having five petals which are clawed so that they fit into the tubular calyx, but differ from them in their bilateral symmetry and marked inequality of their petals; they will be considered later.

Familiar examples of flowers with united petals as well as united sepals are the primrose (*Primula vulgaris*) and its relatives (Fig. 12.1) and the periwinkles (*Vinca* spp.; Fig. 6.6). The corolla in these consists of a slender tube within the calyx and a flat 'limb' or disk on which visiting insects can alight. There are many flowers of this form in the olive family (Oleaceae), including the jasmines (*Jasminum* spp.), *Forsythia* (Fig. 12.2) and lilac (*Syringa vulgaris*), and in the gentian family (Gentianaceae), such as centaury (*Centaurium erythraea*) and spring gentian (*Gentiana verna*). The borage family (Boraginaceae) shows great variation in the relative proportions of the tube and limb. The hanging, tubular bumblebee flowers of comfrey (*Symphytum officinale*;

Fig. 6.49) are at one extreme; at the other are the bright-blue 'rotate' flowers of green alkanet (*Pentaglottis sempervirens*) and the forget-me-nots (*Myosotis* spp.), with a very short tube and a broad flat limb. Between these two extremes lies the houndstongue (*Cynoglossum officinale*), with its dingy purple cup surmounting a short corolla tube.

A variation on these tubular flowers is seen in those species with broad corollas contracted to a narrow mouth. Flowers with 'urceolate' corollas of this kind are common in the heather family (Ericaceae). The small flowers of bell heather (*Erica cinerea*) are probably mainly pollinated by bees. Butterflies can reach the nectar in intact flowers but are probably not effective pollinators; bumblebees commonly perforate the corollas. Cross-leaved heath (*Erica tetralix*), with slightly larger flowers, is also visited mainly by bees, including the long-tongued bumblebee *Bombus pascuorum*. Hagerup &

Fig. 6.7 Bumblebee (*Bombus terrestris*, female), feeding at flower of bilberry (*Vaccinium myrtillus*).

Hagerup (1953) believed that *E. tetralix* is usually pollinated by thrips or selfed; this may be true in some places, but is certainly not generally so (Haslerud, 1974). Many larger flowers of this type, for instance bilberry (*Vaccinium myrtillus*) (Fig. 6.9) and St Dabeoc's heath (*Daboecia cantabrica*), are visited by bumblebees, which 'buzz' the flowers for pollen (pp.125, 179).

Sometimes flowers with tubular corollas have the sepals free or much reduced. Examples are sweet woodruff (*Galium odoratum*), the honeysuckles (*Lonicera* spp.), many of the valerian family (Valerianaceae) and the bellflowers (Campanulaceae). It is noteworthy that these are all flowers with inferior ovaries and all are flowers which (for

Fig. 6.8 Bumblebee (*Bombus pascuorum*), visiting flower of bluebell (*Hyacinthoides non-scripta*).

Fig. 6.9 Honeybee stealing nectar from bluebell (*Hyacinthoides non-scripta*).

various reasons) have a less evident need than many for the extra support at the base of the corolla or protection from nectar robbing that a tubular calyx might help to provide (see p.171).

There are some interesting instances of functionally tubular flowers among the monocotyledons, comparable with the wallflower and herb robert described above. The bluebell (*Hyacinthoides non-scripta*) (Fig. 6.8) has six free perianth segments, which are pressed tightly together to form a tube about 1 cm long. The flowers are pendent, and are especially visited by bumblebees; another pollinator is the long-tongued hoverfly *Rhingia campestris*. Honeybees, on the other hand, sometimes alight on the outside of flowers and push their tongues between the perianth segments, thereby evading the pollination mechanism (Knight, 1961) (Fig. 6.9). In the nodding star-of-Bethlehem (*Ornithogalum nutans*), the stamen filaments are very broad, and overlap to form a tube similar in shape to the flower of the bluebell, while the perianth segments are held well away from the staminal tube and serve for display. The long-tongued bee *Anthophora* and the bumblebee *Bombus pratorum* have been seen visiting the flowers in cultivation (PFY). Related plants in which the perianth segments are united are the grape hyacinths (*Muscari* spp.), with small, usually blue, urceolate flowers in a tight raceme, and the Solomon's seals (*Polygonatum* spp.), which have flowers functionally similar to those of the bluebell.

Generally speaking, the form of these tubular flowers makes for greater precision in the behaviour of visiting insects than the flowers illustrated as examples at the end of Chapter 2. However, as they are radially symmetrical the orientation of the visiting insect in relation to the flower matters little. Apart from the special case of the pendulous flowers (largely visited by bumblebees), the flowers are variously disposed on the plant – often erect or inclined more-or-less indiscriminately at various angles.

Trumpet- and bell-shaped flowers

A rather different line of development is shown by the trumpet- and bell-shaped flowers, which an insect must crawl inside to feed and to bring about pollination. Here the emphasis is on adaptation to the body-form of the pollinator rather than simply to the length of its mouth-parts, though often there are also adaptations preventing short-tongued insects from reaching the nectar. Of course many simple cup- and bowl-shaped flowers are adapted in a general way to the size of their usual pollinators, and no completely sharp line can be drawn between these flowers and those with a deeper trumpet-shaped or bell-shaped corolla. Nevertheless, among the dicotyledons most of the flowers of the kind we are considering here have corollas in which the petals are fused. Among the monocotyledons this generalisation does not hold, and trumpet- and bell-shaped flowers include (among other examples) lilies and fritillaries with free perianth segments.

A common trumpet-shaped flower is the field bindweed (*Convolvulus arvensis*) (Figs. 6.10, 11). The short-lived white or pinkish flowers are about 2 cm long and the corolla, which forms a narrow tube at the extreme base, flares out to a diameter of up to 3 cm. The five stamens closely surround the style so that, in effect, the style and stamens form a short column in the centre of the flower. Nectar is secreted at the base of the ovary, but can be reached only through five narrow passages between the broad bases of the stamens. The anthers dehisce outwards, so that their pollen immediately comes into contact with the body of a visitor; the two stigma lobes project beyond the stamens,

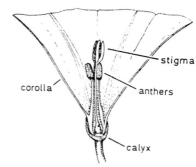

Fig. 6.10 Flower of field bindweed (*Convolvulus arvensis*) with half of calyx and corolla removed.

so that an insect bearing pollen from another flower will readily bring about cross-pollination as it enters the corolla. Bees and a variety of flies are the chief pollinators. The degree of variation in corolla colour and other characters suggests that much cross-pollination must take place.

The structure and mechanism of the larger hedge bindweed (*Calystegia sepium*) is similar. The often-quoted supposition that this species is adapted to pollination by hawkmoths has no more than a tenuous foundation. None of the great nineteenth-century floral biologists ever seems to have observed such visits himself. Müller writes,

a b

Fig. 6.11 **a–b** Insects visiting field bindweed (*Convolvulus arvensis*); **a**, solitary bee (*Lasioglossum* sp.), foraging for nectar. **b**, hoverfly (*Episyrphus balteatus*), feeding on pollen; note passages to nectary between bases of stamens.

Fig. 6.12 Flower of trumpet gentian (*Gentiana acaulis*), with half of calyx and corolla removed, lit from behind to show translucent corolla-tube. The corolla is divided at the base into five deep tubes by the insertions of the stamens.

'Delpino mentions *Sphinx convolvuli* as a fertiliser of *C. sepium*; he tells me by letter that one of his friends catches this insect in numbers, standing by a hedge overgrown with the plant, holding thumb and forefinger over a flower and closing its orifice when the insect has entered!' The convolvulus hawkmoth is so large an insect relative to the size of the bindweed flowers that it seems much more likely that some smaller moth was observed – even though in North America the large trumpet-shaped flowers of *Datura meteloides* are, indeed, visited by the hawkmoth *Manduca sexta* (Baker, 1961). Certainly the large size and pure white colour of the flowers of the hedge bindweed suggest a crepuscular pollinator, and daytime visitors are relatively scarce considering how conspicuous the flowers are. However, the flowers open soon after sunrise, and are visited by considerable numbers of insects, especially bees and hoverflies; on sunny days the pollen is often removed by mid-morning. Some flowers close in the evening, others remain open through the night and close the following day (Stace, 1965). Observations of visits by smaller moths seem to be conspicuously lacking. There can be no doubt that in Britain the flowers are effectively pollinated by hoverflies (Baker, 1957) or bumblebees, particularly *Bombus pascuorum* (Stace, 1965). Generally, failure to set seed is due to self-incompatibility within vegetatively-reproduced populations (Stace, 1961) rather than lack of pollinators.

Flowers of essentially the same type, though with a more bell-shaped corolla, are found in the Alpine trumpet gentians or 'stemless gentians' (*Gentiana acaulis* [Fig. 6.12], *G. clusii* and their allies) and the swallow-wort gentian (*G. asclepiadea*) of Alpine forest margins; the marsh gentian (*G. pneumonanthe*) of lowland damp heaths is similar (Petanidou *et al.*, 1995). These are bumblebee flowers. Unlike the bindweeds, the gentians open for a number of days in succession, closing at night and in dull weather, and they are protandrous which further favours cross-pollination. The interior of the throat of the flower is whitish, in contrast to the deep blue of the lobes at the periphery, so the interior looks light to a bee entering the mouth of the flower. This feature, in various forms, is characteristic of many flowers with tubes or bell-shaped corollas. In *G. pneumonanthe*, and even more strikingly in the commonly-cultivated east-Himalayan *G. sino-ornata*, translucent 'window-panes' form prominent stripes up the side of the tube (Fig. 6.13).

The bellflowers (*Campanula*) have much in common with the flowers we have just considered, but show some interesting differences in detail. Many, like the common harebell (*C. rotundifolia*) (Fig. 6.14), have pendulous flowers; this in itself favours the agile and specialised bees as pollinators, and bees are much the most frequently-ob-

Fig. 6.13 Flower of *Gentiana sino-ornata* lit from behind to show translucent corolla-tube.

served visitors. The flowers are protandrous, and functionally similar to those of the bindweeds or trumpet gentians, with the nectar concealed in the same way by the expanded bases of the stamens. However, the pollen is not transferred direct from the stamens to visiting insects. In the young bud the anthers closely surround the hairy tip of the style, onto which they shed their pollen before the flower opens. In the newly-opened flower the slender filaments of the stamens have already shrivelled, and the pollen adhering to the tip of the style is ready to be picked up on the body of a bee crawling into the bell – an elegant example of 'secondary pollen presentation' (Yeo, 1993). After some days the style branches diverge, exposing the three receptive stigmas which take up the position hitherto occupied by the pollen. Before the flower withers, the stigmas curve back so far that they touch any remaining pollen, so self-pollination may take place if insect visitors fail.

Little need be said about the mechanism of the large bell-shaped flowers among the monocotyledons. Most of these are in the lily family (Liliaceae), and generally function in much the same way as the bindweeds, though the stamens and stigmas are often farther apart. Nectar is secreted in grooves near the bases of the perianth segments. The flowers are usually protogynous, with the style generally slightly longer

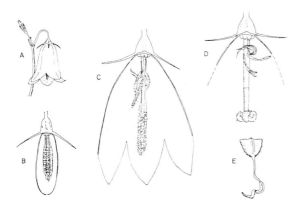

Fig. 6.14 Harebell (*Campanula rotundifolia*). **A**, part of inflorescence. **B**, bud at time of dehiscence of the stamens. **C**, newly-opened flower. **D**, style and stamens of an older flower; the style-branches have reflexed and the stigmas are now receptive. **E**, a single stamen from an open flower, showing the expanded base and the shrivelled filament and anther.

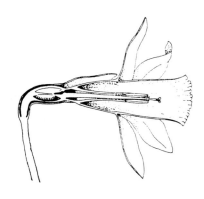

Fig. 6.15 Daffodil (*Narcissus pseudonarcissus*). Half-section of flower. After A.H. Church (from a cultivated form).

than the stamens, and are typically visited by various bees. Richard Bradley's classic early-eighteenth-century observations on cultivated tulips (*Tulipa*) were mentioned in Chapter 1. The flower of the fritillary or snake's head (*Fritillaria meleagris*) is a hanging bell of six free perianth segments, pink in colour, chequered with darker purple. The flowers are visited by bumblebees. The flowers of *Crocus* species (Iridaceae) superficially resemble those of *Colchicum*, but like other members of their family, have only three stamens. They open widely in warm sunshine, and bees often alight on the column formed by the three stamens and the style in a way reminiscent of their behaviour on the open bowl-shaped flowers of meadow cranesbill (Fig. 2.15).

The wild daffodil (*Narcissus pseudonarcissus*) (Fig. 6.15) is another flower pollinated by early-flying bumblebees, but in this case the tubular part of the nodding flower is made up partly of the perianth tube, about 18 mm long, and partly by a tubular outgrowth, the corona, extending for some 30 mm above the level of the spreading perianth lobes. The large queens of *Bombus terrestris* just fill the bell, forcing the anther column and the style to one side, while the rather short tongue of the bee can just reach the nectar around the base of the style through the narrow spaces between the broad bases of the filaments. Various other bees, for example the long-tongued *Anthophora plumipes*, and drone flies (*Eristalis*) can also reach the nectar and can pollinate the flowers. The flowers are fertile to their own pollen, but apparently few seeds are set in the absence of insect visits, so the amount of seed depends greatly on the weather in March and early April when the plants are flowering (Caldwell & Wallace, 1955).

Several of the flowers described in this section are what Kerner called 'revolver flowers'. An insect entering the flower is faced with a ring of narrow tubes – like the barrels in the chamber of a revolver – through which it must probe to reach the nectar. As it sucks the nectar, it may take up any one of the corresponding positions within the flower, and will often move round to feed at several positions in succession. The classic examples are such flowers as the bindweeds and trumpet gentians, but the same principle can be seen in the columbines (*Aquilegia*) and in incipient form even in many simple tubular flowers, where the 'barrels' of the 'revolver' are separated by the stamens or ridges on the corolla, as in herb robert or wallflower. Possibly such a construction serves to detain the insect longer and leads to more effective pollination. All these flowers are tending, in a sense, to become multiple pollination units.

There are some more extreme developments of multiple pollination units from single flowers. In the turk's-cap lily (*Lilium martagon*) there is an open tube on each of the perianth segments, formed by a furrow covered in by flanges (Fig. 6.16) and leading to the nectary at its base. These tubes are very narrow, and suit the slender tongues of Lepidoptera, such as the hummingbird hawkmoth (*Macroglossum stellatarum*), which probe for nectar as they hover close to the exserted stamens and stigmas. The yellow flag (*Iris pseudacorus*) resembles the martagon lily in having multiple tubes giving access

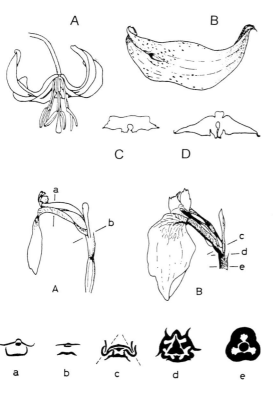

Fig. 6.16 Turk's-cap lily (*Lilium martagon*). **A**, side view of young flower with one perianth segment removed. **B**, single perianth segment, showing entry to nectar-groove. **C**, cross section of perianth segment, one-third of the way up the groove. **D**, the same, three-quarters of the way up the groove. The stamens, and later the style, bend upwards towards the perianth segments.

Fig. 6.17 Yellow flag (*Iris pseudacorus*). **A**, side view of one of the three floral units. **B**, three-quarter front view of same. a–e, sections at the levels indicated in A and B.

Fig. 6.18 Bumblebee (*Bombus hortorum*), sucking nectar from flower of yellow flag (*Iris pseudacorus*).

to the nectar. Here, however, there are three tubes instead of six, three of the perianth segments (the 'standards') serving for display only (Fig. 6.17). The lower side of each tube is formed by the narrow claw or 'haft' of one of the other perianth segments (the 'falls'), each of which has a large free blade on which insects can alight. Lying over the haft of each fall is a greatly expanded and flattened style, looking like a petal and forming the upper side of the tube, and arching over a single anther. The tube is large enough for the bumblebees or long-tongued flies which pollinate the flowers to crawl right in (Fig. 6.18). At the level of the section in Fig. 6.17d, the tube is divided by the filament into two narrow channels which contain nectar. It is clear that the functional unit is not the whole *Iris* flower but a third of it; what is particularly interesting is the remarkable functional analogy between each of the three individual tubes and the whole of one of the bilaterally-symmetrical 'zygomorphic' flowers considered in the next section.

Zygomorphy

The flowers that have achieved the closest adaptation in form to their pollinators are those that have become, like their pollinators, bilaterally symmetrical. These *zygomorphic* flowers have evidently evolved quite independently in many different families of flowering plants. Perhaps zygomorphy has most often arisen, in the first place, as an adaptation to pollination by bees. However, as we shall see, by no means all zygomorphic flowers are bee-pollinated.

In a radially-symmetrical flower an insect can take up any one of a number of positions in relation to the axis of the flower. In a zygomorphic flower the insect tends always to take up a single position, and it is this which allows much more precise adaptation of the flower to particular pollinators. By contrast with radially-symmetrical flowers, zygomorphic flowers are usually placed more-or-less horizontally, and their orientation on the plant generally varies little.

In a zygomorphic flower the stamens and style may be so placed that they come into contact with the underside or with the upperside of the visitor. The first arrangement, called *sternotribic*, is found in the peaflowers (Fabaceae) – the vetches, peas and their allies – and in a number of smaller groups of zygomorphic flowers. Most of these come into the 'flag-blossom' category of Faegri & van der Pijl (1979). The second arrangement, with the pollen normally transferred on the upperside of the visitor, is called *nototribic*. It is particularly characteristic of the deadnettle family (Lamiaceae) and several other large and important families with gamopetalous corollas, such as the figwort family (Scrophulariaceae), the largely tropical families Acanthaceae and Gesneriaceae, and of the orchids. Many of these fall into Faegri & van der Pijl's category of 'gullet blossoms'.

Monkshoods, vetches and other sternotribic flowers

The monkshoods (*Aconitum* spp.) are particularly interesting in this context because they belong to the buttercup family (Ranunculaceae), most of which have radially-symmetrical flowers, and which are by common consent regarded as evolutionarily relatively primitive. The flowers of monkshoods are beautifully adapted in size and form to pollination by bumblebees, and in fact are a classic case of complete dependence on bumblebees for pollination. The monkshoods are closely related to the hellebores (Fig. 2.10), and, although superficially so unlike them, the real differences betwen the flowers are few. The five sepals are strongly coloured, the uppermost forming a large erect helmet-shaped hood, covering the two upper nectaries, which are greatly enlarged (Fig. 6.19). The remaining nectaries are small, or absent altogether. The flowers are protandrous. The young stamens are bent downwards, but stand erect as they mature and dehisce, and then bend back out of the way, so that the underside of a bee landing on the lower sepals and clambering up to reach the nectaries is well dusted with pollen. After the stamens have all dehisced the maturing stigmas are exposed and pollination can take place. The mechanism of the monkshoods is seen at its most highly developed in the beautiful yellow wolfsbane (*A. vulparia*) of the Alps and northern Europe. The helmet is tall and narrow, and the nectaries are some 20 mm long, with spirally-coiled spurs secreting copious nectar which can be exploited only by a few of the longest-tongued bumblebees.

The legumes (Fabaceae), with their familiar pea-flower corollas, are a large, varied, and very important family. Usually the flowers have a more-or-less tubular calyx, divided at the tip into five longer or shorter lobes. The corolla consists of five free petals, of which the uppermost, known as the *standard*, is usually large and conspicuous. The two petals below this are called the *wings*, and the two remaining petals below and between them are pressed together and folded over one another to form a boat-shaped structure en-

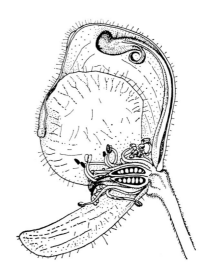

Fig. 6.19 Monkshood (*Aconitum napellus*); half-section of flower.

closing the stamens and ovary, known as the *keel* (Figs. 6.20, 28). The petals are more or less firmly interlocked at the bases of their blades by folds, projections, or zones where the cells interlock or adhere, the details varing from one species to another. The lower nine stamens are fused to form a tube, within which the nectar is accumulated; the tenth stamen is usually free, so allowing access to the nectar from the upper side. A visiting insect usually clings to the wings, and inserts its proboscis between the standard and the upper edges of the keel, sliding it down to reach the nectar in the base of the stamen tube.

The simplest type of mechanism is found in flowers like those of sainfoin (*Onobrychis viciifolia*) or the melilots (*Melilotus* spp.) (Figs. 6.20, 5.14). The upper edge of the keel is open, so that when a visitor forces its way into the flower the wings and keel are pressed down, uncovering the relatively rigid stamens and style which come into contact with the underside of the body of the insect. As the insect leaves the flower the wings and keel spring back into place again, once more covering the stamens and stigma. The flowers of the bee-pollinated clovers (Figs. 6.21, 5.12) work similarly, but the lower parts of the petals and the stamen tube adhere strongly, forming a long

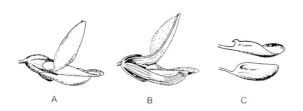

A   B   C

Fig. 6.20 Common melilot (*Melilotus officinalis*). **A**, single flower seen from the side. **B**, flower with half of calyx and corolla removed to show stamens and stigma. **C**, wing and keel petal, seen from outside.

Fig. 6.21 **a–b** Bumblebees (*Bombus* sp.), on red clover, *Trifolium pratense*; notice the length of the tongue as the bee leaves the flower in the second picture.

a

b

Fig. 6.22 Bush vetch (*Vicia sepium*). **A**, flower with half of calyx and corolla removed. **B**, detail of base of stamen-tube, showing the openings to the nectary on either side of the uppermost filament. **C**, tip of ovary and style, showing stylar brush and stigma. **D**, wing petal seen from outside; the broken line indicates the outline of the lower part of the keel. **E**, keel petal seen from outside; the region of adhesion to the wing is shaded.

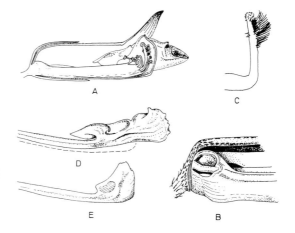

narrow tube through which the bee must probe to reach the nectar and leaving only the upper parts of the wings and keel free to move.

The vetches (*Vicia* and *Lathyrus*) are similar, but have a secondary pollen-presentation mechanism reminiscent of that in *Campanula*. The style is bent up sharply from the tip of the ovary and carries a dense brush of fine hairs below the stigma. The anthers dehisce in the bud, and the pollen is shed onto the hairs of the brush, or into the tip of the keel where the brush sweeps it out as the keel is depressed. By the time the flower opens the anthers have retracted, but the stigma brush is fully charged with pollen as it comes into contact with the underside of a visiting bee. The mechanism is well shown in the bush vetch (*Vicia epium*) (Figs. 6.22, 23). The petals are relatively large and stiff and their bases form a rather long tube; it takes a powerful, long-tongued insect to reach the nectar and the

Fig. 6.23 Bumblebee (*Bombus pascuorum*) on bush vetch (*Vicia sepium*).

flowers are mainly visited by bumblebees. Although the flowers are normally held more-or-less horizontally, they often hang vertically as a heavy bumblebee clings to them.

In some common genera, the two keel petals do not part to expose the style and stamens when the keel is depressed by a visitor. Instead, the pollen is shed into the conical end of the keel, whose edges adhere except for a small hole or slit at the tip. When the wings and keel are pressed down by a visiting insect, the stamens beneath act as a piston, forcing out a string or ribbon of pollen like toothpaste from a tube onto

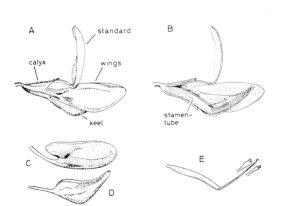

Fig. 6.24 Birdsfoot trefoil (*Lotus corniculatus*). **A**, flower seen from the side.
B, flower with half of calyx and corolla removed.
C, wing petal. **D**, keel petal. **E**, ovary and style, and the tips of two of the longer stamens.

Fig. 6.25 **a–b** Bees visiting birdsfoot trefoil (*Lotus corniculatus*); **a**, bumblebee, *Bombus lapidarius* (worker). **b**, honeybee; the bee has approached the flower slightly from one side, and the keel of the flower with the projecting tip of the style can be seen in front of the bee's abdomen.

a

b

A

B

C

D

Fig. 6.26 Lucerne (*Medicago sativa*). **A**, young flower seen from the side. **B**, a 'tripped' flower with half of the calyx and corolla removed to show the stamens and stigma. **C**, wing petal. **D**, keel petal.

the underside of the visitor. In due course, the stigma protrudes through the slit and pollination can take place. The details of the mechanism in birdsfoot trefoil (*Lotus corniculatus*) are shown in Figs. 6.24, 25. A similar arrangement is found in kidney vetch (*Anthyllis vulneraria*), horseshoe vetch (*Hippocrepis comosa*), the restharrows (*Ononis* spp.) and the lupins (*Lupinus* spp.).

In several genera, there is an explosive pollen-presentation mechanism. Lucerne (*Medicago sativa*) (Fig. 6.26) is the most similar to the simple type of flower of the melilots and clovers. The stamen tube is held under tension between the keel petals by a pair of hollow projections on their upper edges, and to some extent by projections from the upper edges of the wings. The pressure exerted by a visiting

Fig. 6.27 **a–c** Honeybees visiting flowers of common gorse (*Ulex europaeus*). In **a**, the bee is forcing an entry into a fresh flower; in **c** it is leaving the flower following the 'explosion' of the stamens and style from the keel.

bee dislodges these projections, releasing the stamen tube and the style, which spring up, striking the underside of the visitor. In gorse (*Ulex* spp.) (Fig. 6.27), the two keel petals of a newly-opened flower adhere lightly together by their upper edges; in this case it is the keel that is held straight by the stamen tube and the style, rather than vice versa. The stamens dehisce just before the flower opens. The flowers are nectarless, but they are often freely visited by bumblebees and honeybees which forcibly enter the flower as though seeking nectar. This action causes the keel petals to break apart, uncovering the stamens and style and bringing them sharply into contact with the insect, so dusting it with pollen on the underside of the abdomen. Once 'exploded'

Fig. 6.28 Broom (Sarothamnus scoparius). **A**, newly-opened flower with half of calyx and corolla removed to show the position of the stamens and style, held under tension by the keel. **B**, flower 'exploded' following an insect visit.

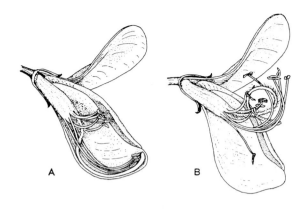

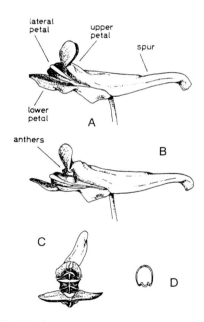

Fig. 6.29 Corydalis cava. **A,** side view. **B**, flower opened as if being probed by an insect, revealing anthers. **C**, front view. **D**, cross-section of spur showing median nectar-groove. The lateral petals enclose the stamens and style like the keel petals of peaflowers (Fabaceae); their claws are fused to the upper edges of the lower petal. The sepals drop when the flower opens.

the spent flower hangs limply open and is seldom visited again by bees. The mechanism of the broom flower (Sarothamnus scoparius) (Fig. 6.28) is similar, but a little more complicated and much more vigorous. Of the ten stamens, five are shorter and strike the bee on the underside, while five are longer and, with the long curved style, commonly strike the bee on the back of the abdomen.

It is inherent in most of these pollination mechanisms in papilionate flowers that the stigma lies close to its own pollen; in some, the stigma is actually embedded in pollen when the flower opens. In some cases, the stigma is not receptive until it has been abraded. Bird's-foot trefoil and a number of the larger-flowered clovers are known to be self-incompatible, and sainfoin is largely so. There are strong indications of self-incompatibility in many other species (although some are known to be self-compatible and regularly selfed), and this is probably the main factor controlling the breeding system throughout the family.

Species of the fumitory family (Fumariaceae) have – apparently quite independently – evolved floral mechanisms remarkably like some of those just described in the Fabaceae. The flower of Corydalis cava, a central and south-European species occasionally found in Britain as a garden escape, is shown in Fig. 6.29. There are four petals, of which the uppermost is the largest and has a long spur at its base. The spur receives the nectar secreted by a long, backward-pointing process from the upper filaments. The two lateral petals are curved inwards at their margins and fused at the tip, so that they form a sheath or hood enclosing the rigid style. The stigma is large and lobed and is covered with pollen by the stamens which dehisce and wither before the flower opens. The bumblebees that visit the flowers depress the hood as they probe for nectar in the spurred upper petal. In young flowers they dust themselves with pollen in the process; in older flowers they may leave pollen on the now-receptive stigma. The flowers appear to be entirely self-incompatible. The whole mechanism of *Corydalis cava* bears a strik-

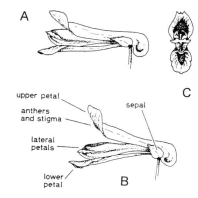

Fig. 6.30 Yellow corydalis (*Pseudofumaria lutea*). **A**, side view of newly-opened flower. **B**, flower 'exploded' after an insect visit; notice the anthers lying against the upper petal. **C**, front view.

ing similarity to that of the vetches. The yellow corydalis (*Pseudofumaria lutea*) (Fig. 6.30), a familar garden plant native in the southern Alps, has an explosive mechanism reminiscent of lucerne or gorse. It is visited by various bees, but is probably self-fertile. The fumitories (*Fumaria*) have similar flowers to the corydalises, but are not much visited and probably almost always self-fertilised.

Deadnettles and other nototribic flowers

The white deadnettle (*Lamium album*) (Fig. 6.31) is a good example of a nototribic flower. The corolla is two-lipped, the upper lip forming a hood over the style and the four stamens, and the lower, marked with a few greenish streaks and dots, forming a landing platform for insects. The lower part of the corolla forms a curved tube about 10 mm long; the nectary lies at the base of the ovary, and nectar accumulates in the narrow part at the base of the tube. The flowers are pollinated by long-tongued bees, especially bumblebees, which accurately fill the space between the lower lip and the stamens and style beneath the hooded upper lip as they suck nectar from the tube. There seems to be no barrier to self-pollination; how much outbreeding actually takes place has apparently not been estab-

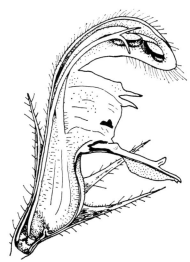

Fig. 6.31 White deadnettle (*Lamium album*). Flower with half of calyx and corolla removed to show the stamens, style and ovary.

162 THE NATURAL HISTORY OF POLLINATION

Fig. 6.32 Bumblebee (*Bombus pascuorum*, female), on flower of yellow archangel (*Lamiastrum galeobdolon*).

Fig. 6.33 Bumblebee (*Bombus pascuorum*, female), on hedge woundwort (*Stachys sylvatica*). The head of the bee comes into contact with the stamens and stigma beneath the hooded upper lip of the flower.

Fig. 6.34 Hoverfly (*Rhingia campestris*) sucking nectar from bugle (*Ajuga reptans*).

lished. Yellow archangel (*Lamiastrum luteum*) has bright yellow flowers very similar in shape, but with a rather shorter tube, so they can be exploited by a greater variety of bees. The speed and precision with which bumblebees visit these flowers and transfer the pollen is impressive to watch (Fig. 6.32).

Many other members of the deadnettle family (Labiatae or Lamiaceae) have similar long-tubed bumblebee flowers, though differing a good deal among themselves in colour and structural details. Some, especially those with smaller flowers and shorter tubes, are commonly visited also by honeybees and the smaller wild bees. There is very noticeable variation from one species to another in the part of the visitor with which the anthers and stigma come into contact (Fig. 6.33). This is easily seen when yellow archangel and bugle (*Ajuga reptans*) (Fig. 6.34) are growing together. A visiting bumblebee transfers the pollen of yellow archangel on its thorax and that of bugle on its head. The stamens of bugle, like those of wood-sage (*Teucrium scorodonia*) (Fig. 6.35), project beyond the very short upper lip, though in the case of bugle they obviously derive some protection from the bracts of the flower above. The functional protandry which is very common in Labiatae is readily

Fig. 6.35 Bumblebee (*Bombus lapidarius*, worker), on wood sage (*Teucrium scorodonia*). The flower that the bee is visiting is in the functionally male stage; the two flowers below are older, with the anthers recurved and the stigma projecting over the entrance to the flower.

seen in these two plants. In a newly-opened flower the stigma stands above the stamens and is not readily touched by a visiting insect. After the stamens have shed their pollen, the style bends down so that the stigma is exposed below the stamens at the entrance to the flower.

An elegant variation on the mechanism of the deadnettle flower is found in the sages (*Salvia*), and is seen at its most highly developed in such species as the bright-blue meadow sage (*S. pratensis*) (Fig. 6.36) and the dull-yellow *S. glutinosa* (Fig. 6.37), both common plants in central Europe. Only two stamens are functional, the other two being reduced to small vestiges. In the two functional stamens, the connective (the tissue between the two anther-lobes) is greatly elongated. One lobe of each stamen is normally developed and lies beneath the hood of the upper lip; the other is abortive, and the connective at its lower end forms an expanded blade partly blocking the entrance to the flower. The anther-filaments are reduced to short, flattened strips joining the connective to the corolla, and providing an elastic torsion joint about which the connective can hinge. When a bee pushes its head into the flower, the connective blades are pushed backwards and upwards, and the fertile anther lobes swing downwards bringing their pollen into contact with the abdomen of the bee. In older flowers, the mature stigmas project in front of the flower so that visiting bees rub against them, usually as they leave the flower. Not all *Salvia* species have the see-saw mecha-

Fig. 6.36 Half-section of flower of meadow sage (*Salvia pratensis*), showing the broadened lower end and the fertile anther cell on the upper end of one of the two fully-developed anthers, and the long style arching over the mouth of the flower.

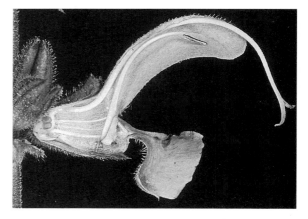

Fig. 6.37 Bumblebee (*Bombus pascuorum*), on flower of *Salvia glutinosa*; the fertile anther-cells have swung down into contact with the abdomen of the bee.

nism devoped to this degree of perfection. In the garden sage (*S. officinalis*), for instance, the connective is much shorter, and both anther-lobes produce pollen, though the lower lobes produce much less than the upper (Fig. 5.17). The mechanism works in the same way as that of *S. pratensis*, but suggests a more primitive condition. On the other hand, in the the scarlet-flowered hummingbird-pollinated *S. splendens* from Brazil, the lever mechanism has evidently been lost. The fertile anther-lobe is still borne on the end of a long connective, but the other end of the connective is not broadened or curved down to block the entrance; the pollen is simply rubbed off onto the head of the hovering visitor as it probes with its bill into the unobstructed mouth of the flower. The lever mechanism is also degenerate in the small-flowered species of *Salvia* that are habitually selfed, like the dingy-purple clary (*S. horminoides*).

Many flowers with comparable 'labiate' corollas are to be found in the figwort family (Scrophulariaceae). There are, in Britain and neighbouring parts of western Europe, two species of lousewort (*Pedicularis*), a scanty representation of a genus rich in species and extraordinarily varied in flower form and colour in North America, Scandinavia, and right through the high mountains of Eurasia, from the Pyrenees and the Alps to the eastern Himalayas, China and Japan. The heath lousewort (*P. sylvatica*) is a common plant of damp heath and moorland, flowering in spring and early sum-

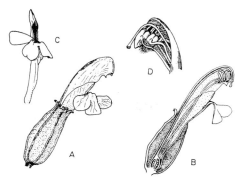

Fig. 6.38 Lousewort (*Pedicularis sylvatica*). **A**, side view of flower. **B**, half-section of flower. **C**, front view of flower. **D**, detail of anthers and upper part of style (semidiagrammatic).

mer. The pink flowers are two-lipped. The upper lip forms a narrow, laterally-flattened hood, enclosing the four stamens, with the stigma just protruding from the underside near the tip. The three-lobed lower lip forms a flat landing platform slightly oblique to the plane of symmetry of the rest of the flower. The two pairs of stamens face one another; the pressure of the sides of the hood keeps them pressed together, preventing the escape of pollen. The entrance to the upper lip forms a narrow slit some 8-10 mm long, usually somewhat widened for about 3 mm at its upper end (where it is separated from the pore through which the style protrudes by two narrow teeth), and with its margins roughened and rolled below. A strong rib on each side of the corolla runs downwards and backwards from the junction of these two parts of the margin. A visiting bee grasps the base of the slanting lip with its forelegs and the corolla tube just below the lip with its middle pair of legs, and inserts its

Fig. 6.39 Bumblebee (*Bombus pascuorum*, female), visiting lousewort (*Pedicularis sylvatica*). Notice the distention of the upper lip of the flower.

head obliquely into the wider part of the entrance to the hood, touching the stigma in the process. As it probes for nectar it prises apart the sides of the hood, at the same time drawing forward the upper part of the hood and releasing the pressure on the stamens, allowing pollen to fall from between them onto its head (Figs. 6.38, 39). The lip of red rattle (*P. palustris*) tends to be more asymmetric, but its pollination is similar (Faegri & van der Pijl, 1979).

The pollination of the louseworts has been studied in Norway by Nordhagen (Lagerberg *et al.*, 1957), and in North America by Sprague (1962) and especially by Macior (1968 onwards). Two widespread North American species studied by Macior (1968a, b) using high-speed cinematography illustrate well the way in which differences in flower shape and other characteristics relate to differences in pollination bi-

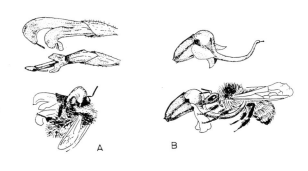

Fig. 6.40 Two American
louseworts. **A**, flower of 'wood
betony' (*Pedicularis canadensis*)
seen from the side and below,
and sketch showing foraging
position of visiting bumblebee.
Iowa (after Macior, 1968b).
B, side view of flower of
'elephant-heads' (*P. groenlandica*),
and sketch showing position
taken up by bumblebee
foraging for pollen. Colorado
(after Macior, 1968a).

ology. *P. canadensis* (known as 'wood betony' in North America) has yellow flowers, similar in shape to those of *P. sylvatica* but with a longer style which projects some distance below the tip of the upper lip. It is visited mainly by the queens of various bumblebee species. Macior's photographs show that, as the bee introduces its head and tongue into the corolla tube to suck the abundant nectar, the stigma sweeps through the pronotum crevice between the head and thorax (Fig. 6.40A). At the same time, pollen is released from the stamens inside the hood, and escapes through the pore surrounding the style, lodging in the pronotum crevice and on neighbouring parts of the bumblebee's body. Most of this pollen is swept away as the bee grooms the hairy surface of its thorax in flying from flower to flower, but the pronotum crevice is inaccessible to the sweeping movements of the middle pair of legs, and the pollen there remains to be swept out by the stigma of another flower. *P. canadensis*, pollinated by queen bumblebees and producing abundant nectar, is typical of early-flowering species in North America and probably throughout the range of the genus. By contrast, 'elephant heads', *P. groenlandica*, is a summer-flowering pink-flowered species, lacking nectar and visited for pollen by worker bumblebees of several species. The onset of flowering coincides with emergence of the worker bees; the queens are too large to operate the mechanism of the flower. The flower resembles the head of a little elephant waving its trunk in the air – the basal part of the hood containing the anthers forming the 'head', the long, upturned beak of the upper lip which ensheaths

Fig. 6.41 Eyebright (*Euphrasia rostkoviana* ssp. *montana*). The mouth of the corolla, with its striking guide-mark pattern, is wide open, but the stamens are in inward-facing pairs as in the larger flowers of *Pedicularis* and *Rhinanthus*.

Fig. 6.42 Bumblebee (*Bombus pascuorum*, female), on yellow rattle (*Rhinanthus minor*).

Fig. 6.43 Bumblebee (*Bombus pascuorum*) visiting flower of foxglove (*Digitalis purpurea*).

the style forming the 'trunk', and the three-lobed lower lip forming the 'ears' and 'jaw'. A visiting bee takes up a position astride the beak, which passes under the thorax and then curves up between the thorax and abdomen so that the stigma is in contact with the front surface of the latter (Fig. 6.40B). The bee grasps the stout central ridge of the hooded basal part of the upper lip with its mandibles, and by rapid movements of its wings shakes out pollen which falls onto the lower lip and is scattered as a yellow cloud enveloping the insect's body (an example of 'buzz-pollination', see pp.125, 179-80). Most of this pollen is groomed from the body and transferred to the corbiculae on the bee's hind legs, but much of that on the front face of the abdomen remains and may pollinate another flower. Bees may sometimes be seen inserting their heads between the sides of the upper lip, as Sprague observed, but the essential contrast between these two species remains clear. In general, the summer-flowering *Pedicularis* species lack nectar and are pollinated by worker bumblebees. In some North American summer-flowering, nectarless mountain species, the pollen-foraging visitors habitually hang from the upper lip (as worker bees sometimes do on late flowers of the nectariferous spring species), and pollen is transferred from flower to flower on the midline of the underside of the bee, a third position on the body from which the pollen cannot be removed by grooming (Macior, 1982).

The louseworts are a particularly varied genus, but their flowers have one feature in common that is found also in all the related semi-parasitic members of the Scrophulariaceae (the Rhinanthoideae) and in the related family Orobanchaceae (broomrapes and toothworts) as well. This is the arrangement of the stamens in two inward-facing pairs, which are normally pressed together preventing release of the dry, powdery pollen until an insect visits the flower. The eyebrights (*Euphrasia*) (Fig. 6.41) have a typically 'labiate' corolla, but in all other members of the Rhinanthoideae the lower

Fig. 6.44 Flowers of *Mimulus glutinosus*; the stigma of the left-hand flower has been touched and the lobes have closed together, exposing the anthers behind it.

lip tends to be reduced and the flower is more-or-less tubular. In the yellow-rattles (*Rhinanthus*) (Fig. 6.42) the flower is strongly flattened laterally, and the stamens function in much the same way as in the louseworts, releasing pollen as the sides of the corolla and the lower parts of the filaments are prised apart by the tongue of a visiting bumblebee (Kwak, 1977). In other genera the anthers have spine-like projections beneath, and pollen is shed when these are touched by a visiting insect.

The remaining zygomorphic flowers in the Scrophulariaceae are a diverse collection. One of the most familiar, the foxglove (*Digitalis purpurea*) (Fig. 6.43, Plate 1h), is in effect a zygomorphic variation of the trumpet- and bell-shaped flowers already described; the orientation of the flowers in their long racemes and the long hairs on the floor of the corolla exclude virtually all insects but the pollinating bumblebees. The monkey flower (*Mimulus guttatus*) is rather similar, but shows an interesting peculiarity in the irritability of the two-lobed stigma. A bee entering the flower brushes first against the stigma, whose two lobes then fold quickly together and lie tightly pressed against the upper surface of the corolla. The insect them comes into contact with the anthers, which lie immediately behind the lower stigma lobe (Fig. 6.44). A very similar mechanism is found in the large-flowered bladderworts (*Utricularia*) of the family Lentibulariaceae (Fig. 6.45). The related butterworts (*Pinguicula*) again have a similar mechanism but here there is no active movement; the movement of the visiting insect suffices to draw back the lower stigma-lobe and uncover the anthers as it leaves the flower. In the Alps, the

Fig. 6.45 Greater bladderwort (*Utricularia vulgaris*). **A**, side view of flower. **B**, side view of flower with half of the lower lip removed. **C**, gynoecium and neighbouring parts of a flower with half the calyx and corolla removed, showing the stigma-lobes closed after being touched. **D**, diagram showing stigma-lobes before and after stimulation.

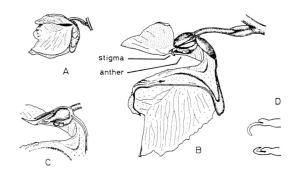

Fig. 6.46 Common toadflax (*Linaria vulgaris*). **A**, front view of flower with lower lip pulled down to show the stigma, stamens, and the proboscis-guide on the lower surface of the corolla-tube. **B**, side view of flower with half of calyx and corolla removed.

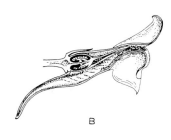

A   B

white-flowered alpine butterwort (*P. alpina*) is visited principally by flies, while the purple-flowered common butterwort (*P. vulgaris*) is visited by bees. However, in north-west Europe the common butterwort is seldom visited by insects and is apparently regularly self-pollinated (Willis & Burkill, 1903b; Hagerup, 1951).

The bladderworts and butterworts have spurred flowers; in the bladderworts, the entry of small insects ineffective for pollination is further impeded by the greatly inflated lower lip, which is pressed against the upper lip, forming a spring-loaded 'door' at the entrance to the flower. The large, bright-yellow flowers of the common bladderwort (*Utricularia vulgaris*) are bee flowers; numerous visits by bumblebees (and two visits by hoverflies) to a population of this species were observed in Gloucestershire (K.G. Preston-Mafham); visits by long-tongued hoverflies, especially *Helophilus lineatus*, were seen by Heinsius (Knuth, 1906-9) and Silén (1906b).

Returning to the the Scrophulariaceae, similarly constructed *personate* flowers are found in the snapdragons (*Antirrhinum*) and toadflaxes (*Linaria* and related genera). The common toadflax (*Linaria vulgaris*) (Fig. 6.46) is a common example. Nectar is secreted by the base of the ovary, and accumulates in the conical spur projecting from the underside of the base of the corolla tube. The personate corolla excludes almost all insects but the strong and 'intelligent' bees; the length of the spur debars short-tongued bees from reaching the nectar. Normally a honeybee or bumblebee visiting the flower lands on the lower lip, which bears a darker yellow or orange guide-mark, and inserts its head between the upper and lower lips, prising open the corolla. The tongue of the bee is guided to the nectar by a smooth channel about 1 mm wide, between two orange hairy ridges on the floor of the corolla tube. As the bee sucks the nectar, its back comes into contact with the stamens and stigma which lie against the upper side of the corolla tube. Bees visiting the flowers for pollen may work the flowers upside down, collecting the pollen directly from the anthers; more than a quarter of a sample of 269 bumblebees foraging on a population of yellow toadflax which Macior (1967) studied in Wisconsin had evidently been working in this way. The flowers of the common toadflax are homogamous, but self-incompatible. The mechanism of the introduced (south European) purple toadflax (*Linaria purpurea*) (Fig. 6.47) is similar. On the other hand, in the ivy-leaved toadflax (*Cymbalaria vulgaris*) and in several of the annual weedy species (e.g. *Kickxia spuria*) the self-incompatibility has been lost and the flowers are regularly self-fertilised.

We cannot leave spurred zygomorphic flowers without mentioning three other, quite unrelated groups – the violets and pansies (*Viola*; Violaceae), the balsams (*Impatiens*; Balsaminaceae), and that most extraordinarily diverse of all flowering-plant

Fig. 6.47 Honeybee visiting flower of purple toadflax (*Linaria purpurea*).

families, the orchids (Orchidaceae), which form the subject of the next chapter. The only species of balsam native to Britain, *Impatiens noli-tangere* (touch-me-not), is an uncommon plant, but the introduced Himalayan balsam (*I. glandulifera*) now grows in profusion along the banks of many rivers and canals. The pink flowers are large enough to enclose a bumblebee completely, bringing its back into contact with the stamens and stigma in the roof of the flower. The busy comings and goings of dozens of bees on a patch of this plant are often a fine sight. Violet flowers (like those of the balsams) have five unequal free petals (Plate 1c). The lowest of these is prolonged at the base into a spur, which accumulates the nectar secreted by the flat, green, tail-like appendages of the two lowermost stamens. The anthers form a cone surrounding the short hooked style, and release pollen onto the tongue of a visitor when the style or the apex of the cone is lightly touched. The style itself is hollow, and flexible at the base. Any pressure on the style causes a small drop of liquid to be exuded from the pore at the tip; it is drawn back again into the cavity of the style, together with any pollen it may have come into contact with, when the pressure is released. This drop of liquid may serve a secondary function in moistening a small area on the tongue of a visiting insect, so causing the powdery pollen to adhere (Beattie, 1969). Pollen grains will germinate only inside the stigmatic cavity, and although the flowers are self-compatible no seed is set in the absence of insect visits. The main visitors to the flowers of violets are bumblebees, other early-flying bees including *Anthophora plumipes* and species of *Osmia*, *Andrena*, *Lasioglossum* and *Halictus*, and hoverflies (Beattie, 1972). It is interesting that many violets produce cleistogamous flowers, self-pollinated in the bud, after the season of the conspicuous chasmogamous flowers is over (p.334). There are also self-pollinated species of *Viola* with normal flowers, such as *V. arvensis*, in which pollen can readily fall into the open stigmatic cavity.

Theft and protection

Many of the flowers we have considered in this chapter regulate the behaviour of particular visitors with considerable precision, and exclude other insects with a fair degree of effectiveness. But with regulation comes the possibility of evasion. In some flowers there is little physical barrier to evasion. Honeybees collecting nectar from *Brassica* crops may approach the flowers from behind and insert their tongues between the sepals and the claws of the petals, instead of entering the flower in the 'legitimate' way and bringing about pollination (Fig. 6.48). We have already seen how honeybees

a   b

Fig. 6.48 **a-b** Nectar-theft: **a,** honeybee sucking nectar 'legitimately' *from* Brassica flower. **b,** honeybee stealing nectar from back of *Brassica* flower ('base working').

may behave similarly on flowers of the bluebell (*Hyacinthoides non-scriptus*). In flowers with the corolla fused into a tube, it is not possible for an insect to get nectar in this manner, unless it can bite a hole through the tube. Many gamopetalous flowers are, in fact, robbed in this way.[1] The usual culprits are the powerful but relatively short-tongued bumblebees *Bombus terrestris* and *B. lucorum* (and other species), which use their mandibles to perforate the corolla tubes or spurs of the long-tubed labiates, comfrey (Fig. 6.49), toadflax, daffodils, columbines (Macior, 1966) and other flowers whose nectar would otherwise be inaccessible to them (Brian, 1957). The holes made by these short-tongued bumblebees are often used subsequently by honeybees and other insects ('secondary robbers'). A flower with a perforated tube may still be visited legitimately for nectar or pollen and pollinated in the normal way – even by flower-robbing bumblebees whose tongues are too short to reach the nectar by that route (A.J.L. & Q.O.N. Kay). Inouye (1983) discusses the ecology of nectar robbing.

A number of flowers show features which may be interpreted as adaptations preventing – or at least minimising – such larceny of nectar. Thus the butterfly flowers of the pinks (*Dianthus*) have a firm leathery calyx, further protected at its base by stout

[1] Inouye (1980) makes a distinction, within the general heading of 'floral larceny', between *nectar theft* in which there is no damage to the flower, and *nectar robbing* which involves 'breaking and entering' by biting a hole through the corolla tube. Inouye calls the kind of behaviour described above on bluebell and brassicas 'base working'.

a   **b**

Fig. 6 49 **a-b** Nectar robbery: **a,** bumblebee *(Bombus pascuorum)*, sucking nectar 'legitimately' from comfrey *(Symphytum officinale)*. **b**, bumblebee *(Bombus* cf. *terrestris)*, robbing nectar from comfrey through hole bitten in corolla-tube.

overlapping bracts (Fig. 6.5). The inflated calyces of the campions *(Silene)* may serve the same function in a different way. In general, the calyx obviously provides some degree of protection against perforation of the base of the corolla tube in many species, as well as protecting the developing bud and providing mechanical support for the mature corolla – which in many flowers will be roughly handled by strong and heavy insects. But of course the calyx evolved in relation to a variety of factors, and the form it takes in a particular instance must reflect a balance between them. It is possible to point to flowers which have tubular corollas apparently with no special protection. Some of these are often perforated, others are not. All that can be said is that different flowers have evolved different patterns of adaptation to the complex environments in which they live – and the continued existence of all of them shows that all of their patterns of adaptation are workable in their own context.

Honeybees often steal nectar from isolated flowers of charlock *(Sinapis arvensis)* in the same way as from brassicas, but seldom do so from flowers in the denser inflorescences (Fogg, 1950; see also *Leonotis*, Chapter 8). This suggests that protection against nectar larceny may have been one factor in the evolution of the dense inflorescences seen, for instance, in clovers *(Trifolium;* Fabaceae), *Buddleja* (Loganiaceae), mints and thymes *(Mentha* and *Thymus;* Lamiaceae), the valerians (Valerianaceae), the Australian blue pincushion flower *(Brunonia;* Brunoniaceae), the teasel and scabious family (Dipsacaceae) and the daisy family (Asteraceae). Grant (1950b) and others have suggested that protection of ovules from damage, especially by beetles, has been an important factor in the evolution of flowers, leading to the development of inferior ovaries, and again favouring dense, head-like inflorescences – a trend whose culmination is seen in the composites (Asteraceae). Dense head-like inflorescences made up of numerous small flowers are so common, and the composites in particular are so manifestly successful, that there must be potent factors of floral ecology working in favour of this form of organisation. It is at first sight a paradox that a family like the labiates (Lamiaceae), so characterised by the freely-visited, beautifully bumblebee-adapted flowers described earlier in this chapter, should have given rise to the 'brush-blossom' heads of the mints, with exposed stamens and stigmas, and short-tubed flowers with nectar accessible to a variety of visitors. But 'brush blossoms' (Faegri & van der Pijl, 1979) are found in many families, and they open up new and different evolutionary possibilities which we examine in the next section and in Chapters 8 and 9.

Umbellifers, composites and others: 'catering for the mass market'

The dense, head-like inflorescences and 'brush blossoms' just considered are often visited by quite a wide range of pollinators. In this they resemble many plants in which numbers of small flowers are massed together in less specialised but more or less compact inflorescences, thus achieving a much more conspicuous floral display than would scattered single flowers. In fact this is a very common pattern of adaptation, a fact that should remind us that the highly specific adaptations to particular insect visitors exemplified by deadnettles, louseworts or orchids are only a limited part of the story, and there are other possibilities for an effective plant-pollinator relationship.

A number of families of flowering plants include members with broad, flat or domed inflorescences of numerous small white flowers. Examples include the elders (*Sambucus*), the guelder-rose (Fig. 15.2), wayfaring tree and their relatives (*Viburnum*) in the Caprifoliaceae, many of the dogwoods (Cornaceae), and members of the saxifrage family (Saxifragaceae) and the related woody Hydrangeaceae. But the prime example of this kind of organisation is the carrot family, the 'umbellifers' (Apiaceae), whose flat white umbels line almost every roadside (and dot many other habitats) in northern Europe in summer (Fig. 12.14).

Umbellifers are visited by a great variety of insects, particularly flies (Diptera), but including also beetles, honeybees and solitary bees, and a range of other Hymenoptera. A total of 334 species of insects, belonging to 37 families, were recorded visiting carrot flowers in Utah (Hawthorn, Bohart & Toole, 1956; see p.358). The umbels themselves are conspicuous from a distance, their flat tops provide a convenient landing platform for insects, and nectar and pollen are both freely exposed to all comers. The plants clearly rely for pollination upon numerous unspecific visits. Different umbellifer species attract rather different assemblages of visitors, and scent probably plays a part in this. Borg-Karlson, Valterová & Nilsson (1993) found different patterns of scent constituents in species from several umbellifer genera and Tollsten & Øvstedal (1994) have shown significant differences in floral-scent chemistry between different populations of pignut (*Conopodium majus*). Most umbellifers tend to be gregarious, so there is a good probability that any insect which finds the flowers attractive will fly to another flower head of the same species. Generally, in this situation it is more important for the visitors to recognise a flower head as such than to differentiate visually between one species and another. Indeed, the recurrence of umbellifer-like heads in different families suggests Muellerian mimicry (p.402) between them – that there is an evolutionary advantage in a common 'advertising style'. The difficulty that we humans have in identifying umbellifers at sight thus has a firm biological basis! One other interesting point about this and the other families that produce similar inflores-

Fig. 6.50 Drone-fly (*Eristalis tenax*), on ox-eye daisy (*Leucanthemum vulgare*). Notice the orchid pollinia (probably of heath spotted-orchid [*Dactylorhiza maculata*] on the proboscis of the insect.

cences is that, by common consent, systematists regard them as evolutionarily rather advanced; they have evidently diverged far from their origins and are the refined product of a long process of evolutionary adaptation to exploit their diverse insect pollinators.

With some 25,000 species, the 'composites' (Asteraceae) are one of the largest families of flowering plants. They grow abundantly in every continent, and are generally regarded as one of the most highly evolved of all flowering-plant families; by any reckoning they are 'successful'. Burtt (1961) has suggested that the particular advantage of the composite head – made up of many small flowers opening over a period of a week or two, each with a single ovule – is that it allows a very wide range of *different* pollinations to take place. In this he contrasts it with the large zygomorphic flowers with their specialised pollination mechanisms and (often) numerous ovules. As he says, 'In these elaborate flowers a very large number of ovules are fertilised by pollen from one, or a few, male parents. This will lead in due course to *intensive* exploration of the possible recombinations between the parents that cross. In the Asteraceae, whose individual ovules are individually pollinated (neighbouring ones often receiving pollen from different plants) the exploration of possible combinations is *extensive* through the population rather than *intensive* between individual plants.' The same argument applies to the umbellifers (producing two seeds from each flower), and in varying degree to all other species with compact flower heads. The Asteraceae are the culmination of a trend – with the orchids (Chapter 7) as their antithesis. More will be said of diversity in pollination relationships later in this chapter.

The Asteraceae are related to the bellflower family (Campanulaceae), and the individual flowers ('florets') may be compared with the *Campanula* flower described on pp.150-51. The florets making up the heads fall rather sharply into two types, well shown in a daisy or ragwort flower (Fig. 6.50). The small tubular 'disc florets' of the middle of the head are actinomorphic, and conspicuous only in so far as many of them are massed together. The 'ray florets' round the edge are zygomorphic, and their strap-shaped (ligulate) corollas are largely responsible for the conspicuousness of the head. In some Asteraceae, such as hemp agrimony (*Eupatorium cannabinum*, Tribe Eupatorieae) and the knapweeds (*Centaurea*) (Fig. 6.54, 15.9) and thistles (*Carduus, Cirsium* and related genera, Tribe Cardueae), only tubular florets occur. On the other hand, in the tribe Lactuceae all the florets are ligulate, resulting in the familiar dande-

Fig. 6.51 Hoverflies (*Syrphus ribesii* and a smaller species) on head of dandelion (*Taraxacum vulgare* agg.).

lion type of flower (Fig. 6.51).

Fig. 6.52 shows three disc florets of the Oxford ragwort (*Senecio squalidus*). The calyx is reduced to a ring of long silky hairs, the *pappus*, around the top of the inferior ovary. The corolla is tubular, very slender below, widening abruptly into a bell shape, shallowly divided like a *Campanula* flower into five lobes. The five stamens are inserted at the top of the narrow part of the corolla tube. Their filaments are free, but the anthers are united into a tube surrounding the style. The pollen is shed into the interior of the tube before the floret opens. At this stage, the style is relatively short, and the tufts of

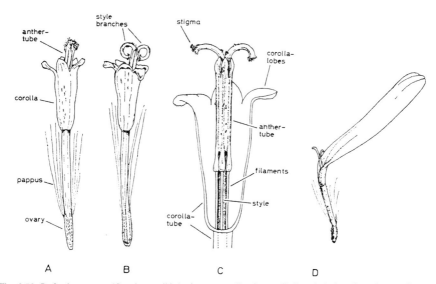

Fig. 6.52 Oxford ragwort (*Senecio squalidus*). **A**, young disc-floret. Pollen is being forced out of the top of the anther-tube as the style grows up through it. **B**, older disc-floret with the divergent branches of the style projecting from the top of the anther-tube. **C**, upper part of disc-floret with half of corolla removed to show details of the anther-tube and filaments surrounding the style. **D**, ray-floret.

hair at the tips of the two style branches fit into the tube like the piston in a cylinder. As the floret opens in the morning, the style grows, forcing pollen out through the top of the anther tube. In fact pollen presentation is not continuous. On stimulation, the filaments contract, measuring out a small amount of pollen to the visiting insect, while the rest of the pollen remains inaccessible inside the anther tube, to be extruded later in the day. Small (1915) remarks on this 'miserly' presentation of pollen; if we accept the importance of repeated visits to the biology of the composite head we may see it as an important and characteristic part of the pollination mechanism. When all the pollen has been swept from the anther tube the style emerges, generally on the second day of opening. The stigma branches begin to diverge, exposing the receptive upper surface which is now ready to receive pollen. The ray florets are purely female, and open before any of the florets of the disc. To this extent, the pollination mechanism of species with both disc and ray florets is more complicated than those with ligulate florets alone. In these plants, for example lesser hawkbit (*Leontodon saxatilis*), the ligulate florets are hermaphrodite (Fig. 6.53), and function in a broadly similar way to the disc florets of the daisy or ragwort, but the pollen-presentation mechanism is different (Yeo, 1993). The upper part of the style is covered with short hairs, forming a stylar brush which lies within the anther tube and becomes loaded with pollen immediately before the floret opens. At anthesis, the whole mass of pollen is carried up on the stylar brush. Later the style branches diverge and can receive pollen. Asteraceae are typically gregarious, and many species are visited by a large variety of insects. Harper & Wood (1957) listed 178 species of visitors to ragwort (*Senecio jacobaea*), including thrips, Hemiptera, beetles, butterflies and moths, and many Hymenoptera and flies. Transport of pollen to another flower of the same species by these largely unselective pollinators depends on the abundance of flowers of the same species in the surroundings. As in the umbellifers, Muellerian mimicry has probably favoured the convergence of composite heads to a few simple visual patterns, in this case especially the yellow-centred white 'daisy' type, and the all-yellow heads of the dandelions, hawkweeds and their relatives, making the Asteraceae another family in which many of the species are notoriously difficult to recognise at sight.

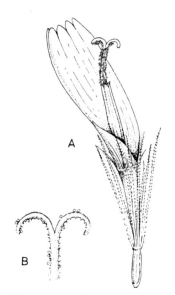

Fig. 6.53 Lesser hawkbit (*Leontodon saxatilis*). **A**, floret from the inner part of a head. The style is exserted through the anther-tube and its branches have begun to diverge, but it still carries a good deal of pollen. **B**, tip of style, showing the stigmatic surface on the upper (inner) side of the style-branches, and the hairs which sweep the pollen from the anther-tube.

Sensitive stamens are found in a number of plant families (p.338), but the functional significance of the movement is seldom as obvious as it is in the Asteraceae. Not all Asteraceae show evident movement. Small (1915) found irritability of the stamens in 64% of the 149 species he observed. The movement is particularly

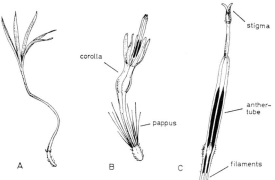

Fig. 6.54 Greater knapweed (*Centaurea scabiosa*). **A**, sterile marginal ('ray') floret. **B**, fertile floret. **C**, detail of anther-tube and style.

striking in the cornflower (*Centaurea cyanus*) and other members of the knapweed genus (Figs. 6.54, 15.9). The conspicuous marginal florets are tubular but sterile and somewhat zygomorphic. The fertile florets have deeply-divided blue corollas, and the long, slender anther tubes stand prominently above the general surface of the head. Pollen presentation has features of both the types described above. The style carries a ring of sweeping hairs below the short stigmatic branches. On stimulation the filaments may contract several millimetres, exposing the pollen-covered style and ejecting a quantity of pollen onto the visitor. After an interval, the filaments recover their original length, and the process can be repeated. Percival (1965) found that in a freshly opened floret of *Centaurea montana*, the amount of pollen delivered on stimulation was about a quarter of the whole amount in the anther tube, but the amount fell progressively on subsequent occasions. Stimulation of the filaments on one side of the floret will cause the anther tube to turn towards the insect causing the disturbance. In due course, the style grows up through the anther tube, and the stigma lobes diverge and can be pollinated.

The Asteraceae are generally thought of as a rather homogeneous family, but they vary a good deal in the details of their florets and capitula (Small, 1917, 1918), and consequently in their floral biology. The purely female ray florets of many species have been mentioned already; these bring an element of protogyny to the the head as a whole, although individual hermaphrodite florets are protandrous. In coltsfoot (*Tussilago farfara*) the ray florets are female, and the disc florets are effectively purely male. In dwarf thistle (*Cirsium acaule*) the plants are gynodioecious, and the smaller female heads can be picked out at a distance from the larger hermaphrodite heads. Both of these arrangements, like the gynodioecy of many Lamiaceae, are evidently adaptations tending to promote a greater degree of outbreeding (Chapter 12, p.345). As in other families, self-incompatibility is often found in the Asteraceae. This would be expected from the pollen-presentation mechanism – as in many legumes (Fabaceae, pp.155-60). Thus self-pollination brought about by the style branches curving back to touch residual pollen at the top of the anther tube, or by the styles picking up pollen from other florets when the heads close at night, will in many cases not result in fertilisation. In fact, there may be a good deal of outbreeding even in self-compatible species; thus Watts (1958) found up to 11.5% outcrossing in an experiment with cultivated lettuce (*Lactuca sativa*), probably brought about mainly by hoverflies.

There are perhaps three main factors limiting the size of the individual heads of Asteraceae. A very large capitulum is mechanically vulnerable; a corymb of smaller heads is less easily damaged, and if a few branches are broken this has little effect on the inflorescence as a whole. Large composite heads provide a rich and compact store of food for insect larvae; the larger the head, the more seeds are put at risk by a single infestation. Here again a corymb of separate smaller heads has an obvious advantage. Thirdly, if the marginal florets are female and the remainder hermaphrodite, selection in favour of a particular ratio of the two types will favour a particular size of head. The balance of selective advantage will be different in different cases, and every gradation exists between the large corymbs of few-flowered capitula of yarrow (*Achillea millefolium*) or hemp agrimony (*Eupatorium cannabinum*) and the massive single heads of the cultivated sunflower (*Helianthus annuus*) or globe artichoke (*Cynara scolymus*).

Insect-pollination syndromes

Regardless of their taxonomic relationships, flowers pollinated by particular visitors tend to show particular features in common, related to the size, behaviour and other biological characteristics of their pollinators; the earlier pages of this chapter have already provided various examples. These patterns of common characters, to which flowers of quite different evolutionary origins may converge, have been called pollination 'syndromes' (van der Pijl, 1961; Baker & Hurd, 1968; Faegri & van der Pijl, 1979). They are not rigid – evolution often finds more ways than one of achieving the same end – but they are clearly expressions of adaptive trends, and it is often not difficult to infer the significant pollinators of a flower from its colour, form and structure. The few examples discussed below are not exhaustive. Syndromes of flower characters associated with pollination by birds and bats are described in Chapter 8, and those associated with pollination by flies which normally breed in decaying animal or vegetable matter ('sapromyiophily') are described in Chapter 10.

Bee-pollinated flowers

Bees (and especially honeybees) visit a very wide range of flowers for nectar and pollen. No single pollination syndrome can be defined for bees in general, but several rather different pollination syndromes are particularly characteristic of them. In general, bees favour yellow and blue or purple flowers with well-marked three-dimensional form, and a large proportion of flowers with well-developed tubular or bell-shaped corollas are mainly or entirely pollinated by bees of various kinds. Zygomorphy is also a common feature of bee-pollinated flowers. This is particularly striking in the large flowers pollinated by bumblebees, such as the monkshoods, deadnettles and louseworts, but the smaller solitary bees of such genera as *Osmia* and *Andrena* are important pollinators of many smaller zygomorphic flowers, including many legumes (Fabaceae) and labiates (Lamiaceae). The sheer diversity of colour of bee-pollinated flowers is in itself a striking and significant feature, which provides an important part of the basis of bees' flower constancy. Menzel & Shmida (1993) have shown that the colours of bee-pollinated flowers in Israel cover virtually the entire range of colour bees can perceive.

Some of the 'brush-blossoms' constitute another pollination syndrome which has undoubtedly often evolved primarily in relation to bees. Several examples have been described already (p.172). There is no sharp distinction to be drawn between typical brush-blossoms and the heads of the rampions (*Phyteuma*, Campanulaceae), scabious species (*Scabiosa* and related genera, Dipsacaceae) and knapweeds (*Centaurea*, Asteraceae) which are also visited mainly by bees. However, the brush-blossom form is also common in relation to bird- and bat-pollination (Chapter 8). It is noteworthy, but perhaps not altogether surprising, that there seems to be no distinctive syndrome of flower characters associated with that super-efficient generalist, the honeybee.

Bee flowers providing pollen as the sole reward

Typical bee flowers provide both nectar and pollen as rewards for the visiting insect. However, there are two important groups of bee-pollinated flowers in which the primary reward is pollen, and there is little or no nectar. The poppies (*Papaver* spp., Papaveraceae) (Fig. 2.12) are a familiar example, the numerous stamens providing abundant pollen avidly collected by visiting bees. The rock roses and sun roses (Cistaceae) are also characteristic pollen flowers, although *Cistus* does produce nectar from around the bases of the stamens (Fahn, 1990; Talavera *et al.*, 1993). The flamboyant pink or white flowers of *Cistus* species, freely visited by bees, bedeck many sunny Mediterranean hillsides in spring; yellow and white-flowered species of the lower-growing genus *Helianthemum* (Fig. 5.17) extend into the dry grasslands of northern Europe. Poppy-like flowers recur in various genera in the Mediterranean region, where they appear to represent a distinct syndrome adapted primarily to pollination by scarabaeid beetles and only secondarily to bees (Dafni *et al.*, 1990). The black-centred red tulips and red or purple-flowered *Anemone* species from the east Mediterranean and neighbouring areas are well known as garden plants; *Ranunculus asiaticus*, with its big, glossy, red or yellow flowers mimics the anemones remarkably closely, although the brightly-coloured perianth-members of the anemones are morphologically sepals, but those of the *Ranunculus* are true petals. In Israel, all the flowers of this type studied by Dafni and his colleagues reflected strongly in the red part of the spectrum but were virtually black in the green, blue and UV regions. However, bee-pollinated corn poppies (*Papaver rhoeas*) in northern Europe reflect UV strongly, so appear as bright 'ultraviolet' flowers to bees (Daumer, 1958) (Chapter 5, p.132).

An interesting and important group of pollen flowers are adapted to vibratory pollen collection – 'buzz pollination' – by bumblebees (Buchmann, 1983; see Chapter 5, p.125). Bittersweet (*Solanum dulcamara*, Solanaceae) (Figs. 6.55, 5.18) is representative of a striking group of (largely unrelated) flowers with rotate or reflexed corollas, and a prominent anther cone in the centre of the flower (Faegri, 1986); it produces no nectar, and is evidently specialised as a pollen flower. The flowers are visited by bees (particularly worker bumblebees) which release the rather small pollen grains (*c.* 14 µm diameter) from the pores at the tips of the anthers by rapid vibration of their thoracic muscles as they hang from the anther cone. Macior (1964, 1970) and Harder & Barclay (1994) have observed precisely similar foraging behaviour on the similarly-formed flowers of the North American 'shooting star' species *Dodecatheon meadia*, *D. amethystinum* and *D. conjugens* (Primulaceae) (Fig. 6.56); the last has

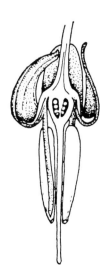

Fig. 6.55 Bittersweet (*Solanum dulcamara*), a buzz-pollinated flower.

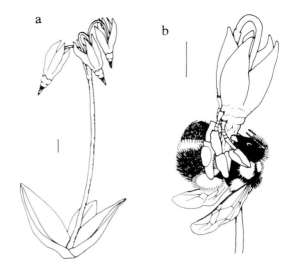

Fig. 6.56 **a**, *Dodecatheon conjugens* (scale bar=1 cm); **b**, position adopted by pollen-collecting bees during flower vibration (*Bombus bifarius* queen; scale bar=5 mm). Drawing by H.C. Proctor, reproduced with permission from Harder & Barclay (1994).

pollen just slightly smaller (*c.* 12.5 μm) than *S. dulcamara*. Borage (*Borago officinalis;* Boraginaceae) is unusual amongst flowers of this form in producing abundant nectar; it is freely visited by honeybees for nectar and by bumblebees for both nectar and pollen. The reflexed petals and exposed anther-cone of the 'shooting-star' flower form may be an adaptation minimising damping of resonance to the visitor's buzz-frequency, or it may be related to floral microclimate, helping to keep the pollen in a dry, powdery, easily-dispersed state (Corbet, Chapman & Saville, 1988). Harder & Barclay found that when *Dodecatheon* flowers were mechanically vibrated at constant energy input, most pollen was shed at 500-1000 Hz. The apparent 'mis-tuning' to the buzz-frequency of the visiting bees (400 Hz), and consequent more miserly meting-out of pollen, may be adaptive as they suggest, but a precisely-tuned resonant frequency could perhaps be missed and more investigation is needed. The same 'solanoid' flower-form recurs in the south-European mountain plant *Ramonda* (Gesneriaceae), in cranberries (*Vaccinium* Sect. *Oxycoccus*; Ericaceae), of which the small-flowered European species are apparently little-visited and usually selfed, and in the Australian genus *Dianella* (Liliaceae). Vibratory pollen release is often important in flowers of various other forms, including many bell-shaped flowers of the heather family (Ericaceae), and the 'shaving-brush' type of which kiwi-fruit (*Actinidia*, p.40) is an example. The buzz pollination of *Pedicularis groenlandica* and similar species was described on p.166. For Australian *Solanum* see Anderson & Symon (1988).

Wasp-pollinated flowers

A diverse assortment of tubular flowers have become adapted primarily to pollination by wasps (Vespidae, p.106). They typically have dingy brownish flowers, and short broad tubes with readily accessible nectar, and do not produce large amounts of pollen. The classic example is the figwort *Scrophularia nodosa* (Fig. 6.57), illustrated with a visiting wasp on the title page of Sprengel's book (p.18). By contrast with the bee-pollinated zygomorphic flowers in the same family, the stamens are held close to the

Fig. 6.57 Common wasp (*Vespula germanica*) visiting flower of figwort (*Scrophularia nodosa*).

lower lip in figworts. The flowers of figworts are protogynous, and the female phase is said to last two days; the stamens then straighten, bringing the anthers to the mouth of the flower above the withered stigma (Fig. 6.58). There seems to be no satisfactory explanation of the function of the staminode in the upper part of the flower (Trelease, 1881). Wasps, mainly *Vespula vulgaris* and *V. germanica*, seem to be by far the commonest visitors to *Scrophularia* species in Britain. However, honeybees and various other bees also visit the flowers commonly, and there is no doubt that in particular

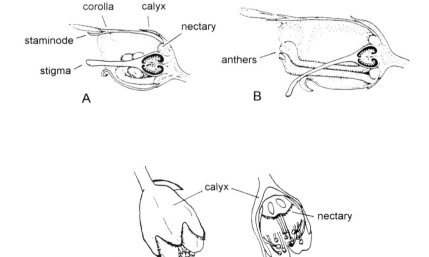

Fig. 6.58 **A**, **B**, Water figwort (*Scrophularia auriculata*). **A**, half-section of newly-opened flower, in functionally female stage. **B**, half section of older flower with dehiscing anthers. **C**, flower of wild cotoneaster (*Cotoneaster integerrimus*), in side view and section, another dull brownish-red wasp-pollinated flower. After Ross-Craig (1956).

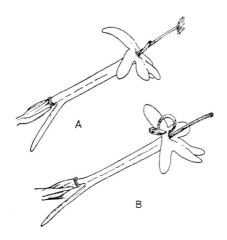

Fig. 6.59 Red valerian (*Centranthus ruber*). **A**, young flower with dehiscing stamen. **B**, older flower, stamen reflexed and style elongated.

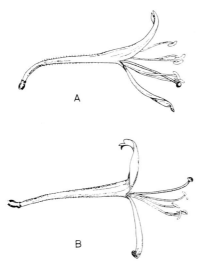

Fig. 6.60 Honeysuckle (*Lonicera periclymenum*). **A**, newly opened flower. **B**, older flower.

places and seasons bees go to the flowers very freely and may even be the dominant visitors. Other flowers showing the same syndrome of characters and commonly visited by wasps include *Cotoneaster* species, with erect dusky brownish-pink petals forming a short tube (Fig. 6.58c), and species of the orchid genus *Epipactis* (pp.190-92; see also Table 5.3).

Butterfly and moth flowers

The butterfly-pollinated pinks (*Dianthus*) were mentioned earlier in this chapter. A good many other flowers are pollinated primarily by butterflies and other day-flying Lepidoptera such as burnet moths (*Zygaena* spp.). They typically combine bright pink or red (or sometimes blue or purple) colours, with sweet (but often not strong) scent, and plentiful nectar accessible only through a slender corolla tube or spur; the stamens and style are often exserted in front of the flower where they can come into contact with the hairy body of the visitor. Examples of flowers showing some or all of the features of this syndrome are red valerian (*Centranthus ruber*) (Fig. 6.59, Plate 3d), species of the common tropical genus *Lantana* (Verbenaceae), hemp agrimony (*Eupatorium cannabinum*), *Buddleja* and the fragrant orchid (*Gymnadenia conopsea*, Fig 7.14) (Chapter 7).

Flowers pollinated primarily by night-flying moths share features with butterfly flowers, but open in the evening and typically produce a heavy sweet fragrance at night. They are generally pale in colour, often with deeply dissected outlines, and are often profuse nectar producers. The long-tubed, pale, sweetly-scented flowers of honeysuckle (*Lonicera periclymenum*, Caprifoliaceae), with their long projecting stamens and stigmas, are visited by hawkmoths and by various other moths including the silver-Y (*Autographa gamma*), which is probably one of the more important visitors in Britain (Figs. 6.60, 61). In the Nottingham catchfly (*Silene nutans*) (Figs. 6.62, 63) the white petals are rolled up during the day, when the flowers could be passed by as no more than withered remnants. In the evening, the petals expand and the flowers become

Fig. 6.61 Silver-Y moth (*Autographa gamma*) visiting flower of honeysuckle (*Lonicera periclymenum*).

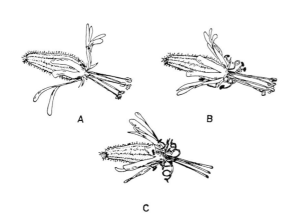

Fig. 6.62 Nottingham catchfly (*Silene nutans*). **A**, flower on first night of opening; the filaments of five stamens have elongated and the anthers are beginning to dehisce. **B**, flower on second night of opening; the stamens which dehisced the previous night have now recurved against the corolla, and the filaments of the remaining five stamens have elongated and their anthers matured. **C**, flower on third night of opening; the filaments are now all recurved and the anthers empty, and the styles have elongated and are receptive.

Fig. 6.63 A noctuid moth visiting Nottingham catchfly (*Silene nutans*, Caryophyllaceae). This creamy-white-flowered species is dependent on dusk- and night-flying pollinators. Many pink 'butterfly' flowers in this and other families are visited by night-flying moths as well.

conspicuous pale stars. The first night, the filaments of the first five stamens elongate, holding the dehiscing anthers in front of the long-tubed flower. The second night, the spent first night's stamens have recurved against the corolla, and the second five have elongated to take their place. On the third night, the stamens have all recurved, their anthers empty, and the styles have elongated and are receptive.

Adaptive patterns and ecological contexts

It is easy to sketch out possible stages in the evolution of specialised and highly adapted flowers like the big bumblebee-pollinated zygomorphic flowers considered earlier in this chapter. As one contemplates the end products of such evolutionary lines, the continued abundance of many simpler, apparently 'primitive' flowers may seem puzzling or even perverse. But it is important to remember that flowers and pollination relationships evolve, not in idealised isolation in which there is a single 'best' solution, but in varied contexts in which the requirements for pollen transport, and the corresponding selection pressures, may be very different.

When the primitive flowering plants first evolved a pollination relationship with insects, they were probably a minor component in a gymnosperm-dominated (and perhaps predominantly wind-pollinated) vegetation. The association with insects increased the probability of pollen being transferred from flower to flower, rather than wasted on other surfaces in the surroundings. A non-specific relationship of this kind is probably hard to better for a reasonably abundant plant. If a buttercup is a primitive generalist flower, the composites (Asteraceae) can be seen as having returned to a similar adaptive niche as highly evolved 'specialised generalists'. Indeed, for plants as gregarious as the dominant trees in a temperate forest canopy, wind may carry pollen better than insects – and the grasses seem to have developed wind pollination secondarily in their evolution as the dominant plants of the expanding treeless plains that accompanied the rise of large grazing mammals (Chapter 14). But at lower densities, especially in a species-rich and diverse community, the flower-constancy of bees can bring about greater efficiency of pollen transfer between flowers of the same species – greater 'dispersal efficiency' (Inouye et al., 1994). This has evidently become a powerful selection pressure on flowers, and a driving force to diversity of colour and form – the sort of diversity seen so strikingly among legumes or labiates. However, flower-constancy is only useful to bees in relation to flowers which are reasonably frequent, and species which are distributed only thinly may be little visited. For flowers visited by bumblebees, the consequences of low density of individual flowers may be offset to some extent by the bees' habit of 'minoring' – of visiting species additional to those to which they are mainly constant. But either of these situations strengthens selection pressure for more precise and longer-lasting placement of pollen on the body of the visitor[1] – for adaptations that favour pollen 'carry-over' – so that pollination can still take place after the insect has visited other flowers of different species.

---

[1] The Australian triggerplants (*Stylidium*) (Fig. 6.64) are a remarkable example of precise placement of the pollen on a visitor. The stamens and style are fused to form a column, which is normally arched back behind the petals on one side of the flower. The column is sensitive at its base. When touched by small bees or bombyliid flies sucking nectar, the column slaps across the flower faster than the eye can follow, striking the visitor and depositing or picking up pollen. Armbruster et al. (1994) found remarkably little overlap among 31 Western Australian species of *Stylidium* in the placing of pollen on visiting insects.

Variations of flower-colour and form influencing the behaviour of visitors and varied structural adaptations exist side by side as important features of the diversity and ecological adaptation of a wide range of flowers. But factors affecting flower-constancy are probably generally more important in species at the commoner end of this range, while structural differentiation (and perhaps also adaptation to unusual pollinators) has probably tended to evolve particularly in flowers which are more widely scattered in the landscape. The culmination of this latter tendency is seen in the orchids. In these plants (as in the unrelated Asclepiadaceae [Figs. 6.65, 66]) the pollen is aggregated into compact 'pollinia' which become firmly and precisely attached to the visitor, so maximising the chance that the pollen-load will reach its destination, even after many visits to flowers of other species. Pollinia may be carried long distances, but if there are many flowers of the same species nearby, the average dispersal distance may be quite

Fig. 6.64 The south-east Australian trigger-plant *Stylidium graminifolium*. The upper flower is in the 'set' position, with the column bearing the anthers and stigma arched back at one side of the flower. The lower flower has been 'triggered' as if by a visiting insect.

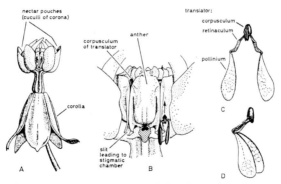

Fig. 6.65 The milkweed family (Asclepiadaceae) are particularly widespread and important in tropical, subtropical and warm-temperate regions, but *Asclepias* species reach as far north as southeastern Canada, and *Vincetoxicum officinale* grows on dry limestone over much of Europe to northern France and south Sweden, but just fails to reach Britain. The drawing (**A**) shows the tropical American *Asclepias curassavica*. In the middle of the flower the stamens and style are fused together to form a column (**B**), which bears a crown (*corona*) of nectar-pouches (*cuculli*), one to each anther. In the drawing, the cut base of one cucullus is hatched; the channel at the top carries nectar to the cucullus from the nectaries in the stigmatic cavities (Galil & Zeroni, 1965). The pollen forms compact pollinia, connected to a wishbone-shaped *translator* (**C, D**); the five translators and anthers alternate, so each translator is joined to one pollinium from each of the two adjacent anthers. Visitors come to the flowers for nectar, and as they fly away their legs, guided by grooves on the column, tend to catch in the clip formed by the corpuscula of the translators. Pollinia are thus withdrawn attached to the claws or hairs on the insects' feet. When the insect visits another flower the pollinia are drawn into the stigmatic slits between the anthers.

Fig. 6.66 Honeybee (*Apis mellifera*), sucking nectar from the cuculli of the corona of *Asclepias verticillata* (in cultivation). Notice the pollinia attached to the bee's front right foot.

short (Pleasants, 1991; Broyles & Wyatt, 1991; Broyles, Schnabel & Wyatt, 1994). A detailed account of the pollination of some prairie species of *Asclepias* in Wisconsin is given by Macior (1965). The remarkable structural and other adaptations involved in the pollination of orchids are the subject of the next chapter.

    Various different kinds of pollination relationships can thus be seen as highly advanced evolutionary end-points, not in competition with one another, but complementary. The pollination system of a particular species reflects both its recent ecology and evolution, and the longer-term constraints of its evolutionary origins. All families of flowering plants show a range of possibilities in their pollination systems – some more than others, but none embraces the whole range of anthecological diversity sketched in the last paragraph. Most families have a particular *métier*. It is significant that three of the largest families of flowering plants, the composites (Asteraceae), the grasses (Poaceae) and the orchids (Orchidaceae) – all three universally regarded as evolutionarily highly advanced – are characterised by quite different pollination systems.

7

The Pollination of Orchids

The orchids display a remarkably diverse range of intriguing and beautiful pollination mechanisms, and include some of the most extraordinary examples of adaptation to insect visitors. Although earlier botanists had described the structure of orchid flowers and observed visits by insects, the nature and variations in detail of pollination mechanisms in orchids were first fully appreciated by Charles Darwin. His book *The various contrivances by which Orchids are fertilised by Insects*, first published in 1862, is the record of a great deal of painstaking and perceptive observation. From 1842 until his death 40 years later, Darwin lived at Down, close to the crest of the North Downs in Kent. Many of the British orchids are plants of chalky soils, and it was in the country around Down that many of Darwin's observations were made. These were the orchids with which he was most familiar. They show well the essential features of pollination in the family, even though they embrace only a fraction of the variation found in other parts of the world, of which many examples are included in his book.

The orchids are, by common consent, one of the most advanced families of flowering plants – in the sense that the structure of their flowers has diverged more than almost any others from the condition of their primitive ancestors of 100 million years ago. They are the ultimate expression of the evolutionary trend of increasingly precise adaptation to particular flower-visiting insects seen in the zygomorphic flowers, considered in Chapter 6. But a very high degree of evolutionary advancement could equally be argued for the Composites (Asteraceae) (Chapter 6) and the grasses (Poaceae) (Chapter 9), which have quite different pollination mechanisms and relationships. It was suggested at the end of Chapter 6 that the very precise adaptations in orchids have evolved to provide specific and effective pollination of flowers which are often widely scattered and typically form only a small part of the bulk of the vegetation in which they grow. Many of the pollination systems that have evolved in this situation are in varying degrees pollinator-limited, which may make adaptive shifts to new pollination systems rather easy. The present chapter may be read with these thoughts in mind.

The structure of orchid flowers

At first sight, an orchid flower has little in common with the flowers of any other family. However, when it is examined in detail it appears that it is, in effect, an exceedingly specialised version of the kind of flower seen in the lilies and their relatives (Liliaceae). A lily or tulip flower has six perianth segments, three outer and three inner, six stamens, again in two whorls of three, and an ovary made up of three fused carpels – though the ovary is superior in the tulip or lily, but inferior in the orchids. The orchid flower also has six perianth segments, though one segment of the inner whorl is usually larger than the others, and is called the lip (or *labellum*). The lip is strictly the uppermost petal, but in most orchids the ovary is twisted through 180° so that the flower is, in fact, upside down. The lip then appears at the bottom, where it forms an alighting platform for insects; it often has a spur or nectary at its base. It is in the

stamens that the greatest modifications of the orchid flower are found. Most of the
stamens have been either completely lost or reduced to sterile vestiges. Two members
of the outer whorl are missing; Darwin believed that they had become fused with the
sides of the lip, but it is generally thought now that they have vanished without trace.
The remaining stamens, together with the stigmas, have become fused into a stout
column which projects in the centre of the flower, above the lip. The small tropical
Asian and Australian genus *Neuwiedia* (usually regarded as a primitive orchid, but
sometimes, with the related two-stamened *Apostasia*, placed in a separate family
Apostasiaceae) is a nicely illustrative 'missing link' with three fertile stamens (Dressler,
1993). In all other orchids, never more than two stamens are fertile, and in most orchids the only fertile stamen is the remaining one in the outer whorl. Only two stigmas
are functional; the third forms the *rostellum*, which generally projects from the top of
the column, and produces sticky matter whose function will be referred to repeatedly
in what follows.

Orchids with two stamens: the Lady's slippers

The orchids fall into two main groups. Most primitive are the Lady's slipper orchids,
Cypripedium and related genera (Fig. 7.1). In these, two stamens of the inner whorl are
fertile, the third forming the front of the column. The single remaining stamen of the
outer whorl is represented by a thick petal-like staminode, overarching the stigma.
The lip forms a deep pouch. The flowers of the European Lady's slipper *Cypripedium
calceolus* (Fig. 7.2) offer no reward, but the bright yellow lip and somewhat fruity scent
attract a variety of bees. In the colonies studied by Nilsson (1979a) on the Swedish
island of Öland, the commonest visitors, and the most important for pollination, were
medium-sized female solitary bees of the genus *Andrena*, especially *A. haemorrhoa*; in
Czechoslovakia, Daumann (1968) found that the main pollinators were larger *Andrena*
species such as *A. tibialis* and *A. nigroaenea*. The bees enter the lip through the obvious
large opening. Large bees, such as bumblebees (and, on Öland, the larger *Andrena*
species), seldom enter the flowers and if they do, they can generally quickly climb out
the same way as they came in. Smaller bees are trapped in the lip. After some minutes
of wing-buzzing and undirected efforts at escape, the bees begin to prise methodically
under the stigma, slightly depressing the elastic lip. In so doing, they free a passage for
themselves out through the back of the flower. From this point, the translucent 'window-panes' in the sides of the lip near its base probably help to guide the bee towards

Fig. 7.1 Lady's slipper
(*Cypripedium calceolus*). Flower
with half of lip removed to
show details of the column
and the path taken by a
visiting insect.

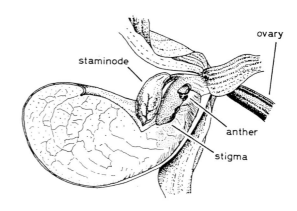

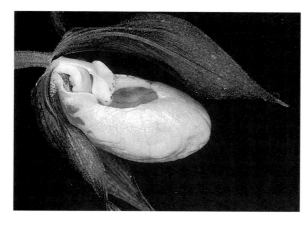

Fig. 7.2 Lady's slipper (*Cypripedium calceolus*); close-up of lip to show the large opening by which a visiting insect enters the flower, the translucent 'window-panes' at the back of the lip, and one of the smaller openings beneath the anthers by which the insect escapes.

one of the two narrow openings on either side of the stigma and staminode, past one of the stamens where some of the sticky pollen is smeared onto the upper side of its thorax. On visiting another flower, the bee will leave pollen from the first flower on the stigma before it squeezes out past one of the two stamens to pick up a further load of pollen. The mechanism does not always work in this neat and tidy way; a bee may have to make several attempts before it finds its way out of the lip. However, it seems to be rare for a flower to receive its own pollen in the course of a single visit. The Lady's slipper flower is generally an effective 'one-way-traffic' device, ensuring that the insect passes the stigma before it comes into contact with the anthers, which in a plant with few, large flowers should favour cross-pollination. This in itself is probably less effective than might be expected, because *Cypripedium* is rhizomatous and often grows in clonal patches. It has been suggested that after a few visits the bees learn to avoid the flowers; this would tend to limit pollinations within a single patch and the bees might then fly some distance before trying another *Cypripedium* flower. There seems to be no firm evidence on this question one way or the other.

The bees visiting *Cypripedium calceolus* are predominantly females. This may be related to the scent of the flowers. The scent of the European Lady's slipper is unusual in being dominated by octyl and decyl acetates, with smaller amounts of various other substances common in flower and fruit scents. These are chemically similar to constituents of pheromones important in odour-marking and other aspects of the behaviour of *Andrena* bees, suggesting that *Cypripedium* has evolved a scent which specifically manipulates the innate behavioural responses of its pollinators. Of two American forms of *C. calceolus*, var. *parviflorum*, pollinated by *Ceratina* bees (Anthophoridae), has a scent dominated by mono and sesquiterpenoids, and var. *pubescens*, pollinated by halictid bees, has a scent dominated by 1,3,5-trimethoxy benzene (Bergström *et al.*, 1992).

The large genus *Paphiopedilum* is the tropical counterpart of the bee-pollinated Lady's slippers. It has the same one-way pollination system, but the flowers are probably generally adapted to lure flies, beetles and perhaps other insects. The colours, prevailingly green, brown, dull red, purple and white in various combinations, the sometimes putrid smell, and other features of the flowers recall the deceptive fly-pollinated aroids and asclepiads (Chapter 10, pp.295ff). According to Atwood (1985) *P. rothschildianum* is pollinated by syrphid flies which are deceived into laying their eggs in the flowers, especially on the staminode (see also p.306).

The mainstream: orchids with a single stamen

In the great majority of orchids there is only a single, much-modified anther. Like most anthers, this is two-lobed, but by contrast with the loose, powdery pollen of most plants, the pollen grains are bound together by slender elastic threads into pollen masses or pollinia. The two anthers which are fertile in *Cypripedium* are reduced to projections on the column, often forming part of the *clinandrium* – the little hood protecting the fertile anther. In the greater number of orchids, the anther is borne by a comparatively slender stalk on the back of the column. Its apex is close to the rostellum, and it is by their apices that the pollinia become attached to a visiting insect. This condition is found in the huge subfamily Epidendroideae, which grow in a wide diversity of habitats and include almost all the epiphytic orchids of the tropics as well as such temperate orchids as the helleborines (*Epipactis, Cephalanthera*) and twayblades (*Listera*), and also in the Lady's tresses and related genera (subfamily Spiranthoideae). In the subfamily Orchidoideae, *Orchis* and its close relatives, the single anther is perched on top of the column, and the two pollinia are furnished with minute stalks or *caudicles* at their bases, attached to viscid discs (*viscidia*) formed of rostellum tissue, but in other members of the subfamily the relation of the pollinia to the viscid matter of the rostellum varies a good deal. The Orchidoideae are typically ground orchids, often grassland plants, and although a minority in the world orchid flora in both species and individuals, they are well represented in temperate regions and include many of the best-known orchid species of Europe and North America.

Insect pollination in some European orchids

The helleborines, Epipactis *and* Cephalanthera

The marsh helleborine *(Epipactis palustris)* is a locally-distributed plant of open calcareous fens and dune slacks over much of Europe. The whitish flowers (Figs. 7.3, 7.4), which have no noticeable scent, are borne in a loose raceme on a stem 10-30 cm or more tall, and are visited by a wide diversity of insects, including various bees, social and solitary wasps, ants and two-winged flies (Nilsson, 1978b). The summit of the column in a newly-opened flower is occupied by the large, projecting, almost globular rostellum, with the broad squarish stigma below it. The anther overhangs the rostellum. Even before the bud opens the anther cells dehisce, releasing the rather friable pollinia which come to lie with their tips touching the rostellum. As the rostellum matures, its outer surface develops into a soft elastic membrane; at a slight touch it becomes viscid, so that the pollinia stick to it. The tissue within the rostellum develops into a lining of sticky matter which, on exposure to air, hardens in a few minutes. An object brushing upwards and backwards against the rostellum easily removes the whole of

Fig. 7.3 Marsh helleborine *(Epipactis palustris).*

the elastic skin of the rostellum as a little cap, which sticks firmly by its adhesive lining. The base of the lip forms a cup, the *hypochile*, containing nectar; its broad flat tip, the *epichile*, is attached to the base by a slender, elastic 'waist'. An insect visiting the flower for nectar and alighting on the lip depresses it, and as long as it is feeding, is well clear of the rostellum and the pollinia. But as soon as it makes to leave the flower and takes its weight from the lip, the epichile returns to its original position. Darwin saw this as an important element in the mechanism of the flower, causing the visitor to fly upwards and strike the rostellum with its head as it left the flower, but Nilsson's observations cast doubt on this, and Darwin himself ex-

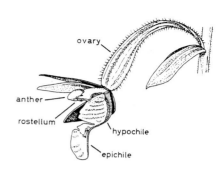

Fig. 7.4 Marsh helleborine (*Epipactis palustris*). Side view of flower with half of perianth removed.

pressed reservations in the second edition of his book. In the Isle of Wight, Darwin's son, William, observed visits by honeybees, but he also saw visits (and pollinia removed) by small solitary wasps and by various Diptera. Honeybees are an introduced species in northern Europe, so they cannot be the original pollinators of *E. palustris*. Nilsson saw very few honeybees on marsh helleborines in Sweden. The flowers are probably primarily adapted to pollination by solitary wasps, perhaps particularly *Eumenes*, but a wide range of insects *can* bring about pollination, including honeybees if they are abundant. If the flowers are not visited by insects, the friable pollinia sooner or later break up, and the loose pollen falls down over the the rostellum and stigma. This may happen even before the flower opens, and the relative importance of cross and self-pollination probably varies greatly at different times and in different localities.

The broad-leaved helleborine (*Epipactis helleborine*), another species very widespread in Europe (and naturalised in eastern North America), is a shade-loving plant perhaps more common along shady roadsides and wood margins than in extensive woods. The flowers (Fig. 7.5) vary in colour from dull purple to greenish, and are pollinated

Fig. 7.5 Flowers of broad-leaved helleborine (*Epipactis helleborine*).

mainly by common wasps (*Vespula* spp.) which are sometimes attracted to the flowers in considerable numbers, perhaps in the first instance by scent. The colour of the flowers is reminiscent of the dingy brownish-purple of the wasp-pollinated figworts (*Scrophularia* spp.) (Chapter 6, pp.106, 181). The mechanism of pollination is similar to that of the marsh helleborine, but the smaller lip is not hinged in the middle and removal of the pollinia depends entirely on the insect's head striking upwards against the rostellum as it backs out of the flower; the more protuberant rostellum no doubt aids in bringing this about. This species can also be self-pollinated. According to Hagerup (1952), pollen grains which fall onto the rostellum are trapped by its viscid secretion, which later spreads out over the stigmas, where the pollen grains germinate and bring about fertilisation; Waite *et al.* (1991) found substantial levels of selfing, especially in short, few-flowered inflorescences. In common with orchids known to be regularly selfed, the broad-leaved helleborine shows remarkably regular production of well-developed capsules, nicely graded in size as they mature in succession from the bottom of a long flower-spike to the top.

The helleborines of the related genus *Cephalanthera* illustrate the way in which the *Epipactis* type of pollination mechanism probably originated. In size and structure the flowers are not unlike those of *Epipactis*, but the centre stigma lobe forms no more than a rudimentary rostellum and the pollen grains are only weakly bound together by a few elastic threads. The narrow-leaved helleborine (*C. longifolia*), which is distributed very widely across Eurasia, is mainly pollinated by small bees. Dafni & Ivri (1981b) observed numerous visits by *Halictus* species in Israel. Its white flowers (fragrant there, but apparently not so in northern Europe) provide no nectar; visiting bees are presumably attracted by the scent and the 'pseudopollen' of the papillose yellow ridges on the lip. The flowers open rather widely, but the narrow tubular space between the lip and the column forces an insect penetrating the flower against the stigma, which is covered with a copious sticky secretion. Leaving the flower, the insect brushes past the anther, which arches over the front of the stigma, and the friable pollen adheres to the stigmatic secretion smeared on its back. The anther has an elastic hinge at its base and springs back to its original position as soon as the insect has gone. It is arguable whether the lack of a developed rostellum, and other features of the *Cephalanthera* flower, are primitive or degenerate, but the flower nicely illustrates how the *Epipactis* condition might have evolved from that seen in ordinary flowers with friable pollen and sticky stigmas. It is particularly interesting that the sticky stigmatic secretion serves to stick the pollen to the visiting insect, because it is the sticky secretion of the rostellum – generally thought to be derived from a vestigial stigma – which takes over this function in all the more advanced orchids. Self-pollination evidently does not take place in *C. longifolia* in Israel, because no seed is set if pollinators are excluded. The red helleborine (*Cephalanthera rubra*) is also visited by bees, and apparently (to insect senses) mimics tall *Campanula* species with which it often grows; the orchid sets more seed in localities where *Campanula* is present. It is curious that the helleborine is pollinated by early-emerging male leaf-cutter bees of the genus *Chelostoma*, while the *Chelostoma* females gather pollen almost exclusively from *Campanula* (Nilsson, 1983c). The self-pollinated white helleborine (*C. damasonium*) is discussed later in this chapter.

The twayblades: precision from unspecialised visitors

The common twayblade (*Listera ovata*) (Figs. 7.6, 7.8) shows a rather similar mechanism to that of *Epipactis*, with some interesting differences in detail. The greenish flowers are borne in a long, slender raceme above the two leaves that give the plant its name.

a   **b**

*Fig. 7.6 **a-b**, Ichneumon wasp (Ichneumon sp.) visiting flowers of twayblade (Listera ovata). **a**, the insect is sucking nectar from the groove up the centre of the lip. **b**, the insect has reached the base of the lip and its head is about to make contact with the pollinia.*

They are particularly attractive to ichneumons (a twayblade with a visiting ichneumon appears on the title page of Sprengel's classic book), especially the males, which visit the flowers in considerable numbers; sawflies and beetles are also frequent visitors (Nilsson, 1981), and these three groups are the principal pollinators. The scent is dominated by two common monoterpenes, linalool and *trans*-β-ocimene; some of the many minor ingredients may play a part in attracting these particular insects. The lip of the flower is broadly strap-shaped, bent sharply downwards from a point near its base, and deeply notched at the tip. It forms a landing platform leading up to the column; a groove, secreting much nectar, runs from the notch up the centre of the lip. The anther lies behind the rostellum, protected by a broad expansion of the back of

*Fig. 7.7 **a-b** Twayblade (Listera ovata):* **a**, newly-opened flower; the pollinia have been removed on the head of a pin. **b**, older flower; the column has curved upwards so that pollinia can now come into contact with the stigmas.

a   b

Fig. 7.8 **a-c** Pollination of twayblade (*Listera ovata*), by the skipjack beetle *Athous haemorrhoidalis*: **a**, beetle with freshly acquired pair of pollinia on its head. **b**, beetle has arrived at another flower and is beginning to suck nectar from the groove on the lip. **c**, pollinia have come into contact with the stigma to which pollen is firmly adhering.

the column. As in *Epipactis*, the anther cells dehisce before the bud opens, and the pollinia are left quite free, supported in front by the concave back of the rostellum. A visiting insect crawls slowly up the narrowing lip, feeding on the copious nectar, which leads it to a point just below the rostellum. On the gentlest touch the tip of the rostellum exudes, almost explosively, a drop of viscid liquid which, coming into contact with the tips of the pollinia and the insect, cements them firmly to its head and sets in a matter of seconds. As the drop of viscid matter is expelled, the rostellum bends sharply downwards, but within 2-3 hours it straightens from its arched position close to the lip, leaving clear the way to the stigma (Fig. 7.7). Now an insect crawling up the nectar groove can pollinate the flower with pollinia brought from another, younger flower. It has been suggested that the almost explosive ejection of the viscid matter from the rostellum may startle the pollinating insects sufficiently to make them fly to another plant before they start feeding again. However, many visiting insects seem little disturbed by the explosion of the rostellum and, as insects generally work up the inflorescence from the bottom, this probably makes little difference to the effectiveness with which cross-pollination is brought about. The most interesting feature of the pollination of twayblade is the way in which a precision mechanism has evolved depending on relatively undiscriminating pollinators; the twayblade probably attracts a greater diversity of insects than any other European orchid. It is striking and curious that an ichneumon or a skipjack beetle (Fig. 7.8) will operate the mechanism neatly and accurately, but the flower evidently does not provide the appropriate cues for orientation of the bees which visit this species casually for nectar and in general are not effective pollinators. The twayblade depends entirely on insects for pollination; self-pollination under natural conditions is apparently rare (Nilsson, 1981). The tiny lesser twayblade (*Listera cordata*), which grows in boreal and mountain coniferous forests and on moist upland heather moors, has a similar floral mechanism. The principal pollinators in North America are fungus gnats (Ackerman & Mesler, 1979); in California, experimentally emasculated flowers showed a capsule set of 72% which must all have been due to pollinia brought from other spikes (Mesler *et al.*, 1980). However, this species seems to be largely autogamous in north and west Europe.

The Lady's tresses orchids, Spiranthes

The Lady's tresses orchids and their relatives (subfamily Spiranthoideae) make up a rather distinctive, largely tropical, group. Autumn Lady's tresses (*Spiranthes spiralis*) is an attractive though inconspicuous little orchid locally common in central, western and southern Europe in short turf in late summer. In the south of England the leaf-rosettes die down about May, and the leafless flower spikes appear in August or early September, just before the new season's leaf-rosette emerges beside them. The small, tubular, sweetly-scented whitish flowers are borne in a spiral on the upper part of the stem, giving a plait-like appearance, hence the English name of the plant. The flowers

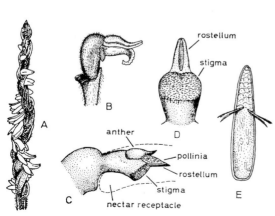

Fig. 7.9 Autumn Lady's tresses (*Spiranthes spiralis*). **A**, inflorescence. **B**, single flower, with lower sepal removed. **C**, detail of column; the broken line indicates the outline of the perianth. **D**, front view of column. **E**, disc with pollinia attached. B–E after Darwin.

Fig. 7.10 Autumn Lady's tresses (*Spiranthes spiralis*), with visiting bumblebee.

project almost horizontally from the stem and never open widely (Fig. 7.9). The rostellum is a slender flattened structure, projecting forwards above the stigma. The central part of its upper surface consists of an elongated mass of thickened cells forming a wedge-shaped viscidium (which Darwin called the 'boat-formed disc') to which the tips of the pollinia are attached; the viscid matter on its underside is protected by the delicate membrane of the lower surface of the rostellum. At a touch, the membrane splits down the middle and around the edges of the viscidium, exposing the viscid matter and leaving the viscidium free but supported between the prongs of a fork formed by the sides of the rostellum. In a newly-opened flower, the column lies close to the lip, leaving only a narrow passage for the tongue of a bee visiting the flower to reach the nectar in the cup-shaped base of the lip. The flower cannot be pollinated, but the bee inevitably touches the lower surface of the rostellum, and the 'boat-formed disc' with its attached pollinia becomes cemented to the upper side of its proboscis. Following removal of the pollinia, the remains of the rostellum wither and the column and lip slowly move apart, leaving the stigma freely exposed to pollen brought by a visiting bee from another flower. The flowers at the bottom of a spike always open first, and those at the top last, so that while the most recently-opened flowers at the top of the spike have pollinia waiting to be removed, those at the bottom of the spike are ready for pollination. The principal visitors are bumblebees (Fig. 7.10), which invariably start at the bottom of a spike and work upwards, visiting flowers in succession until they reach the top and fly off to repeat the process on another flower-spike, so cross-pollination is practically assured. The flowers appear to be quite freely visited. In southern England, Darwin observed visits by bumblebees at Torquay, and visits by bumblebees are frequent at a population on the university campus in Exeter (MCFP), where a good deal of seed is set. Visits by honeybees have been observed at Exeter and in Gloucestershire (K.G. Preston-Mafham), where the flowers were also visited by solitary bees (probably *Andrena* sp.); and pollinia of *S. spiralis* were seen on an unidentified small solitary bee in Dorset (MCFP). The floral mechanism of temperate North American species of *Spiranthes* is essentially similar. Bumblebees are the principal pollinators of most of the species, with leaf-cutter bees (Megachilidae) playing a minor role (Catling, 1983).

Orchis *and related genera: adaptive radiation – and deception*

The species which Darwin took as the first example in his book, the early purple orchid (*Orchis mascula*) occurs almost throughout Europe and is surely one of the best known of all orchids. In Britain, it is often common in woods and pastures and along roadsides in spring and early summer, with its bright purple flowers borne above rosettes of dark-spotted leaves in a rather loose spike which may be anything from 5 cm to 40 cm or more tall. The flowers emit a rather strong and, to our senses, unpleasant scent, due mainly to monoterpenes, especially pinenes, myrcene and *trans*-β-ocimene

(Nilsson, 1983a). The lip is broad, flat or somewhat reflexed at the sides, slightly lobed, and with a long stout spur at its base (Fig. 7.11). The two lateral sepals spread widely, while the upper sepal and the two upper petals form a hood over the single anther, which stands erect just above the wide entrance to the spur. The pollen grains are aggregated into small compact masses (*massulae*) which are bound together by slender elastic threads into a pair of club-shaped pollinia; the elastic threads run together at the base to form the slender stalks (*caudicles*) by which the pollinia are attached to a pair of sticky discs, the viscidia, formed of rostellum tissue. The rostellum forms a protective pouch, the bursicle, enclosing the viscidia immediately over the entrance to the spur. Just behind this, forming the upper side of the throat of the spur, is the sticky stigmatic area.

The anther cells open even before the flower expands, so that the pollinia are quite free within them. An insect visiting the flower lands on the lip, and inserts its proboscis into the spur. In doing so, it can hardly avoid touching the pouch-like rostellum. At the slightest touch this ruptures along the front, and the bursicle is easily pushed back by the insect's movements, exposing the sticky discs attached to the bases of the pollinia. Almost infallibly, one or both will touch the insect, and stick firmly to it. Curiously the spur contains no free nectar. Darwin thought that visiting insects pierced the cells of the wall of the spur to feed on the abundant cell sap, but it is now generally accepted that the spur is, as Sprengel believed, merely a sham nectary (Daumann, 1941; Dafni, 1984). Deception, as we shall see, is a recurrent theme amongst the ground orchids. In the few seconds that the insect remains at the *O. mascula* flower, the viscid matter sets hard and dry, and the insect leaves the flower with a pollinium, or a pair of pollinia, cemented like horns to its head. To begin with, the pollinia lie in much the same direction as they occupied in the flower from which they came. In this position, if the insect were to visit another flower, they would simply be pushed against the pollinia there. But about half-a-minute after their removal, as the membrane forming the top

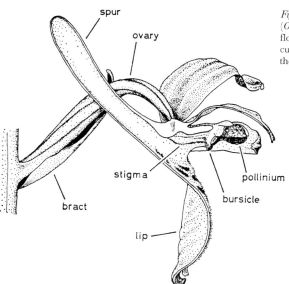

Fig. 7.11 Early purple orchid (*Orchis mascula*). Side view of flower with half of perianth cut away to show details of the column.

a   **b**

Fig. 7.12 **a–b** Common spotted orchid *(Dactylorhiza fuchsii)*. **a**, pollinia freshly removed from a
flower on the point of a pencil. **b**, pollinia about half-a-minute after removal, now in a
position to strike the stigma.

of the viscidium dries out, each pollinium swings forward through an angle of about
90°. This movement, completed in a time which would allow the insect to fly to another flower, brings the pollinia on the head of a suitably-sized insect (face about 3.2
mm wide) into exactly the right position to strike the sticky stigmas, leaving a layer of
pollen massulae on the surface. The remainder of the pollinium remains firmly attached to the insect's head, and a single pollinium can pollinate several flowers. The
whole process is easily reproduced if a well-sharpened pencil is substituted for the
tongue of the insect (Fig. 7.12). However, many visitors are larger or smaller than the
optimum size, so in practice the mechanism works with less than perfect precision
(Nilsson, 1983a).

    In Britain, Sweden and other parts of northern Europe, the main visitors are queen
bumblebees and cuckoo bees *(Psithyrus)* recently emerged from hibernation, and males
of the solitary bee *Eucera longicornis*, with occasional visits from other solitary bees and
Diptera. Farther south in Europe solitary bees are probably the main pollinators; in
fact the form of the flowers seems better adapted to these than to bumblebees. The
plant appears to be exploiting its superior floral display at a time when the bees are
generally inexperienced and have not yet established nests and regular foraging rou-
tines, and food flowers of any kind are rather few. Bumblebees and *Eucera* males alight
at the bottom of the spike and visit only one or a very few flowers before flying off.
This results in a characteristic rapid decline in fruit-set from the bottom of the spike
upwards (Nilsson 1983a). Nilsson found enormous variation in the number of flowers
in a spike setting capsules, with population means varying from about 3% to 20%.

    A mechanism identical in all its essentials is found in many other members of the
genus *Orchis*, and in the marsh and spotted orchids *(Dactylorhiza* spp.) (Figs. 7.12, 13).

These plants are variously pollinated by social and solitary bees and Diptera. The widespread continental European elder-flower, or 'Adam and Eve' orchid (*Dactylorhiza sambucina*), with its apple-scented flowers and striking red-purple/yellow colour dimorphism, is pollinated on the Baltic island of Öland almost entirely by bumblebees; the proportion of flowers setting seed varied between localities and years from just over 2% to nearly 50% (Nilsson, 1980). The common and heath spotted orchids (*D. fuchsii* and *D. maculata*) appear to be pollinated mainly by bumblebees and honeybees, but various Diptera and beetles (Gutowski, 1990) also visit the flowers and may locally be significant pollinators; according to Hagerup (1951), *D. maculata* depends for pollination mainly on the drone-fly *Eristalis intricarius* in the Faroes and Iceland. Honeybees and the bumblebees *Bombus lapidarius* and *B. terrestris* were the predominant visitors to a large colony of *D. fuchsii* near Tring in Buckinghamshire

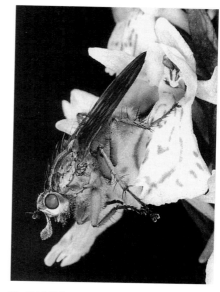

Fig. 7.13 Yellow dung fly (*Scathophaga stercoraria*), bearing a pair of pollinia, on heath spotted orchid (*Dactylorhiza maculata*).

(Dafni & Woodell, 1986). At this site, the bees appeared to be exploiting the copious stigmatic secretion, which contains glucose and amino acids, as a true 'reward'. On average, 53.7% of the flowers produced capsules, and the proportion in some plants was as high as 90%. As in *Orchis mascula*, most seed was set in the lower part of the spikes, but this could at least in part reflect nutrient limitation for capsule development. In a population of *D. fuchsii* at Leith Hill in Surrey, the proportion of pollinia removed, and the proportion of flowers pollinated, was extremely variable (Waite *et al.*, 1991); the proportion of flowers producing capsules ranged from zero to 97%, with a mean of 46.6% in grassland but only 10.6% in woodland. The examples just quoted are instances of rather generalised food deception; the orchid simply has the kind of floral display and scent that would normally be associated with a nectar-producing flower. It is probably significant that many of these orchids are variable in flower colour and form, thus making it more difficult for insects to learn to avoid them (Dukas & Real, 1993). Sometimes the orchid appears to mimic a particular nectar-producing species, as in the case of *Orchis israelitica* which shares the same flowering season and pollinators – mainly solitary bees – with the common east Mediterranean spring-flowering bulb *Bellevalia flexuosa*, with which it often grows (Dafni & Ivri, 1981a). A different category of deception is apparent in the orchids that specifically attract males of one particular pollinator; in general this specificity must be due to scent. An example is the east Mediterranean *Orchis galilaea*, common in Israel, which is pollinated exclusively by the males of the bee *Halictus marginatus* (Bino *et al.*, 1982); more extreme cases of sexual deception are described later in this chapter. The gaudy pink Mediterranean *Orchis papilionacea* is pollinated by patrolling males of the anthophorid bee *Eucera tuberculata* (Vogel, 1972).

These deceptive orchids must certainly be derived from nectar-producing ancestors.

Only two species of *Orchis* are known to produce nectar, the widespread European bug orchid (*Orchis coriophora*) and the similar east Mediterranean *O. sancta*. In Israel, Dafni & Ivri (1979) found the fragrant flowers of *O. coriophora* freely visited, especially by honeybees and *Nomada*, and most of the flowers produced capsules. There are many nectar-producing flowers in genera related to *Orchis*, and there has been a good deal of adaptive radiation to different pollinators amongst them.

Some butterfly- and moth-pollinated orchids: Gymnadenia, Platanthera, Nigritella *and* Anacamptis

The fragrant orchid *(Gymnadenia conopsea)*, with its slender spikes of heavily scented long-spurred pink flowers (Fig. 7.14), occurs throughout Europe; the smaller, shorter-spurred and even more sweetly-scented *G. odoratissima* is mainly central European. Both produce copious nectar and are pollinated by Lepidoptera. The flowers of both species are rather small, with a short three-lobed lip. The pollinia are placed so that their elongated viscidia form part of the arched roof of the entrance to the slender spur. The rostellum does not form a bursicle, so the two viscidia are freely exposed to the air. The viscidia become fixed lengthwise to the tongue of a visiting insect, and stick sufficiently firmly even though the viscid matter does not set hard as in *Orchis* and its near allies – though, as if to compensate for this, the pollinia are more friable and the massulae more easily detached than in *Orchis*. After removal of the pollinia, the caudicles bend forward and downwards, so that they come to lie almost parallel with the tongue of the insect. In this position they readily strike the two protuberant stigmas, to right and left of the entrance to the spur, when the insect visits another flower. Darwin saw visits by a number of noctuid moths to *G. conopsea*, and this species seems to set abundant seed; day-flying forester moths (*Adscita* sp.) are among the visitors to *G. odoratissima* in Switzerland (MCFP).

Fig. 7.14 Fragrant orchid (*Gymnadenia conopsea*).

The butterfly orchids *Platanthera bifolia* and *P. chlorantha* resemble the fragrant orchids in their naked viscidia. In fact, the pollination mechanism of the lesser butterfly orchid (*P. bifolia*) (Fig. 7.15a) is very like that of the fragrant orchid, but the rather spidery fragrant white flowers are more obviously adapted to attracting night-flying moths. Pine and small elephant hawkmoths are major visitors to this species in Sweden, but noctuid moths are probably the main pollinators of the shorter-spurred races of *P. bifolia* that occur in oceanic regions of Europe such as the British Isles (Nilsson, 1983b). The small, round viscidia are placed facing each other close together over the mouth of the spur and the stigmas, and become attached to the tongues of visiting moths. The greater butterfly orchid (*P. chlorantha*) (Fig. 7.15b), though closely related to *P. bifolia* and very like it in the superficial form of its flowers, is strikingly different in the form of the column and the disposition of

Fig. 7.15 **a–b** The common European butterfly orchids, **a**, lesser butterfly orchid (*Platanthera bifolia*; south-west England form); the pollinia, close together over mouth of spur, are carried on the tongues of visitors.
b, greater butterfly orchid (*Platanthera chlorantha*); the pollinia, widely spaced at the base, are carried on the compound eyes of visiting moths.

a

b

the pollinia and viscidia. The viscidia are placed wide apart, at either side of the entrance to the spur, with the pollinia forming an arch over the large confluent stigmas. The flowers are visited largely by night-flying noctuid moths, but the pollinia become attached to the insect's head, usually to its compound eyes – which may become so plastered with viscidia that the insect can hardly see. This difference in the structure of the column and spacing of the viscidia places an effective breeding barrier between the two species, and, although they are freely interfertile and overlap both geographically and in flowering time, hybrids are rare. The 'mechanical isolation' is reinforced by a difference in scent. On Öland, Nilsson (1983b) found about 22% linalool and 60% methyl benzoate (long known to be attractive to hawkmoths) in the scent of *P. bifolia*, whereas the scent of *P. chlorantha* was dominated by lilac alcohols (*c.* 70%) with *c.* 25% methyl benzoate.

Of 333 visitors to *P. chlorantha* recorded by Nilsson (1978b) on Öland, 280 were noctuids, belonging to 22 species. The most frequent (106 individuals), and the species carrying the most pollinia, was the plain golden-Y moth (*Autographa jota*). The tongue length of this species almost exactly matched the most frequent depth of accumulated

Fig. 7.16 **a–b** Pyramidal orchid (*Anacamptis pyramidalis*). **a**, five-spot burnet moth (*Zygaena* cf. *trifolii*), visiting a flower; the proboscis of the moth bears several pairs of pollinia. **b**, another five-spot burnet moth, probing for nectar in an inverted position; the collar-like viscidium encircling the proboscis is clearly visible.

nectar in the spurs, and its head width was just slightly less than the mean distance between the viscidia. The large and small elephant hawkmoths (*Deilephila elpenor* and *D. porcellus*) were also rather frequent visitors on Öland, but with their larger heads are probably less effective pollinators than the noctuids. Pollinium removal (and pollination) can only take place if the insect brings its head up to the mouth of the spur, so longer-tongued moths such as the larger hawkmoths can suck nectar without bringing about pollination. This must impose a selection pressure for adaptation to the longest-tongued of the plant's major visitors (Nilsson, 1988). Judging from the amount of seed set, the butterfly orchids have an efficient means of pollination. In counts made over three seasons on Öland, Nilsson found seed-set in *P. chlorantha* ranging from 31% (probably explained by bad weather) to 78%.

The sweetly vanilla-scented *Nigritella nigra* of the high pastures of the Alps, with its small, dark red-purple flower-head, is also a butterfly flower, producing nectar in a short, narrow-mouthed spur. As in other head-like inflorescences visited by butterflies (e.g. red valerian and hemp agrimony, p.182), the form of the perianth is of little consequence to the pollination mechanism; there is, in fact, little differentiation between the lip and the other perianth members, and in *Nigritella* (unlike most other orchids) the ovary is not twisted, so the lip of the flower is at the top. Otherwise, the flower works in much the same way as the fragrant orchids, except that the pollinia become attached to the lower side of the visitor's proboscis.

After these nectar-providing moth and butterfly flowers, the pyramidal orchid (*Anacamptis pyramidalis*) poses something of an enigma. It has all the marks of a beautifully-adapted butterfly flower. The rather small, pink, sweet-scented, long-spurred flowers are borne in a dense pyramidal spike. The three-lobed lip bears two conspicuous projecting ridges, forming a guide like half a funnel leading into the narrow mouth of the slender spur; Darwin compares them to the sides of a bird decoy. The pollinia are borne on a single saddle-shaped viscidium, placed very low on the column over the mouth of the spur, so that the two stigmas (which are confluent in many orchids) are here widely separated. Butterflies and moths, including both day-flying burnet moths (*Zygaena* spp.) (Fig. 7.16) and night-flying noctuids, visit the flowers in large numbers. The proboscis of a visiting insect is guided straight into the mouth of the spur by the converging ridges on the lip. As it is inserted into the spur it brushes past the bursicle, which moves back at a slight touch and exposes the viscidium. As soon as this is exposed to the air it begins to curl inwards, clasping the insect's proboscis around which it fits like a collar. Indeed, if the proboscis is slender, the two ends of

Fig. 7.17 Glanville fritillary butterfly (*Melitaea cinxia*), on pyramidal orchid (*Anacamptis pyramidalis*), with a pair of pollinia on its proboscis.

the viscidium may encircle it completely. Within a few seconds the viscid matter has set and the pollinia are firmly cemented in place, though owing to the curling of the viscidium they now diverge more widely than they did in the anther. After a short interval they begin to swing forward, and soon come to project one on either side of the insect's proboscis, exactly placed to contact the two stigmas when the insect visits another flower.

Both the number of pollinia removed and the number of capsules produced by the pyramidal orchid suggest that this finely-coordinated mechanism is generally highly effective. Many species visit the flowers, and individual moths often visit this species repeatedly. Darwin remarks on a noctuid moth bearing eleven pairs of pollinia of *A. pyramidalis* on its proboscis, 'The proboscis of this latter moth presented an extraordinary arborescent appearance!' Yet Darwin was unable to find even a trace of free nectar in the spurs of the pyramidal orchid, and was driven to conclude that visiting insects must suck from the intercellular spaces nectar secreted within the tissue of the spur. This is clearly not so in the early purple orchid, in which recent study has confirmed that there is no nectar and the flower depends on deception for pollination, as Sprengel suggested two centuries ago (Nilsson, 1983a). But insects visit the pyramidal orchid so persistently that one can only echo Darwin's comment (after remarking on a spike of *A. pyramidalis* which had produced twice as many capsules in the upper as in the lower half), '... it appears to me quite incredible that the same insect should go on visiting flower after flower of these Orchids, although it never obtains any nectar,' (Darwin, 1877). The relation of this orchid to its pollinators calls for critical experimental study.

Beetles and wasps, and parallels with the twayblades: Coeloglossum *and* Herminium

The frog orchid (*Coeloglossum viride*) (Fig. 7.18) and the musk orchid (*Herminium monorchis*) have inconspicuous flowers, secreting nectar and pollinated by various small crawling and flying insects. They represent a further variation on the *Orchis* theme, and provide interesting parallels with the twayblade, which has a similar range of pollinators. The green to brownish-tinged flowers of the frog orchid are like those of

Fig. 7.18 Frog orchid (*Coeloglossum viride*).

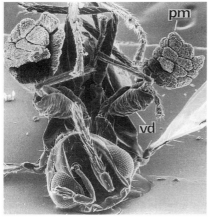

Fig. 7.19 Chalcid wasp (*Tetrastichus conon*, female) with a pollinium of musk orchid (*Herminium monorchis*) attached to each front femur. Scanning electron micrograph, × 64.

Orchis in structure, but the lip is broadly strap shaped with a short broad spur at its base and the remaining perianth segments form a helmet over the column. The two viscidia are rather widely spaced; the stigma is small and lies in the centre of the flower between them. Nectar is secreted in the spur, but in addition there are two small nectaries on either side of the lip close to its base, and almost beneath the viscidia. The lip has a median ridge, which tends to make an insect landing on it crawl up one side or the other, towards one of the drops of nectar beneath the viscidia, rather than up the middle. Feeding at one of these nectaries the insect can easily remove a single pollinium on its head. The forward movement of the pollinium, completed within a minute in *Orchis*, takes 20 minutes or half-an-hour in the frog orchid – time enough for its rather slow-moving pollinators to visit another spike. The pollen is then readily transferred to the central stigma as the insect explores the nectaries at the base of the lip. Beetles are probably among the commonest pollinators of the frog orchid; soldier-beetles (*Rhagonycha fulva*) and a small black sawfly have been seen on the inflorescences, bearing pollinia, in Sussex (K.G. Preston-Mafham). Silén (1906a) observed many visits by beetles of the genus *Cantharis* in northern Finland, and some visits by ichneumons and other insects. The smaller yellowish-green flowers of the musk orchid do not expand widely, and the lip, which has a very short spur at its base, does not differ greatly from the other petals. The flowers are visited by a variety of small Diptera and Hymenoptera; the main pollinators are female parasitic wasps of the genus *Tetrastichus* (Nilsson, 1979b). Attracted by the characteristic fragrance (probably mainly *p*-methoxybenzaldehyde, with common monoterpenes), these crawl into the flowers on either side between the perianth segments to seek nectar in the spur. So placed in a semi-inverted position in the corner of the flower, the insect's leg is immediately below one of the relatively large saddle-shaped viscidia, which become transversely attached to the femur, usually near its base (Fig. 7.19). The stigmas are transversely orientated with their broadest parts just below the viscidia, where they receive pollen when the insect visits another flower. Insect visits are essential for pollination, but as *Tetrastichus* wasps are virtually ubiquitous, seed-set is generally good; Nilsson found that about 70% of flowers formed capsules in southern Sweden.

Sexual deception in *Ophrys* and other genera: 'pseudocopulation'

The discovery of sexually-deceptive pollination in the genus Ophrys

Quite the most remarkable pollination mechanisms among European orchids – and indeed among the most remarkable to be found in any plants – are those of the 'insect orchids' of the genus *Ophrys*. These orchids are well known for the fancied resemblance of their flowers to various insects. What function, if any, this resemblance served was long a matter for conjecture; Darwin was plainly puzzled by it. It was not until the early decades of the twentieth century that it was discovered that pollination in most species is brought about by insects going through part at least of their mating behaviour in response to the flower. This process, often called *pseudocopulation*, was first elucidated by Pouyanne (Correvon & Pouyanne, 1916; Pouyanne, 1917), who observed the common Mediterranean mirror orchid (*Ophrys speculum*) for many years in Algeria, where he was *Président du tribunal de Sidi-Bel-Abbès*. His observations on *O. speculum* and other species were soon confirmed by those of Col. M.J. Godfery (1925 onwards) in the south of France. Not long afterwards, a similar sexually-deceptive mechanism was described by Mrs Edith Coleman (1927 onwards) in the south-east Australian tongue-orchid *Cryptostylis leptochila*, pollinated by males of the ichneumon *Lissopimpla excelsa*. Since then, pseudocopulation has been found in a number of other southern Australian genera, and it may well occur in more genera and species of orchids, and involve a greater diversity of pollinators, in southern Australia than anywhere else in the world. The discovery that two species of *Disa* in the Cape Province of South Africa are pollinated similarly, *D. atricapillata* by a sphecid wasp *Podalonia canescens*, and *D. bivalvata* by a pompilid wasp *Hemipepsis hilaris*, adds yet further examples of sexual deception from a different, yet climatically similar, part of the world (Steiner, Whitehead & Johnson, 1994).

Ophrys is represented by many species in southern Europe, north Africa and the Levant. The flowers are similar in general plan to *Orchis*, but the lip is thick, brownish and velvety-textured, often with metallic bluish markings, there is no spur, and the two viscidia are covered by separate bursicles. Much of our knowledge of *Ophrys* pollination biology is due to the researches of Kullenberg (1961).

The mirror orchid, Ophrys speculum (O. vernixia)

The mirror orchid (Plate 6b; Fig. 7.20B) is a widespread and common Mediterranean species which will be familiar to many people who have spent spring holidays anywhere between Portugal and the Aegean islands. The lip is like an oval convex mirror, of a curious glistening metallic violet-blue colour, with a narrow yellow border thickly fringed with long reddish brown hairs. The thread-like dark red upper petals can be imagined as simulating an insect's antennae. Pouyanne found, from 20 years' observa-

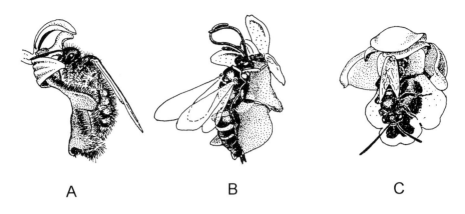

A   B   C

Fig. 7.20 Insect visits to *Ophrys* flowers. **A**, male of the solitary wasp *Campsoscolia ciliata* on a flower of *O. speculum*. **B**, male of the solitary wasp *Argogorytes mystaceus* visiting a flower of the fly orchid, *Ophrys insectifera*. **C**, male of the bee *Andrena maculipes* visiting a flower of *O. lutea*. After photographs by Kullenberg (1956a, 1961).

tion, that *O. speculum* is visited by one insect only, the scoliid wasp *Campsoscolia ciliata*, and of that species only by the males. The females ignore the flowers, although both sexes visit other flowers for nectar.

Campsoscolia ciliata is rather larger than a honeybee; each segment of the abdomen is fringed with long red hairs. The males appear several weeks before the females, and Pouyanne often saw them during March skimming with a swift zig-zag flight over the dry sunny banks where the wasps make their burrows. The females spend much of their lives underground, hunting for the beetle larvae with which they provision the burrows for their own progeny, and scarcely leave the soil except to mate and feed.

The flowers of *Ophrys speculum* are eagerly sought out and visited by the males,[1] though the insect neither seeks nor finds nectar or other food. Although the flowers have no appreciable scent to us, the males can detect their presence at some distance; Pouanne remarked that if one sat in the sun 'un petit bouquet d'*O. speculum* à la main', the flowers soon attracted the insects, sometimes several hustling one another on the same flower, and apparently oblivious of the observer. The attraction of the flowers resides in the lip; flowers with the lip cut off were completely ignored. Detached flowers laid face-upwards on the ground were as attractive as if they were on the flower spike. If they were laid face-downwards, with the 'mirror' hidden, the insects were still attracted but had difficulty in finding the flowers. The wasps were clearly aware of the presence of *O. speculum* flowers even when they were hidden from sight.

Alighting on the flower of *O. speculum*, the male *Campsoscolia* sits lengthwise on the lip, with his head just beneath the rostellum (Fig. 7.20A), and plunges the tip of his

[1] Kullenberg (1961) saw 19 visits during 11 hours' watching in Morocco in late February and early March 1948. MCFP saw two visits in late March in Mallorca, 1982, but none in April in Mallorca (1983) or Greece (1986), even where the orchid was abundant.

abdomen into the fringe of long reddish hairs at the end of the lip with brisk, tremulous, almost convulsive movements, in the course of which he rarely fails to carry off the two pollinia on his head. Pouyanne was struck by the resemblance of the behaviour of the wasp to copulatory movements; later, when he was able to observe the males pursuing the females, he described them alighting on their backs and performing exactly the same movements as they did on the flowers.

Even to our eyes, the lip of *O. speculum* bears a certain resemblance to the female of *Campsoscolia ciliata*, with her broad abdomen likewise fringed with red hairs. At first sight the mirror is puzzling, but as Pouyanne realised when he was able to observe the females of *Campsoscolia* at close quarters, it corresponds exactly with the position of the bluish shimmering reflection on the wings of the wasp when she is resting or crawling over the ground (Correvon & Pouyanne, 1923). To us the resemblance between the wasp and flower may seem crude, but combined with scent and tactile stimuli it attracts the male wasps and elicits the copulation behaviour effectively enough for some 40% of the flowers to produce capsules.

The fly orchid, Ophrys insectifera

The fly orchid (*O. insectifera*), which occurs widely from the west Mediterranean countries northwards to Britain and southern Scandinavia, is related to *O. speculum*. Its pollination, first observed by Godfery (1929) and since studied in detail by Wolff (1950) in Denmark and Kullenberg (1950, 1961) in Sweden, bears many points of resemblance to that species. The flower-spike is slender, typically about 20-40 cm high, and bears up about ten rather widely-spaced flowers. The lip is rather long and narrow, dark reddish brown with a metallic bluish patch in the centre, and shallowly lobed at the tip (Fig. 7.21). The two upper petals are small, narrow and blackish, forming the 'antennae' of the 'fly'. Perhaps its most characteristic habitat is about wood margins on calcareous soils, but it also occurs in woods, in chalk and limestone grassland, and in open calcareous fens.

The only insects known as regular pollinators of the fly orchid are the solitary wasps *Argogorytes mystaceus* and *A. fargei*. As in the mirror orchid, the wasps are attracted to the flowers in the first place by scent. Upon settling, the wasp sits lengthwise on the lip – like *Campsoscolia* on *O. speculum* – with its head close to the column (Plate 6a; Fig. 7.20B). Often it remains on the flower for many minutes, every now and then restlessly changing its position before settling down again and performing movements which look like an abnormally vigorous and prolonged attempt at copulation. While it is on the flower, the wasp seems quite oblivious of the observer's presence. Very similar accounts of visits by *A. mystaceus* are given by

Fig. 7.21 Close-up of single flower of fly orchid (*Ophrys insectifera*).

Godfery from the south of France and by Wolff and Kullenberg in Scandinavia, and similar visits have been photographed in south Germany (Baumann & Künkele, 1982) and in Surrey (G.H. Knight, C. Johnson) and Wiltshire (H. Jones) in southern England. Darwin found that pollinia had been removed from 88 of the 207 flowers he examined; in a small Wiltshire colony visited in early June, 1969, of 13 flowers (on six plants) all but three of the oldest had at least one pollinium removed and eight had been pollinated (MCFP). But it is unusual for more than a quarter of the flowers to be pollinated, and sometimes the proportion is much lower than that, especially in large colonies and in seasons when the orchid is particularly numerous. The flowers are self compatible with their own pollen, but most of the flowers which are pollinated at all must receive pollen from other plants.

Godfery remarked that with only one, apparently accidental, exception, he never saw *Argogorytes* visit any orchid but *O. insectifera*; and apart from the two *Argogorytes* species, *O. insectifera* received no more than casual visits from other insects. In fact, the *Argogorytes* species regularly visit umbellifers (Apiaceae) for nectar, and also twayblade (*Listera ovata*) (Nilsson, 1981). Nevertheless, the relationship between the fly orchid and *Argogorytes* is so specific that crosses with other species of *Ophrys* cannot be more than rare accidents. Obviously the fly orchid is closely adapted to *Argogorytes*. The upper surface of the lip bears a remarkable general resemblance to the back of the female in contour and the nature of its hair covering. The fly orchid has a long flowering season, from early May to the latter part of June, broadly spanning the period when the male wasps emerge. Kullenberg noticed the interesting fact that in a Swedish locality the flowers were visited by *A. mystaceus* in the early part of the flowering season, but a fortnight later they were being visited by *A. fargei*.

Other Ophrys species

There is much diversity in detail in the pollination of *Ophrys*. One of the species studied by Pouyanne and by Godfery (1930) was the widespread Mediterranean orchid *O. lutea* (Plate 6c). In this plant the lip is brilliant yellow, with a dark raised area in the centre, and a pair of narrow metallic bluish patches on either side of a dark marking near the base. *O. lutea* flowers in March around Algiers, when even in North Africa calm sunny days are few and far between, and its visitors are much harder to observe than those of *O. speculum*. The number of capsules produced varies enormously from place to place; in the localities Pouyanne examined it ranged from as few as 3% to as many as 70-80% of the flowers produced. It was in this last favourable locality that Pouyanne was able to witness repeated visits to the flowers by small bees, males of *Andrena nigro-olivacea* and *A. senecionis*. The bees made the same kind of movements as *Campsoscolia* on *O. speculum*, but in contrast to the insects visiting that species and the fly orchid, the visitors to *O. lutea* always took up a position with their heads outwards on the flower, so that the pollinia were borne away on the tip of the abdomen (Fig. 7.20C). Evidently to the male bees, the 'decoy' represents a female bee sitting head downwards on a large yellow flower. Kullenberg found that by cutting off and reversing the lip of *O. lutea* he could induce the males to visit the flowers the 'normal' way round with their heads next to the column! *Ophrys fusca*, also a widespread Mediterranean species, works similarly; visiting bees carry the pollinia on the tip of the abdomen (Correvon & Pouyanne, 1916; Godfery, 1927, 1930; Vogel, 1976a; Paulus & Gack, 1981, 1990a).

Another species Godfery observed in the south of France was the late spider orchid, *O. fuciflora*, which extends from the east Mediterranean to France, and just reaches the chalk of south-east England. The flowers were visited by the large grey males of the bee *Eucera tuberculata*. The bees became aware of the flowers remarkably promptly and pounced on them, staying only momentarily but quickly and neatly removing the pollinia as they flew away. Kullenberg (1961) observed many visits by *Eucera longicornis* to plants in experimental cultivation in Sweden, and this bee also effectively pollinated the flowers. Many of the Mediterranean *Ophrys* species with flowers of the same general form as *O. fuciflora* are pollinated by bees in a similar way, but these brief visits are easily missed (Paulus & Gack, 1990a, 1990b). A population of the early spider orchid (*O. sphegodes*; Fig. 7.22), which is very local on chalk and limestone at its northern limit on the south coast of England, showed surprisingly rapid population turnover; individuals lived on average not more than a couple of years, and (at least under good grazing conditions) there was abundant recruitment to the population from seed (Hutchings 1987a, b; Waite & Hutchings 1991). But pollinator visits to this species appear to be infrequent. Godfery (1933) found that of 27 flowers he examined near Swanage, in Dorset, four had both pollinia removed and six had pollen on the stigma; a visit by a bee already bearing pollinia was observed in Dorset in 1975 by Mr J. Moore.

Fig. 7.22 A group of plants of early spider orchid (*Ophrys sphegodes*).

Self-pollination in the bee orchid, Ophrys apifera

The commonest *Ophrys* in western Europe, the bee orchid, *O. apifera*, is regularly self-pollinated, at least in the northern part of its range. Structurally, the bee orchid is very like other members of the genus. Its two significant differences are that the anther cells open a little more widely, and that the caudicles are a little longer and more flexible. Apparently insects occasionally visit the flowers; such visits are probably commoner in the Mediterranean region than in Britain or Ireland. In Morocco, Kullenberg observed visits by *Eucera* and *Tetralonia* males (the main species visiting the allied species *O. scolopax*, *O. bombyliflora*, *O. tenthredinifera* and *O. fuciflora*), but these insects seem often to fail to come into contact with the pollinia and are therefore often not effective pollinators. However, some of the *Eucera* males observed by Kullenberg visiting the sawfly orchid (*O. tenthredinifera*) attempted copulation with the labellum but here too failed to reach the pollinia, so in the Mediterranean region there may not be a sharp difference between the bee orchid and some of the related species in this respect. In Britain, too, pollinia are sometimes removed from flowers of the bee orchid; hybrids with the two spider orchids and the fly orchid have been reported, and a correspondent of Darwin's saw a bee 'attacking' a bee orchid flower. However, these are uncommon occurrences, and normally, when the flowers have been open for a day or two, the pollinia fall out of the anther and hang down in front of the stigma. With the spike shaking in the wind, the pollinia swing against the sticky stigma and are held fast. Kullenberg, who made most of his observations on the bee orchid in Morocco, doubted whether self-pollination would take place regularly without the disturbances caused by insect visits, and in north Africa this may be so. The experience of many observers confirms that in the south of England the bee orchid is self-pollinated with a very high degree of regularity. Often a spike of *O. apifera* can be found with the uppermost flower freshly opened and the pollinia still in their cells, the next flower with the pollinia dangling freely from the column (Plate 6d), and the lowest flower faded, with the pollinia caught against the stigma and a plump capsule developing beneath the flower. In contrast to the other species, almost every flower produces a capsule. *O. apifera* is often a notably early colonist of suitable disturbed calcareous habitats, but contrary to common belief it is quite a long-lived plant; despite the heavy investment in seed production, the same individual may flower repeatedly in successive years (Wells & Cox, 1991).

The scent chemistry of Ophrys

The remarkable pollination relationships in *Ophrys* have stimulated much research into the chemistry of the fragrances produced by the flowers and their insect visitors (Borg-Karlson, 1990). Several conclusions stand out. First, the range of compounds produced by the flowers and insects is similar; both include aliphatic hydrocarbons, alcohols, ketones and esters, benzenoid substances, various common monoterpenes, and sesquiterpenes, especially farnesol and farnesyl esters (but there are compounds in some insect fragrances not matched in *Ophrys* flowers). Second, there are recognisable broad correspondences between the fragrances produced by individual *Ophrys* species and those of the insects that visit them. Thus the scents of typical *O. insectifera* and of *Argogorytes* are both rich in aliphatic hydrocarbons (*O. insectifera* subsp. *aymonii* contains less hydrocarbons and more aliphatic alcohols and terpenes, and is also pollinated by *Andrena*). In general, the *Ophrys* species pollinated by *Andrena* bees have scents rich in aliphatic alcohols, ketones, esters and terpenes, and amongst these there is fair correspondence between the scents of particular groups of *Ophrys* species and of the particular groups of *Andrena* species pollinating them. However, these correspondences are by no means exact, and it is clear that the mimicry of insect fragrances by *Ophrys* is more subtle than straight chemical duplication – and also less than perfectly effective. In field experiments, males were always less attracted to the *Ophrys* flowers than to their own females. *Ophrys* pollination thus depends heavily on the newly emerged males early in the flying season before the females appear.

Sexual deception among southern Australian orchids

There are many biological parallels between the orchids of the Mediterranean region and those growing in similar climates in the southern parts of Australia, despite the fact that the two floras are not closely related (Dafni & Bernhardt, 1990). Little more than a decade after Pouyanne first recognised pseudocopulation in *Ophrys*, a similar pollination relationship was found in the south-east Australian small tongue-orchid *Cryptostylis leptochila*, in this case involving males of the ichneumon wasp *Lissopimpla excelsa* (*L. semipunctata*) (Coleman, 1927, 1928a, 1928b, 1929a). The spidery-looking flowers of *Cryptostylis* are 'resupinate', with the narrow, warty reddish-brown lip at the top and curved over the back of the flower (Plate 6e). The male wasps are strongly

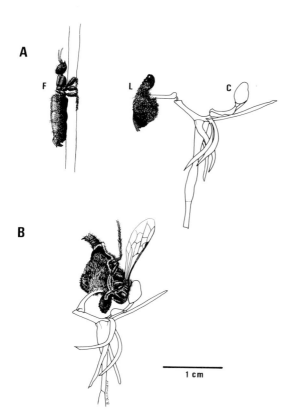

Fig. 7.23 Pollination of the West Australian orchid *Drakaea glyptodon* by the thynnine wasp *Zaspilothynnus trilobatus*.
A, The pre-mating posture of the wingless female wasp when calling for a mate, compared with the orchid flower.
B, Male wasp tipped against the column in the position required for pollination.
C, column; L, labellum; F, female wasp. Reproduced with permission from Peakall (1990). Compare Plate 7f.

1 cm

attracted to the flowers by a scent which is imperceptible to us, and alight on the lip with the tip of the abdomen towards the column. As it attempts to mate with the lip, the visiting insect probes the area around the column with its genitalia, often presenting the flower with its sperm-packet. When the end of the insect's abdomen comes into contact with a viscidium it carries away a pollinium, which is then brought into contact with the stigma of the next *Cryptostylis* flower with which the wasp tries to mate. At least five species of *Cryptostylis* are pollinated by males of the same species of wasp, but the species are apparently incompatible with one another's pollen and hybrids between them are unknown (Coleman, 1929b, 1930, 1931, 1938).

Some very striking Australian examples of pollination by sexual deception depend on thynnine wasps (Hymenoptera Aculeata: Tiphiidae subfamily Thynninae). The female wasps are wingless and, apart from mating (which can take place several times), spend most of their lives underground, searching for the root-feeding scarabaeid beetle larvae upon which they lay their eggs. When ready to mate, the female climbs to a vantage point from which she releases a pheromone that attracts the patrolling, winged males. Final recognition is by sight; the male seizes the female and carries her off. Copulation takes place in flight and is prolonged, the pair visiting flowers at which both sexes feed on nectar while mating. In the west-Australian hammer orchids (*Drakaea*), the solitary flowers are borne on a slender stem 10-25 cm high. The end of the lip forms a dummy thynnine female, maroon in colour, glossy and usually variously decorated with protuberances or warty excrescences. This dummy is connected to the base of the lip by a slender stalk with a flexible hinge in the middle (Fig. 7.23). The flowers attract the male wasps by odour. Stoutamire (1974) decribes how on several occasions thynnine males followed his car down the road and flew in through the open windows to locate *Drakaea* flowers on the floor behind the driver's seat. Peakall (1990) found that flowers of *Drakaea glyptodon* introduced experimentally into a habitat were discovered by the males of *Zaspilothynnus trilobatus* within a minute. Once close to the flower, the visitors locate the dummy female by sight, and most alight. Some of the visitors seize the tethered dummy female and attempt to fly off with 'her', in doing so swinging back against the column and receiving a pair of pollinia on the top of the thorax – or bringing previously acquired pollinia into contact with the stigma. Peakall found that, following one attempt to carry off a dummy female, a wasp generally did not repeat the attempt with another flower in the immediate vicinity. Different species of *Drakaea* are pollinated by different wasp species; Stoutamire found that *D. elastica* was visited by *Z. nigripes*. As in *Ophrys*, the number of flowers pollinated varies greatly from place to place; in nine populations examined in 1985 and 1986, the proportion of flowers pollinated ranged from zero to 58% (Peakall, 1990).

Other Australian orchids are pollinated by thynnine wasps in a similar way, including the east-Australian elbow orchids (*Spiculaea*), and a number of species of the large and varied genus *Caladenia*, which occurs throughout the southern parts of Australia. Some *Caladenia* species have colourful sweet-scented flowers visited by small bees in apparent search for pollen or nectar. Other species pollinated by male thynnines have dull-coloured and (to us) odourless flowers; they have mobile, more-or-less lobed lips, bearing a dark raised mark (or a series of dark protuberances) about the size and shape of a female thynnine in the centre, sometimes with leg-like dark markings on either side. In eastern Australia, pseudocopulation with thynnine wasps has also been observed in the bird-orchids *Chiloglottis* (Stoutamire, 1974, 1975), and in the copper beard-orchid (*Calochilus campestris*) in Victoria.

Several other instances of sexual deception have been described from Australian orchids. These include the sawfly *Lophyrotoma leachii*, visiting the large duck-orchid (*Caleana major*) (Cady, 1965), and winged male ants (*Myrmecia urens*) which are the exclusive pollinators of the fringed hare-orchid (*Leporella fimbriata*) (Peakall, Beattie & James, 1987). A curious case occurs among the greenhood orchids (*Pterostylis*), where the species of the *P. rufa* group are pollinated by male fungus gnats which appear to be attracted sexually to the oddly insect-like small brown lip. The lip is irritable, and when triggered by an insect touching the sensitive base it snaps smartly upwards (Fig. 7.24). This traps the visitor with its back against the column, and if the insect is carrying pollinia these are thrown into contact with the stigma. The column has two wings near the tip which project towards the lip, and the only way the insect can escape is by pushing between these wings with its back towards the column, picking up

Fig. 7.24 **a–b** Flower of the rufous greenhood orchid (*Pterostylis rufa*); Eltham, Victoria. In **a** the lip is in its normal position; in **b**, following stimulation, it has swung up against the column × 3.5.

a   **b**

pollinia on its thorax as it does so. A different species of insect appears to be responsible for the pollination of each species of *Pterostylis* (Coleman, 1934; Sargent, 1909, 1934; Beardsell & Bernhardt, 1983).

A tropical miscellany

There are perhaps 20,000 species of orchids. Most of them grow in the tropics, where probably nearly three-quarters of them are epiphytes on the branches of the forest trees, especially in the tropical mountains. It is clear that our knowledge has only begun to scratch the surface of the pollination biology of these plants. However, a few generalisations can probably be made. First, the kind of highly specific relationships that have figured prominently in the earlier parts of this chapter do occur throughout the orchids (for example, the Andean *Trichoceros antennifera*, pollinated by tachinid flies, and its relatives are tropical American instances of sexual deception), but they are probably the exception rather than the rule. Many orchids provide nectar or other reward and are 'normally' pollinated by bees, moths or other visitors (van der Pijl & Dodson, 1966); indeed, some are quite promiscuous and are pollinated effectively by a range of insects. Second, as we have already seen among European orchids, a remarkably large proportion of orchid species, perhaps more than a quarter, practise 'false advertisement' and do not reward their pollinators (Dressler, 1993). Third, it is clear that the same sort of pressures for floral diversification seen at work among the zygomorphic flowers described in the last chapter have also operated in the orchids. The paragraphs that follow highlight only a few selected examples.

'*Angraecum sesqipedale*, of which the large six-rayed flowers, like stars of snow-white wax, have excited the admiration of travellers in Madagascar, must not be passed over. A green whip-like nectary of astonishing length hangs down beneath the labellum,' (Darwin, 1862b). In specimens sent to him, Darwin measured spurs 11½ inches

Fig. 7.25 The long-spurred hawkmoth-pollinated orchid *Angraecum arachnites*, growing as an epiphyte in primary forest in central Madagascar. Scale bar = 1 cm. Reproduced with permission from Nilsson *et al.* (1987).

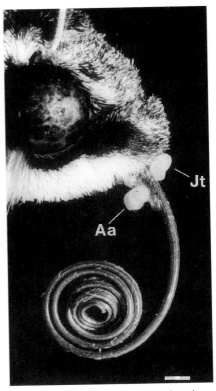

(29 cm) long, 'with only the lower inch and a half filled with nectar. What can be the use ... of a nectary of such disproportionate length?' Darwin found when he examined the flowers that for a moth to reach deeply into the spur, it would be forced to push its proboscis through the deep notch in the centre of the rostellum, and only if the moth did this would the pollinia become attached to the thick base of its proboscis. It is thus essential to the

Fig. 7.26 Head of long-tongued hawkmoth (*Panogena lingens*), with pollinia of two angraecoid orchid species attached to the base of the proboscis, *Angraecum arachnites* on the ventral and *Jumellia teretifolia* on the dorsal side. Scale bar = 1 mm. Reproduced with permission from Nilsson *et al.* (1987).

pollination mechanism (as in the north-temperate butterfly orchids [*Platanthera*, p.200]) that the flower's spur should be a little longer than the tongue of the pollinator. Even so, this still implies a remarkably long-tongued pollinator, and Darwin concluded that ' ... in Madagascar there must be moths with proboscides capable of extension to a length of between ten and eleven inches!' (25-28 cm). His surmise was vindicated by the discovery 40 years later of the hawkmoth *Xanthopan morgani* subsp. *praedicta* (Rothschild & Jordan, 1903), which has an average tongue-length of about 20 cm, reaching over 24 cm in some individuals. In fact, visits of this moth to *A. sesquipedale* have yet to be observed in the field, and there is at least one other hawkmoth in Madagascar, *Coelonia solani*, with a tongue of comparable length (Nilsson *et al.*, 1985). Several long-spurred orchids in the forests of central Madagascar are pollinated by a long-tongued form of the hawkmoth *Panogena lingens*, with a tongue averaging about 12 cm; the spurs of the orchids are a centimetre or two longer. *Angraecum arachnites* (Fig. 7.25) places its pollinia near the base of the visitor's proboscis on the under side (Fig. 7.26); the pollinia of *A. compactum*, *Jumella teretifolia* and *Neobathiea gran-didierana* also become attached to the base of the proboscis but on the upper side, while those of *Aerangis fuscata* are generally carried on the head and the palps (Nilsson *et al.*, 1987). The orchids and the hawkmoth have almost certainly interacted evolutionarily over a long span of time – an instance of diffuse co-evolution, where the orchids have generated a selective pressure as a group. The individual orchid species may compete for pollination, but also benefit from common support of the nectar requirements of their shared pollinator. A hawkmoth the size of *P. lingens* probably requires about 1.3 mg sugar (22 J) per minute for hovering flight. Nilsson *et al.* (1985) calculated that for the long-tongued form of this moth the nectar from a single visit to a flower of *Angraecum arachnites* would yield enough energy to maintain hovering for about 70 seconds.

Several groups of tropical orchids produce oil as a reward, and are visited by oil-collecting bees (pp.43, 115). These include species of *Disperis*, pollinated by bees of the genus *Rediviva* (Steiner, 1989), and related African ground orchids, the dwarf tropical American epiphytes of the subtribe Ornithocephalinae, and others (Dressler, 1993). A good many tropical American orchids are visited by hummingbirds, and various genera of diverse systematic affinity (e.g. *Stenorrhynchos* [Spiranthoideae] and many Epidendreae), have red or yellow flowers which appear to be adapted primarily to bird pollination.

Other pollination systems are based on deception of one sort or another. Two species of *Oncidium* are pollinated by males of the solitary bee *Centris* (Dodson & Frymire, 1961b; Dodson, 1962). The bees usually have a favourite perch near an *Oncidium* plant, and from time to time take off and hover near the orchid. The flowers are borne in long racemes, and when they are moved by the wind the bee darts in and buffets one of the flowers. The *Centris* bees appear to hold territories by chasing off all insects which fly nearby. Possibly this is a case of 'aggressive mimicry', the bees seeing the flowers as flying insects. As a result of repeated buffeting flights, all the flowers of an inflorescence may be pollinated in a short space of time – but often the bees appear to ignore the flowers altogether. However, the flowers are open for about three weeks, which gives them a fair chance of being visited, though investigators have little chance of seeing pollination take place. The plants are apparently self-incompatible, and often only one flower of a pollinated plant produces a fruit.

Various orchids in both the Old and New World floras are pollinated by flies, attracted to the brownish or dull reddish flowers by foul odours ('sapromyophily', Chapter 10). In *Bulbophyllum macranthum*, the two lateral sepals are directed upwards and meet near their tips. Here the flies alight and spend much time licking the surface, holding onto the outside of the otherwise slippery sepals. Crawling towards the centre of the flower, they find the sepals parted so they can no longer straddle them. As they slip on the surface of the sepals they clutch at the solid tongue-like lip, which affords a good grip. On transferring their weight to the lip in a head-upwards attitude, they are suddenly flung backwards and downwards, for the lip is hinged and delicately balanced, and tips with the weight of the fly. Two springy arms near the tip of the column embrace the fly, while the lip returns to its original position. The fly soon escapes, but in its struggles removes the pollinia on its abdomen (Ridley, 1890). Different species of *Bulbophyllum* exploit flies of different sizes, but all are characterised by the finely balanced lip tipping to throw the visitor against the column. Similar deception is seen in the tropical American *Masdevallia fractiflexa*, which by its colour and smell of carrion lures flesh-flies to effect pollination, without offering any tangible reward (Dodson, 1962).

Fig. 7.27 A pendent flower of the bucket orchid *Coryanthes.* The column descends vertically, almost closing the left-hand side of the 'bucket' formed by the descending arm of the large concave lip; C, column. Arrows show the course of visitng bees, falling into the 'bucket' after brushing perfume from the basal parts of the lip, and then, following immersion, forcing their way out past the stigma and anther. Funcionally, this parallels the one-way mechanism of *Cypripedium,* but its details and structural basis are quite different. After Lindley, redrawn from Darwin (1877).

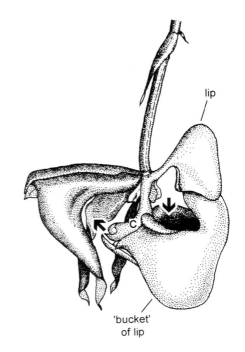

lip

'bucket'
of lip

Orchids and scent-gathering by euglossine bees

A particularly interesting group are the tropical American orchids pollinated by euglossine bees – large solitary, communal or primitively social bees, related to the bumblebees. Some species have only a thin coat of hairs, and are often brilliant metallic blue, green or bronze in colour. Fast fliers, capable of covering long distances, the euglossines are widespread in mainland tropical South America. They are typically long tongued, and feed on nectar from a wide variety of often deeply-tubular flowers. The males also gather scent (which they mop up with the feathery brushes on their front tarsi and carry in their inflated hind tibiae), for reasons generally thought to have some connection with mating, but not fully understood (Dressler, 1982; Williams, 1982). Many genera of orchids are pollinated mainly or exclusively by scent-gathering male euglossine bees. The plants provide no reward other than profuse scent (containing monoterpenes such as 1,8-cineole, limonene, linalool, myrcene, *trans*-β-ocimene and pinenes, and benzenoids such as benzaldehyde, methyl benzoate and methyl cinnamate), and they attract no other visitors. Williams (1982) has written an excellent detailed review of the pollination of these plants.

Among the most bizarre of all pollination mechanisms is that of the genus *Coryanthes* (Fig. 7.27), first described by Crüger (1865; summarised by Darwin, 1877) from his observations on *C. macrantha.* The big, waxy-looking flower hangs down and part of the lip forms a bucket into which fall drops of watery liquid secreted by a pair of knobs on the column. Male bees of the genus *Eulaema* eagerly visit the flowers, attracted by the strong scent. This leads them to an area at the base of the lip, which they scratch with their forelegs to collect the liquid scent from the surface. In doing so,

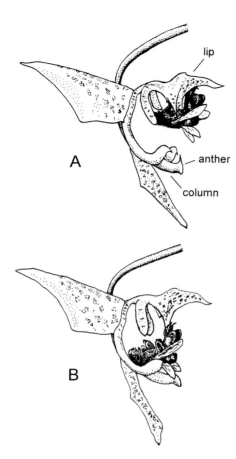

Fig. 7.28 Two stages in the visit of a bee, *Euglossa viridissima*, to the South American orchid *Gongora maculata*. **A**, bee upside down, brushing the lip. **B**, bee loses grip and falls with back to column, down which it slides. After photographs by Dr C.H. Dodson.

a bee will often slip and fall into the bucket. They swim around in the water but cannot climb the sides. The only way out is through the narrow opening at the apex of the lip, close to the tip of the column. In squeezing through this, the bee passes first the stigma, then the anther. Dodson (1965) found that the first bees to enter a flower usually took 15 to 30 minutes to find their way out, because of the resistance of the finger-like rostellum which slips between the thorax and abdomen, fixing the pollinia to the base of the abdomen of the insect. Once the pollinia are removed, bees can escape much more quickly, leaving any pollinia they are carrying on the stigma as they pass the column. Two related orchid genera, *Gongora* (Fig. 7.28) and *Stanhopea* (Dodson & Frymire, 1961a), have 'fall-through flowers.' The scent-collecting male bee falls from the lip in such a way that his back touches the column and pollinia are deposited on his thorax. In the massive, intensely fragrant, waxy-looking flowers of *Stanhopea* there are usually large prongs on either side of the lip which guide the falling bee past the tip of the column (Fig. 7.29). Some species of *Stanhopea* are pollinated by *Euglossa* but others, with a larger gap between the tip of the column and the lip, are pollinated by the larger *Eulaema*.

Fig. 7.29 Single flower of the
tropical South American
orchid *Stanhopea wardii*. The
visiting bee falls down the
'chute' between the
channelled upper surface of
the lip (below the prominent
eye-spot) and the column to
its right, striking the pollinia
close to the tip of the
column in the process.
Compare *Gongora*, Fig. 7.28.

Catasetum and its relatives are another group of orchids pollinated by scent-gathering male euglossine bees; Darwin considered them 'the most remarkable of all Orchids'. In the swan orchid (*Cycnoches lehmannii*), the unisexual flowers are pendent (Fig. 7.30) and the visiting *Eulaema* bees alight upside-down on the lip. To reach the scent-producing area the bee is forced by a projection of the lip to let go with its hind legs, whereupon the body swings down and, if the flower is a male, touches the anther-

Fig. 7.30 '*Cycnoches ventricosum.*
Flower viewed in its natural
dependent position. *c*, column,
after the ejection of the
pollinium together with the
anther. *f*, filament of anther. *s*,
stigmatic cavity. *L*, labellum.
pet, the two lateral petals. *sep*,
sepals.' From Darwin (1877).

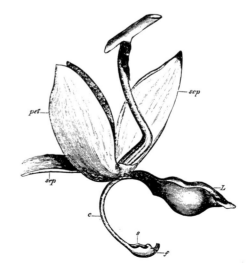

cover. This triggers discharge of the pollinia, which are forcibly ejected onto the under side of the bee's abdomen near the tip. Female flowers are slightly smaller, and there are hooks on the column which catch the pollinia carried by the bee as its abdomen swings past them. In *C. egertonianum*, pollinated by *Euglossa*, the male and female flowers are very dissimilar, the female functioning in the way described, while in the male the slender lip (which is uppermost in the flower) bends under the weight of the insect, bringing it into contact with the anther, when the pollinia are shot onto the visitor in the same way. *Catasetum* also has unisexual flowers. The male flowers are showy and vary greatly between different species; some have the lip uppermost, some below. The female flowers are greenish and all have the hooded lip uppermost; careful examination is needed to distinguish the species (but male and female flowers of the same species produce the same floral fragrances [Hills *et al.*, 1972]). Indeed, the flowers look so different that they were at first placed in separate genera; Darwin sets out at length his reasons for concluding that *Catasetum tridentatum* (= *C. macrocarpum*) and *Monachanthus viridis* are the male and female plants of the same species – a conclusion confirmed in the field by Crüger, then Director of the Botanical Garden in Trinidad. The pollinia of *Catasetum* are attached to a remarkably large viscid disc. In the newly opened flower, this viscidium is turned towards the back of the column, away from visiting insects, and the pedicel joining it to the pollinia is strongly curved and under tension. The rostellum is prolonged downwards into two slender tapering 'antennae', one of which projects towards the lip. If anything touches this, the viscidium is released from

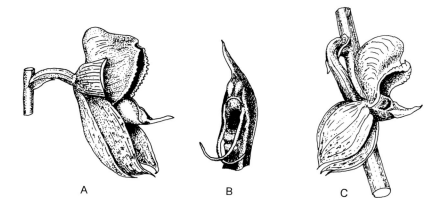

Fig. 7.31 Male and female flowers of *Catasetum macrocarpum*, redrawn from Darwin (1877). **A**, male flower, with the two lateral sepals cut away for clarity; the helmet-shaped lip is uppermost in the flower, withe the column projecting in the centre of the flower below it. **B**, detail of the column. The two pollinia are joined by the broad stipe to the viscid disc above the otherwise functionless stigmatic cavity (pollinia, stipe and viscid disc together make up the 'pollinarium'). Expulsion of the pollinarium is triggered by a touch to the antenna which projects forward near the base of the column. (Darwin, p.194, as *C. tridentatum*). **C**, female flower; the sepals are reflexed and there are many other detail differences, but the lip is uppermost s in the male flower. The pollinia are rudimentary and functionless; pollinia carried by a visiting bee readily slip into the narrow transverse stigmatic cleft (Darwin, p.199, as *Monachanthus viridis*).

the surrounding membrane of the rostellum, the pedicel straightens, and 'the pollinia are shot forth like an arrow, not barbed however, but having a blunt and excessively adhesive point,' (Darwin, 1877). Crüger wrote, 'a great number of [bees] ... may be seen every morning for a few hours disputing with each other for a place in the interior of the labellum, for the purpose of ...[brushing scent from] ... the cellular tissue on the side opposite to the column, so that they turn their backs to the latter. As soon as they touch the upper antenna of the male flower, the pollen mass with its disc and gland, is fixed on their back, and they are often seen flying about with this peculiar-looking ornament on them. I have never seen it attached except to the very middle of the thorax. When ... the insect enters a female flower, always with the labellum turned upwards, the pollinium, which is hinged to the gland by elastic tissue, falls back by its own weight and rests on the anterior face of the column. When the insect returns backwards from the flower, the pollinia are caught by the upper margin of the stig-matic cavity which projects a little beyond the face of the column ...' Darwin (1877) noted that the bees sent to him by Crüger were two species of *Euglossa*.

Conclusion: some general thoughts on orchid pollination

It is easy to marvel at the extraordinary and often unexpected intricacy and perfection of the adaptations and pollination relationships of orchid flowers, and in so doing, not to enquire how these precise and intricate flower structures and relationships came about, how they relate to more commonplace pollination mechanisms, or how they fit into the larger biological picture. There are perhaps three features of orchid flowers which particularly invite comment. Why have various species of orchids become so precisely and specifically adapted to such a remarkable range of different pollinators? Why do so many orchids rely upon deception of one sort or another? With so many apparently hazardous pollination relationships in the family, why are not more orchids self-pollinated?

Adaptive radiation and specificity to minority pollinators

From the point of view of the plants, pollination is rather like an animal population ecologist's 'mark-release-recapture' experiment. An insect picking up a pollen load is an insect 'marked'; a successful pollination is a 'recapture'. Nowhere is this clearer than among the orchids, where it is easy to see whether the pollinia have been re-moved from a flower, and whether the stigmas have received pollen; pollinia are gen-erally removed from more flowers than are pollinated.[1] How should an experimenter (or a plant) maximise the number of 'recaptures'? In fact, the most important factor is the proportion of the population 'marked'. A very abundant flower can thus evolve a pollination relationship with an abundant insect, because many individuals will have visited flowers of the same species previously, and will carry their pollen. For minority flowers like most orchids, it is neither possible to produce enough pollen to 'mark' a sufficient proportion of the population of a common insect, nor to attract a big enough proportion of the insects to the flowers to pick up or deposit pollen. Added to

---

[1] A fascinating exception is the Madagascan orchid (*Cynorkis uniflora*) (Nilsson *et al.*, 1992). Attachment of pollinia, which have extraordinarily long (16 mm) rigid caudicles, to the eyes of visiting hawkmoths is very chancy, but the pollinia that do attach bring about numerous pollinations. Often more flowers are pollinated than pollinia removed, especially at high flowering densities.

this, to maximise 'recaptures' it is important that the 'mark' should be durable. Orchids have achieved this by evolving pollinia, but that in itself makes it impossible to spread pollen thinly among a large number of individual insects. In this situation it is better for the plant to rely on a minority pollinator, which it can attract specifically and with high effectiveness. This kind of selection pressure must be one factor in the remarkable diversity of pollen vectors used by different orchid flowers, and the specificity with which these are attracted. The situation is also one in which there are probably heavy selection pressures on both the male and the female functions of the flower (Willson, 1994), and in which adaptive shifts to new pollinators may be relatively easy.

Thus although there *are* many orchids which are pollinated by bees visiting the flowers for the nectar reward they receive, the exceptions to this 'mainstream' pattern of floral adaptation are very numerous. The orchids include various unrelated groups of precisely-adapted butterfly and moth flowers. Diverse instances of pollination by beetles, by social and solitary wasps, by ichneumons and other parasitic wasps, have been described earlier in this chapter. The Australian *Microtis parviflora* is pollinated by flightless worker ants (Peakall & Beattie, 1989); the extraordinary subterranean orchid *Rhizanthella gardneri* of Western Australia, which never appears above ground but produces its heads of flowers beneath the leaf-litter around bushes of the shrub *Melaleuca uncinata* (with which it shares a mycorrhizal fungus), is probably pollinated by termites (Dixon, Pate & Kuo, 1990). The orchids dependent for pollination on pseudocopulation, and the tropical orchids pollinated by scent-gathering euglossine bees, are the most spectacular examples of a pattern of close adaptation to one or a very few pollinators which is a very widespread feature of the Orchidaceae. In almost every case, the initial specific attraction of the pollinator to the flower is by scent.

The evolution of pseudocopulation has often been seen in the past as one of the major enigmas of orchid evolution. We now know so many instances amongst the orchids of flowers producing a fragrance which specifically attracts only males or females of a particular species or group of insects, that there can be little doubt that in genera like *Ophrys*, *Cryptostylis* or *Drakaea* it was a specific scent attraction for the males of a particular insect which arose first, the visual resemblance to a female insect evolving later. The evolution by flowers of scents mimicking insect pheromones is probably less surprising than it appears at first sight. All plant and animal cells have much of their biochemical mechanism in common. The 'metabolic pathways' leading to the formation of insect pheromones and flower scents are largely the same, so insects and flowers have a similar repertory of possibilities and a degree of apparent 'mimicry' is quite likely to arise by chance. Once a specific attraction was established, of the kind apparent in *Orchis galilaea* or *Catasetum*, the way was open for the development of the sort of visual and tactile resemblances we see in *Drakaea* or *Ophrys* (Bergström, 1978; Harborne, 1993).

Deception and the pollination of orchids

As Abraham Lincoln is reputed to have said, 'You can fool all the people some of the time, and some of the people all the time, but you can not fool all the people all of the time.' Much the same applies to insects, and the orchids have been well placed, in evolutionary terms, to profit from Lincoln's dictum. For an abundant species that is dependent for pollination on a 'mainstream' pollinator, for which it provides a staple food source, the continued viability of the pollination relationship is tied to the continued provision of the reward – usually nectar or pollen – at least by the majority of flowers. For minority-species, more options are open, and pollination systems based on deception are very widespread amongst flowering plants (Dafni, 1984) (Chapter 9).

They are particularly prevalent amongst the orchids, where pollinia make possible deceptive pollination systems which would not be viable with loose pollen. 'Deception' generally implies a degree of mimicry, exploiting the normal behavioural responses of insects. This may be quite general or very precise.

Many orchids simply have conspicuous and colourful flowers, but provide no reward. They may be said to mimic in a general way the broad characteristics of nectariferous flowers. Most of them are pollinated by bees or flies that habitually visit nectariferous flowers; there are many examples amongst the European species of *Orchis* (Dafni, 1990). Pollination may depend in varying degrees on visits by 'naïve' newly-emerged bees, or on exploratory behaviour or imperfect discrimination by pollinators mainly visiting rewarding flowers. It is probably often advantageous to deceptive flowers of this kind to be varied in appearance, making it harder for visitors to learn to avoid them (Dukas & Real, 1993); populations of such non-rewarding species as the military and lady-orchids (*Orchis militaris* and *O. purpurea*), the heath spotted-orchid (*Dactylorhiza maculata*) and the beautiful *Calypso bulbosa* of Eurasian and North American boreal forests (Wollin, 1975; Ackerman, 1981; Boyden, 1982) are often notably variable in flower colour, form and pattern. Many orchid flowers provide more specific deceptive attractions to food-gathering pollinators, especially in the form of 'pseudopollen', which may be simply a roughened yellow patch of the lip (*Cephalanthera*) (p.192), stamen-like tufts or patches of hairs, as in the North American *Calopogon*, or powdery tissue on the lip surface, as in the saprophytic Australian orchid *Gastrodia sesamoides* (Beardsell & Bernhardt, 1983). The first two types are purely deceptive, the last may at least in some degree be providing a real reward. There are a number of instances where orchids appear to mimic some more-or-less specific reward-providing model, and to share its pollinators. The examples of *Cephalanthera rubra/ Campanula* and *Orchis israelitica/ Bellevalia* have already been mentioned; another probable case is *Orchis pallens*, mimicking the flower spikes of *Lathyrus vernus* (Vöth, 1982). The curiously unorchid-like yellow flowers of some Australian species of *Diuris* mimic the flowers of leguminous subshrubs of such genera as *Daviesia* and *Dillwynia* (Beardsell & Bernhardt, 1983).

The orchids discussed in the previous paragraph are imposing a small proportion of unrewarding visits on insects that habitually visit flowers for nectar or pollen. They are 'fooling all the insects some of the time.' The very highly specific deceptive mechanisms exemplified especially by pseudocopulation come close to 'fooling some of the insects all the time.' In both cases, it is important to the long-term stability of the relationship that the orchid should not place too heavy a burden on the pollinator. As long as the deception does not matter significantly to the pollinator population, natural selection will not operate significantly against it.

What advantage does deception bring the orchids? The most obvious is that it saves resources; orchids are seldom gregarious enough to provide a worthwhile resource for flower-constant pollinators, so there is probably little selective advantage to be had from providing a reward (Cohen & Shmida, 1993). Indeed, it has been suggested that the lack of a reward may be advantageous to the orchid by discouraging insects from visiting successive flowers on the same spike. Various factors that may be involved are discussed by Nilsson (1992b).

Self- and cross-pollination in orchids

The orchids are so pre-eminently an insect-pollinated group that at first sight it seems surprising to encounter self-pollinated members among them. But orchids are subject to the same conflicting selection pressures as other plants. Self-pollination has obvious advantages in maintaining a high and reliable seed production even if insect visitors are scarce – and isolated individuals can still set a full crop of seed. Its genetic effects may also be beneficial in the short run, by preserving adaptively favourable gene combinations, even if at the price of accumulating deleterious mutations, and longer-term loss of genetic flexibility. So we should ask not only 'Why are some orchids regularly self-pollinated?' but also 'Why are not more orchids self-pollinated?' What determines the balance between cross- and self-fertilisation in the family?

We have considered the case of the bee orchid (*Ophrys apifera*) already (pp.209-10) – regularly selfed in the north-west European part of its range, but perhaps at least sometimes insect pollinated in the Mediterranean part of its area. In southern England, a very much commoner plant than the insect-pollinated narrow-leaved helleborine (*Cephalanthera longifolia*) (p.192) is the nearly related white helleborine (*C. damasonium*), a very characteristic plant of the 'hanger' beechwoods of the chalk escarpments. In this species, the flowers open for only a short time, and then never widely, yet almost every flower produces a well-developed capsule. The flowers are occasionally visited by bees and other insects, so some outcrossing may possibly take place, but there is no doubt that most seed is produced by selfing. Self-pollination in the broad-leaved helleborine (*Epipactis helleborine*) probably supplements cross-pollination by wasps, though the relative importance of selfing and crossing probably varies from place to place and from time to time. The narrow-lipped and green-flowered helleborines (*E. leptochila* and *E. phyllanthes*) are regularly self-pollinated. They are all very local plants, and much less variable than *E. helleborine*, from which some or all of them probably originated. Perhaps all of these regularly self-pollinated species are products of similar selection pressures to those that have favoured apomixis in genera such as *Sorbus* (whitebeams) and *Rubus* (brambles), probably related to establishment in particular kinds of habitats which arise repeatedly but unpredictably and are of limited duration.

Many orchids with pollination mechanisms based on deception ripen only a small proportion of capsules, so seed-set appears to be pollinator-limited; figures have been quoted earlier in this chapter for species of *Orchis*, *Dactylorhiza*, *Ophrys* and *Drakaea*. In populations of the pink Lady's slipper orchid or moccasin flower (*Cypripedium acaule*) in north-eastern North America, the proportion of flowers producing capsules ranged from zero to 23%; the overall average is probably less than 10% (Davis, 1986; Gill, 1989; Primack & Hall, 1990). It may be argued (as Gill does) that a pollination system as 'inefficient' as this cannot possibly be adaptively stable – that it should be immediately vulnerable to takeover by a mutant capable of regular self-pollination. But the widespread occurrence of deceptive pollination systems, and the fact that these have clearly arisen many times independently and ultimately from nectar-providing ancestors, demonstrate that they *must* be adaptive and 'evolutionarily stable' under some conditions. Darwin's experiment with *Cephalanthera damasonium* suggests that there may be quite severe inbreeding depression of seed-set even in an habitually self-pollinated species. Perhaps the crucial point is that seed-set, although a simple measure of 'fitness' in a glasshouse or garden experiment, does not measure fitness in a field situation – where representation in the next generation depends not only on seed-set but also longevity of the parent plants and on seedling establishment. It has been shown that in many situations 'fitness' in this sense is maximised at moderate or even low levels of seed-set – which require correspondingly low levels of pollination (Calvo & Horvitz, 1990; Calvo, 1993). In a four-year study of *Cypripedium acaule* in Massachusetts, Primack & Hall (1990) found that an average-sized plant that produced a fruit in the current year showed an estimated 10-13% decrease in leaf area and a 5-16% decrease in the probability of flowering in the following year. Perhaps our preconceptions are at fault. We accept readily that few pollen grains can reach a stigma, and that most pollen must be lost and 'wasted'. We are less ready to accept apparent 'waste' of flowers, but in some situations that may be the price for getting the 'right' number of pollinations and producing the 'right' amount of seed. The demographic data of Gill for the moccasin flower and of Hutchings (p.209) for the early spider orchid both show plentiful recruitment from seed. Generally, deceptive orchids show quite good seed-set at least in *some* localities – and, of course, a single orchid capsule contains a great many seeds. Probably the plants of any one generation are drawn from only a relatively small proportion of the individuals in the generation before (Calvo, 1990). This is true for birds (Newton, 1995) and it may be a common feature of many plants and animals. The deceptive pollination systems seen in so many orchids are (up to a point) manifestly successful, but their particular pattern of adaptation probably limits their further evolutionary options – on the principle that, 'If I were going there, I wouldn't be starting from here!' We need to know much more about the demography and reproductive biology of orchid populations before we can hope to understand them fully.

8

Birds, Bats and Other Vertebrates

Although insects are undoubtedly the most important animal pollinators, a significant role in pollination is played by vertebrates. Of these, birds are important in many parts of the world, while bats are an element in the pollinating fauna of the tropics and some subtropical regions. Beyond these groups there are non-flying mammals that pollinate flowers, and flowers that are clearly adapted to pollination by them.

Bird-pollination: the birds

In latitude, bird-pollination extends from the southern tip of South America to Alaska in North America, while its nearest occurrence to Britain is in Israel. There are no bird-pollinated flowers in Europe, nor in Asia north of the Himalayas. Bird-pollination is known to occur up to an altitude of about 4,000 m in the mountains of East Africa and South America, the birds migrating locally to these levels. Flower-visiting has been recorded for about 50 bird families and there is a whole range of birds showing different degrees of adaptation to a floral diet. Some of the adaptations seen in bird-pollinated flowers are parallel to those found in insect-pollinated flowers. Examples are: food supply, conspicuousness, guide-marks and the size, shape and positioning of flowers. Some of the bird-flowers are extremely large, while the smallest are no bigger than typical bumblebee flowers. We shall return to bird-flowers when we have reviewed the flower-visiting birds.

The floral foods that are taken by birds are nectar, pollen and solid food bodies. The food that is most generally offered and accepted is nectar, the carbohydrate energy-food. Birds using nectar as their energy source usually get their protein from insects, sometimes picking these up from the flowers they visit for nectar. Birds are so active and inquisitive that the evolution of their habit of feeding on nectar was perhaps to have been expected. There are two ways in which they might have been led to this source of food. One is by visiting foliage to drink raindrops, a habit common among birds of tropical forests, and the other is by going to flowers to look for small insects. Some nectar-feeding birds show great speed and precision of movement, together with considerable constancy to flowers of a single species. They can therefore be highly efficient pollinators, and individual birds may visit many thousands of flowers in a day (Porsch, 1933; Scott Elliot, 1890b).

In all regions where there are nectar-feeding birds there is, however, a tendency for the birds to steal nectar by piercing the sides of tubular flowers of all sizes. It was considered by Sargent (1918) that the prevalence of destructive nectar-robbing by birds in Western Australia created a strong evolutionary pressure upon insect-pollinated plants to become adapted to bird-pollination and so increase the chances of legitimate visitation. Swynnerton (1916a) made a special study of nectar-thieving by birds in what is now Zimbabwe. His most interesting results concerned thieving from bird-pollinated flowers by their legitimate pollinators. For example, in *Leonotis mollissima* (Lamiaceae) the flowers are in dense spherical clusters placed at intervals up erect stems, and they radiate horizontally or slightly downwards, being reached legitimately by birds perched just below. Some of the birds preferred to do this and were thus valuable pollinators, only piercing the calyces or using previously made punctures when the flowers were disarranged and so had ceased to protect one another (compare pp.172-3). Other birds of the same two sunbird species (*Anthothreptes hypodilus* and *Cinnyris niassae*), however, always pierced the flower or used punctures, usually approaching from above; but even if approaching from below, when legitimate visiting would have been more convenient, they went to some trouble to steal the nectar through the tough sides of the flower, probably owing to their having a dislike of getting pollen on their plumage. In *Leonotis* and other genera, damage depended much on whether the plants were included in the 'beat' of destructive birds. Birds which are not regular nectar-feeders may sometimes steal nectar too, and this occurs even in Europe, the main genera concerned being *Sylvia* (warblers) and *Parus* (tits); in Britain, the tits attack *Mahonia*, American currant (*Ribes sanguineum*), gooseberry (*R. grossularia*) and cherry and almond (*Prunus* spp.) (Swynnerton, 1916b; Zucchi, 1989; PFY). However, blue-tits (*P. caeruleus*) can actually pollinate sallows, *Salix caprea* and *S. cinerea* (Kay, 1985; AJL) and the crown-imperial, *Fritillaria imperialis* (Búrquez, 1989).

Table 8.1 The main flower-visiting families of birds, with names of the genera mentioned in this book (from Porsch, 1924; Stiles, 1981).

Family	No. of species	Distribution	Genera
Trochilidae Hummingbirds	300+	North and South America	*Glaucis* (Plate 7a), *Mellisuga*, *Patagona*, *Selasphorus* (Plate 7b) *Trochilus*
Thraupidae (part) Honeycreepers	15	Tropical America	
Nectariniidae Sunbirds	106	Africa, South-west Asia to Philippines	*Anthothreptes*, *Arachnothera*, *Cinnyris*, *Nectarinia* (Plate 7c)
Zosteropidae White-eyes	c. 85	Africa, Asia, Australia	*Zosterops* (Fig. 8.5)
Promeropidae Sugarbirds	2	South Africa	*Promerops* (Plate 7d)
Meliphagidae Honeyeaters	160+	Australasia	*Acanthorhynchus*, *Anthochaera*, *Anthornis*, *Phylidonyris* (Plate 7e), *Prosthemadera*
Dicaeidae Flower-peckers	55	Asia, Australasia	
Fringillidae subfam. Drepanidinae Hawaiian finches (Hawaiian honeycreepers)	23	Hawaiian Islands	
Icteridae American orioles	90, few flower-visiting	North and South America	*Icterus*
Psittacidae subfam. Loriinae Lorikeets subfam. Loriculinae Hanging-parrots	c. 60 ? no.	South-east Asia, Australasia South-east Asia	

The nine most important families of flower-visiting birds are listed in Table 8.1, but there is much variation in the proportion of nectar-feeding species in the different families.

Hummingbirds

There are over 300 species of hummingbirds and, although the area which is richest in species is the northern part of the Andes, studies of their behaviour have been made mainly in Central and North America. Most hummingbirds are very tiny – they may weigh less than 3 g, and few weigh more than 10 g (Wolf, Hainsworth & Gill, 1975). The smallest (the Cuban bee hummingbird [*Mellisuga helenae*]) is only 5 cm long, and half of this length is accounted for by bill and tail. Their rapidly vibrating wings make the hum which gives them their name, but the largest species (the great hummingbird [*Patagona gigas*]) is over 20 cm long and flaps its wings with a slow, butterfly-like motion when hovering at flowers (Ridgway, 1891). In North America the hummingbirds are migratory, and there are notable coincidences between the movements of the birds and the flowering periods of the plants they visit (Pickens, 1936; Bené, 1946; Grant & Grant, 1968).

Hummingbirds nearly always feed while in flight, though at least some will perch if they get the chance (Stiles, 1981, p.331; Westerkamp, 1990). Nectar is taken on a large scale by many species of hummingbird and even when feeding on insects these birds usually prefer to find them in the flowers which are most suited to the size and shape of their bills. They may dive right inside large trumpet-shaped flowers such as those of the trumpet creeper (*Campsis radicans*), which is generally considered to be primarily adapted to hummingbirds.

The needle-like bills of hummingbirds are usually straight or slightly decurved (Fig. 8.1A-C); in rare instances, they have a stronger downward curvature or are curved upwards. They are unlike those of most other birds in that there is a strong overlap between the upper and lower mandibles (Fig. 8.1D). During nectar-feeding the tip of the bill is opened sufficiently to allow the tongue to move rapidly in and out (Moller, 1930, 1931a). The tongues of birds have few internal muscles. They consist of a horny skin ensheathing bones and cartilages that extend back into the head, where

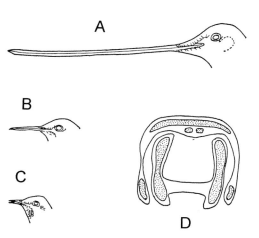

Fig. 8.1 **A–C**, heads of hummingbirds to show extremes of bill-length and a commonly occurring intermediate size; **D**, transverse section of bill of hummingbird showing overlap of upper and lower mandibles, greatly magnified; bone shaded, horn unshaded. A–C after Ridgway (1891), D after Moller (1930).

Fig 8.2 **A**, bones and mantle of an unspecialised avian tongue; **B**, the same parts of the rufous hummingbird (extremities omitted), with the tongue longitudinally sectioned; **C**, mantle and bones of tongue of another hummingbird (*Eulampis holosericeus*), with transverse sections of the mantle. In hummingbirds the protruding bones are coiled round the cranium. Bone and horn unshaded, cartilage shaded; corresponding bones bear the same numbers. Based on Lucas (1897), Moller (1930) and Ridgway (1891).

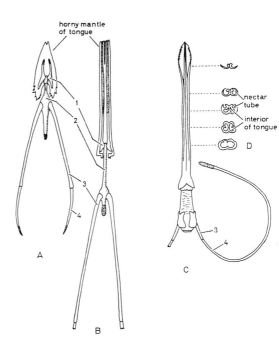

they are activated by muscles (Lucas, 1897; Weymouth, Lasiewski & Berger, 1964) (Fig. 8.2). The tongue of the hummingbird is deeply bifurcated and each lobe is produced into a thin lamina rolled up lengthways to form a slender tube. Nectar is probably held in this double tube by capillarity until the tongue is brought back into the beak, at which stage the nectar can be sucked back into the mouth and swallowed. The end of each tube is papery and frayed, giving a brush-like effect, which may increase the capillarity and which helps in the capture of small insects.

The behaviour of hummingbirds

Hummingbirds readily learn to feed from tubes and small vessels containing liquid food, even without any floral adornments (Ridgway, 1891). They are thus easily subjected to feeding experiments, from which it has been concluded that two North American species, the black-chinned and ruby-throated hummingbirds, like bees, have no inherited colour preference, though they have temporary preferences capable of being modified by conditioning (Bené, 1941, 1946).

Bené noted that individual hummingbirds had differing preferences for particular flower species in the same area, and if the favourite flowers were sufficiently abundant others were rarely visited. This also recalls the behaviour of bees and indicates a degree of constancy valuable for pollination. In Bené's list (1946) of the flowers most visited by hummingbirds, whether for nectar or for insects, and including many species not specialised to hummingbird pollination, the commonest colour is red. In addition, all the flowers in Table 8.2 are red, or red with orange or yellow. Thus it seems that there is selection in favour of red in hummingbird flowers, and this is presumably

Table 8.2 The eight most important hummingbird flowers east of the Mississippi (James, 1948).

Plant	Family	American name
Aesculus pavia	Hippocastanaceae	Red buckeye
Aquilegia canadensis	Ranunculaceae	Wild columbine
Campsis radicans	Bignoniaceae	Trumpet creeper
Impatiens capensis	Balsaminaceae	Jewelweed
Lobelia cardinalis	Campanulaceae	Cardinal flower
Lonicera sempervirens	Caprifoliaceae	Trumpet honeysuckle
Macranthera flammea	Scrophulariaceae	—
Monarda didyma	Lamiaceae	Oswego tea

more because most insects do not perceive it than because the birds actually prefer it. As bird-pollinated flowers have to provide much larger carbohydrate (sugar) rewards to their pollinators than insect-pollinated plants (see Stiles, 1981, p.326), it is important that the nectar should not be removed by insects, and inconspicuous colouring is one factor in preventing this (Raven, 1973). K.A. Grant (1966) has suggested that the uniform colouring of the North American hummingbird flowers is related to the migratory habits of the birds, which involve immigration from the south at breeding time followed (in the west) by ascent to higher altitudes after breeding. The hummingbirds thus have only to seek red flowers on entering a new area with a different flora, since the hummingbird flowers have all evolved the same colour as those from which the birds have already learnt to feed. The North American hummingbirds are all much alike in bill-length, so that there is no likelihood of plants specialising to suit particular types of hummingbird. In tropical America, on the other hand, the birds vary in bill-length, so there is opportunity here for specialisation, and as the birds are resident they learn to distinguish individual plant species. Consequently the hummingbird flowers of this area are much less uniform in shape, size and colour. This theory and others are further expounded in a book on hummingbird pollination in western North America (Grant & Grant, 1968).

Aggressive behaviour in the defence of small feeding territories is very conspicuous in hummingbirds, and these territories are sometimes only a few metres across (Bené, 1946; Pickens, 1944). Indeed, a single flowering tree in Costa Rica was seen by Moller (1931b) to be parcelled out territorially between numerous hummingbirds of three species. Such behaviour evidently limits the dispersal of pollen and thus restricts the range over which outcrossing, if any, takes place. Feeding territories may be held for very short times compared with the nesting territories of temperate passerine birds. Not all hummingbirds are territory-holders, however. In fact the various species can be classified into ecological groups which have different flower-visiting habits. In a study of hummingbirds in Trinidad, Snow & Snow (1972) found that the hermit-hummingbirds (Phaethorninae) differed ecologically from the 'non-hermits' (Trochilinae) which we may call exhibitionists. Within each group there was a sharp division into two size ranges, comprising respectively birds weighing over 6 g and those weighing under 5 g. The hermits had long decurved beaks and drab colouring (Plate 7a) and they spent their time in the forest, remaining within 3-5 m of the ground and visiting only large-flowered shade-tolerant and usually herbaceous plants. The exhibitionists were typically found in more open places than hermits (though they varied in their habitat preferences) and they were the ones that held territories. They were dressed in the bright iridescent plumage for which hummingbirds are famous (Plate 7b) and they

Fig. 8.3 Heliconia bihai in
montane forest in Dominica;
the plant is small compared
with species of open
habitats (which can be many
metres tall). The large
attractive bracts are orange
and the small flowers white
with a green spot near the
tip. A,J.L.

had straight beaks. The two species of large exhibitionists fed at large red flowers,
while the four small ones were often at pink, white or yellow flowers, most of which
were probably insect-adapted. These birds were apparently also visiting hummingbird-adapted flowers, but only when the nectar levels in them were high.

In a wider survey Stiles (1981) found it possible to divide the exhibitionists into three
groups; apart from the small species with bills only 10-15 mm long, there were medium-sized species weighing $3\frac{1}{2}$-7 g with straight bills about 20 mm long, and medium-sized to large species ($5\frac{1}{2}$-12 g) with curved bills over 30 mm long (the ones that
Snow & Snow found visiting hummingbird-adapted flowers). The hermits and the
curved-billed exhibitionists show high degrees of co-adaptation with the flowers they
visit. At the opposite extreme are some short-billed species that actually concentrate
on piercing flowers, and while this is so, co-evolutionary relationships with them are
impossible (Stiles, 1981).

Interesting studies of the interactions between the ecology of plants and hummingbirds were carried out in Costa Rica by Linhart (1973) and Stiles (1975). A single
bird-pollinated plant genus, *Heliconia* (family Heliconiaceae, closely related to the banana family, Musaceae), was considered. *Heliconia* consists of large herbs, often producing massive inflorescences; these have conspicuously coloured boat-shaped bracts
each subtending a condensed branch on which is produced a succession of tubular
flowers, differing between species in size and curvature (Fig. 8.3). Different ecological
tendencies were evident: some species grew in the forest interior, sometimes in clearings and light gaps, and formed small plants with few inflorescences (as in Fig. 8.3),
while others grew on forest margins and river banks, by sunny forest streams or in
unforested swamps, and formed large clumps (probably by vegetative spread) producing hundreds of inflorescences; these also invaded second-growth forest. The flowers
of the first group were visited by hermit hummingbirds and those of the second by
exhibitionist territory holders. In Linhart's experiments, dye powders were put on the
flowers and from the dispersal of the particles the likely dispersal of pollen was inferred. It was found that in two forest-margin species there was a high rate of transfer
near the source and an abrupt drop in dispersal outside the clump. The two species of
the forest interior showed a lower percentage of visits to flowers near the dye source,
but the decline with distance was more gradual and the maximum observed dispersal
distance was greater.

Output of energy in the nectar of a *Heliconia* clump was low in two of the forest species studied by Stiles and highest in two species of open habitats. In Stiles's area there were five *Heliconia* species in which a clump could supply the whole of a hummingbird's daily energy requirement, and three forest species that could not.

The forest species are visited by hermit hummingbirds that have to move from clump to clump. In fact, the birds practise the 'traplining' strategy already described for bees (pp.134-5). The plant must supply enough energy to be worth revisiting, but not enough to be worth defending. (Hermits do occasionally hold territories but do not defend them consistently.) The exhibitionists were again found to vary in their habitat-preferences but they were again territory holders on the *Heliconia* species that produce large clumps. These plants are also characterised by synchronised flowering, high rate of flower-production, more or less high rate of nectar-production and possession of a straight flower-tube. Forest *Heliconia* species tend to have the opposite attributes. However, over the whole group of *Heliconia* species there are some inconsistencies and anomalies.

The kind of floral strategy that is open to the plant depends on its habitat. Forest *Heliconia* species would not be expected to achieve large size, so floral biology will tend towards adaptation to hermits. Second-growth habitats initially have high light intensity; this allows vigorous growth and also intense competition. This may have favoured the ability to form large clumps by vegetative spread, and this in turn confers the ability to produce enough flowers to attract territory-holding hummingbirds. However, as the hummingbirds do not need to move much from clump to clump, cross-pollination will be at a low level (as implied for tree species, p.229). The situation suggests that the plant has been forced into a compromise. As most species of *Heliconia* are self-compatible (Kress, 1983) the consequence is not low seed-set but a low level of outbreeding. However, such *Heliconia* species are the exception; most are in fact hermit-pollinated and hermits in turn are almost always associated with plants in the order Zingiberales, which includes the Heliconiaceae (Stiles, 1981, p.342), (see also pp.409-10).

The two main categories of hummingbird also differ in the extent of their movements over the longer term. Thus Arizmendi & Ornelas (1990) found that, in a part of Mexico with a dry season, medium-sized territory-holding species were residents, while traplining species moved from habitat to habitat following richly rewarding nectar-sources. The small hummingbirds could not hold territories against the medium-sized birds and were forced into non-territorial foraging, particularly at insect-adapted flowers.

Knowledge of energy-production in nectar and energy-consumption in pollinators can be used to investigate feeding efficiency and pollination strategies. An example of such a study is that of Wolf, Hainsworth & Stiles (1972). Three species of hummingbird visiting three species of *Heliconia* at one locality were observed. After a flower visit some nectar remained in the flower, the amount being constant for a given bird species. The hourly rate of nectar production was known, so the time between visits could be used to find the amount of nectar taken.

It was found that the nectar extraction efficiency for the largest of the three hummingbirds was significantly higher on one of the plant species than on the other two. This was due to a higher sugar concentration in the nectar of this plant. Also, the smallest hummingbird showed a higher extraction efficiency at one species of plant than was achieved by the other two birds; by volume it took nectar at the same rate as the largest bird but, being smaller, it used less energy in collecting it. Because of the hummingbirds' speed of flight and the conspicuousness of the flowers in this study,

Table 8.3 Energy expenditures of some bird and hawkmoth pollinators.

Pollinator (source)	Weight (g)	Activity	Rate of energy use (Joules/g body wt/h)	Power consumption by this pollinator (Joules/s=Watts)	Sugar used by this pollinator (mg/h)
Hawkmoth (Heinrich, 1975a)	1	hovering	1250	0.35	75
Hummingbird (Heinrich, 1975a)	5	hovering	900	1.25	270
Hummingbird (Wolf *et al.*, 1975)	5	forward flight	760	1.06	230
Hummingbird (Wolf & Hainsworth, 1971)	5	sitting	250	0.35	75
Hummingbird (Wolf *et al.*, 1975)	13.5	hovering	900	3.39	730
Sunbird (Wolf *et al.*, 1975)	13.5	forward flight	836	3.14	680
Sunbird (Wolf *et al.*, 1975)	13.5	foraging[1]	322	1.21	260

[1] estimate, based on 10% of time flying, remainder hopping and sitting

the time spent searching was short and the cost believed to be low. In this situation it might be advantageous for the birds to specialise but, if lower flower density caused an increase in search time, and so reduced foraging efficiency, it might be better to forage promiscuously.

Table 8.3 shows some energetic costs for hummingbirds compared with a hovering insect (hawkmoth) and a perching bird (sunbird). A large hummingbird has been included for comparison with a sunbird of the same weight. Hovering is the most energy-demanding of all methods of foraging but it makes possible the highest rates of flower visitation (see also remarks under 'Sunbirds', p.234). However, like hawkmoths, hummingbirds need high-reward flowers (see p.244). Hovering for nectar-feeding is only feasible for birds up to the size of the largest hummingbirds. When not active, hummingbirds can reduce energy consumption by becoming torpid (small homeothermic animals need high food intake to avoid starvation in cool conditions). When torpid they regulate their temperature to a lower setting than when active, unlike bees or hawkmoths which abandon temperature control when torpid.

Other American flower-visiting birds

The honeycreepers (part of Thraupidae) and the bananaquit (*Coereba flaveola*; of uncertain affinity) are also important flower visitors in the Americas (Porsch, 1930; Moller, 1931a,b; Campbell & Lack, 1985; Stiles, 1981). Some of them have peculiar bills, adapted to piercing flowers and stealing their nectar; these, in fact, are parasitic on the relationship between flowers and hummingbirds. A few, however, feed on nectar in a legitimate manner, and are tit-like in their movements. They feed when perched, performing difficult contortions in order to reach flowers, and seem to probe the flowers farther from them instead of those most conveniently situated (Moller, 1931b). Their bills are sharply pointed and commonly slightly decurved and fairly short. Their tongues (Fig. 8.4a) are rather similar to those of hummingbirds. Honeycreepers feed

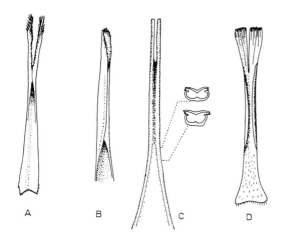

Fig. 8.4 Birds' tongues, from above. **A**, honeycreeper (Thraupidae), with each lobe twisted to form an incomplete tube; **B**, Hawaiian honeycreeper (Drepanidinae); **C**, sunbird (Nectariniidae) with transverse sections (internal muscles draw the upper surface downwards, widening the space between the tongue and upper mandible and producing suction); **D**, tongue of New Zealand bell bird, *Anthornis melanura* (Meliphagidae). A, B after Lucas (1897), C, D after Moller (1930).

on insects and fruit juices as well as nectar.

The New World passerine family Icteridae includes species that are quite important nectar-feeders, especially in the dry tropics, but they are relatively unspecialised, even compared with Old World flower-visiting birds (Stiles, 1981).

Old World flower-visitors: the sunbirds

The main flower-visitors in Africa are the sunbirds (family Nectariniidae), which are also important in Asia; these are mostly small birds with very slender decurved bills, similar to those of some hummingbirds (Plate 7c). The smallest species weigh about 5 g (Wolf, Hainsworth & Gill, 1975), while four species studied by Gill & Wolf (1978) ranged from 7.5 to 17 g. The most important African genera are *Nectarinia* and *Cynniris*, while *Cynniris osea* is a pollinator in Israel. A well-known representative in Asia is the long-billed spider-hunter (*Arachnothera longirostris*), in which the bill is enormous in relation to the size of the body. This bird is a regular flower-visitor as well as specialising in catching spiders while hovering; it is dull in colour and very strongly associated with *Musa superba* (a wild banana) which occurs as widely dispersed riverside clumps. Thus it presents a parallel to hermit hummingbirds visiting the related *Heliconia* (Stiles, 1981). Sunbirds may hover while feeding at flowers, but usually they perch. Hovering is energy-demanding but it seems to be a good option for small animals where the food object is stationary (as in nectarivory and taking spiders from webs) but it is not suitable for mobile food objects (as in insectivory) where hopping and making short flights is effective and more economical of energy (Wolf, Hainsworth & Stiles, 1972). Van der Pijl's account (1937) of the sunbirds' acrobatic habits while drinking from flowers is almost identical with Moller's description of the honeycreepers. The tongue of a sunbird is channelled, but divides into two separate rods near the tip; when the rods are pressed against the upper mandible, the resulting channel forms a sucking tube (Fig. 8.4c).

When the visits of four different species of *Nectarinia* to a single flower species (*Leonotis nepetifolia*, family Lamiaceae) were compared in East Africa, the results obtained were comparable with those for hummingbirds visiting *Heliconia* (Gill & Wolf, 1978). Differences between species in the average duration of a flower visit corresponded principally to differences in bill morphology and the consequent ease of inserting the bill into the corolla. The bills differed in length, depth and curvature. The slowest and least efficient bird was a small species (7.5 g) with a short curved bill that always fed by piercing the calyx. Two large species (weight 15.5 g and 17 g respectively) with strongly curved bills of intermediate length that could easily be inserted into the corollas of the *Leonotis* were the most efficient foragers. The fourth species, weighing 13.8 g, and with a long nearly straight bill which was difficult to insert into the flower, was intermediate in efficiency. The smaller foraging costs of the two smaller species compensated only partially for their reduced nectar-extraction rate. However, on account of its lower flight costs, the smallest bird could achieve a net energy gain when flower-density was half that required by the medium-sized species. Thus a small bird might be able to co-exist in the same habitat as a large one by adopting a strategy of ranging widely and visiting scattered flowers (compare with report on small hummingbirds by Arizmendi & Ornelas cited on p.231). The relative success of the second- and third-largest species was reversed when they visited an *Aloe* species which has a straight flower; the four sunbird species presumably have co-adaptations to different flowers or flower types but were apparently all able to feed profitably together on the dense stands of *Leonotis nepetifolia* growing in fallow fields. (An estimate of the cost of foraging by a sunbird is included in Table 8.3).

It has been pointed out by Westerkamp (1990) that exclusiveness in a pollinator-plant relationship cannot be achieved by lengthening the bill and the flower-tube in a straight line, as to do this would make it difficult for a perched bird to probe the flowers. So, as a rule, the bills of sunbirds and the flowers adapted to them are curved. The same problem is overcome in Lepidoptera by the 'knee-bend' in the proboscis. The curved flower is zygomorphic, restricting direction of approach, so that its stamens and stigmas contact the pollinator with greater precision than those of an actinomorphic flower (see Chapter 6).

Other flower-visiting birds of the Old World, Australasia and the Pacific

The white-eyes are all very similar small warbler-like birds, weighing about 11 g (Paton & Ford, 1977), with very slender, but usually short, bills (Fig. 8.5). They resemble honeycreepers in their gutter-shaped tongues, each lobe of which ends in a brush of flattened hairs, and also in their habit of sucking the juices of fruits as well as the

Fig. 8.5 Indian white-eye,
Zosterops palpebrosa
(Zosteropidae).

Fig. 8.6 Eastern spinebill, (*Acanthorhynchus tenuirostris*) (Meliphagidae, Australia), visiting flower-head of *Banksia spinulosa* (Proteaceae). S.M. Carthew.

nectar of flowers (Moller, 1931a; Holm, 1988).

The sugarbirds of southern Africa (Plate 7d) comprise only the two species of *Promerops* (Promeropidae, perhaps allied to starlings, Sturnidae). They breed only in *Protea* vegetation in various habitats. Their peak of breeding coincides with the flowering of certain species of *Protea* and, as well as feeding on the nectar of the flowers, they use the fluff from the inflorescences for their nests (Campbell & Lack, 1985; Rebelo, 1987).

The honeyeaters of Australia, New Zealand and the Pacific islands make up the family Meliphagidae. Together with the lorikeets, the Meliphagidae are the main pollinators of the important Australian tree and shrub families Proteaceae and Epacridaceae and of some specialised bird-pollinated members of the Myrtaceae and Fabaceae-Faboideae, among others. They are numerous in species, and their size ranges from about the same as a goldcrest to as large as a jay. The moderately small *Phylidonyris* (Plate 7e) and *Acanthorhynchus* (spinebills) (Fig. 8.6) weigh about 11 g, medium-sized species weigh 13-20 g and the largest (the red wattlebird [*Anthochaera*]) weighs about 110 g (Paton & Ford, 1977). The regular consumption of nectar by such large birds occurs also among the Icteridae of South America. Most of the honeyeaters feed on insects as well as nectar (Campbell & Lack, 1985).

The bills of large species of Meliphagidae can be stout and similar to that of the European blackbird as, for example, in the New Zealand bell bird (*Anthornis melanura*). The tongue of this bird (Fig. 8.4D) is channelled, bifurcated and slightly brush-tipped. Holm (1988) described the tongues of several Meliphagidae. Most are channelled and bifurcated (in one instance split into four) and bear a brush of flattened hairs of varying length (in relation to the length of the whole tongue) but better developed than that of *A. melanura*. The flattening of the hairs minimises the width of capillary spaces within the brush (compare the glossa of the wasp *Vespula* and the bee *Eucera*, Chapter 5). This is important because many of the flowers from which the birds feed hold their nectar in the capillary spaces between hairs, but these hairs are cylindrical, so that capillary traction is less than that of the bird's tongue. Whereas the beaks of these brush-tongued species are not particularly slender, the spinebills have long slender bills like those of the sunbirds, and they can hover while feeding. The beak of the eastern spinebill (*Acanthorhynchus tenuirostris*) (Fig. 8.6) is 2 cm long; at a distance of 12 mm back from its tip its thickness is only 1.5 mm. This species is described as tube-tongued because towards the tip the edges overlap; the extreme tip is made of four rods, each with a fringe of flat hairs on one side which are curved to form an open-work continuation of the tube.

The tongues of Meliphagidae can extend 10-37 mm beyond the beak and can lick 6-17 times per second. The bill is closed between each lick and during closure the nectar is presumably squeezed from the brush and into the channel of the tongue, whence it is drawn back into the throat. In the case of the spinebill, capillarity will draw nectar into the brush and the tubular part of the tongue.

Modern work suggests that Meliphagidae do not eat pollen; if they do ingest it, they cannot digest it (Holm, 1988). It has been demonstrated in many investigations that they carry pollen of supposedly bird-pollinated plants and in some that they transfer it to the stigmas of other flowers of the same species (Collins, Newland & Briffa, 1984; Paton & Ford, 1977; Ford, Paton & Forde, 1979; Hopper, 1980a). Species of Meliphagidae and other Australian nectar-feeding birds tend to visit a wider range of plant species than do species of hummingbirds, and the plants themselves receive visits from a wide range of birds. This means that effective cross-pollination is most likely to occur where plants are clumped so that a bird leaving a plant will have a high probability that the next suitable plant it meets will be of the same species.

The flowerpeckers of Asia and Australasia frequently have bills well adapted to flower-visiting, and are particularly associated with the mistletoe family (Loranthaceae) both as pollinators and as distributors of seed (Davidar, 1983). In this family of plants bird-pollination is very common, and is also carried out by a variety of other birds including the sunbirds in Asia and Africa and the white-eyes in Asia (Evans, 1895; Davidar, 1983).

The various species of Hawaiian honeycreeper have bills of different lengths, adapted to the various species of tree belonging to the subfamily Lobelioideae of the Campanulaceae, which they pollinate. However, they also visit for nectar the open flowers of a tree of the family Myrtaceae which is believed not to have reached Hawaii until long after the tubular-flowered tree lobelias (Perkins, 1903). The tongue of these honeycreepers is rolled to form a tube, and has a fringed tip (Fig. 8.4B). The most highly specialised birds of this group have become extinct as a result of changes in the fauna and flora brought about by man (Stiles, 1981).

The last birds to be mentioned here are specialised parrots, the lorikeets. They have the usual short hooked bill of the parrots and a tongue that is very short for a flower-visiting bird. The tip of the tongue appears obliquely cut off (truncate) and the end is provided with a patch of papilliform hairs which is horseshoe-shaped but can be opened out to a greater width. Both pollen and nectar are gathered by the tip of the tongue which is then withdrawn and forced forward against a U-shaped lobed ridge on the palate. This removes the food from the tongue and pushes or drops it further back on the tongue. On the next withdrawal the back of the tongue forces that food into the pharynx. Pollen is the main food but nectar, insects, fruit and grain are eaten (Holm, 1988). These birds move in flocks and visit eucalyptus flowers, frequently becoming thoroughly dusted with pollen.

Bird-pollination: the flowers

A wide range of plant families is involved in bird-pollination (Tables 8.2 & 8.4), which means that the basic structure of bird-pollinated plants is diverse. It is also true that flower-form is varied (Table 8.4).

Bird-pollinated flowers: form and structure

A rough grouping of bird-flowers into five classes according to their general shape is given in Table 8.4. The overhung mouth and reflexed lobes of flowers of Group 1 produce the appearance of a dogfish or shark (Fig. 8.7). An interesting feature of some members of Group 2 is that the flower-tube remains closed at the top until probed by a bird, when it opens explosively, scattering a cloud of pollen over the visitor. It may also be noted that some families with a bilaterally symmetric flower structure (Acanthaceae, Gesneriaceae and Scrophulariaceae) have produced species of plants having almost regular tubular flowers, as one of their forms of adaptation to bird-pollination, in conformity with the syndrome (Table 8.5). In Group 2 (Plate 7c,f, Fig. 8.8), the stamens and style (or the style only) may protrude from the tube. Another example is *Macranthera flammea*, which has a special arrangement for ensuring contact of the stigma and anthers with the neck of a hummingbird visitor. The narrow tubular flowers stand erect on springy pedicels, while the style and stamens in turn protrude from the tube. The hovering bird inserts its bill and then drops down so as to bring the flower into a more conven-

Fig. 8.7 Bird-pollinated gullet blossom, *Columnea microphylla* (Gesneriaceae), with a bright red and orange corolla.

ient horizontal position. In this way, either the stigma or the anthers are levered into contact with a precise spot on the nape of the bird's neck (Pickens, 1927). Typical brush flowers (Group 3) are shown in Plate 7e and Fig 8.6. Flowers with protruding stamens are particularly common in Australia, partly because the dry climate makes protection of the pollen from rain unnecessary. The flowers in Group 4 all face one way, in line with the peduncle and usually upwards, and the heads are closely surrounded by bracts. They often resemble Asteraceae, particularly thistles, but in fact there are very few Asteraceae that are bird-pollinated. By holding their nectar in spurs

Fig. 8.8 A radially symmetric hummingbird-pollinated flower, *Manettia inflata* (Rubiaceae); coloured orange-red with a yellow mouth to the corolla.

Table 8.4 Flower forms of bird-pollinated species, with their family and distribution.

1. Gullet flowers (*'Rachenblumen'* in German)
Long-tubed, zygomorphic, upper lip overhangs mouth, lower lip reduced/recurved

Columnea microphylla	Gesneriaceae	Central America (Fig. 8.7)
Rechsteineria cardinalis	Gesneriaceae	Tropical America
Mimulus cardinalis	Scrophulariaceae	N America
Salvia splendens	Lamiaceae	Tropical S America
Eremophila spp.	Lamiaceae	Australia
Leonotis leonurus	Lamiaceae	S Africa
Vitex lucens	Verbenaceae	New Zealand
Antholyza spp.	Iridaceae	Africa

2. Tubular flowers
Long-tubed, radially symmetric or nearly so, mouth slightly flared

Abutilon megapotamicum	Malvaceae	Tropical S. America
Bruguiera spp.	Rhizophoraceae	Tropical mangrove swamps[1]
Fuchsia spp.	Onagraceae	Central to temperate S America
Ipomopsis aggregata	Polemoniaceae	N America (Plate 7f)
Manettia inflata	Rubiaceae	Tropical S America (Fig. 8.8)
Lonicera sempervirens	Caprifoliaceae	N America
Erica spp.	Ericaceae	S Africa
Macleania spp.	Ericaceae	Tropical America
Astroloma humifusum	Epacridaceae	Australia
Russelia juncea	Scrophulariaceae	Central America
Penstemon centranthifolius	Scrophulariaceae	N America
Macranthera flammea	Scrophulariaceae	N America
Rechsteineria lineata	Gesneriaceae	Tropical S America
Aeschynanthus radicans	Gesneriaceae	Indonesia
Odontonema schomburgkianum	Acanthaceae	Tropical S America
Loranthus kraussianus and *L. dregei*	Loranthaceae	S Africa[1]
	Bromeliaceae (many)	Central to temperate S America
Kniphofia spp.	Liliaceae	Africa
Aloe spp.	Liliaceae	Africa (Plate 7c)
Lachenalia spp.	Liliaceae	Africa
Phormium spp.	Liliaceae	New Zealand

3. Brush-flowers
Clustered in spheres or cylinders, often densely, with numerous protruding coloured stamens

Greyia sutherlandii	Melianthaceae	S Africa (flowers well spaced)
Callistemon spp.	Myrtaceae	Australia
Beaufortia sparsa	Myrtaceae	Australia
Eucalyptus spp.	Myrtaceae	Australia (flowers well spaced)
Acacia celastrifolia	Fabaceae sub-family Mimosoideae	Australia
Calliandra fulgens	Fabaceae sub-family Mimosoideae	C America
Banksia spp.	Proteaceae	Australia (Fig. 8.6)
Nuytsia floribunda	Loranthaceae	Australia
Amyema spp.	Loranthaceae	Australia (flowers well spaced)

4. Capitula
Flowers clustered in disc-like or hemispherical heads, parallel or radiating, often zygomorphic

Dryandra spp.	Proteaceae	Australia
Protea spp.	Proteaceae	S Africa (Plate 7d)
Mutisia spp.	Asteraceae	S America (flower stalks pendent)
Haemanthus natalensis	Amaryllidaceae	S Africa

Table 8.4 (continued)

5. Spurred flowers

Petals separate; nectar-tube in the form of a spur formed by a single sepal or petal

Aquilegia canadensis	Ranunculaceae	N America (5 spurs)
Delphinium cardinale	Ranunculaceae	N America (1 spur)
Tropaeolum pentaphyllum	Tropaeolaceae	Tropical S America (1 spur)
Impatiens niamniamensis	Balsaminaceae	Tropical Africa (1 spur)
Impatiens capensis	Balsaminaceae	N America (1 spur)

[1] Flowers open explosively when probed

formed from a single perianth member, the flowers of Group 5 direct birds' bills away from the ovaries (Grant, 1950b). These spurred flowers belong to mainly insect-pollinated families, and it seems that the possession of a spur has made it easy for some species to evolve into bird-flowers.

Good examples of adaptation to different types of bird are seen in the genus *Erythrina* (Fabaceae-Faboideae). In Indonesia, different species are adapted to birds of different sizes (Fig. 8.9A-C) which feed while perched on the peduncles, and the flowers face inwards so that they can be easily reached from this position (Docters van Leeuwen, 1931). On the other hand, there are many American species of *Erythrina* that are highly adapted to hummingbirds, having closed tubular flowers providing no

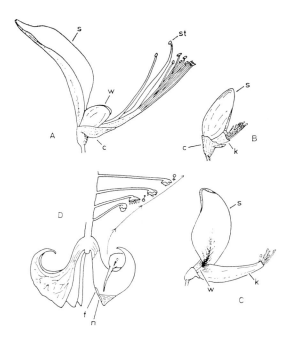

Fig. 8.9 **A**, Flower of *Erythrina variegata* var. *orientalis* which is pollinated by starlings, *Sturnopastor* sp., and other birds of similar size (orioles, drongos, thrushes etc.); **B**, *E. subumbrans* pollinated by smaller perching birds (bulbuls, sunbirds and white-eyes); **C**, *E. umbrosa*, pollinated by humming-birds. These flowers are all red and the shape of the calyx suggests that it is important in keeping petals and stamens in position. **D**, inflorescence of a hummingbird-pollinated species of *Marcgravia* (incomplete, and with one of the flattened nectar- vessels cut open). c, calyx, f, flower rudiment, k, keel, n, nectary, s, standard, st, stigma, w, wings, k flower in male stage, j flower in female stage; arrows show course of departing bird's head. A-C after Docters van Leeuwen (1931), D after Wagner (1946).

Fig. 8.10 Bird-of-paradise flower, *Strelitzia reginae* (Strelitziaceae), a very large flower pollinated by perching birds.

alighting place and facing outwards to suit hovering birds (Raven, 1977; Cruden & Toledo, 1977). American bird-flowers are generally longer-tubed than Asiatic ones, in correspondence with the generally longer bills of the American pollinating birds (van der Pijl, 1937).

Fig. 8.11 Inflorescence of *Marcgravia umbellata* (Marcgraviaceae): the brim-full nectar-pouches hang down in the centre, while the unopened flowers stand above them at an angle to their pedicels. When the flowers open the pedicels bend further and the cap-like corollas are shed; the knob on the central nectary is a rudimentary flower.

A South African bird-pollinated plant which does not easily fall into any of these groups is the bird-of-paradise flower, *Strelitzia reginae* (family Strelitziaceae, closely related to the banana family, Musaceae). It has very unequal perianth segments; the outer three are orange and serve for display, while the inner three are blue, two being large and concealing the stamens, and one smaller, concealing the nectary (Fig. 8.10). The two segments that conceal the stamens have lobes on which a bird must stand in order to reach the nectar; the weight of the bird pushes the lobes apart so that the anthers are then exposed and dust the underside of the bird with pollen (Scott Elliot, 1890a). The much larger flowers of the giant strelitzia (*S. nicolai*) are effectively pollinated by sunbirds that perch crosswise on the blue perianth segments (the outer ones are white) and on the projecting tip of the style. The pollen is very sticky and hangs together in threads, so that although birds do not often perch on the style, a good seed-set is achieved. This plant flowers all the year round and produces very dilute nectar (about 10% sucrose equivalent). Several species of sunbirds visit it (Frost & Frost, 1981).

The tropical American genus *Marcgravia*, some species of which are bird-pollinated, is another very distinctive plant, having its nectar contained in modified bracts. In an unnamed species studied by Wagner (1946), these are arranged in a whorl which terminates the pendent inflorescence, and above them a series of flowers develops on stalks which lengthen with age (Fig. 8.9D). The shape of the nectar vessel forces the hummingbird visitors to depart in such a way that their heads touch the flowers. A similar species (*M. sintenisii*) was observed by Howard (1970) to be visited by hummingbirds, honeycreepers and todies (*Todus*, family Todidae, gaudy little insectivorous birds, 11.5 cm long, confined to the Greater Antilles); its nectaries are at first yellow but turn bright red when the flowers open. The extraordinary appearance of the *Marcgravia* inflorescence is well shown in Fig. 8.11.

One further unusual plant is the screw-pine (*Freycinetia*, family Pandanaceae), of the Pacific region, which provides fleshy, sugary bracts surrounding the flowers. These are eaten by birds or bats. Those of the Hawaiian *F. arborea* were formerly eaten by the Hawaiian crow (*Corvus tropicus*) and other extinct endemic birds which presumably pollinated the plant; currently it is pollinated by the introduced *Zosterops japonica* (Cox, 1983a). In the Samoan islands, *F. reineckei* is pollinated by the Samoan starling (*Aplonis atrifuscus*, family Sturnidae) and a day-flying fruit bat, both of which eat the bracts (Cox, 1984).

The syndrome of bird-pollination

In spite of the variation in bird-pollinated flowers, it is possible to draw up a list of reasonably constant features that characterise them (Table 8.5) (for the syndrome concept, see pp.142, 178.)

The prevalence of red as a colour in hummingbird flowers has already been noted and discussed (pp.228-9). In all the main areas of the world where bird-pollination is significant, bird-adapted flowers are frequently red (syndrome, item 3). Apart from the red-blindness of insects, another factor in this may be that birds have their greatest spectral sensitivity and finest hue discrimination towards the long-wavelength (red) end of the spectrum (references in Stiles, 1981). Hummingbirds can respond to near-ultraviolet light which is invisible to man, but it is not known how important this is in

Table 8.5 Features of the floral syndrome of bird-pollination (ornithophily) (mainly from Faegri & van der Pijl, 1979 – presentation modified).

1) daytime flowering
2) flower-opening concentrated in the early morning[1]
3) vivid colours, often scarlet or with contrasting 'parrot-colours'
4) colour-signal often long-lasting[2]
5) lip on flower-margin absent or curved back
6) flower tubular and/or pendent or nodding
7) external wall of flower firm
8) stamen-filaments stiff and/or united
9) ovary protected
10) scent absent
11) nectar abundant
12) peak of nectar-production in early morning[1]
13) nectar held in flower by capillarity
14) flower tube or spur deep and wider than in butterfly flowers
15) separation of nectar reservoir from receptive stigmas and polliniferous anthers relatively great.

[1] Rebelo (1987) [2] Stiles (1981)

Fig. 8.12 A little hermit hummingbird, *Phaethornis longuemareus*, hovering in front of the red-spotted leaves of *Dalbergaria florida* (Gesneriaceae) behind which the flowers are concealed. After Susan Payne Smith in Jones & Rich (1972).

bird-flower relationships (Goldsmith, 1980). Bird-flowers can often be recognised by their harsh primary colours (Plate 7), frequently arranged to form a simple pattern in a single flower or inflorescence, and then referred to as 'parrot-coloration'. This is particularly common in the pineapple family (Bromeliaceae) (Porsch, 1924) and the *Heliconia* family (see pp.230-1), in both of which it often extends to the bracts. By the involvement of bracts, and sometimes stems, the plant is able to present a long-lasting colour-signal (syndrome, item 4). Several species of the epiphytic hummingbird-pollinated *Dalbergaria*, (formerly included in *Columnea*, family Gesneriaceae), have inconspicuous pale yellow tubular flowers hidden under large spreading or drooping leaves. The signal to the birds that flowers are present takes the form of red blotches on the leaves. Usually there is an irregular blotch on the underside of the leaf, but *D. florida* produces, in addition, two red spots on the upper surface near the tip; these only appear just before flowering begins. As the leaves are drooping at their tips, the positioning of these spots is highly significant (Jones & Rich, 1972) (Fig. 8.12).

Guide-marks are frequent in bird-flowers but are not usually as well defined as those of insect-pollinated flowers (Plate 1); the ultra-violet component of these guide-marks is very weak (Kugler, 1966).

The projecting styles and stamens found in many Australian bird-pollinated flowers or inflorescences of the brush type (see pp.173 and 237) are hardened but also elastic (Holm, 1988) (syndrome, item 8) and birds may perch on them when feeding at the flowers. Except in extremely large flowers, petals (and/or sepals) do not usually provide any alighting place (syndrome, item 5), but many plants have portions of bare stem, leaf-stalks, bracts or flower-buds in a convenient position for birds to perch on. For example, *Kniphofia* (red-hot poker), *Aloe* (Plate 7c) and many other South African plants have rather short inflorescences surmounting a stout bare stem, while in the genus *Puya*, a South American member of the pineapple family (Bromeliaceae), the rigid inflorescence-branches are prolonged beyond the flowers (Baker & Baker, 1990). The flowers of *Erythrina* that have been mentioned as being directed back along the inflorescence-stalk represent an asymmetric version of the same thing, with the flowers along the upper side of an inclined and basally bare axis. This arrangement occurs in other genera, particularly *Combretum* (Combretaceae) (Gruyj *et al.*, 1990). Such an inflorescence may lie along the ground, as in *Pitcairnia corallina*, a terrestrial member of the Bromeliaceae from Colombia. Some Australian herbs and undershrubs are also adapted to pollination by birds standing on the ground (for example, *Brachysema* and its allies [Fabaceae-Faboideae]; see Porsch, 1927; Keighery, 1982), and the habit occurs again in South Africa.

Robustness (syndrome, item 7) may serve to prevent nectar-thieving by both insects and birds, as well as enabling birds to perch on part of the inflorescence while probing its flowers (Rebelo, 1987). Flexible pedicels, as in many *Fuchsia* species, are also a hindrance to attempts to pierce the perianth for nectar (Swynnerton, 1916a) and the same may be true of nectaries in spurs (see p.239 and syndrome, item 15). Flower-adapted birds are thought unlikely to be the cause of accidental damage (Stiles, 1981).

Bird flowers are scentless (syndrome, item 10), for flower-visiting birds have little or no sense of smell.

In relation to their size, the flowers of plants adapted to bird-pollination secrete large quantities of nectar (syndrome, item 11) which, as we have already seen, is relatively dilute and sometimes slimy. Species of *Banksia* in Australia produce so much nectar that it is used as food by aboriginals (Werth, 1956), while showers of nectar can be brought down by shaking the branches of *Erythrina* and *Grevillea* (Swynnerton, 1916a). Typical nectar concentrations (weight of sucrose or equivalent in a given weight of solution) of hummingbird-pollinated flowers have been assembled by Baker (1975) from his own and others' observations. The extremes reported are 10% and 34%. Means calculated from the plant species growing in particular areas and habitats were however, remarkably close to one another, varying only from 20% to 24%. These concentrations are weak compared with those of insect-pollinated flowers; the figures available for bee-pollinated flowers in the same areas are: 10% to more than 75%, with means of 30% to 48%.

Baker (1975) has suggested that the relatively low concentration is necessary to keep the viscosity down to a level that permits rapid uptake by the tongue, since this would be important in a bird that hovers while feeding. This assumes that uptake is firstly by capillary movement into the two troughs of the tongue, as already described, rather than by suction. The viscosity of a sugar solution rises steeply with concentration, so that at 20°C a 40% sugar solution is over three times as viscous as a 20% solution. The suggestion that viscosity is important is supported by the observation that the viscosity of nectar of about 20% concentration found in lowland Costa Rica is about the same as that at high tropical mountain sites with about 12% concentration. The temperature difference, from 30°C in the lowlands to about 10°C in the tropical highlands, accounts for the similar viscosities. Possibly the concentration is adjusted so that a certain viscosity is not exceeded. Another possible factor in the low concentration of the nectar of hummingbird-pollinated flowers is the need for replacement of water supplies which may be more pronounced in birds than in insects.

Bolten & Feinsinger (1978) argued against Baker's theory on the ground that, given a choice, hummingbirds always prefer the strongest sugar solutions, even up to 49% and even to the point where the higher viscosity reduces foraging efficiency. If, however, nectars below about 20% concentration are unprofitable for collection by bees but not by hummingbirds, they might have been selected for as a deterrent to the collection of the nectar by bees. Bolten & Feinsinger compared a set of hummingbird-flowers according to flower depth. It was found that those with nectar hidden too deeply to be reached by any bees averaged a distinctly higher concentration (though the sample was very small: 9 species). The reasoning is the same as that already quoted to account for the red coloration of bird-pollinated flowers. The fact that plants have both red flowers *and* dilute nectar might be due to selection for stronger nectar by hummingbirds constantly pushing the concentration to the edge of the bee-acceptable range, and/or that the red signal is valuable in flower location in its own right.

The proportions of the three nectar sugars (see p.40) vary according to the type of bird attracted to the flowers. Hummingbird-flowers show a great preponderance of sucrose; Freeman *et al.* (1985) reported this as averaging 70% in a sample of Mexican plants. However, plants apparently adapted to other avian pollinators and producing low-sucrose nectars are sometimes heavily exploited by hummingbirds (Freeman *et al.*, 1985).

The nectar of the most widely open bird-pollinated flowers is easily visible, and some tubular flowers, for example *Antholyza bicolor*, also store their nectar where it can be seen, with the result that the birds by-pass those flowers which are empty. Many other tubular flowers have a swelling at the base in which the nectar accumulates. Arrangements that hinder the accidental loss of the relatively large volumes of nectar in the flowers are well developed (Holm, 1988). The simplest is the narrowing of the spaces between the floral tube, the stamens and the style, or the provision of hairs at the base of the flowers, thus creating capillary interstices to hold the nectar. If there is an enlarged nectar store, its mouth may be protected by incurved hairs or stamen-filaments. The result is that nectar cannot be shaken from these flowers; at the same time, change of concentration through evaporation is minimised.

Nectar sugar production per flower per 24 hours shows enormous variation in hummingbird-pollinated plants. Figures compiled by Carpenter (1983) range from 0.22 to 80 mg.

Pollination by bats

Visits to flowers by bats are known to have been observed as early as 1772, but it was only towards the end of the nineteenth century that bats were recognised as significant pollinators (Jaeger, 1954b; Baker & Harris, 1957; Cox, 1983a). Now the subject is firmly 'on the map' with the publication of a book devoted to it (Dobat & Peikert-Holle, 1985).

Bat-pollination occurs mainly in the tropics, but in the Andes it occurs up to 3,400 m above sea level, where frosts are frequent at night. Flower-visiting bats are found mainly where there is a succession of suitable flowers for them all the year round, but those that pollinate cacti in southern Arizona are migratory (McGregor *et al.*, 1962). The latitudinal limits extend about 30° either side of the equator. Most bat-pollinated plants are trees or woody climbers, but some are dwarf shrubs, herbs or herbaceous climbers. Observation of bat-pollination is difficult because it takes place mainly in high trees and at dusk or in darkness. However, bat-pollination can often be deduced from the presence of claw-marks on the flowers or the fallen corollas. Photography and night-viewing aids can show how the bats behave during their visits to flowers, and the animals can be netted so that their pollen loads can be examined. Pollen grains can also be identified in faeces deposited at roosts.

Table 8.6 The floral syndrome of bat-pollination (chiropterophily) (from van der Pijl, 1936; Faegri & van der Pijl, 1979; Vogel, 1968-9; and Skog, 1976).

1) flowers are open and pollen and nectar are available at night
2) flower-opening takes place late in the day
3) colour white, creamy, ochre-yellow or dingy shades of green and/or purple
4) flowers usually last only one night
5) scent nocturnal, strong, unlike that of insect-pollinated flowers, often fruity, sour, musty, cabbagy or suggestive of fermentation
6) flowers robust
7) flowers often widely bell-shaped or scuttle-shaped
8) large quantity of dilute nectar (sometimes mucilaginous)
9) large quantity of pollen
10) flowers are in exposed positions (accessible to bats)

Bat-pollinated flowers

The chief characteristics of bat-pollinated flowers are listed in Table 8.6. The flowers may open just before dark, but if the plant is pollinated by bats living in huge roosts in caves and ranging widely for food the flowers may not open until some hours after sunset, when the bats may have had time to reach them (Gould, 1978). The colours are more or less like those of hawkmoth-pollinated flowers (Chapter 4) but dingy colours are more prevalent, and these are perhaps adapted to discouraging visitation by other classes of pollinator. The scents are sometimes like those of the bats themselves. Placement of the flowers clear of the foliage may be achieved by having branches in layers with space between them ('pagoda-style'), by bearing them on trunks and limbs ('cauliflory'), by holding the flowers aloft on stiff stalks or by letting them hang below the leaves on rope-like stalks ('flagelliflory'). The first two types are more or less exclusive to the Old World. In addition, if the plant is small it is likely to grow epiphytically or on rocks, so that approach is possible by an ascending flight. Herbaceous bat-pollinated plants, which are mainly characteristic of the New World, are always larger than their herbaceous relatives with other pollination arrangements. While more nectar is produced by the blossoms of bird-pollinated plants than insect-pollinated, bat-pollinated flowers produce still more. The nectar provision is related to the greater size of the bats: for example, 10-45 g in the New World, and 16-120 g in the Old, although the mainly fruit-eating flying foxes (*Pteropus* spp.) can be up to 1,500 g and may visit flowers occasionally. Some bat-pollinated flowers last two or three nights but many (especially brush-flowers, see below) last only one, which is remarkable in view of the heavy investment by the plant in massive flowers. It has been suggested that retaining these parts after they have performed their function might induce animals to eat them and incidentally damage the ovaries (Cox, 1982).

As with bird-pollinated plants, the flower food offered may be nectar, pollen or, rarely, fleshy sugary bracts. The latter occur in *Freycinetia* (Pandanaceae), already mentioned in connection with bird pollination. The Indonesian species *F. insignis*, pollinated by flying foxes, differs from the bird-pollinated members of the genus (p.08.19) in that the inflorescences open in the evening and produce a musty fruity odour. The nectar provided by bat-pollinated flowers in the New World is glucose-rich or glucose-dominated, whereas in the Old World sucrose-rich and sucrose-dominated nectars occur as well (Baker & Baker, 1983a, b; Freeman *et al.*, 1991). The pollen (or stamens) of certain brush-like flowers (which also provide nectar) is eaten by bats in both the Old and the New Worlds. Examples are found in the families Bombacaceae, Myrtaceae, Sapotaceae, Fabaceae-Mimosoideae, Agavaceae and Cactaceae (van der Pijl, 1936, 1956; Vogel, 1958; Howell, 1979). Large amounts of pollen may be produced through having many stamens or extra-large anthers or by mixing male flowers in with the hermaphrodite ones (Vogel, 1968-69; Heithaus, Opler & Baker, 1974). A striking case is provided by *Pseudobombax* (Bombacaceae), where there are hundreds of stamens. *P. munguba*, of Brazil, offers no nectar (Gribel, 1995). Apparently the need to produce a large amount of pollen arises from the large surface area of a bat in relation to the stigmatic area of the flower (Heithaus, Opler & Baker, 1974), as well as the bats' appetites. Nevertheless, the work of Howell (1974) confirms that bat-pollinated plants have become adapted to the nutritional needs of the bats by having more protein in their pollen than related insect-pollinated or unspecialised flowers. Howell demonstrated that the bat *Leptonycteris sanborni* had protein requirements similar to those of many other animals. However, it ate very few insects but large quantities of pollen; captive bats remained in good health when fed on artificial diets based on the

Fig. 8.13 African bat-pollinated
flowers. **A**, the baobab
(Adansonia digitata);
B, *Parkia clappertoniana*,
diagrammatic section through
the inflorescence.
A after Jaeger (1954b),
B after Baker & Harris (1957).

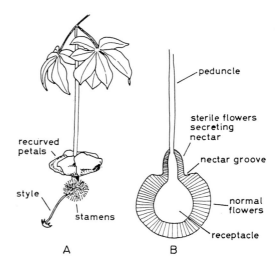

pollen of bat-pollinated species of *Agave* and *Carnegiea* (the saguaro cactus). Provision of some pollen purely for the bats appears to occur in *Parkia*, since it is released directly into the nectar (Hopkins, 1984).

The commonest type of bat-pollinated flower in the Old World is the separate-petalled brush-flower (Table 8.7). One such species, the kapok tree, has American affinities but is believed to have reached Africa naturally. Its creamy-white flowers occur in spherical clusters where there are no leaves. The flowers are 5 cm in diameter and their pedicels are 8 cm long, so the clusters are quite large objects. Other examples are the durian and the baobab tree (Fig. 8.13A); bats alight on the tuft of stamens of the baobab and lap up nectar which has run out on to the petals. A distinctive form of inflorescence is found in the mimosoid tree *Parkia biglobosa*; it consists of tiny flowers densely packed on a solid body of the shape shown in Fig. 8.13B. These flowers are at first red, then purple and finally salmon pink, and produce a weak fruity scent. Fruit-sucking bats cling to the flower-heads when drinking the nectar, which is produced by sterile flowers near the peduncle and collects in the trough just below. Harris & Baker (1959) indicated that in Ghana, in the form then called *P. clappertoniana*, the head lasts two nights, being male the first night and female the second, but Hopkins (1983) found that it lasted only one night. Four species of *Parkia* are recognised in Africa, of which three are known to be bat-pollinated (Hopkins, 1983). Species of *Parkia* are also pollinated by bats in south-east Asia (Docters van Leeuwen, 1938; Gould, 1978) and in tropical America (Hopkins, 1984).

Wild bananas (*Musa* spp., family Musaceae) are either bird-pollinated or bat-pollinated. In the Malay Peninsula, the bat *Macroglossus sobrinus* is strongly associated with *Musa* and, in places, perhaps totally dependent on it (Start & Marshall, 1976). The flowers do not fit any of the usual patterns, being rather narrowly tubular and situated side by side with their mouths downwards in a row against the most recently exposed bract, which is very large and supports the bats (Plate 8b).

A good example of an Old World bat-pollinated gullet-flower is *Kigelia africana*, the sausage tree (Table 8.7). The flowers are dark-coloured, fleshy, sour-smelling and nocturnal, and they hang down on long stalks. The corolla is 7 cm long and 12 cm wide,

Table 8.7 Some bat-pollinated members of Old-World plant families.

BRUSH FLOWERS

Families	Species	Figures & references
All dicotyledonous with separate petals*		
Bombacaceae	*Adansonia digitata* (baobab)	Fig. 8.13A; Harris & Baker, 1959; Jaeger 1954a & b
	Adansonia gregorii	Armstrong, 1979
	Ceiba pentandra (kapok)	Vogel, 1958; Baker & Harris, 1959
	Durio zibethinus (durian)	van der Pijl, 1936
	Gossampinus valetonii	van der Pijl, 1936
Chrysobalanaceae	*Maranthes polyandra*	Lack, 1978
Lecythidaceae (incl. Barringtoniaceae)	*Barringtonia acutangula*	Armstrong, 1979
	Planchonia careya	Armstrong, 1979
Fabaceae subfamily Mimosoideae	*Parkia* spp.	Fig. 8.13B; Baker & Harris, 1957; Gould, 1978; Hopkins, 1983
Sonneratiaceae	*Duabanga* spp.	Start & Marshall, 1976
	Sonneratia spp.	Start & Marshall, 1976
Myrtaceae	*Angophora* spp.	Armstrong, 1979
	Eucalyptus spp.	Armstrong, 1979
	Melaleuca spp.	Armstrong, 1979
Proteaceae	*Banksia* spp.	Armstrong, 1979
	Grevillea spp.	Armstrong, 1979
	Protea elliotii	Lack, 1978

GULLET FLOWERS

Dicotyledonous families with petals separate		
Fabaceae subfamily Faboideae	*Mucuna macropoda*	Plate 8a; Hopkins & Hopkins, 1993
Dicotyledonous families with petals joined		
Bignoniaceae	*Kigelia africana*	Vogel, 1958
	Oroxylum indicum	Gould, 1978

OTHER FLOWER FORMS

Monocotyledonous family		
Musaceae	*Musa* spp.	Gould, 1978; Start & Marshall, 1976

*sequence according to Heywood (1978)

Fig. 8.14 Flowers of South
American bat-pollinated plants.
I, *Trianaea speciosa* (Solanaceae);
II, *Symbolanthus latifolius*
(Gentianaceae); **III**, *Cayaponia*
sp. (Cucurbitaceae); **IV**, *Cobaea
scandens* (Polemoniaceae);
V, *Campanaea grandiflora*
(Gesneriaceae); **VI**, *Cheirostemon
platanoides* (Malvaceae). From
Vogel (1958).

and has a lower lip with a wrinkled surface, affording a grip for alighting bats which crawl inside.

The bat-pollinated flowers of the New World have been surveyed by Vogel (1958, 1968-69). They fall into two main groups, the bell-shaped or wide gullet-like flowers (Fig. 8.14) and the brush-like form, both of which, as we have just seen, also occur in the Old World. Brush-blossoms are made up of one or more flowers, in the latter case forming a dense cluster. The blossom is usually held above the foliage and the brush effect is produced by the numerous stamens that are much longer than the sepals and petals. All the dicotyledons with brush-flowers in Table 8.8 belong to families that are primitive in having the petals and/or sepals free or nearly so, and the same is true of the monocotyledon family Agavaceae. In the Marcgraviaceae, the nectary is formed by a bract and so separated from the flower, but the stamens tend to occur in a protruding brush, so the plants may be classed as having brush-flowers. *Marcgravia* includes bat-pollinated species in which the inflorescences hang down on rope-like peduncles or on leafy stems, like those pollinated by hummingbirds (Figs. 8.9D, 8.11) (Vogel, 1958; Sazima & Sazima, 1980). In the bat-pollinated *Norantea macrocarpa*, however, the inflorescence is an erect umbel and each floret has a nectary of its own formed by a bract near the base of the pedicel. All the nectaries are therefore near the centre of the umbel, where there is an open space giving easy access to them (Vogel, 1958, using the name *Marcgravia* cf. *rectiflora*). Bat-pollination has been discovered more recently in a member of the Euphorbiaceae, *Mabea occidentalis*, a small tropical American pagoda-style tree, the inflorescences of which show parallels with those of

Table 8.8 New-World plant families that include bat-pollinated members.

Monocotyledons	Dicotyledons with petals separate	Dicotyledons with petals joined
Brush flowers		
Agavaceae	Bombacaceae	Cactaceae
	Capparidaceae	*Carnegiea*
	Caryocaraceae	
	Chrysobalanaceae	
	Euphorbiaceae	
	Fabaceae-Mimosoideae	
	Marcgraviaceae	
Bell-shaped or gullet-like flowers		
Bromeliaceae	Bombacaceae	Acanthaceae
Cannaceae	Fabaceae-	*Trichantha, Louteridium*
	Caesalpinioideae	Bignoniaceae
	Fabaceae-	*Parmentiera*
	Faboideae	Cactaceae
	Malvaceae (Fig. 8.14)	Cucurbitaceae (Fig. 8.14)
	Melastomaceae	Gentianaceae (Fig. 8.14)
		Gesneriaceae (Fig. 8.14)
		Campanulaceae-Lobeloideae
		Burmeisteria,
		Siphocampylus
		Polemoniaceae (Fig. 8.14)
		Rubiaceae
		Hillia
		Solanaceae (Fig. 8.14)

(Vogel, 1958, 1968-69; Sazima & Sazima, 1980; Steiner, 1983)

Marcgravia (Steiner, 1983).

Some of the bell-shaped or gullet-like bat-pollinated flowers belong to the more primitive families that have free or almost free petals (Table 8.8). Here the petals are usually just firmly overlapped, but in *Eperua leucantha* (Fabaceae-Caesalpinioideae) the large upper petal is modified to form a bell on its own. This flower-form is, however, more common in the more advanced families that have united petals; some of these families include a strong representation of hummingbird-pollinated species. Typically the bell or gullet leads into a narrow tube, at the base of which the nectar is held, while the corolla-lobes are turned sharply back. The stamens and style most often lie along the upper or lower side of the bell and the flower posture ranges from pendent to upwardly inclined.

A bat's tongue, unlike a bird's beak, is not available for the transfer of pollen, so the anthers are always at the throat or outside it, so as to deposit pollen on the head or fore-body. The flowers themselves fall into two groups on size: larger ones which the bats partially enter or cling to, holding the reflexed petals or the sepals, and smaller ones that fit them like a face-mask, from which nectar is taken by the bats in flight. About the maximum size for a flower of the large-sized group, 12 cm long and 8 cm across the mouth, is found in *Ochroma pyramidata* (*O. lagopus*), the balsa-wood tree, in family Bombacaceae. The mask-like flowers actually offer much less nectar than most bat-pollinated flowers (only a few microlitres) (Frankie & Baker, 1974).

A New World bat-pollinated plant well known in cool-temperate gardens is *Cobaea scandens* (the cup-and-saucer vine, family Polemoniaceae). It is a climber with pendent shoots, from which hang long pedicels supporting the bell-shaped flowers in a nearly horizontal posture (Fig. 8.14IV). The flowers are greenish when they open and dull purple later; their scent is somewhat sweet but also includes a cabbagy element. (Pollination records of *C. scandens* are summarised by Dobat & Peikert-Holle, 1985.)

Flower-visiting bats

The classification and main features of vegetarian bats are shown in Table 8.9. It will be seen that the ability to fly and feed in the dark using echo-location (sonar), for which bats are famous, is restricted to the Microchiroptera, and that the flower-visiting bats of this sub-order are confined to the New World. The Old World flower visitors (belonging to the Megachiroptera) are, on average, much heavier.

The Pteropidae include large fruit-eating species of *Pteropus* (flying foxes) and *Cynopterus*. *Pteropus* has already been mentioned as a pollinator of *Freycinetia* (p.241). The

Table 8.9 Classification of fruit-eating and flower-feeding bats (order Chiroptera) with distribution and size (weight) where known.

Classification	Distribution	Weight (in grams)
A. MEGACHIROPTERA		
Absent from the New World; with good eyesight and sense of smell, poor navigation in the dark		
Pteropidae (only family)		
Subfamily Pteropinae (fruit-bats)		
Cynopterus	S. E. Asia, Australasia	
Eidolon helvum	Africa	110-330
Epomophorus gambianus	Africa	
Epomops franqueti	Africa	
Nanonycteris veldkampii	Africa	
Pteropus (flying fox)	S. E. Asia, Australasia	up to 1500
Rousettus aegypticus	Africa	95-160
Subfamily Macroglossinae (nectar-feeding bats)		
Eonycteris spelaea	S. E. Asia	40-65
Macroglossus minimus	S. E. Asia	15-25
Macroglossus sobrinus	S. E. Asia	15-25
Syconycteris australis (Plate 8a)	Australasia	very small
B. MICROCHIROPTERA		
World-wide distribution; poor eyesight, good navigation in dark using sonar; mostly insectivorous		
Phyllostomidae (confined to New World; the only family of flower-visiting bats in this suborder)		
Subfamily Phyllostominae (less specialised flower-visitors)		
Phyllostomus discolor		22-40
Subfamily Stenoderminae		
Artibeus jamaicensis		45
Subfamily Carolliinae		
Carollia perspicillata (mainly fruit-eating)		15
Subfamily Glossophaginae (28 spp., all flower-visitors, relatively small)		
Glossophaga soricina		10
Leptonycteris sanborni		15-20
Musonycteris		

day-flying *Pteropus samoensis* (perhaps the largest of all bats, with a wingspan of at least 1.8 m) visits and pollinates the Samoan species *F. reineckei* in the day time, like the bird mentioned on p.241 (Cox, 1984). (*Cynopterus* also consumes the edible bracts of *Frey-cinetia*, but is thought not to effect pollination.) Smaller species of *Pteropus* feed on and probably pollinate *Eucalyptus* and *Melaleuca* (Myrtaceae) in tropical Australia (Beardsell *et al.*, 1993). The exotic fruits eaten by the fruit-bats are specially adapted to them by their peculiar smells and by growing in such a position (often on the trunk or main branches) that the bats can easily get at them. These same scents are reproduced by the flowers that attract Old World nectar-feeding bats of subfamilies Pteropinae and Macroglossinae. These bats have reduced teeth, a well developed snout, and a long slender tongue with a brush of backwardly directed hairs at the tip. The lower size limit of the Pteropidae is reached in African flower-visitors like *Nanonycteris* (Lack, 1978) and the Queensland blossom bat of Australia and New Guinea (*Syconycteris aus-tralis*) with a body-length of 5 cm (Plate 8a); the latter is totally dependent on flowers (Armstrong, 1979). The tongue of another species, which has a body length of 7.5 cm, can be protruded to a length of 5-6 cm, making it comparable with the most highly specialised New World bats (below). The flower bats take substantial quantities of pollen, mainly by grooming the fur. Claw marks indicate that *Eonycteris* clings to the outside of flowers, whereas *Macroglossus* crawls inside when possible. However, visits are often fleeting, as in the New World (van der Pijl, 1936, 1956; Start & Marshall, 1976).

Flower-feeding species of the Microchiroptera belong to the family Phyllostomidae (Table 8.9). However, the tongue of these American nectar-feeders is similar to that of the Old World nectar-bats, having a brush of bristles and scales at the tip (Fig. 8.15) which hold nectar by capillarity. The lower incisors are greatly reduced, allowing the tongue to move in and out without the jaws being opened, and the tongue extension can be several centimetres. The genus *Phyllostomus* itself (subfamily Phyllostominae) consists of comparatively large species that show a relatively low level of adaptation to flower-visiting and confine themselves to brush-blossoms; the snout-length, tongue-length and dentition correspond to frugivory and carnivory. These bats always alight on the flowers but keep the wings spread. On pendent flowers they cling with their hind feet and feed head down. They show some awkwardness in alighting and taking off from flowers. Their flower visits last up to a few seconds. The tongue and snout of *Carollia* (subfamily Carolliinae) are not developed for flower-visiting (Steiner, 1983). All 28 species of the subfamily Glossophaginae are flower visitors. Within the group, increasing adaptation to flower feeding is expressed in the development of a slender snout, involving modification of the skull, and a lengthening of the tongue, which reaches its limit in *Musonycteris harrisonii*, with a body-length of 8 cm and a tongue-length of 7.6 cm. The coat is composed of hairs bearing scales (Dobat & Peikert-Holle, 1985, p.111-2) which are sometimes drawn out into points to give a plumose effect like the hairs of bees. This makes them good at holding pollen. Bats of this group visit brush-flowers, but they are mainly visitors to bell-shaped and gullet-like flowers.

It has sometimes been held that bats do not hover and that the fluttering of New World bats in the face of the smaller flowers, at which they cannot alight, is mainly the effect of braking to a standstill (followed by a quick probe of the tongue) and then getting into motion again. However, although the flower-visit may be from a fraction of a second up to two seconds in duration, the wings can be beating 16 times per second (as in *Glossophaga soricina*), so it is now considered that these flower-bats genu-inely hover. In fact, the small Old World *Macroglossus* can also hover, but it uses this for

Fig. 8.15 The New World
flower-visiting bat
Glossophaga soricina: head
with extended tongue from
above (A) and from the side
(B). St Vogel.

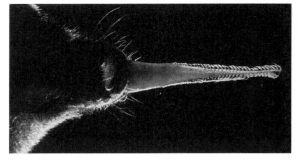

a

b

a quick inspection of the flower before alighting. At some flowers, Glossophaginae
hover and grasp the flower with the hind claws at the same time (Dobat & Peikert-
Holle, 1985).

The behaviour of flower-visiting bats

The foraging behaviour of bats in both hemispheres appears to be flexible and determined by the interplay of the bats' own attributes (particularly size), their roosting
sites and the distribution of food sources. Marshall (1983), dealing with Megachiroptera, cited observations from Africa that the large *Eidolon helvum* feeds on peripheral
branches while the smaller *Epomops franqueti* feeds deeper in the foliage, and that *Epo-
mophorus gambianus* arrives earlier to feed than *Rousettus aegypticus*[1] or *Nanonycteris veld-
kampii*. The West Malaysian *Eonycteris spelaea* roosts in caves, and the only large roost in
Selangor State contains tens of thousands of bats. Such large numbers cannot find
food near the roost and the bats travel nightly to feeding sites up to 38 km away,
probably flying for 1½ hours to get there. They break up into groups of five to 20

---

[1] Late feeding in this species is facilitated by its power of echo-sounding (Dobat & Peikert-Holle, 1985) using
clicking noises made by the tongue (Pye, 1983).

(rarely to 50) for foraging. They are known to feed on flowers of at least 31 different plant species, most of which are spatially widely scattered or seasonal in flowering. In the same area, two species of the still smaller *Macroglossus* roost in trees, either alone or in small scattered groups, near their major food sources which are restricted as to species (*Sonneratia* spp. for *M. minimus*, *Musa* spp. for *M. sobrinus*). These two bats forage singly, remain near the roost, and often return repeatedly to one small area or even a single plant (Start & Marshall, 1976). Their food-plants remain in flower throughout the year.

New World flower-visiting bats studied by Vogel (1958) began to visit flowers at sunset but did not necessarily arrive at a particular feeding place at the same time on successive nights. They always fed in flocks and moved from tree to tree, any one tree usually being invaded repeatedly during the course of an evening ('pulsed visitation'). Visits to flowers lasted only a fraction of a second but, by counting the number of wing-claw marks on the flowers of the sausage tree (introduced into South America), he estimated that a flower might receive up to 50 visits in one night. Although the bats normally alighted, they fed from *Marcgravia* and *Purpurella* (Melastomaceae) almost entirely while hovering. They became well dusted with pollen and their stomachs contained nectar but no pollen.

New World bats that probably forage singly are *Sturnira lilium*, *Carollia perspicillata* and *Glossophaga soricina*. The larger *Artibeus jamaicensis* and *Phyllostomus discolor* probably forage in groups (Heithaus, Opler & Baker, 1974; Heithaus, Fleming & Opler, 1975). Heithaus, Fleming & Opler (1975) estimated flight distances in Costa Rican seasonally dry forest by 'capture-mark-recapture', and found that they were proportional to size of bat (except that one small species, *Carollia perspicillata*, travelled further than expected). It was suggested that the larger bats (with mean flight distances of more than 400 m) were the ones that could utilise those plant species that were widely spaced in the forest and that group-foraging may be a good arrangement for exploiting these plants. In the dry forest, the trees flower mainly in the dry season and the species flower in some sort of succession which covers its whole length. In the wet season fruits are available and most nectar-eating bat species then become frugivorous. It was found that bats commonly carried mixed pollen loads (though the frequency of these varied according to the species of bat). This might create selection pressure for successional flowering of different plant species, which would reduce the effects of the low visitor-constancy.

On the tree *Bauhinia pauletia*, Heithaus, Opler & Baker (1974) found that the larger bats almost confined themselves to flowers high up, whereas the smaller ones visited the lower flowers nearly as much as the upper. This behaviour, and that already quoted for Old World bats entering the canopy, is probably related to the greater manoeuvrability of smaller bats (Dobat & Peikert-Holle, 1985). This is also associated with a lower aspect-ratio of the wings and consequent relatively higher flight costs, leading to the need for frequent rests.

Flexibility of foraging behaviour seems to be exemplified by *Artibeus jamaicensis*, as this species was probably a group-forager in Costa Rica (as mentioned above) but was found foraging singly in South America on *Lafoensia* and apparently trap-lining (pp.134-5) on these well-spaced plants (Sazima & Sazima, 1975). Flexibility was also shown by the observed variation in the size of the foraging group of *Phyllostomus discolor*, which was graded according to the number of flowers open on individual plants of *Lafoensia glyptocarpa* and species of *Bauhinia* in the foraging area. Pulsed visitation lasted through the evening, and while bats were present individual flowers normally received only one visit, indicating that the bats had the means of recognising recently visited flowers (Sazima & Sazima, 1977). The trap-lining technique may be used equally by solitary or group foragers (Frankie & Baker, 1974).

Heithaus, Opler & Baker (1974) considered that the resource-utilisation patterns of bats visiting *Bauhinia pauletia* were related to aspects of social behaviour. The flowers were clumped and visitation by the group-foraging *Phyllostomus discolor* was pulsed here too. The bats always drained all the nectar from the flowers. A return visit after an interval (average 21 minutes) allowed time for fresh nectar to accumulate in the flowers. The organisation of the bats into groups provides an effective response to the pattern of nectar availability. In contrast, the small bat *Glossophaga soricina* forages singly and could not drain a flower at one visit; a single individual appeared to come repeatedly to the same flower. A detailed analysis of a flock-foraging situation in Arizona given by Howell (1979) led to similar conclusions about this habit. The only food resource here was *Agave palmeri* and the bats were migrant summer residents, *Leptonycteris sanborni*. The plant had a clumped distribution and unpredictable flowering. Yet again, the visiting was pulsed; between foraging bouts the bats rested while digesting the food. The analysis indicated that flocks search more efficiently than individuals and that by their special foraging behaviour they exploit plant patches more efficiently. During resting the bats conserve heat in the cool night air and speed up digestion by hanging in clusters. Several features of the bats' behaviour and the flowering behaviour of the plant indicated co-evolution between the two.

Insectivorous bats may hold territories (Bradbury & Emmons, 1974; Rydell, 1986) and in 1970 Baker raised the question of territory holding by flower-visiting bats, but reported examples are few (Gould [1978] observed that *Pteropus* took up territory on durian trees). The success of systematic group-foraging would seem to depend on there not being more than one party of bats in the same area, otherwise a second party could pre-empt a patch of flowers while the first was deliberately abstaining from a visit. This suggests that a group may be able to exercise territorial control by some means.

The evolution of bat-pollination

Bats were in existence in the Eocene (50 million years ago) and were then already highly developed Microchiroptera. They had a dentition typical of insectivorous mammals, suggesting that their non-flying ancestors were insectivorous. By this time, Africa and America were well separated as a result of continental drift. However, the New World plant-feeding Microchiroptera are thought to have originated no more than 15 million years ago (reviews by Marshall, 1983, and Dobat & Peikert-Holle, 1985). As pointed out by Wilson (1974), they radiated into an environment containing pollinating insects and fruit-eating birds, so that both floral nectar and the flesh of soft fruits were already available. These were to be found in some structurally primitive plant groups as well as in more advanced ones. The same would have applied to the Old World bats, although the first fossil Megachiropteran is dated between 15 and 25 million years ago.

Some genera of bat-pollinated plants are native to both the Old and New Worlds. These are *Parkia*, *Mucuna*, *Eperua*, *Ceiba* and perhaps *Crataeva* (Vogel, 1968-69); they might have evolved before the separation of Africa and South America, but if so they must have evolved very similar bat-pollinated species subsequently. This seems possible in view of the fact that the whole syndrome of bat-pollination is very similar in the two hemispheres. So much so that when plants are transported from the East to the West and vice versa, they receive bat visits in their new territory. (Examples are *Musa* [banana] [Marshall, 1983], *Kigelia africana* [Vogel, 1958] and *Durio zibethinus* [Baker,

1970], when introduced into the New World, and *Agave angustifolia* and *Crescentia cujete* when introduced into the Old World [Marshall, 1983]). Perhaps the bats of the two hemispheres are sufficiently similar to call forth similar adaptations on the part of the plants. Van der Pijl (1936) suggested that the unpleasant smell of the glandular secretion that most bats produce, and by means of which they find each other when they gather in flocks, is similar to the scent of bat-pollinated flowers, which may have achieved adaptation to bat-pollination by imitating the smell of the bats, as proposed for butterflies by Müller (p.87). Thus bat-pollinated flowers would acquire similar scents wherever they evolved.

Against the hypothesis of independent evolution of bat-pollination within genera after the separation of Africa and South America are the facts that in *Mucuna* and *Parkia* the bat-pollinated species in the two hemispheres are considered to be closely related, and that two bat-pollinated species, *Ceiba pentandra* and *Mucuna pruriens*, are found on both sides of the Atlantic. Here one must suppose that migration across the Atlantic occurred at some stage. The problem is extensively discussed by Dobat & Peikert-Holle (1985), who mention that the seeds of *Parkia* are heavy and unlikely to cross the ocean. However, many very heavy seeds of Fabaceae drift from West to East across the Atlantic today (Nelson, 1978) and *Ceiba* has light, wind-dispersed seeds.

It is possible to speculate on the origin of bat-pollinated flowers by looking at closely related species pollinated by other animals. Vogel (1968-69) lists bat-pollinated groups with sister-groups that are respectively hawkmoth-pollinated, bird-pollinated and bee-pollinated, the last group being the least common. Hawkmoth-pollinated flowers resemble bat-pollinated flowers in their nocturnal flower-opening and in their colouring, while bird-flowers resemble them in their high nectar-production and firm consistency. Stiles (1981) points out that plants adapted to passerine birds (flower visitors other than hummingbirds) are like bat-pollinated species in the prevalence of the tree habit and frequency of brush-flowers, while their visitors are often group-foragers; to this may be added glucose-rich or glucose-dominated nectar (Baker & Baker, 1983a, b), while Stiles (1981) also points out that evolution could in some cases have proceeded from bat-pollination to bird-pollination.

Pollination by non-flying mammals

It has been known for some time that there are a number of small arboreal marsupials in Australia that feed on a mixed diet of vegetable matter, insects and nectar, or even chiefly on nectar, which they find in flowers generally regarded as bird-pollinated. However, 25 years ago it was possible to say that no plants primarily adapted to the visits of non-flying mammals ('NF-mammals') had been recognised, even though the latter might act as effective pollinators. Now the picture is substantially changed; plants adapted to pollination by these animals are known from Australia, South Africa and tropical America. Flowers adapted to pollination by birds or, particularly, bats are the most likely to be already suited to pollination by NF-mammals. When this happens, the flowers may be regarded as accidentally pre-adapted to pollination by this kind of visitor. Apart from this, some flowers appear to be ambivalent in their adaptation. We shall consider these last two categories before going on to describe the flowers that seem to be fully and exclusively adapted to pollination by NF-mammals; in each case we shall take the New World plants first, and then those of the Old World.

Flowers incidentally pre-adapted to NF-mammal pollination

In New World lowland tropical forests, the small bat-pollinated tree *Mabea occidentalis* (family Euphorbiaceae) is almost certainly effectively pollinated also by a marsupial,

Table 8.10 New-World plants the flowers of which are visited by NF-mammals (NFM).

Plant	Supposed legitimate pollinator	NFM visitors	Dimensions/ weight
Acanthaceae			
Trichanthera gigantea	Bats Glossophaga	Caluromys[1]	180-290mm (+long tail)/300±70g
Bombacaceae			
Ceiba pentandra	Bats	Cebus[3], Saimiri[3], Ateles[3], Potos[2]	/Ateles 7kg
Ochroma pyramidata (syn. O. lagopus)	Bats	Didelphis[1]	
Quararibea cordata	NFM?	Cebus[3], Aotus[3], Potos[2], Bassaricyon[2], Caluromys[1], Didelphis[1], Caluromysiops[1]	
Pseudobombax tomentosum[4]	NFM and large bats	Caluromys[1]	see above
Combretaceae			
Combretum fruticosum	Birds, especially passerines	8+ spp. of primate	/100g-7kg
Euphorbiaceae			
Mabea occidentalis	Bats	Caluromys[1]	see above
Fabaceae-Mimosoideae			
Inga thibaudioides, Inga ingoides	?	Caluromys[1]	see above
Fabaceae-Caesalpinioideae			
Hymenaea courbaril	Bats	Caluromys[1]	see above
Melastomaceae			
Blakea austin-smithii, chlorantha penduliflora	NFM	5 spp. of rodents, in 4 genera	

[1]Marsupials [2]Procyonids (olingos & kinkajous, Carnivora) [3]Primates [4]see p.245

(Lumer, 1980; Steiner, 1981; Janson, Terborgh & Emmons, 1981; Lumer & Schoer, 1986; Gribel, 1988)

the red woolly opossum (*Caluromys derbianus*) (Table 8.10). The importance of the opossum as a pollinator may vary over the flowering period, depending on the prevailing bat-visitation rate, while bat-visitation will affect the profitability of the flowers to *Caluromys* (Steiner, 1981). Calculations showed that the animal could get a substantial proportion of its daily energy requirement from the nectar of *M. occidentalis*. However, it is far from dependent on the tree as it is more or less omnivorous. *Caluromys* takes in pollen as it feeds on nectar and it eats the pollen that it grooms from its coat.

Comparable observations have been made in Africa. In the tropics, Lack (1977) reported genets (*Genetta tigrina*, family Viverridae), which seem to be omnivorous, visiting the flowers of *Maranthes polyandra* (family Chrysobalanaceae). This plant has clustered flowers forming brush-blossoms that display the syndrome of bat-pollination and are effectively pollinated by *Nanonycteris* bats (Lack, 1978). Similarly, Coe & Isaac (1965) saw bush-babies (*Galago* sp., family Lorisidae) feeding on the flowers of the baobab, *Adansonia digitata* (p.246 & Fig. 8.13A).

In Australia the waratah, the national flower of New South Wales, *Telopea speciosissima* (Proteaceae), falls into the pre-adapted category since it shows the syndrome of

Plate 1 Flowers and guide-mark patterns. **a**, alpine clover,
Trifolium alpinum; **b**, a sun rose, *Halimium lasianthum*;
c, half section of flower of common dog violet,
Viola riviniana; **d**, *Androsace sarmentosa*: the flowers have
a yellow ring round the mouth of the corolla tube when
newly opened, which changes to red as the flower ages;
e, bittersweet, *Solanum dulcamara*, a flower adapted to
vibratory pollen collection: the anther-cone contrasts in
colour with the rest of the flower; **f**, bastard balm, *Melittis
melissophyllum*; **g**, a lousewort, *Pedicularis oederi*;
h, half section of flower of foxglove, *Digitalis purpurea*.

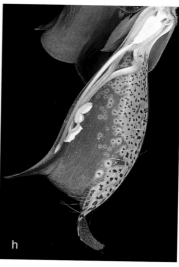

Plate 2 **a**, Sawfly, *Tenthredo* sp., on marsh marigold, *Caltha palustris*; **b**, solitary bee, *Colletes daviesanus*, on *Anthemis tinctoria* (PFY); **c**, solitary bee, *Andrena haemorrhoa* on pear blossom; **d**, solitary bee, *Anthophora plumipes*, approaching a garden comfrey, *Symphytum* sp. (PFY); **e**, bumblebee, *Bombus pratorum*, on rayed form of common knapweed, *Centaurea nigra*; **f**, bee-fly, *Bombylius major*, visiting primrose, *Primula vulgaris* (J E Bebbington).

Plate 3 **a**, Hoverfly, *Rhingia campestris*, visiting red campion, *Silene dioica*; **b**, common blue butterfly, *Polyommatus icarus*, on fleabane, *Pulicaria dysenterica*; **c**, red admiral butterfly, *Vanessa atalanta*, on hemp agrimony, *Eupatorium cannabinum*; **d**, Lulworth skipper butterfly, *Thymelicus acteon*, sucking nectar of red valerian, *Centranthus ruber*; **e**, broad-bordered bee hawkmoth, *Hemaris fuciformis*, hovering at field scabious, *Knautia* sp., Dordogne, (J E Bebbington).

Plate 4 Wind pollination. **a**, Scots pine, *Pinus sylvestris*: male cones shedding pollen; **b**, Scots pine, young female cones at receptive stage; **c**, Norway spruce, *Picea abies*: young female cones at the receptive stage (the cones turn brown and hang down from the branches at maturity); **d**, the yellow male catkins and the red female flowers of hazel, *Corylus avellana*; **e**, a spikelet of couch grass, *Elymus repens*, showing the large freely-exposed stamens and the feathery stigmas.

a

b

d

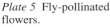

Plate 5 Fly-pollinated flowers.
a, *Aristolochia elegans* (Aristolochiaceae), a liana, with a U-shaped tube and large entrance funnel;
b, *A. cretica*, cut open to show entrance to prison guarded like a lobster pot, receptive stigmas and undehisced anthers (PFY);
c, *Stapelia hirsuta* (Asclepiadaceae):, a carrion-like flower borne near the ground, with blowfly eggs;
d, *Arisaema propinquum* (Araceae), W. Himalayas: the translucent veins of the spathe show up brightly behind the entrance;
e, 'mouse plant' *Arisarum proboscideum* (Araceae): inflorescence with half of the spathe removed, showing the fungus-like spadix appendage, a female flower and several male flowers;
f, umbels of hogweed, *Heracleum sphondylium* (Apiaceae), with flies and an ichneumon wasp (PFY).

c

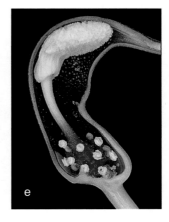

e

f

Plate 6 Sexually-deceptive orchids.
a, male solitary wasp, *Argogorytes mystaceus,* on fly orchid, *Ophrys insectifera*, Wiltshire (H Jones); **b**, mirror orchid, *Ophrys speculum*; **c**, 'yellow bee orchid', *Ophrys lutea*; **d**, self-pollination of bee orchid, *Ophrys apifera* one pollinium is hanging from the anther, a second is adhering to the stigma; **e**, ichneumon wasp, *Lissopimpla excelsa*, visiting Australian tongue-orchid, *Cryptostylis leptochila* (J A L Cooke/OSF); **f**, thynnine wasp visiting W. Australian hammer orchid, *Drakaea glyptodon* (Babs and Bert Wells/OSF).

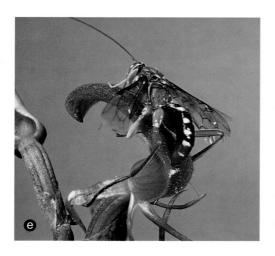

Plate 7 Flower-visiting birds.
a, bronzy hermit hummingbird, *Glaucis aenea*, at passion flower, *Passiflora vitifolia*, Costa Rica (Michael Fogden/OSF); **b**, broad-tailed humming-bird, *Selasphorus platycercus* (f), at scarlet gilia, *Ipomopsis aggregata*, USA (Claude Steelman/OSF); **c**, male purple sunbird, *Nectarinia asiatica*, at aloe, Oman (Mike Brown/OSF); **d**, Cape sugar bird, *Promerops cafer*, on *Protea*, S. Africa (E & D Hosking/FLPA); **e**, tawny-crowned honeyeater, *Phylidonyris melanops*, at *Melaleuca* sp.,W. Australia (Babs and Bert Wells/OSF).

a

b

c

Plate 8 Flower-visiting mammals.
a, Queensland blossom bat,
Syconycteris australis, on *Mucuna macropoda*, Papua New Guinea
(M J G & H C F Hopkins);
b, short-nosed fruit bat, *Cynopterus sphinx*, visiting wild banana
inflorescence (Merlin D Tuttle/OSF); **c**, honey possum,
Tarsipes rostratus, on *Banksia coccinea*, W. Australia (Babs and
Bert Wells/OSF); **d**, Namaqua rock mouse, *Aethomys namaquensis*,
feeding at *Protea* sp, South Africa
(J A L Cooke/OSF).

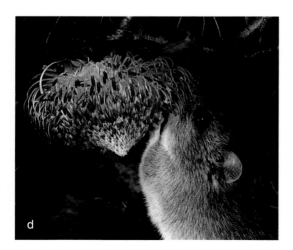

d

bird-pollination but is regularly visited by the eastern pygmy possum (*Cercartetus nanus*), which carries significant loads of its pollen (Goldingay *et al.*, 1991). There are probably other Proteaceae in the same situation, such as *Banksia coccinea* (Plate 8c).

Flowers ambivalently adapted to birds and NF-mammals

In southern Africa, certain NF-mammal-pollinated species of *Protea* receive visits also from birds and are perhaps ambivalently adapted (Rebelo & Breytenbach, 1987). These species have exposed flower heads, unlike other South African species that are highly adapted to NF-mammals and have concealed flowers.

A wide-ranging survey of Australian vertebrate-pollinated plants has been carried out by Holm (1988). Here it was found that in *Banksia* and *Dryandra* there are typical bird-pollinated species and typical NF-mammal-pollinated species, but in addition other species that could seemingly be pollinated by either group (Table 8.12).

These apparently ambivalent species are of three kinds. There are ground-flowering ones that are stemless, with large clustered leaves and large flower-heads borne at the edge of the clump on underground rhizomes; the colour is more or less dull and the scent is of a sort preferred by mammals. There is a second group in which the plant forms a bush within which the inflorescences are concealed. Two of the four species known to Holm had dull-coloured inflorescences and one had a mammal-preferred scent. The third intermediate group, described by Hopper (1980b), had completely exposed heads but one of the two species had a strong odour at all times (and produced new flowers steadily through the 24 hours), while the other opened all its new flowers at night (but had only a faint odour). These groups thus show signs of adaptation to non-flying mammals but they do not prevent access of birds to the flowers, and might be pollinated by either. (The primarily bird-adapted species have inflorescences that are red, orange or yellow in colour and easily visible from a distance; the stiff styles of the immature flowers are arched, thereby providing a firm perch for access to the mature flowers. There is copious nectar and the flowers are scentless. Observations of bird-visits exist in Holm's work or in earlier publications for four of the five species named in Table 8.12, column 1.)

In a study of two Australian Proteaceae, *Banksia ericifolia* (reddish-orange inflorescences) and *B. spinulosa* (more golden, Fig 8.16), 31 individual birds and three native rats (*Rattus fuscipes*, southern bush-rat) were caught near the plants in New South Wales. Two birds carried a light load of pollen of these plants, while two rats carried a heavy load and one a light load. It was found that the style remained sharply hooked so that the pollen presenter was rather near the base of the flower. This resulted in deposition of pollen on the snouts of the rats but not on the much longer bills of the birds (Carpenter, 1978a). However, the association of hooked styles with NF-mammal-pollination has not been found in any subsequent study of *Banksia* (for example, Hopper, 1980b). (In most Proteaceae, the anthers deposit their pollen on the tip of the style before the stigma becomes receptive, and this part is termed the pollen-presenter.)

In many *Banksia* species, especially where the flowers in the cylindrical heads are arranged in double rows with grooves between them, the nectar overflows the perianth and accumulates in the grooves, whence it is gathered by the pollinators. However, in *B. ericifolia* and *B. spinulosa* the nectar, which has a pungent odour, is produced in such quantity that it flows out of the grooves and trickles down the stems to the ground. Carpenter suggested that this could lead to rats on the ground discovering the nectar and being induced to climb the shrubs to find the flowers. Other features of these two banksias consistent with NF-mammal-pollination are a preponderance of flower-opening and nectar-secretion at night, and a far higher sugar-energy production than in normal bird-pollinated flowers. Carpenter suggests, however, that the plants have kept open the option of being pollinated by birds, and so are ambivalent. In some locations birds, but no mammals, visit the flowers (Paton & Turner, 1985), possibly because the mammals are absent, so ambivalence could be important. Another study of these and other species at various localities in eastern New South Wales showed that similar amounts of seed-set resulted from night-time pollination as from day-time pollination, and that two nocturnal flower-visiting marsupials, the sugar-glider (Fig. 8.16) and the brown antechinus (see Table 8.13), regularly carried large pollen loads from the banksias. This is claimed to be the first proof that mammals visiting flowers in Australia actually cause seed-set (Goldingay *et al.*, 1991). Carthew (1993) found that the sugar-glider behaved on the inflorescences in a manner more likely to cause pollination than the eastern spinebill (*Acanthorhynchus tenuirostris*). She also found that these marsupials, plus the eastern pygmy possum, made at least as many inter-plant movements as birds, and that distances exceeding 10 m between plants visited were more frequent (Carthew, 1994).

Flowers exclusively adapted to pollination by NF-mammals

These may receive visits from both opportunistic species and from those for which flower-food is the primary source of nourishment. Adaptation to NF-mammals is probable in some tropical South American plants previously regarded as adapted to bat-pollination; these can be effectively pollinated by a wide variety of mammals (monkeys, kinkajous and opossums – listed against Bombacaceae in Table 8.10). Monkeys are mostly active by day and opossums by night. Monkeys sometimes damage or destroy flowers, but the plants produce many more than could be expected to ripen into fruits, so there could easily be a net benefit from the monkeys' visits. The features that adapt the plants to visits by these animals are conspicuous, upright blossoms borne in stalkless clusters a little way back from the tips of leafless branches, a tough, partially fused perianth forming a shallow cup and protruding stamens. The flowers usually open all at once, giving an immediate substantial food supply. True New World bat-flowers are contrasted as being white or pale green, scented, long-stalked and borne at the very tips of branches, with usually only a few open at a time on each plant (Janson, Terborgh & Emmons, 1981). Just such a divergence in flower characters and pollinators was later found between *Pseudobombax tomentosum* and *P. longiflorum* growing in the same area (Gribel, 1988) (Table 8.10). However, Gribel suggested that some of the plants adapted to NF-mammals are equally adapted to Phyllostomine (but not Glossophagine) bats, and so are ambivalent or possibly pre-adapted.

The rodent-pollinated species of *Blakea* listed in Table 8.10 are 'hemi-epiphytic' shrubs growing in cloud forest at middle and upper altitudes in the American tropics (Lumer, 1980; Lumer & Schoer, 1986). Their flowers are bell-shaped, nodding and hidden within the foliage. Their petals and sepals are green and their anthers usually purple. The flowers are nocturnal, the anthers ripening during the first day of opening and becoming ready to release pollen in the evening. A sucrose-rich nectar is secreted in the first and second nights but not during the day. Pollen is released explosively by slight pressure on the outside of the petals when the animals grasp the flowers with their fore-paws, and by pressure at the bases of the stamen-filaments, as when the animal's snout reaches the base of the floral bell. If there is any odour, it is not detectable by the human nose. The rodents move rapidly among the flowers and do not damage them. No bats and few birds and insects visited the flowers during the periods when they were observed. The rodents' visits take place either through the

Table 8.11 Small mammals visiting *Protea amplexicaulis* and *P. humiflora* in South Africa (Wiens & Rourke, 1978).

Family and Scientific name	English name	Remarks
RODENTS		
Muridae		
Aethomys namaquensis	Namaqua rock-mouse	Rocky habitat
Praomys verreauxii	Verreaux's mouse	Rocky habitat
Rhabdomys punilio	Cape striped field-mouse	Deep-soil habitat (burrower), feeding by day
Mus minutoides	(house-mouse relative)	
Gliridae		
Graphiurus occidentalis		
SHREWS		
Soricidae		
Crocidura spp.[1]		
Macroscelididae		
Elephantulus edwardii[1]	Elephant shrew	

Animals weigh 20-70 g and are crepuscular to nocturnal, unless stated.
[1] do not carry pollen

night or at dawn and dusk.

These *Blakea* species live in a harsh environment of strong winds and wind-driven rain, and the vegetation of the cloud forest is dense. These conditions are unfavourable to other types of pollinator and may have led to the adoption of NF-mammals as pollinators. Some species of *Blakea*, like the majority of the family Melastomaceae, are insect-pollinated, and they have white or pink flowers with widely spread petals; they offer pollen, but not nectar, as a reward to their visitors and they are sweetly scented. *Blakea* therefore seems to have become adapted to rodent-pollination; whether the rodents are adapted in structure or inherited behaviour patterns to feeding on *Blakea* remains to be seen.

The flower-feeding activities of six species of nocturnal lemurs (the primates of Madagascar) were summarised by Sussman & Raven (1978). Three had been reported as eating whole flowers, but one of these and three others were seen taking floral nectar from native and introduced plants (including the usually bat-pollinated *Ceiba pentandra*) without destroying the flowers. The weights of these four species of lemur range from 2.5 kg down to between 50 and 150 g. Probably some of the native Malagasy flowers visited by lemurs (such as *Strongylodon craveniae*) are adapted to them (Nilsson *et al.*, 1993), but there is as yet no suggestion of reciprocal adaptation in morphology on the part of the lemurs.

Adaptation to exclusive NF-mammal pollination takes a different form in the Cape fynbos vegetation of South Africa (Rourke & Wiens, 1977; Wiens & Rourke, 1978; Wiens *et al.*, 1983; Wiens, 1985). The fynbos is a low scrub formation in a region of Mediterranean climate (cool moist winters and hot dry summers). Many species of *Protea* and a few of *Leucospermum* (also in Proteaceae) share a syndrome which has been proved in some cases to secure pollination by ground-living rodents. The flower-heads are wider and flatter than in the bird-pollinated members of the same genera and are

mostly dingy in colour, or dull outside and white within. Large quantities of concentrated nectar (about 36% sugar) with a high sucrose content (40-70% of the sugars)
are provided, and the nectar has a yeasty smell, sometimes with a sweet or rancid
scent superimposed. The amount of nectar available in a single head is 2-3 ml. The
peak hour for opening new flowers is 21.00 hrs (for *P. humiflora*). The heads themselves
are borne inside the bushes and near the ground, on short stout peduncles that enable
them to bear the weight of a mouse (Table 8.11), though sometimes they face downwards and may be reached by the animal from below (Plate 8d). Although it is not
usually possible to observe the visitors' activities in the field, animals in captivity
eagerly visited *Protea* heads for the nectar and responded to the scent of concealed
heads; those trapped near the plants regularly carried *Protea* pollen on their snouts.
Tracks in the vegetation made by the rodents ran under the bushes.

When the *Protea* flower-head is upright, the visitor is forced by the inward curvature
of the stiff styles to dip into the middle of the head and probe outwards. From this
position it reaches the flowers in such a way that they can be pollinated. Each flower
presents its nectar in a bowl-like structure formed by the perianth just above the point
where it splits, and this is over-arched by the style, which presents the flower's pollen
and receives pollen from the visitor. There is always a separation of about 10 mm
between the nectar and the style-head, so that the latter makes contact with the animal's snout in front of the eyes and about 10 mm from its tip. The stigmatic slit near
the tip of the style never gapes, and pollination is apparently effected only when pollen is forcibly pressed into the slit. The rodents (Table 8.11) have no special adaptation
to nectar feeding and the tongue cannot be projected beyond the tip of the snout.

Flowering takes place mainly in late winter, 'typically a low point in rodent food
cycles'. The nectar is 'an important community food resource for rodents'. *Protea humiflora* has an average flowering period of 45 days. For a sample area it was calculated,
assuming that half the nectar is lost to thieves, that the *P. humiflora* plants could supply
the energy requirements for all the small mammals present for eight days.

The effectiveness of the rodents as pollinators compared with other visitors was
assessed by bagging the heads to keep the rodents out. Seed-set in preliminary experiments was thereby reduced by 50% in *P. humiflora* and by 95% in *P. amplexicaulis*. Insects
foraging for nectar do not normally touch the pollen presenter. In other experiments,
in which powder that fluoresces in ultra-violet light was added to flower-heads, it was
found that it was spread along rodent runways and on to conspecific flower-heads for
15 m. (Early reports of rodents chewing the bracts refer to behaviour that is unusual.)
Conservative calculations, based on trapping records, indicate a home-range for the
mice of about 25-60 m.

Probably adaptation of *Protea* to pollination by NF-mammals arose several times;
there are many such species and it seems likely that their ancestors were mostly bird-
pollinated, since these have the most in common with them. In *Leucospermum*, however,
NF-mammal-pollination may have evolved from insect-pollination (Rebelo &
Breytenbach, 1987). There are a number of different growth habits by which the
flower-heads are concealed. Often the habitats of the rodent-pollinated proteas are
specialised, the species are of extremely limited distributional range and their occurrence is in isolated colonies.

Unlike the plants, the rodent visitors have very wide geographical ranges. Thus the
facts that the proteas have a smaller range than the mice and only feed them for part
of the year, mean that the animals cannot have developed any special adaptations to
flower-visiting. Rather, the plants have found the key to attracting them. Wiens (1985)
suggests a parallel with the marketing of 'junk-food' for humans; the rodents can eas-

ily live without floral sugar, as many of them have to when not living in *Protea* habitats, but if it is available they seek it out. It has been suggested that concealment of flower-heads is needed to reduce nectar-robbery by birds and insects; the large quantity of nectar and its high concentration make it doubly attractive. Nevertheless, such robbery does occur, though robbery by insects is much less than it is from bird-pollinated *Protea* species. As rodents are particularly vulnerable to predatory birds, it would seem that protected feeding sites are necessary for this reason also.

One other mammal has, extraordinarily, turned out to be a potential pollinator. In South Africa there is evidence that *Acacia nigrescens* is adapted to pollination by giraffes (*Giraffa camelopardalis*) in the savanna vegetation of the Kruger National Park (du Toit, 1990). Flowers of riverine species of plants that bloom in the dry season are a signifi-cant part of the giraffes' diet; these plants include *Acacia tortilis*, which has rounded, short-stalked inflorescences surrounded by straight spines which, though not a com-plete defence, do protect the greater part of each flower-cluster. *A. nigrescens*, on the other hand, has bottle-brush inflorescences held clear of the plant's prickles, which are recurved. The giraffes spend much time eating them and get pollen loads on their heads as a result. It is therefore suspected that enough flowers that receive pollen transferred by giraffes survive to form pods. In fact, all the *Acacia* species in the area produce far more flowers than ever form fruits. Du Toit notes that there are many other *Acacia* species differing in the same ways as the two above-named, and that they also differ consistently in flower colour, those with capitate inflorescences being bright yellow and those with cylindrical ones being pale yellow. The other pale ones are thus also likely to be adapted to giraffe-pollination.

Holm's work (1988) showed that as well as pre-adapted or ambivalent NF-mammal-pollinated plants, there are species of Proteaceae in Australia as fully adapted to small-mammal-pollination as those of South Africa, though he did not demonstrate pollinator visits. The close parallels between the two regions in NF-mammal-pollina-tion are emphasised by Rourke & Wiens (1977). However, in Australia there are mam-mals truly adapted to flower-feeding.

The strictly NF-mammal-adapted species of south-west Western Australia (Table

Table 8.12 Pollinatory adaptation of *Banksia* and *Dryandra* in south-west Australia.

Bird-pollinated	*NFM-Pollinated*	*NFM-Pollinated but accessible to birds*
Banksia ashbyi	*Banksia candolleana*	a. ground-flowering
Banksia coccinea	*Banksia dryandroides*	*Banksia blechnifolia*
Banksia grandis	*Dryandra mucronulata*	*Banksia petiolaris*
Banksia prionotes		*Banksia prostrata*
Dryandra formosa		b. flowers concealed in bush
		Banksia baueri
		Banksia pilostylis
		Banksia prostrata
		c. flowers exposed; some
		'NFM' characters present
		Banksia baxteri[1]
		Banksia ericifolia[2]
		Banksia occidentalis[1]

NFM=non-flying mammal
[1]Hopper, 1980b [2]Carpenter, 1978a; otherwise based on Holm, 1988.

8.12) have the inflorescences concealed within the bush and amazingly well protected by the leaves. The leaves in these plant genera are variable, but in the two species of *Banksia* and one of *Dryandra* that we are considering, they are almost identical. They are linear in outline and cut to the midrib on either side forming triangular lobes, broadest at the base; the lobes are more or less horny and are further strengthened either by inrolled edges or by thickening of the veins. If any attempt is made to part the leaves, their saw-like edges interlock and to separate them forcibly may then require all of a man's strength. Other features are that the inflorescence colour is dull, the odour sour (known to Holm in one species only but reported as 'yeastlike', as in *Protea*, in one *Dryandra* by Rourke & Wiens, 1977), and the styles are not strongly hooked and are presumably not required to support a pollinator's weight.

Australian flower-adapted mammals

Specific adaptations in NF-mammals for flower feeding are rare, and are at present known only in Australian marsupials. These have been surveyed by Armstrong (1979) and Turner (1982). Out of a total of 25 species recorded at flowers, the nine most significant for pollination are included in Table 8.13. The most famous of them is the honey-possum (*Tarsipes rostratus* [syn. *T. spenserae*]) (Plate 8c). It is very small and has a prehensile tail (like some of the New World flower-visiting opossums) as long as its body. The snout is slender and tubular. There are only a few teeth which provide guidance for the tongue. The tongue is extensible and has a brush of large papillae at the tip, and smaller hair-like papillae in its middle section (Holm, 1988). The palate is transversely ribbed and these ribs, together with the teeth, act as scrapers, helping to squeeze nectar and pollen out of the tongue as it moves in and out of the mouth. The animal is thus structurally highly adapted to flower feeding, though it can eat insects (Vose, 1973, cited by Holm, 1988). The honey-possum is confined to the south-west corner of Western Australia and is normally nocturnal, but it feeds in daylight in cool, cloudy weather at the end of the winter. It is only distantly related to other marsupials and is therefore placed in a family of its own. Pollen carried by *Tarsipes* comes from

Table 8.13 Australian NF-mammals that feed non-destructively on flowers (marsupials included are only those for which pollen and/or nectar form a moderate to large component of the diet).

Group	Name	Weight in grams
Marsupials		
Dasyuridae –		
marsupial 'mice'	*Antechinus apicalis* (dibbler)	55
	Antechinus stuartii (brown antechinus)	20-50
Burramyidae	*Acrobates pygmaeus* (feathertail glider)	12
	Cercartetus caudatus	
	Cercartetus concinnus	20
	Cercartetus nanus (eastern pygmy possum)	15-30
Petauridae	*Petaurus breviceps* (sugar-glider) (Fig. 8.16)	90-140
Tarsipedidae	*Tarsipes rostratus* (honey-possum) (Plate 8c)	15
Placentals		
Muridae	*Mus* species	
	Rattus fuscipes (southern bush-rat)	90-130

(Authors: Carpenter, 1978a; Armstrong, 1979; references in Hopper, 1980; and Collins & Rebelo, 1987, p.49; weights partly from Goldingay *et al.*, 1991.)

Fig. 8.16 Sugar-glider, *Petaurus breviceps*, visiting *Banksia spinulosa*; the extended body-wall of the animal that allows for gliding is visible. S.M. Carthew.

Banksia and *Adenanthos* in Proteaceae, and *Beaufortia* and *Calothamnus* in Myrtaceae. Because the animals are so small in relation to the brush-blossoms that they visit, the pollen gets all over the body.

Some of the other marsupials are phalangers (gliders) which are nocturnal partial insectivores, sometimes very small (Fig. 8.16). The flowers they visit are again Proteaceae and Myrtaceae, which are more or less clearly brush-blossomed. Holm (1988) found that the south-western pygmy possum (*Cercartetus concinnus*) has a brush tongue and a ribbed palate (like *Tarsipes* and the lorikeets), indicating that it is a pollen-feeder. The feathertail glider (*Acrobates pygmaeus*) has a slightly brush-tipped tongue and is frequently seen in flowering *Eucalyptus* trees.

Turner (1982) suggests that flower-visiting marsupials, together with parrots (lorikeets), co-evolved in the Cretaceous with *Banksia*, as many species of the latter possess characters that fit them for pollination by both these groups of vertebrates. The animals can digest pollen in huge quantities and some have specialised gut morphology. Thus in Australia there is a true co-evolution of both the plants and their pollinators among the NF-mammals. Pollen of *Banksia* (and presumably of Myrtaceae) is an important source of protein for flower-visiting NF-mammals in Australia and can be present all the year round, as in habitats in New South Wales (Turner, 1984; Goldingay et al., 1991) and Western Australia (Collins & Rebelo, 1987, p.391). The plants in this relationship appear to be adapted to NF-mammals, but most of them in such a way as not to exclude bird-pollination. We see here an extension of the occurrence of the relatively low level of pollinator-plant specificity which characterises bird-pollination in Australia.

Is there a syndrome of NF-mammal pollination?

There has been much discussion as to whether a syndrome of NF-mammal-pollination ('therophily') can be recognised. It seems that one could characterise such a syndrome as follows:

1) blossoms large (compared with most insect-pollinated blossoms)
2) blossoms of firm construction
3) blossoms borne on robust and/or short stems
4) blossoms dull in colour

5) nectar abundant
6) nectar rather concentrated
7) nectar sucrose-rich
8) odour yeast-like or musky, sometimes like that of bat-pollinated flowers

As concealment of the blossom is absent at least as often as present, it is better not to list it. From our presentation of NF-mammal-pollination, it is clear that there are several rather sharply defined subsyndromes which are characteristic of specific eco-geographical situations, namely the upright-flowered lowland tropical syndrome and the mossy forest syndrome of the New World, and the south temperate scrubland and savanna syndromes of the Old World. As regards the report of probable giraffe-pollination, we can assure readers that it was not published on 1 April. And as a postscript, we may mention that there are occasional reports in the literature of lizards causing pollination; an example from the West Indies concerns a cactus that is sometimes deserted by its regular hummingbird pollinators and is then visited by lizards, which appear to effect pollination (Scott, 1991).

9

Pollination by Wind and Water

All of the plants we have considered so far depend for pollination on insects, or other animal visitors such as hummingbirds and bats. However, many plants rely for pollination on other agencies, of which by far the most important is wind. The clouds of pollen blowing like yellow smoke from pines and other conifers are a familiar sight in early summer. Other wind-pollinated plants include many of the commonest forest trees of temperate climates, almost all the grasses, sedges and rushes (Poaceae, Cyperaceae and Juncaceae), many seashore plants and weeds belonging to the goosefoot and dock families (Chenopodiaceae and Polygonaceae) and a diverse assortment of other species.

It may at first sight seem surprising that wind pollination is so common if, as we tend to assume, pollination by insects is so much more efficient. However, as already pointed out in Chapter 6, wind may transfer pollen more efficiently than insects for such highly gregarious plants as the dominant trees of a temperate forest canopy, or the grasses of a savanna, steppe or temperate grassland. On the one hand, these are the plants for which wind pollination is most efficient. On the other hand, effective insect pollination of such very abundant plants would probably require a greater population of insects than most temperate – or indeed many tropical – ecosystems could support during a limited flowering season (Whitehead, 1969, 1983; Regal, 1982). Species-poor situations in which insects are relatively sparse, such as saltmarshes and some semi-arid habitats, also tend to show a large proportion of wind-pollinated plants (Cox, 1991). By contrast, floras in true deserts often show a high incidence of insect pollination, probably because plants grow and flower only after heavy rains, when conditions are not in fact arid, and there is often intense insect activity.

Wind pollination (or *anemophily*) has the obvious advantage of being independent of the possibly erratic occurrence and capricious behaviour of insects, and is effective when insects are scarce or absent. On the other hand, effective wind pollination requires the production and dissemination of very large amounts of pollen. If effective pollination requires no more than one pollen grain to reach a stigma with an area of one square millimetre (about the area of an oak stigma), every square metre of the plant's habitat must receive around a million pollen grains to make pollination reasonably certain. In fact, pollen production is ample to achieve this sort of density. It has been estimated that a single birch catkin produces about five-and-a-half million pollen grains and a hazel catkin nearly four million; a single floret of rye produces over 50,000 grains (Pohl, 1937). In general, wind-pollinated plants produce more pollen than the insect-pollinated plants, whether per stamen, per flower, per inflorescence or per plant, but the relationship is by no means clear cut. Some insect-pollinated plants produce very large amounts of pollen, especially those such as the corn poppy (*Papaver rhoeas*, p.47) in which pollen is the main or only reward to insect visitors.

Wind-pollinated plants tend to produce small numbers of ovules. In many cases, each flower produces only a single seed, as in oak, hazel or the grasses and sedges.

Table 9.1 The pollen rain over Great Britain, 1943, (data recalculated from Hyde, 1950).

Pollen type	Average catch (grains cm^{-2} y^{-1})	Least and greatest catch (grains cm^{-2} y^{-1})	
Poaceae (grasses)	2106	725 (Paddington)	4455 (Chesterfield)
Fraxinus (ash)	271	89 (Aberdeen)	505 (Llandough)
Quercus (oak)	178	10 (Aberdeen)	504 (Cambridge)
Ulmus (elm)	146	24 (Chesterfield)	773 (Cardiff)
Plantago (plantain)	134	52.8 (Paddington)	238 (Llandough)
Platanus (plane)	96	0 (Aberdeen)	428 (Paddington)
Urtica (nettle)	88	22 (Aberdeen)	156 (Llandough)
Rumex (dock)	62	34 (Paddington)	117 (Chesterfield)
Pinus (pine)	60	14 (Chesterfield)	105 (Aberdeen)
Corylus (hazel)	47	18 (Edinburgh)	94 (Llandough)
Betula (birch)	41	22 (Chesterfield)	71 (Cardiff)
Taxus (yew)	36	9 (Aberdeen)	138 (Cambridge)
Populus (poplar)	30	0 (Aberystwyth)	133 (Cambridge)
Salix (willow)	28	6 (Aberdeen)	67 (Chesterfield)
Chenopodiaceae	27	8 (Edinburgh)	93 (Cambridge)
Sambucus (elder)	17	3 (Aberystwyth)	59 (Chesterfield)
Tilia (lime)	17	0	84 (Cambridge)
Alnus (alder)	16	6 (Aberdeen)	35 (Cardiff)
Ericaceae	14	2 (Cambridge)	43 (Edinburgh)
Aesculus (horse chestnut)	14	0 (Edinburgh)	73 (Cambridge)
Juncaceae (rushes)		5 (Paddington)	33 (Aberystwyth)
Luzula (woodrush)	10		
Juncus (rush)	5		
Fagus (beech)	12	1 (Chesterfield)	47 (Edinburgh)
Apiaceae (umbellifers)	12	3 (Aberystwyth)	36 (Chesterfield)
Artemisia (mugwort)	11	5 (Chesterfield)	27 (Cardiff)
Ranunculaceae	11	1 (Aberdeen)	19 (Cambridge)
Asteraceae (composites), misc.	11	8 (Paddington)	22 (Cardiff)
Acer (sycamore, maples)	8	1 (Aberystwyth)	19 (Edinburgh)
Rosaceous trees	8	2 (Aberystwyth)	17 (Cardiff)
Brassicaceae (crucifers)	7	2 (Aberystwyth)	31 (Llandough)
Asteraceae (composites) Cichorieae	7	2 (Edinburgh)	15 (Llandough)
Castanea (sweet chestnut)	6	0	28 (Cardiff)
Carpinus (hornbeam)	6	0	17 (Paddington)
Cyperaceae (sedges)	5	2 (Chesterfield)	12 (Aberystwyth)

The table includes all the pollen types of which the average catch at eight sites during the year was more than 5 grains per sq. cm. Comparable pollen-counts for the 1990s would certainly show less elm and more coniferous pollen than 50 years ago, and there would be other more local differences, but otherwise the table is probably reasonably representative of the long-term picture. These are mostly town sites, so the pollen of planted trees is prominent. The sites were: Llandough Hospital, Penarth, Glamorgan; National Museum of Wales, Cardiff; University College of Wales, Aberystwyth; St Mary's Hospital Paddington, London; Botany School, Cambridge; Derbyshire Sanatorium, Chesterfield; King's Buildings, Edinburgh; City Hospital, Aberdeen. No station is quoted for minimum catch if this was zero at more than one station.

Climax forest trees tend to produce large single seeds, because a large food store is required for the establishment of seedlings in deep shade on the forest floor, where they may remain for years in a suppressed state among the ground flora until death of an old tree creates a gap that allows them to grow up to the canopy. For such trees

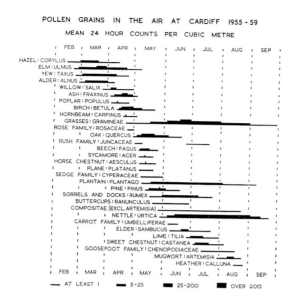

POLLEN GRAINS IN THE AIR AT CARDIFF 1955 - 59

MEAN 24 HOUR COUNTS PER CUBIC METRE

Fig. 9.1 Pollen calendar based on counts obtained with a Hirst automatic volumetric spore trap on the roof of the National Museum of Wales, Cardiff, from 1955 to 1959. The concentrations indicated were attained on average during the periods shown except that certain trees (poplar, willow, hornbeam, beech, *Acer*, and horse chestnut) reached a mean 24-hour count of over 5 grains per cubic metre in certain years only; elm would nowadays give lower figures than recorded here. Figure drawn by Mrs K.F. Adams, reproduced by permission from Hyde & Williams (1961). Stix & Grosse-Brauckmann (1970) and Solomon (1979) give comparable diagrams for Darmstadt and for New York City.

pollen limitation may thus be of little consequence. The rushes (Juncaceae) produce many-seeded capsules, but this is unusual among wind-pollinated plants. It is noteworthy that in rushes the pollen grains remain together in tetrads, and in many species there is a good deal of self-pollination.

The 'pollen rain' falling from the air during 1943 at eight sites widely scattered over Britain was studied by H.A. Hyde of the National Museum of Wales, by identifying and counting the pollen grains caught on gelatine-coated glass slides. He found that the total annual catch of grass pollen averaged about 2,100 grains/cm^2; the total tree-pollen count averaged just over half that number. Taking the eight sites together, the most abundant tree-pollen types were ash, oak and elm (elm would be much less prominent now, since the Dutch elm disease epidemic of the 1970s). Some of Hyde's results are summarised in Table 9.1. For various reasons these must be nearly minimum figures. Horizontal 'gravity slides' are inefficient in trapping pollen under ordinary windy conditions, especially for the smaller grains. The slides were all exposed well above the surrounding vegetation, usually on a building. Most of the sites were in built-up areas – and even rural Britain is only thinly and sporadically wooded. Taking these factors together, there is no doubt that much higher figures for particular tree-pollen types would be found using more efficient trapping surfaces in and around woods; and similar considerations apply to herbaceous plants. In the much more heavily-forested landscape of Sweden, Erdtman estimated that the total annual pollen-rain may amount to 30,000 grains/cm^2; at this sort of density, pollen deposition can be several grams per square metre – several tons per square kilometre (Solomon, 1979). Wind-pollinated plants have well-defined flowering seasons (Fig. 9.1). The anemophilous deciduous trees mostly flower in early spring while the branches are still

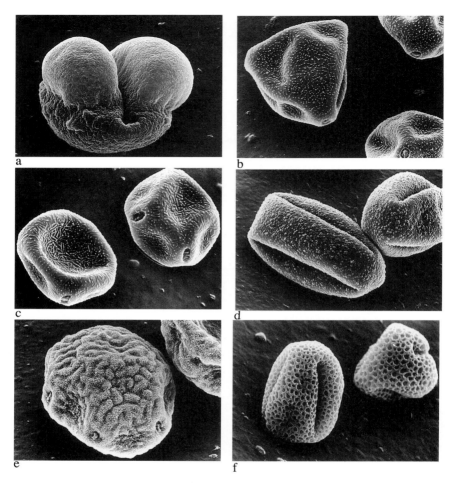

Fig. 9.2 Some wind-borne tree-pollen grains. **a**, jack pine, *Pinus banksiana*; typical pine grain with two air-sacs. **b**, hazel, *Corylus avellana*; three-pored. **c**, alder, *Alnus glutinosa*; the pollen wall is thickened betwen the four pores. **d**, pedunculate oak, *Quercus robur*; three longitudinal furrows. **e**, English elm, *Ulmus procera*; about six equatorial pores. **f**, ash, *Fraxinus excelsior*, three longitudinal furrows, grains in nearly equatorial and polar view. Scanning electron micrographs of air-dry pollen, × 1000. See also Nilsson *et al.* (1977).

bare (hazel, elm, alder, ash), or as the leaves unfold (birch, oak); the evergreen conifers tend to flower rather later. The summer pollen-rain is dominated by grasses and such herbaceous species as the plantains (*Plantago* spp.), stinging nettle (*Urtica dioica*) and in late summer in North America, ragweed (*Ambrosia artemisiifolia*).

Wind-dispersed pollen grains usually have a smooth dry surface, in contrast with the sticky and often highly ornamented grains that are often common in insect-pollinated

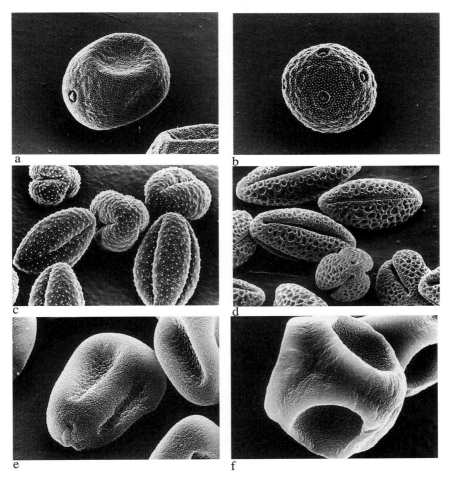

Fig. 9.3 Some wind-borne herbaceous pollen grains. **a**, perennial rye-grass, *Lolium perenne*; typical grass pollen with a single pore. **b**, ribwort plantain, *Plantago lanceolata*; pores scattered over the grain. **c**, mugwort, *Artemisia vulgaris*; compare the insect-pollinated Asteraceae in Fig. 2.8c–f. **d**, dog's mercury, *Mercurialis perennis*. **e**, hare's-tail cottongrass, *Eriophorum vaginatum*; pear-shaped grain with no sharply defined apertures, but with four thin areas in the wall, one at the blunt end and three round the sides. **f**, great wood-rush, *Luzula sylvatica*; grains shed in tetrads, each individual grain with a diffuse thin area at its distal pole. All × 1000.

species. Consequently, they are dispersed singly or in twos and threes, rather than sticking together in larger groups. The pollen grains of insect-pollinated species are of very varied sizes; those of anemophilous species vary much less, and are commonly between 25 and 40 μm in diameter in angiosperms and between 30 and 60 μm in diameter in conifers (Figs. 9.2, 9.3). This is probably because, on the one hand, larger grains are trapped more efficiently from a moving airstream than smaller grains while, on the other hand, the high rate of fall of very large grains will limit their range of dispersal. Rates of fall of wind-borne pollen grains in calm air range from between 1 and 2 cm/s for the smallest and lightest, to about 40 cm/s for the largest and heaviest. The rates of fall for many common anemophilous species are around 2-6 cm/s (Table 9.2).

Table 9.2 Rates of fall of pollen grains in still air (for authors see Gregory, 1973).

Species		Rate of fall (cm s^{-1})
Abies alba	silver fir	38.7
Betula pendula	silver birch	2.4
Corylus avellana	hazel	2.5
Dactylis glomerata	cocksfoot	3.1
Fagus sylvatica	beech	5.5
Larix decidua	European larch	9.9–22.0
Picea abies	Norway spruce	8.7
Pinus sylvestris	Scots pine	2.5
Quercus robur	pedunculate oak	2.9
Salix caprea	goat willow	2.2
Secale cereale	rye	6.0–8.8
Tilia cordata	small-leaved lime	3.2
Ulmus glabra	wych elm	3.2

The dispersal and deposition of wind-borne pollen

In fact, under normal conditions, the rates of fall of pollen grains in still air are of only secondary importance. Most wind-pollinated species possess adaptations which prevent the release of pollen under perfectly calm conditions. For instance, in most catkins the pollen shed by one flower lodges on the horizontal surface of the bract of the flower below, and in grasses the pollen is held by the curved, spoon-shaped lower ends of the anther loculi (Fig. 9.15, 9.16). If there is even slight wind to dislodge them,

Table 9.3 Standard deviation pollen-dispersal distance for some wind-pollinated trees (rounded to nearest 5 m).

Species		Standard deviation (m)	Author
Fraxinus americana and *F. pennsylvanica*	ash	15–45	Wright, 1953
Pseudotsuga menziesii	Douglas fir	20	Wright, 1953
Populus deltoides and *P. nigra* var. *italica*	poplar	300 or more	Wright, 1953
Ulmus americana	elm	300 or more	Wright, 1953
Picea abies	Norway spruce	40	Wright, 1953
Cedrus atlantica	Atlas cedar	75	Wright, 1953
Cedrus libani	cedar-of-Lebanon	45	Wright, 1953
Pinus cembroides var. *edulis*	pinyon	15	Wright, 1953
Pinus elliottii	ash pine	65	Wang, Perry & Johnson, 1960

Box 9.1 The transport and deposition of wind-borne pollen

Pollen grains are dispersed by atmospheric turbulence in the same way as the plume of smoke from a chimney (Gregory, 1973; Mason, 1979) or a volatile scent (Murlis, Elkinton & Cardé, 1992), with average density showing a 'normal distribution' across the plume, but falling off downwind at a rate rather less than the inverse-square law which it approaches under extremely turbulent conditions; dispersal across the axis of the plume is always greater horizontally than vertically, especially if the pollen source is close to the ground.

If pollen were dispersed only by completely random air movements, its density would be expected to show a normal distribution around the source; an appropriate measure of dispersal is then the root-mean-square (or 'standard deviation') distance moved from the source. Investigations using sticky microscope slides as pollen traps to study the deposition of pollen at different distances from isolated source plants, or genetic markers to follow the dispersal of particular genes from a test population, show that, on average, pollen deposition is greater very close to the source and at substantial distances from it than would be predicted from a normal distribution – the distribution is of the kind known as *leptokurtic* (Bateman, 1947c; Wright, 1953) (and indeed insect dispersal of pollen is always leptokurtic too). The diagrams below show the relative number of pollen grains caught on sticky slides at various distances from an isolated tree of slash pine (*Pinus elliottii*) in Florida (Wang, Perry & Johnson, 1960), and the percentage pollination of test plants of perennial rye grass (*Lolium perenne*) at various distances from a 'contaminating' plot of a variety of the same species with red-based shoots (Griffiths, 1950). Estimates of the root-mean-square dispersal distance for some American trees are given in Table 9.3 (see also Adams *et al.*, 1992). For herbaceous plants, pollen dispersal distances are probably equally diverse, but smaller – perhaps typically a tenth to a hundredth of these figures. The data of Bateman (1947b) suggest figures of about 2-4 m for cultivated beet, and about 6-9 m for maize. Tonsor (1985) found a standard deviation pollen-dispersal distance of 58 cm for *Plantago lanceolata*; he points out that for several reasons gene-dispersal may not closely match pollen dispersal.

Pinus elliottii
pollen caught on sticky slides
at distances from an isolated tree

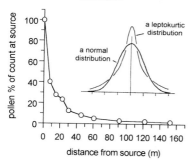

Lolium perenne
'contamination' of seed by pollen from a test plot
of plants with red shoot bases

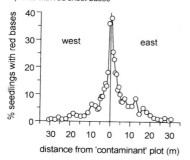

These pollen-dispersal curves are of course averages, resulting from the interaction of various, constantly changing factors over a period of time. What happens on particular occasions will depend on such factors as settling velocity, the height of release of the pollen, windspeed and turbulence (Okubo & Levin, 1989). Particular weather conditions may sometimes lead to local deposition of large amounts of pollen at a

Box 9.1 continued

considerable distance from the source. Rempe (1937) investigated the distribution of pollen in the course of a series of flights over Göttingen. He found that under conditions of strong convection pollen concentration was almost unchanged up to an altitude of 1000 m, and that considerable amounts of pollen were still present at 2000 m; he found larger amounts of pollen in cumulus clouds (up-currents) than outside. By contrast, under stratus cloud the pollen concentration fell off sharply with altitude. In the course of one flight, Rempe encountered a heavy fall of spruce pollen associated with the base of a dissolving cumulus cloud over Göttingen, at a time when spruce was past flowering in the lowlands. The wind was north-easterly, and the source of the pollen was evidently in the Harz Mountains, where spruce was still flowering and cumulus cloud building up, some 34 km to the north-east. In this instance, with a windspeed of 14 km/h, the pollen would have taken about 2½ hours to reach Göttingen. As Rempe pointed out, transport of pollen for distances up to some 300 km/day is readily explained. It is probably for reasons of this kind that 89 per cent. of the spruce pollen counted by Hesselman (1919) on Västra Banken lightship, 30 km from the Swedish coast in the Gulf of Bothnia, was trapped during 2 of the 40 days of observation (16 May-26 June). Cumulus clouds break up and dissolve at the end of the day, so pollen carried high into the atmosphere during the day largely settles to the ground during the night. Some pollen may be caught up and carried again to high levels on subsequent days, but probably little

remains viable beyond the first day or two, because at least some pollen grains, and probably all, are quickly damaged by more than a few hours' exposure to the ultra-violet radiation in sunlight (Werfft, 1951).

The mechanisms of deposition of airborne particles are complex; they are discussed more fully by Gregory (1973), Edmonds (1979), Harrington (1979), Chamberlain & Little (1981), Niklas (1985) and Monteith & Unsworth (1990). Gravity is of little significance except in conditions of complete calm, rarely found except on clear, still nights. The most important mechanism for pollination – for the entrapment of pollen grains on stigmas – is aerodynamic impact. As the airstream diverges round a solid object, suspended particles tend by virtue of their momentum to follow a straighter path and collide with the surface; the heavier the particle the more likely this is to happen. Impaction is most likely with surfaces at right angles to the airstream. A high degree of atmospheric turbulence increases the likelihood of pollen being deposited on surfaces parallel to the airstream, and tends to increase the rate of deposition of pollen to the ground. It has been suggested that electrostatic attraction may play a part in the deposition of pollen onto stigmas (Corbet, Beament & Eisikowitch, 1981; Erickson & Buchmann, 1983). There is usually some separation of electrostatic charge between the air and objects on the ground surface, which will generally work in a direction which favours the dispersal and later deposition of pollen, but its effects are probably always secondary to aerodynamic impact.

the pollen grains will be kept in suspension by the turbulence of the air. Turbulence arises in two ways. The interaction of wind with solid objects on the ground generates eddies which break up the smooth flow of the air. Also, on clear days turbulence is produced by convection currents, as the ground and the air above it are warmed by the sun – a phenomenon familiar to air travellers. In scale, turbulence varies from the eddying around the leaves and stems of plants on a windy day to the great weather-systems so dramatically shown in photographs of the Earth taken from spacecraft.

The great majority of wind-borne pollen grains of all species are deposited quite close to their source. Some pollen is carried for very long distances. At the beginning of June 1937 Erdtman demonstrated by a sophisticated vacuum-cleaner technique the presence of pollen grains in the air even in mid-Atlantic, though less by a factor of ten or twenty thousand than the average over southern Sweden for April and May (Erdtman, 1969). Such long-distance transport would explain the presence in moss samples from Spitzbergen of Scots pine (*Pinus sylvestris*) pollen which must have come from Scandinavian forests 750 km to the south, and *Nothofagus* pollen in peat on Tristan da Cunha, 5000 km from the nearest source in South America (Hafsten, 1960). However, spectacular long-distance transport of this kind is of little significance for pollination. The relatively few grains carried to such distances, even if still viable, would be swamped by vast quantities of locally produced pollen. Normal dispersal distances are typically from tens to a few hundreds of metres for wind-pollinated trees, and from a few decimetres to a few metres for herbaceous plants (see box and Table 9.3). Compared with insect-pollinated species, rates of gene flow in wind-pollinated plants are notably high (Hamrick, Godt & Sherman-Broyles, 1995).

Pollen grains are captured by the stigmas mainly through impact – by collision – as the air streams past. The aerodynamic effects on airflow of the architecture of the shoots and the form of the flowers are undoubtedly important for both dispersal and capture of pollen (Niklas, 1986, 1987).

Such physical considerations account for many of the features commonly found in wind-pollinated flowers. The stamens are usually large and well exposed, often hanging freely on long filaments or in catkins. The stigmas too are large and well exposed, and often finely divided and feathery. A narrow surface is more efficient in trapping particles from moving air than a broad surface of the same area. However, even large feathery stigmas are generally small enough to behave as single objects in relation to the boundary layer of the airflow, so their form may largely reflect selection for large receptive surface with minimal expenditure of material. The functions served by the perianth in insect-pollinated flowers are irrelevant to wind pollination; indeed, a well developed perianth would be a hindrance to free transfer of pollen, and accordingly the perianth is generally much reduced or absent altogether. Free from the constraint of having to flower when insects are active, most of the wind-pollinated deciduous forest trees flower very early in the year (Fig. 9.1) when the trees are bare of leaves. The flowers are then most freely exposed to disperse and receive pollen, and the surrounding area 'competing' with the stigmas for pollen is least. The concentration of pollen is very high in the immediate neighbourhood of the dehiscing anthers; stigmas close to them would be so thickly covered with their own pollen that there would be little chance of fertilisation by pollen from other individuals. It is not surprising, therefore, that many wind-pollinated plants have the sexes in separate flowers, or if the stamens and stigmas are borne close together in space they are separated in time by strongly marked dichogamy.

The conifers

The conifers, which dominate the boreal forests of Eurasia and North America, are among the few groups of plants that are consistently wind-pollinated – and may be seen ecologically as a gymnosperm analogue of the later-evolved catkin-bearing trees which dominate the deciduous forests to the south. Their 'flowers' are very different from those of true flowering plants. The reproductive parts are typically aggregated into strobili or 'cones', as in Scots pine (*Pinus sylvestris*) (Plate 4a,b). The male cones are grouped around the bases of the elongating new shoots, and mature about the end of May or early June. Each is about 5-8 mm long, and made up of numerous 'stamens', each consisting of a scale (microsporophyll) with a narrow upturned crest, bearing two pollen sacs on its under side. The pollen grains each have two wings or bladder-like expansions of the wall (Fig. 9.2a) which serve to reduce the density and rate of fall

Fig. 9.4 Ovules of yew, *Taxus baccata*, showing the pollination drop.

of the grain – though their main significance may be rather in orienting the grain relative to the nucellus as it is drawn into the micropyle by the pollination fluid (Doyle, 1945). The young female cones are formed near the tip of the current year's shoots. At the time of pollination, in June, the cone is around 8 mm long. The axis of the cone bears small bract scales and much larger and thicker ovuliferous scales, each with two ovules near the axis on its upper surface. Each ovule is covered by a single integument, with a rather wide micropyle facing the axis. The tips of the ovuliferous scales, which make up the outside of the young cone, gape slightly apart, so that pollen can reach the ovules. Pollen grains which lodge around the mouth of the micropyle are drawn into contact with the nucellus by a drop of liquid exuded into the micropylar canal and then reabsorbed. After pollination, the scales thicken and seal the exterior of the cone. Fertilisation does not take place until the following summer. The cones ripen in the second year after pollination, when the scales of the now dry, woody 'fir cone' gape apart once more to release the winged seeds.

Most of the other common conifers differ only in minor particulars from the Scots pine, but the seeds usually ripen in the same year as the cones are pollinated as in the spruces (*Picea*) (Plate 4c). In some conifers, such as the Douglas fir (*Pseutotsuga menziesii*), the bract scales are long and project from between the ovuliferous scales of the ripe cone. In the larches (*Larix*) the bract scales are large and brightly coloured in the young female cones, forming the attractive 'larch roses'. The cedars (*Cedrus*) are unusual in flowering in autumn rather than spring.

Yew (*Taxus baccata*) is related to the conifers but its ovules are solitary. They are produced from buds borne in the leaf axils during the winter. The ovule itself is borne at the tip of a minute shoot in the axil of a scale just below the apex of the bud. The tip of the ovule emerges from between the bud scales very early in the year. By February or March it is ready for pollination. A sticky drop of liquid is exuded from the micropyle (Fig. 9.4), in which the pollen grains are trapped and find their way down the canal of the micropyle to the nucellus. The male cones of yew are also produced from buds formed in the leaf axils, but on separate trees. They are smaller than those of the Scots pine and the scales are umbrella-shaped, with 5-9 pollen sacs on the lower surface. The pollen grains lack the characteristic 'wings' of the pine and its close relatives, and the large seed with its fleshy red aril is ripened within the year.

The catkin-bearing trees, and some other trees

The trees of the birch and hazel families (Betulaceae and Corylaceae) bear their flowers in catkins rather comparable in their general arrangement with the strobili of conifers. In the birches, of which we have two common species in Britain (*Betula pendula* [Fig. 9.5] and *B. pubescens*) the male catkins are formed in the autumn, and as they mature in March and April they expand and hang freely, dangling from the tips of the twigs. The individual flowers are borne in threes, with a few bracteoles, in the axil of each catkin scale. Each consists of a pair of deeply divided stamens, with a small bract-like perianth. The short carpellary catkins are borne on the same tree, and remain stiff and erect as they expand. The flowers lie in the axils of the catkin scales as

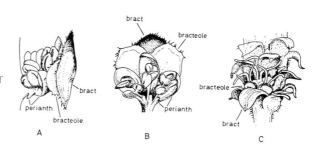

Fig. 9.5 Silver birch (*Betula pendula*). **A**, side view of a group of male flowers in the axil of a single catkin scale. **B**, a similar group of flowers seen from the adaxial surface (i.e. the side towards the tip of the catkin). **C**, part of a female catkin.

in the male catkins. There is no perianth; each flower consists of an ovary bearing two styles, and containing two ovules. The fruit, ripened in late summer, is a small winged nutlet with a single seed. Alder (*Alnus glutinosa*) is similar in essentials to birch, but there are four stamens to each male flower, and the female catkins become woody and cone-like in fruit (Fig. 12.11). Between late February and the end of March, the long reddish male catkins expand and shed their yellow pollen, and the small dark female catkins, in clusters a few inches back from the tips of the shoots, are enlivened by the red of their stigmas.

The long yellow 'lamb's tail' catkins of hazel (*Corylus avellana*) are among the first signs of approaching spring as they expand in the first mild weather of the new year (Plate 4d). If the air is perfectly calm, little pollen escapes from the catkin, but pollen blows out in clouds as the catkins bob and dangle in the wind. The male flowers have no perianth, and are borne in pairs in the axils of the catkin scales (Fig. 9.6). The female flowers are also borne in pairs, and possess a small perianth. The catkin is reduced to a plump bud containing only a few flowers, which at the time of pollination are so undevel-

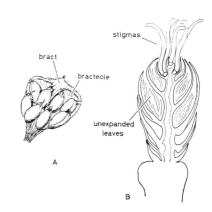

Fig. 9.6 Hazel (*Corylus avellana*). **A**, pair of male flowers in the axil of a single catkin scale. **B**, vertical section of female catkin.

Fig. 9.7 Pedunculate oak (*Quercus robur*) with catkins and expanding leaves; arrows indicate female flowers.

oped that they consist of little more than the crimson stigmas projecting from the scales at the tip. After fertilisation, the ovary develops into a one-seeded nut, with a leafy involucre formed from the bracteoles at the base of the flower. Hazel has a particular place in the history of pollination biology as the subject of Richard Bradley's experiments in the early eighteenth century (see p.14).

The family which includes the oaks and beeches (Fagaceae) is related to the two families just considered. There are two native oaks in northern Europe. In the pedunculate oak (*Quercus robur*) the ripe acorns are borne on a stalk a few cm long, while in the sessile oak (*Q. petraea*) the acorns are only shortly stalked or almost sessile on the twigs. The slender yellowish-green catkins of both species appear with the opening leaf-buds in April and May (Fig. 9.7). The catkin scales are very small, and the individual flowers have a rather larger perianth and more stamens than those of the catkin-bearing trees we have discussed so far (Fig. 9.8). About six stamens is usual, but the number is variable. The female flowers are borne a few together in short spikes. Each is surrounded at the base by a scaly involucre which later develops to form the cupule or 'acorn cup'; the minute green perianth forms a toothed border at the top of the ovary surrounding the three styles. The ovary contains six ovules, but only one develops.

Beech (*Fagus sylvatica*) has its male flowers in long-stalked tassel-like heads. The female flowers are borne in pairs, surrounded by the involucre which grows to form the four-valved cupule enclosing the beech nuts. The individual flowers are very much like those of oak and, as in the oaks, they appear as the foliage begins to expand.

Beech and the European and North American oaks are regularly wind-pollinated and are visited only casually by insects for their pollen. By contrast, the sweet chestnut (*Castanea sativa*) is largely insect pollinated; so too are some tropical species of oak. The

Fig. 9.8 Pedunculate oak (*Quercus robur*). **A**, side view of a single male flower. **B**, two female flowers. **C**, young developing acorn, showing perianth and cupule.
A and B, 22 May; C, 18 June.

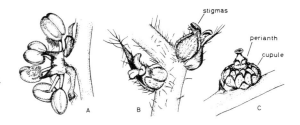

Fig. 9.9 Sallow (*Salix cinerea*);
a, solitary bee (*Andrena* sp.)
on male catkin. **b**, hoverfly
(*Syrphus* sp.) feeding on
pollen (see also Fig 4.4).

a

b

chestnut has stiff, erect catkins up to 15-20 cm long, which appear in July. Usually the catkin is male in its upper part, with a small number of female flowers at the base, but some catkins are entirely male. The male flowers have 10-20 stamens; the female flowers have a six-celled ovary and are borne in groups of three in each cupule. Pollination is brought about by bees and other insects which visit the catkins for nectar and pollen; wind pollination can take place later as the pollen dries and is blown from the male flowers.

The same kind of relationship between wind-pollinated and insect-pollinated species is shown in another family of catkin-bearing trees, the willow family (Salicaceae). The willows and sallows (*Salix*) are insect pollinated; sallows are favoured collecting sites for entomologists in early spring. All the willows and sallows are dioecious; the male and female catkins are borne on separate plants. The catkins are stiff and usually erect. In the larger willows they are about 6 cm long and a little less than 1 cm thick, and appear with the leaves in April and May. In the sallows ('palm', 'pussy willow') they are shorter and broader and appear on the bare twigs in March and April (Fig. 9.9). The individual flowers, one in the axil of each catkin scale, are very simple. There is no perianth, but the flowers possess a nectary at the base; male flowers usu-

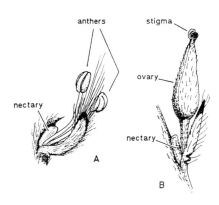

Fig. 9.10 Sallow (*Salix cinerea*). **A**, single male flower. **B**, single female flower. Compare Fig. 9.9.

ally have two stamens (more in some species) while female flowers have a two-celled ovary with a short style and two stigmas (Fig. 9.10). The catkins are freely visited by insects in fine weather. Although the pollen is quite sticky (Hesse, 1979a), some pollen is dispersed by wind, but how significant this is for pollination is questionable. Sacchi & Price (1988) found that at two sites near Flagstaff, Arizona, almost all pollination of *S. leptolepis* was by insects; Vroege & Stelleman (1990) in the Netherlands concluded that wind pollination is probably important for both the goat willow (*S. caprea*) and for the creeping willow (*S. repens*), but that the former is somewhat better adapted to pollination by insects and the latter by wind. The poplars (*Populus*) are closely related to the willows, but are entirely wind-pollinated. The catkins are up to 10 cm long, and are flexible and pendulous like the male catkins of oak and hazel; they appear in early spring before the leaves. The individual flowers lack the nectaries of the willows (but have a cup-like disc at the base), and the stamens of the male flowers are more numerous.

The remaining important wind-pollinated trees in the woods of northern Europe

Fig. 9.11 Flowers of English elm (*Ulmus procera*).

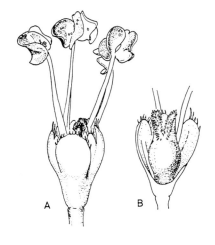

Fig. 9.12 English elm (*Ulmus procera*). **A**, single flower. **B**, flower with part of perianth cut away to show ovary and stigmas.

are the elms (*Ulmus*) and ash (*Fraxinus excelsior*). The elm family are all wind-pollinated, and our native elms are among the earliest trees to flower in spring; the flowers are often expanded by the end of January. Before Dutch elm disease killed most of the sizeable elms in southern England in the 1970s, the crowns of the hedgerow trees of *Ulmus procera* outlined against the sky by the tight clusters of dark reddish flowers on the twigs (Fig. 9.11) were a characteristic feature of the English landscape in late winter. The individual flowers of elms are bisexual, unlike those of any of the trees we have considered so far (Fig. 9.12). There is a small bell-shaped perianth divided into four or five lobes, as many stamens, and a one-celled ovary with two styles which in due course develops into the one-seeded winged fruit. The flowers are strongly protandrous, and at first sight may appear to be unisexual, especially in the male phase when the stamens project far out of the flowers on their long filaments.

Ash also has bisexual flowers, which appear in coarse, dark greenish masses on the naked twigs in April (Fig. 9.13). Most members of the olive family (Oleaceae), to which ash belongs, are entomophilous; familiar examples are privet (*Ligustrum*), lilac (*Syringa vulgaris*) and jasmine (*Jasminum*). Ash flowers have no corolla; there are two stamens as in other members of the family, and the long ovary which later develops into the flat, one-seeded 'ash key' bears two rather large blackish stigmas at the tip (Fig. 9.14). Some trees bear hermaphrodite flowers, others may bear purely male or purely female flowers or a mixture of hermaphrodite and unisexual flowers; the same tree may vary somewhat in behaviour from year to year. The common ash is an obvious example of a plant of entomophilous ancestry which has quite recently (in evolutionary terms) become adapted to wind pollination. It is particularly interesting that the Mediterranean manna ash (*Fraxinus ornus*) still possesses a white corolla, and the fragrant flowers, borne when the tree is in full leaf, are pollinated by insects. A parallel case is the American silver maple (*Acer saccharinum*), a precocious-flowering wind-pollinated member of an otherwise generally insect-pollinated genus.

Fig. 9.13 Flowers of ash (*Fraxinus excelsior*).

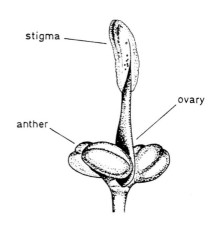

stigma

ovary

anther

Fig. 9.14 Ash (*Fraxinus excelsior*); single hermaphrodite flower.

The grasses and sedges

The grasses (Poaceae) are by far the most important family of wind-pollinated herbs. The main exceptions to wind pollination in the family are the species which are self-pollinated (often cleistogamous) or apomictic (Campbell *et al.*, 1983) (but see p.392). The small, greenish flowers of grasses are grouped in *spikelets* (Figs. 9.15, 19.16; Plate

4e), each enclosed at its base by a pair of chaffy *glumes*. The individual flowers or 'florets' in the spikelet are borne alternately on either side of the slender axis or *rachilla*. Each flower is enclosed by a glume-like *lemma*, usually with a prominent mid-nerve and often with a slender awn at the back or tip, and *palea* usually with two prominent nerves. The flower itself consists of an ovary with two long feathery stigmas, three stamens with slender filaments and large versatile anthers, and a pair of small swollen scales called lodicules – perhaps the last vestiges of the perianth – which swell to open the floret when the anthers and stigmas are mature. The spikelets are variously arranged into slender wiry inflorescences, which hold the flowers above the turf and the leaves of neighbouring plants; most grasses flower in June and July, at the height of the growing season, a fact of which hay-fever sufferers are keenly aware. A glance at a few grasses in flower will convey better than any description how effectively the stamens and stigmas are presented to the wind. It has been suggested that the constant movement of grass stems in even a gentle breeze may be important in in-

Fig. 9.15 Spikelet of false-oat grass (*Arrhenatherum elatius*). The spikelet contains one male floret with three stamens, and one hermaphrodite florest with three stamens and two feathery stigmas; the tips of the long lodicules are clearly visible. Compare with Fig. 9.16.

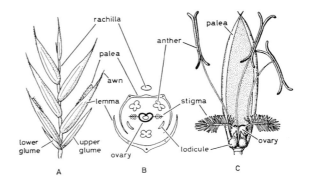

Fig. 9.16 The floral structure of grasses. **A**, spikelet of red fescue (*Festuca rubra*). **B**, diagram of a grass flower. **C**, flower of meadow fescue (*Festuca pratensis*) with lemma removed. A, after Hubbard (1954); C, after Rendle (1930). Compare with Fig. 9.15.

creasing the volume of air swept for pollen by the stigmas (Niklas, 1985).

The sedges (Cyperaceae) are superficially similar to the grasses and, like them, have much-reduced greenish or brownish flowers with prominent stigmas and large stamens. They differ from the grasses in the structure of the individual flowers and inflorescences. The flowers are borne singly in the axils of scales or 'glumes' forming catkin-like spikes. In most genera the flowers are hermaphrodite, with (usually) three stamens and an ovary containing a single ovule, with a style divided into two or three long rough stigmas. The perianth may be represented by bristles (which form the 'cotton' of the cotton-grasses, *Eriophorum*) or may be absent altogether. In the large and common genus *Carex* the flowers are unisexual, and are often grouped into separate male and female spikes (Fig. 9.17), an arrangement recalling the wind-pollinated catkin-bearing trees.

Fig. 9.17 Inflorescence of a sedge, *Carex demissa*. The terminal spike is male, with conspicuous exserted stamens; below it are three short female spikes, each flower with three spreading stigmas.

Other wind-pollinated herbs

The flowers of grasses and sedges are so specialised for wind pollination that they bear little resemblance to insect-pollinated flowers. Among the remaining wind-pollinated herbs, an entomophilous ancestry is often obvious. Thus the rushes (Juncaceae) have a small chaffy perianth, large stamens and large rough stigmas (Fig. 9.18), but the arrangement of the parts of the flower leaves no doubt that they are closely related to

Fig. 9.18 Flowers of a woodrush, *Luzula forsteri*. Each flower has six chaffy brown perianth segments, six stamens, and three large slightly feathery stigmas.

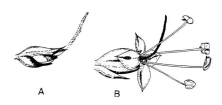

A   B

Fig. 9.20 Ribwort plantain (*Plantago lanceolata*).
A, young flower, with receptive stigma.
B, older flower; stigma withered, anthers
dehiscing. Compare Fig. 9.20.

the lily family (Liliaceae). The plantains
(*Plantago*, family Plantaginaceae) (Fig.
9.19) have no close typically entomophilous relatives, but are comparable with
the rushes in their adaptations to wind
pollination – though in no way related to
them. The corolla is small and membranous, and the strongly protogynous flowers have long, rough stigmas and large
versatile anthers borne on long filaments
(Fig. 9.20). Some species, such as the
hoary plantain (*P. media*), which has con-

Fig. 9.19 Inflorescence of ribwort plantain
(*Plantago lanceolata*). The stigmas expand first
(visible in the younger flowers near the tip
of the spike), succeeded later by the large
anthers freely exposed on their long
filaments. See Fig. 9.21.

spicuous heads and is slightly scented, are visited by insects; they are probably another
case (like the willows and some further examples considered below) in which there is
a balance between insect and wind pollination (Stelleman, 1981).

The pollen of the common stinging nettle (*Urtica dioica*) is among the most abundant
in the pollen rain of late summer. The stinging nettle has small, greenish unisexual
flowers, borne in catkin-like inflorescences hanging from the leaf axils; usually male
and female flowers are found on separate plants (Fig. 9.21). The female flowers have

Fig. 9.21 A female and a male
plant of stinging nettle (*Urtica
dioica*). Because the nettle is a
perennial plant with vigorous
spread by rhizomes, large
patches of one sex are
common.

two smaller and two larger perianth segments, and a one-celled ovary with a sessile tufted stigma. The male flowers have
four perianth segments and four stamens;
the stamens are incurved and under tension in the bud and spring back, scattering the pollen explosively, after the flowers
open (Fig. 9.22).

Fig. 9.22 Stinging nettle (*Urtica dioica*). **A**,
newly opened male flower. **B**, male flower
after dehiscence of anthers. **C**, female flower.

The docks (*Rumex*) are also abundantly
represented in the pollen rain. Of the
European genera in the dock family (Polygonaceae), *Polygonum* and its allies are
mainly pollinated by insects or self-pollinated. Thus bistort (*Persicaria bistorta*) and
the introduced Japanese knotweed (*Fal-
lopia japonica*) have conspicuous inflorescences and well developed perianths;
bistort has sticky pollen (Hesse, 1979b).
Others such as the common knotgrass
(*Polygonum aviculare* agg.) have small inconspicuous flowers in the leaf axils. The
wind-pollinated docks and sorrels, on the
other hand, have bulky lax inflorescences
well exposed to the wind, dry pollen,
rather small perianths and tufted stigmas
(Fig. 9.23).

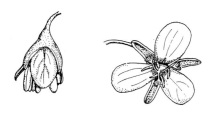

Fig. 9.23 Broad-leaved dock (*Rumex
obtusifolius*). **A**, newly-opened flower, in
functionally male phase. **B**, older flower with
anthers shed and stigmas expanded.

The members of the goosefoot family
(Chenopodiaceae) are largely plants of
exposed maritime habitats or salt steppes.
Their insect-pollinated relatives are to be
sought in such (largely tropical) families as
the Amaranthaceae and Nyctaginaceae.
The flowers of Chenopodiaceae are small
and greenish, with no corolla, and the
short stamens and stigma are usually
quite freely exposed (Fig. 9.24). Pollen of
Chenopodiaceae appears abundantly in
the pollen rain, and wind pollination is
certainly very effective in, for instance,
beet (*Beta vulgaris*), where pollen from wild
populations of sea beet (subsp. *maritima*)
can lead to troublesome genetic contamination of seed crops. There is also a good
deal of evidence that some other Chenopodiaceae are commonly self-pollinated (for example *Atriplex patula*).
Probably various degrees of balance between anemophily and self-pollination
are to be found in the family.

Many aquatic and waterside plants be-

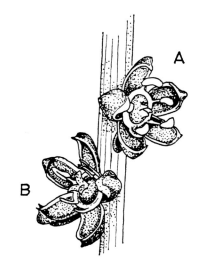

Fig. 9.24 Sea beet (*Beta vulgaris* ssp *maritima*).
A, young flower about time of dehiscence of
anthers. **B**, older flower; anthers mostly
shed, and stigmas mature.

Fig. 9.25 Wind-pollinated
aquatic plants; **a**, mare's tail
(*Hippuris vulgaris*).
b, broad-leaved pondweed
(*Potamogeton natans*).

longing to various families are wind-pollinated, for example mare's tail (*Hippuris*) (Fig. 9.25a), water milfoil (*Myriophyllum*), some of the 'pondweeds' (*Potamogeton*) (Fig. 9.25b), and the bur-reeds (*Sparganium*) (Fig. 12.12) and bulrushes (*Typha*). Most of the remaining anemophilous herbs are scattered species belonging to predominantly insect-pollinated families, for example the wind-pollinated species of *Thalictrum* (meadow-rues) in the Ranunculaceae, salad burnet (*Sanguisorba minor*) (Fig. 12.8) in the Rosaceae (Hesse, 1979c), and dog's mercury (*Mercurialis perennis*) (Fig. 12.9) in the Euphorbiaceae. Even so specialised an entomophilous family as the daisy family (Asteraceae) includes such anemophilous genera as the mugworts (*Artemisia*); the North American ragweed (*Ambrosia artemisiifolia*) is a troublesome cause of hay fever in the United States in late summer. The interesting general point is the way in which the combination of characters associated with wind pollination – the 'syndrome of anemophily' – has appeared independently in flowers of widely varied basic structure.

A striking feature of studies of the present-day pollen rain is the amount of pollen found of types we ordinarily think of as purely entomophilous. This should not be a cause for surprise. There are probably few entomophilous species of which no pollen is ever shed into the air, and there must be many for which this pollen brings about a small amount of local wind pollination. Even a very small proportion of wind pollination will provide a basis for natural selection to work on under conditions where wind pollination may be advantageous, and so lead to the evolution of anemophily. We tend to think of plants in terms of clear-cut categories. However, it is a common assumption that many insect-pollinated species may be self-pollinated if insect pollination fails, and a balance between two or more pollination mechanisms is probably common in other cases as well. It is not easy to demonstrate and evaluate a balance of this kind, but there can be little doubt that there is such a balance between entomophily and anemophily in, for example lime (*Tilia*), ling (*Calluna vulgaris*) and the rock-roses (*Helianthemum*), as well as in the sweet chestnut and willows mentioned already; in *Tilia* the pollen is only slightly sticky, and in *Calluna* it is not sticky at all (Hesse, 1979a). All of these flowers are visited by insects, often in large numbers, but pollen of all is released plentifully into the air, and all show at least some indication of the syndrome of anemophily. However, in three possible instances which have been analysed, Free (1964) concluded that wind-dispersed pollen is of negligible importance in the pollination of apple trees, Hatton (1965) concluded that mistletoe (*Viscum album*) is predominantly anemophilous, while the experiments of Anderson (1976) suggest that *Tilia* in North America is pollinated mainly by a wide range of day- and night-flying insect visitors, with anemophily playing a subsidiary, but not insignificant, role.

Water pollination

Pollination by wind is common and important, but pollination by water is surprisingly rare. As Agnes Arber wrote in her classic book on *Water Plants* (1920), 'The most notable characteristic of the flowers of the majority of aquatic angiosperms is that they make singularly little concession to the aquatic medium ...' And indeed, insect-pollinated and wind-pollinated aquatics abound (Cook, 1988), but water-pollinated species are relatively few and their structure and modes of pollination suggest scattered and diverse origins. Among the dicotyledons, the only genera with water-pollinated species are the hornworts (*Ceratophyllum*) and the water starworts (*Callitriche*). Water pollination is known (or probable) in 29 genera of monocotyledons, from nine families. Six of these families are strictly hydrophilous, the fresh or brackish-water Ruppiaceae, Najadaceae and Zannichelliaceae, and the seagrass families Posidoniaceae, Cymodoceaceae and Zosteraceae (Cox, 1988).

Adaptations to water pollination are diverse (Sculthorpe, 1967; Cox, 1988, 1993); there is no single 'syndrome of hydrophily'. Most often, pollination takes place at or above the water surface ('epihydrophily'). In some cases, the surface film provides the means of transport but pollination clearly takes place in the air above the surface. In others, the pollen and stigmas actually float in the surface film, so that pollen transfer takes place at the water surface itself. Relatively few species are pollinated completely under water ('hypohydrophily'). In water-pollinated plants, pollen dispersal is thus often confined to a two-dimensional plane, within which the stigmas also lie. It can be shown theoretically that this should increase greatly the chances of successful pollination, and it is probably particularly important adaptively in view of the much lower rates of mixing in water than in air. Wind-pollinated plants typically maximise the probability of pollination by increasing the size of the 'target'; large plumose stigmas are common in anemophilous species. Water is a much denser medium, imposing much greater 'form drag' (Monteith & Unsworth, 1990), so stigmas of water-pollinated plants are rigid and simple in outline, though often elongated. But in water-pollinated plants, natural selection seems often to have favoured increase in size of the 'search-vehicle', through elongation of the pollen grains, or through a tendency for the pollen to stick together in space-filling rafts in the surface film, or loose, submerged, fluffy or gelatinous masses (Cox, 1983b, 1988, 1993).

Some members of the frog-bit family (Hydrocharitaceae) have conspicuous insect-pollinated flowers, and others are wind-pollinated, but the family includes also one of the best-known examples of pollination at the water surface, the ribbon-weed (*Vallisneria spiralis*) (Fig. 9.26). This is a plant of warm climates, but is naturalised at a num-

Fig. 9.26 Tapegrass (*Vallisneria spiralis*). **A**, semi-diagrammatic sketch to illustrate habit of plant (the size of the leaves and depth of the water are more than proportionately reduced). **B**, female flower. **C**, male flowers. **D**, stigma-lobes of female flower. A, based on Kausik (1939) and Sculthorpe (1967); B–D after Kausik.

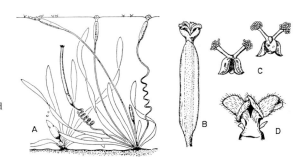

ber of places in Britain where the water is warmed by industrial effluents, and is commonly grown in aquaria. The plants are dioecious. The minute male flowers are borne many together in a tubular spathe near the base of the plant. Each has two stamens, tightly enclosed by the three sepals. At maturity, the flowers break free and float to the surface of the water, where they open and the stamens dehisce, exposing the pollen which adheres in a globular mass to the tip of each stamen. The female flower is borne to the surface of the water on a slender flexible peduncle, where it lies more or less horizontally in a shallow dimple in the surface film. When the sepals open, exposing the three large fleshy stigmas, the flower is some 3-4 mm across; the tubular spathe enclosing it at the base is a little over a centimetre long. The stigmas are unwettable, like the leaves of a number of floating aquatics, owing to the dense velvety pile of water-repellent hairs with which they are covered. The male flowers are carried about by water currents and the wind. If one chances to encounter a female flower, it slides down the depression in the surface film and comes to rest with the projecting stamens in contact with the stigmas. In *V. americana*, in which the stamens are often erect and do not project beyond the sepals of the male flower, pollination probably depends on the male flowers being toppled into the female flower when the latter is momentarily submerged.

In many species of the related genus *Elodea*, including the familiar Canadian waterweed (*E. canadensis*), it is the pollen grains themselves which are dispersed across the water-film to reach the stigmas. As in *Vallisneria*, the plants are dioecious. Both male and female flowers have six perianth segments, and open at the surface of the water, carried up by the slender perianth tube (from which fragile attachment the male flow-

Fig. 9.27 Canadian waterweed (*Elodea canadensis*). **A**, habit of plant. **B**, male flower. **C**, female flower. B, after H. St John (1965).

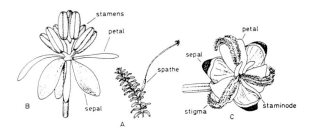

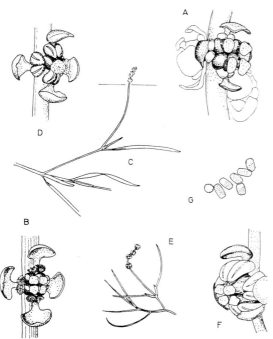

Fig. 9.28 Pondweeds (*Potamogeton*). **A**, single flower of *P. obtusifolius*. **B**, single flower of *P. berchtoldii*. **C**, habit of *P.trichoides*. **D**, single flower of *P. trichoides*. **E**, habit of *P. pectinatus*. **F**, single flower of *P. pectinatus*. **G**, pollen of *P. pectinatus*.

ers often break free). The male flowers open suddenly, the water-repellent perianth segments reflexing against the surface film and holding erect the nine stamens. These dehisce explosively, scattering the large pollen grains over the surrounding water. The rather densely-set spines of the exine hold back the surface film and prevent the pollen grains from being wetted, so they can be freely moved over the water surface by wind and other disturbances. The female flower lies in a shallow depression in the surface film as in *Vallisneria*, usually resting on two of the three water-repellent stigmas which project beyond the perianth (Fig. 9.27). Male plants are exceedingly rare in Britain, and are apparently uncommon within the species' native range in North America.

There are a number of other examples of surface hydrophily in the flora of northwest Europe. In some species of *Potamogeton* (pondweeds) (Figs. 9.25b, 9.28) the protogynous flowers are wind-pollinated. However, in *P. filiformis* and *P. pectinatus* the lax interrupted flower-spikes float at the water surface (Fig 9.29), and the pollen grains are carried on the surface film to the stigmas. According to the late J.E. Dandy, for many years a taxonomic authority on this genus, *P. pectinatus* when growing in deep water may be pollinated and set fruit without the spikes reaching the surface; Philbrick & Anderson (1987) suggest that in submerged *Potamogeton* species, bubbles formed as the anthers dehisce provide an air-water interface bringing pollen to the stigma of the same flower. Probably all gradations between regular anemophily and regular surface or submerged hydrophily are to be found in the genus, with at one extreme the dense spikes of *P. natans*, standing stiffly above the floating leaves, and at the other the lax

Fig. 9.29 Inflorescences of *Potamogeton pectinatus* floating at the surface in a slow-flowing river, Dorset. In Switzerland, Guo & Cook (1989) found an average seed set of 4% (2.5–7.5%) in permanently submerged inflorescences, which increased to an average of 23% (6.5–40%) in inflorescences that could reach the surface.

floating inflorescences of *P. pectinatus*. Daumann (1963) showed that the pollen of *P. natans* rapidly loses its power of germination on contact with water; viability fell to only 10% after four hours' wetting. On the other hand, the pollen of *P. lucens* – a large, submerged species on which only the flower-spikes appear above the surface – remained 45% viable after a day in water. Daumann noticed quantities of pollen floating in the neighbourhood of inflorescences, and showed in aquarium experiments that pollen grains could in this way reach the stigmas and germinate. The small grass-leaved species such as *P. pusillus* and *P. berchtoldii* have small few-flowered spikes, which project only a little above the surface, and it is likely that they too are usually pollinated in the surface film. All the possibilities are realised in the opposite-leaved pond-weed, *Groenlandia densa*, which may be pollinated by wind, in the surface film, or on either the water or the air side of the surface film in bubbles under water (Guo & Cook, 1990).

In the species of *Ruppia* (tassel pondweeds), which grow in brackish pools and ditches, the pollen is similarly liberated at or just above the water surface, upon which it floats to the flat, shield-shaped stigmas (Fig. 9.30, 9.31). According to Gamerro (1968), who has given a detailed account of the pollination of *R. cirrhosa*, the pollen is

Fig. 9.30 Tassel pondweed (*Ruppia maritima*); inflorescence shedding pollen just above the water surface.

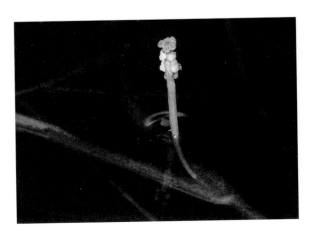

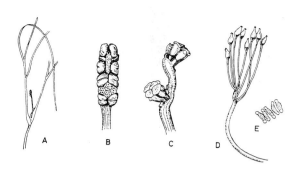

Fig. 9.31 Tassel pondweed (*Ruppia maritima*). **A**, habit of plant. **B**, inflorescence at time of dehiscence of anthers (compare Fig. 9.33). **C**, inflorescence after pollination and shedding of stamens. **D**, shoot with ripe fruits developed from two flowers. **E**, pollen grains.

released into bubbles of gas expelled from the dehiscing anthers, spreading over the water surface as the bubbles burst. The shortly sausage-shaped pollen grains tend to line up side by side, forming irregular snowflake-like patches a few millimetres across in the surface film (Fig. 9.32). Patches a centimetre or so across are more effective 'search vehicles' and have a better chance of striking a stigma and bringing about pollination than smaller groups (Cox & Knox, 1988, 1989), but whether this increased probability outweighs the larger number of pollen grains that go to make large 'search vehicles' is an open question. *R. cirrhosa* flowers are protandrous. After dehiscence the anthers are shed, and the inflorescence remains floating for a time, exposing the stigmas in the surface film where they can be pollinated, before being withdrawn beneath the surface by the coiling of the peduncle, as in *Vallisneria*. The other European species, *R. maritima*, is pollinated in the same way, but there is no spiral coiling of the peduncle after flowering. The weight of evidence (including our own observations) is that, despite some earlier statements to the contrary, pollination of *Ruppia* always takes place in the surface film. Little or no pollination occurs if the water level is high enough to keep the stigmas submerged.

Submerged hydrophily is found in only a few genera. The hornworts (*Ceratophyllum*)

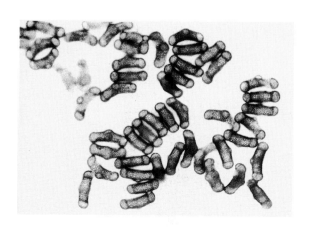

Fig. 9.32 Pollen of *Ruppia cirrhosa* forming raft in surface-film.

Fig. 9.33 Hornwort
(*Ceratophylum demersum*).
A, immature male flower.
B, mature stamen releasing
pollen grains. **C**, female flower.

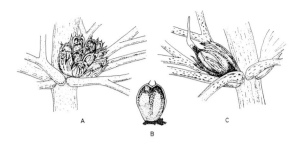

belong to an ancient and isolated angiosperm lineage, and have probably always been water pollinated; a fossil *Ceratophyllum*-like plant is known from Australian Cretaceous deposits pre-dating the main diversification of flowering plants (Dilcher, 1995). The flowers are borne in the axils of the finely-divided leaves (Fig. 9.33). The female flower consists of a one-celled ovary, containing a single ovule, with a slender, oblique style, surrounded by a small cup-shaped perianth divided into 10-15 lobes. In the male flower, a similar perianth surrounds 10-20 stamens. Each consists of a large anther with an expansion of the connective at the tip which acts as a float. As they mature, the stamens break away and float to the surface where they dehisce, releasing the pollen grains which sink slowly through the water as they are wafted among the submerged stems. Like many other water plants, *Ceratophyllum* spreads mainly by vegetative growth, and most seed is probably self-pollinated. In a study of isoenzyme distribution in *Ceratophyllum* populations, Les (1991) found *C. demersum* 'structured genetically like inbreeding terrestrial plants.'

Various pollination systems have been attributed to the water starworts (*Callitriche*), including hydrophily. These small amphibious and aquatic plants have minute, unisexual flowers borne singly or in pairs in the leaf axils (Fig. 9.34); the pollination of the American species has been examined critically by Philbrick & Anderson (1992). The species with flowers borne on the aerial parts of the shoots (e.g. *C. stagnalis*) are probably almost always selfed by pollen from stamens of adjacent flowers. In the American amphibious species *C. heterophylla*, *C. trochlearis* and *C. verna*, which produce flowers both above and below water, seed-set results from selfing brought about either by normal transfer of pollen, or by growth of pollen tubes through the vegetative tissues of the plant to reach the

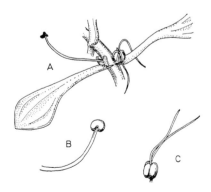

Fig. 9.34 Water starwort (*Callitriche obtusangula*). **A**, part of flowering shoot, showing a male and a female flower. **B**, undehisced stamen. **C**, ovary and stigmas.

ovules (Philbrick & Bernardello, 1992). In the wholly submerged *C. hermaphroditica* (which like most of the more terrestrial species) fruits freely, pollination takes place by growth of pollen tubes through the water from the anthers to adjacent styles, rather than by dispersal of the pollen itself. Usually the result will be pollination of another flower on the same plant, but the occurrence of occasional outcrossing has been demonstrated using DNA (RAPD) genetic markers (Philbrick, 1993).

The genera which remain to be considered all belong to the monocotyledons. *Zannichellia palustris* (horned pondweed) (Fig. 9.35) looks at first sight very like *Ruppia* or one of the narrow-leaved *Potamogeton* species. The flowers are borne in small axillary clusters, each cluster usually comprising one male flower, and a few female flowers surrounded at the base by a cup-shaped spathe. The male flower consists of a single stamen, with a rather long filament raising it well above the female flowers. The female flowers each consist of a single carpel with one ovule and a more-or-less funnel-shaped stigma. The pollen grains are somewhat sticky, and are initially released into the water in small

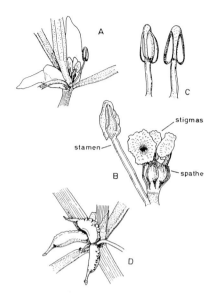

Fig. 9.35 Horned pondweed (*Zannichellia palustris*). **A**, node showing a male flower and a group of female flowers. **B**, enlarged view of male and female flowers. **C**, two views of an undehisced stamen. **D**, node with ripe fruits.

stringy or flaky masses. Within minutes these break up and set free the individual grains, but the initial dispersal in groups may increase the chance of collision with a stigma. On reaching a stigma, the pollen grains germinate and produce pollen tubes; occasional prematurely-germinated grains may drift freely, but these do not seem to be important for pollination. Although underwater pollination appears a chancy process, surprisingly good fruit set has been recorded in this species, ranging from 56-91% (Guo *et al.*, 1990). Most seed is selfed, but some outcrossing occurs. The north-west European species of *Najas* are rare and local; *N. flexilis* is scattered in clear base-poor lakes from Kerry, the English Lake District and the Hebrides, across northern Europe to Finland and Russia, and *N. marina* (one of the few dioecious members of the genus) occurs in clear, base-rich, fresh or slightly brackish waters northwards to the Norfolk Broads and the Baltic. They are probably pollinated in a similar way to *Zannichellia*. The male flower consists of a single stamen and a small perianth enclosed in a short tubular spathe; the female flower consists of a single carpel with one ovule, and three long stigmas at the apex. The pollen grains are rich in starch and have a greatly reduced exine; they have often begun to germinate before they escape from the envelopes of the male flower, and the developing pollen tubes may increase their chance of being caught by the stigmas.

Among the most specialised of hydrophilous species are the various genera of marine seagrasses that occur in different parts of the world. In the dioecious *Halodule*

Fig. 9.36 Eel-grass, *Zostera angustifolia*, on an intertidal mud-flat, with flowering shoots.

Fig. 9.37 Eelgrass (*Zostera angustifolia*). **A**, habit of plant, × 0.3. **B**, sheath of flowering shoot, showing projecting stigmas, × 2. **C**, part of flowering spike removed from sheath × 5. **D**, part of older spike; the ovaries have begun to swell and the stigmas are withering and falling, the lowest anther is just dehiscing, two dehisced anthers project beyond the translucent margins of the sheath, a fourth is shown broken from its fragile attachment, × 5.Compare Fig. 9.36.

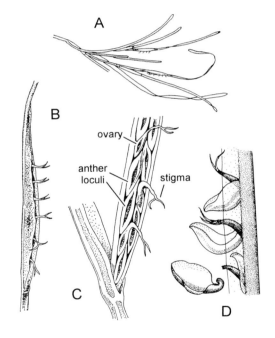

pinifolia (Cymodoceaceae) studied in Fiji by Cox & Knox (1989), pollination takes place very much as in *Ruppia*, but the pollen is elongated into spaghetti-like threads which, on release from the anther, assemble themselves into spidery rafts up to a few mm across in the surface film. The female plants produce remarkably long (22 mm) straight stigmas, which float in the water surface, providing long linear targets for the drifting pollen (Cox, 1988). Three species of estuarine and marine eelgrasses (*Zostera*) occur commonly on west-European coasts. All three produce shoots bearing linear grass-like leaves from rhizomes rooted in sand or mud. *Z. noltii* and *Z. angustifolia* (Fig.

9.36) grow between the tidemarks on estuarine mud-flats and so are exposed for
much of the time; *Z. marina* occupies a
zone several metres in depth from about
low water of spring tides downwards in
sandy bays and estuaries, and all but the
upper fringes of the *Z. marina* beds are
constantly submerged. A vivid account of
the pollination of *Z. marina* is given by
Clavaud (1878). Male and female flowers
alternate on a small flattened axis more-
or-less enclosed in a leaf sheath (Fig.
9.37). The female flowers consist of an
ovary, containing a solitary ovule, with
two long stigmas at the apex; the male
flowers, which mature 1-3 days later, consist of a single anther. As in *Halodule*, the
pollen 'grains' are spaghetti-like, about
0.25 mm long, and of the same density as
sea water; the exine is extremely reduced.
They are released in cloudy masses,
which are drifted through the eelgrass
beds by the tides. If a pollen grain comes
into contact with a narrow object, it rapidly becomes curled around it; in this way
the pollen is securely anchored to any
stigmas it may reach. The intertidal species of *Zostera* are structurally similar to *Z.*

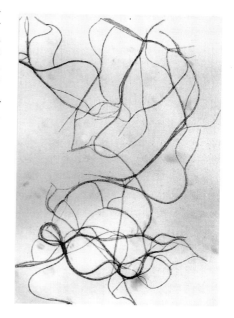

Fig. 9.38 Eel-grass (*Zostera angustifolia*). Pollen
floating in the surface-film.

marina, but smaller. The ripe anthers are full of air, and the newly released filamentous
pollen is water-repellent, readily forming spider's-web-like rafts in the surface film
(Fig. 9.38). These intertidal eelgrasses are probably generally pollinated in the surface
film as the tides ebb and flow over their sheltered mud-flat habitat (the north-Pacific
genus *Phyllospadix*, also in the Zosteraceae, is pollinated in the same way [Cox, 1993;
Cox, Tomlinson & Nieznanski, 1992]). *Z. marina* too may be pollinated in the surface
film at the upper end of its range on the shore, but, if submerged, the pollen is neu-
trally buoyant and pollination can also take place under water (De Cock, 1980; Cox,
Laushman & Ruckelshaus, 1992). Submerged hydrophily is characteristic of other
permanently submerged seagrasses such as *Posidonia*, widespread in the Mediterranean, which has elongated pollen like *Zostera*, and the 'Turtle-grass' *Thalassia testudinum*
of the Gulf of Mexico and the Caribbean, in which the round pollen grains are arranged like pearls on a necklace in long mucilaginous strands (Cox & Tomlinson,
1988).

10

Deception and Diptera: 'Sapromyiophily'

We have already seen in Chapter 7 how, among orchids, there are many unconventional ways of securing pollination, including various kinds of deception. Here we look at further cases of deception in the cause of pollination, including more examples from the Orchidaceae.

Pollen production partly disguised

Deceit creeps in by small steps. A mild form of it occurs in some pollen flowers (not providing nectar) that have two types of stamen, one of which provides food while the other dusts visiting insects with pollen. This is a general feature of the larger-flowered members of the family Melastomaceae (Fig. 10.1). Here the flowers are usually purple or pink and the pollination anthers are of a similar colour, while the food-anthers are yellow. The pollination anthers are carried on jointed filaments on the lower half of the sideways-facing flower, serving as a support for an insect collecting pollen from the conspicuous food-stamens while being rather inaccessible themselves. Pollen is released by vibration (buzz-pollination, see below and pp.125, 179-80 and 167). The flowers are sacrificing some of their pollen to the bees in exchange for pollination. This happens in many plants but here the allocation of pollen to bee-food is clearly defined. Similarly, in many species of the spiderwort family (Commelinaceae) some of the stamens bear modified anthers of conspicuous colour; in some species, these produce no pollen or only a minute quantity. Differentiation of stamens also occurs in *Cassia* (family Fabaceae) in which the arrangement of floral parts is similar to that of Melastomaceae. Here the pollen is dry and is shed in clouds from pores at the tips of the pollination anthers when these are vibrated by an insect, or as they spring up when the departing visitor takes its weight off them (van der Pijl, 1954). Species of *Lecythis* and *Couroupita* in the brazil-nut family, Lecythidaceae, which have hundreds of stamens in the flower, also have two types, one for pollination and one for the pollinators to forage at. The pollination stamens are in the part of the flower where stamens are usually to be found,

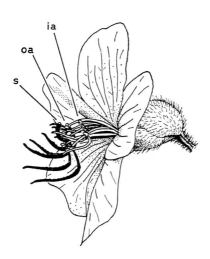

Fig. 10.1 Flower of *Tibouchina* (family Melastomaceae). The large purple pollination anthers (**oa**) and the style (**s**) are shown in black, the smaller yellow food anthers (**ia**) in white. Pointers to anthers indicate where they join their filaments.

while the feeding stamens are attached to a lateral appendage which over-arches the flower. The visitor thus has stamens beneath it and stamens above it and its back rubs the pollination stamens. In one species it is known that the feeding stamens produce larger pollen grains than the pollinating stamens (Mori, Prance & Bolten, 1978).

In plants such as these, one can argue that there is an element of deception because the plant is making some of its anthers attractive and others inconspicuous, as well as difficult to manipulate. The insect is not supposed to find all the pollen that is there. The less pollen there is in the food-stamens, the deeper the deception. Even when all the stamens are the same, some trickery may be practised. Several genera of various families have bushy hairs on the stamen filaments which may give the impression of pollen-richness and deflect the insects' efforts to collect pollen to the wrong place (for example *Narthecium* and *Verbascum*) (Vogel, 1978a). In other plants, rather small anthers are borne on enlarged coloured filaments that look like anthers; alternatively the connective (the sterile piece that lies between the pollen-containing parts of the anther) can be blown up into a dummy anther. When pollen is dry and released from holes at the tips of the anthers ('poricidal') by the buzzing activity of bees (p.125), the anther takes the form of a rigid bottle that does not collapse; this can retain its bright yellow colouring and so continue to lure insects long after it is empty. In the dioecious Begoniaceae and Cucurbitaceae, there are bulky bright yellow stigmas looking similar to the anthers of the male flowers (Ågren & Schemske, 1991). The female flowers in these and several other families are rewardless and receive comparatively few pollinator-visits. They rely for pollination on 'mistakes' by animals that have already visited male flowers (Baker, 1976; Little, 1983; see also p.340).

Brood-site imitation

The relationship between the plant species and the insect, in the cases just described, is usually not totally one-sided. On the other hand, as we have seen earlier (Chapter 7), some orchids, particularly those that are flower-mimics – either generalised or specialised – do create a totally one-sided relationship with their pollinators. But deception does not necessarily involve the false promise of food. Deceitful attraction of insects that are seeking an appropriate place to lay their eggs is an option taken up by a whole range of plants. The victims are mostly Diptera and Coleoptera. (Some plants actually do provide tissues in which their pollinators breed: these are described in Chapter 11.)

The special features involved in this kind of pollination recur again and again in unrelated plants; they constitute the syndrome of 'sapromyiophily' – pollination by insects associated with decaying organic matter. Along with these, we shall now describe also pollination by insects that breed in living fungi.

The main families of plants practising deceit by imitating the 'brood-place' are Aristolochiaceae, Asclepiadaceae, Araceae and Orchidaceae. There are two levels of adaptation. In plants on the first level, insects are lured to the flower but are not detained or, if they are, they are released after a short interval. All four families have representatives of this type. Flowers on the second level are more complex; insects are imprisoned for a considerable time, usually 24 hours or more, and the prisons can hold not just one or a few insects, but many. The orchids are not represented on this level; they are dealt with after the other families.

Brood-site imitation without prolonged imprisonment

An example on the first level is the genus *Asarum* in the birthwort family, Aristolochiaceae (Vogel, 1978b, c). In *A. caudatum*, from North America, the perianth forms an open cup with three brownish flesh-coloured lobes that are drawn out into long points. Tail-like points and dingy purple to brown perianths are features of the syndrome; often the tails are the source of scent and the colour suggests carrion, although here there is no evidence for imitation of flesh. In fact the pollinators are fungus-gnats (Diptera, family Mycetophilidae). Male and female flies come to the flowers and sometimes mate there. In the floral cup below each perianth-lobe are two translucent whitish patches, edged with dark red. Here the female flies lay eggs. These may hatch but the resulting larvae never get beyond the first instar. While the female is laying the eggs, her back touches the plant's sexual organs in the centre of the flower and causes pollination. There is no scent perceptible to humans but there is evidence that scent attracts the flies. A special feature of the pale patches in the flower is their dampness, resulting from a much higher rate of transpiration than from the neighbouring surfaces. When laying eggs on a mushroom-type fungus, the fly pushes between the gills which are also very damp and may be translucent towards the stalk. Thus the flower seems to imitate the smell, dampness and illumination of the fungus. The number of flies in the habitat builds up in autumn when fungi are abundant. In spring, when *Asarum caudatum* flowers, there may be plenty of flies but few fungi, and this is the plant's opportunity to get pollinated by deceiving the flies. It has a creeping habit, grows in shade and bears its flowers underneath the leaves.

This system works efficiently even when there is no exact imitation of the form of a fungus. However, in spite of this some species of *Asarum* have longitudinal folds or rectangular chambers formed by a grid of ridges inside the flower, thus giving a closer resemblance to the 'damp crypts' between the gills of mushrooms (Vogel, 1978b, c).

The milkweed family, Asclepiadaceae (see Chapter 6), has a very elaborate arrangement for attaching pollinia to the hairs, feet or mouth-parts of insects that come for nectar. In such flowers there is a strong tendency to attract Diptera, which has perhaps led to the evolution of the many deceit flowers in the family. In *Stapelia* and allied genera, which are cactus-like plants of Africa and southern Asia, the flowers look and smell like bad meat or carcasses, characteristically being flesh-coloured or dark purplish red, often covered with hairs, and often of large size – up to 40 cm across (Plate 5c). The largest genera of this type, *Stapelia*, *Caralluma* and *Huernia*, comprise about 250 species, among which there is extensive variation in colour, patterning, type and distribution of hairs, and surface sculpture of the corolla (White & Sloane, 1937). Much of this variation suggests adaptation to pollinators with specific requirements, and there would seem to be a possibility of interesting research into the instinctive requirements and sensory discrimination of the flies that pollinate these strange flowers. When cultivated in Britain the flowers attract muscid and calliphorid flies, which lay eggs in them. These are easily seen, usually near the centre of the flower where presumably the pollinia clip on to the insects.

In the *Arum* family (Araceae) there are again non-trapping deceit flowers that attract fungus-gnats. In the Mediterranean genus *Arisarum* there is one species that does not closely imitate a fungus, and one that clearly imitates one in a most perfect way. The first is *A. vulgare*, which has a striped hood and a club-shaped spadix appendage (see later for details of the inflorescence of Araceae); the second is *A. proboscideum*, the mouse plant, in which the spadix appendage is the fungus-mimic (Plate 5d). Vogel (1978b, c) found that here, too, the chambered surface of the 'fungus-cap' and the neighbouring internal surface of the spathe were moist. Below the moist zone, the surface carries a powdery wax on which the flies cannot walk. The base of the spathe is brightly lit, so flies apparently walk downwards and then fall. The bright light may delay their departure somewhat, but flies are rarely found inside the chamber so it seems they are not effectively imprisoned. The flies' eggs are found on the spadix appendage of *A. proboscideum*. This plant also has a tail (hence 'mouse-plant'), here developed from the tip of the spathe. The flies found by Vogel were Mycetophilidae and, in *A. vulgare*, also Sciaridae. Strangely, it has been found that *A. vulgare* attracts hardly any pollinators and reproduces mainly by small tubers (Kroach & Galil, 1986).

Brood-site imitation with long-term imprisonment

We can now move on to look at members of these same families that are on the second level of adaptation, imprisoning their pollinators for a time. In this group, most flowers of Aristolochiaceae and inflorescences of Araceae are protogynous and depend for pollination on insects arriving with pollen; later they shed their own pollen all over the insects and then release them. Protogyny is not needed by the Asclepiadaceae, with their precise pollination arrangements.

In the large genus *Aristolochia* (a more advanced member of its family than *Asarum*) the plants trap insects in their specially modified tubular perianths. In the European *A. clematitis* (birthwort) the perianth is greenish-yellow and 3.5 cm long, including the tongue-like lobe at one side of the mouth (Fig. 10.2). Biting-midges and other small flies are the pollinators, being attracted by the smell of the flower and alighting on the tongue. If they enter the tube they fall because the cells that line it form downwardly-directed conical lubricated papillae; pinching a papilla with the claws forces the foot

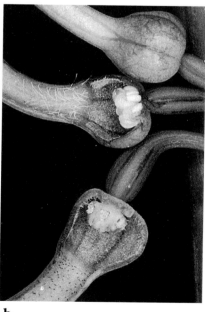

a   **b**

Fig. 10.2 Aristolochia clematitis (birthwort, Aristolochiaceae); a straight-tubed species. **a**, erect stem with flower clusters at the nodes; **b**, part of a flower cluster, with two flowers dissected. The upper flower is in the receptive female stage, with the anthers still undehisced; note the stiff downward-pointing hairs in the tube. The lower flower is in the male stage (or later), with the anthers dehisced and the hairs in the tube shrivelled and brown.

Fig.10.3 Flower of *Aristolochia lindneri*, another straight-tubed species. **A**, flower in first stage of anthesis with half the perianth removed; **B**, flower seen from above (tail-like lobe of perianth foreshortened in this view; as the flowers are borne near the ground the other lobes suggest a faecal deposit); **C**, a multicellular trap hair seen from above; **D**, the same from the side, with adjoining papillate cells of trap wall. Shading represents purple colouring; the prison is bright, but an aggregation of dots sets off the 'window pane'. After Cammerloher (1933) and Lindner (1928).

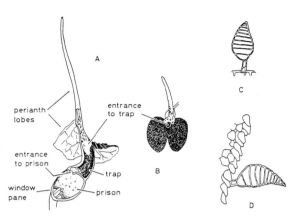

off the tip. Once the insects have fallen, their escape is prevented by long hairs that can bend downwards but not upwards. A clear translucent ring round the reproductive organs falsely suggests a way of escape and ensures that the flies effect pollination. When the prisoners are due for release, the tube becomes horizontal or drooping and the hairs and papillae shrivel (Knoll, 1956; PFY). More complex structures are found in two South American species of this genus, *A. lindneri* and *A. grandiflora* (Figs. 10.3, 10.4). In both species there are conspicuous perianth lobes, a long, tail-like append-age, a dark antechamber or trap, a brighter prison with which the trap connects by a

Fig. 10.4 Flower of *Aristolochia grandiflora*, with U-shaped tube. **A**, flower in first stage of anthesis with half the perianth cut away; **B**, part of a trap hair from above (the swollen base limits movement); **C**, the same, from the side (the swollen base prevents bending upwards but not downwards). The trap is blackish purple inside and the prison lighter purple with darker freckles; as the hairs in the right-hand arm of the trap are upside down they help flies to climb into the prison. This species is a liana. After Cammerloher (1923).

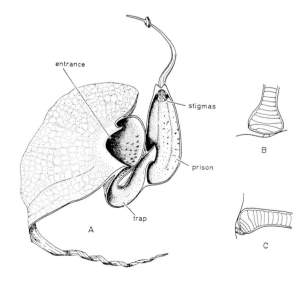

funnel-shaped passage, and a 'window-pane' encircling the reproductive organs (better seen in the photograph of *A. sipho*, Fig. 10.5). They too have internal downwardly pointing lubricated papillae, together with larger hairs. On the day the flowers open, a foul smell is produced by the perianth lobes, and flies are attracted to them and trapped in the prison where nectar is secreted. On the second day no smell is produced, and the stigmas bend together so that they cannot receive pollen. The anthers then open and the prisoners, newly dusted with pollen, are allowed to escape by the widening of the entrance and the shrivelling of the trap hairs. In *A. lindneri* the purple colour of the trap, which is confined to its inner surface, disappears on the second day, brightening this part of the flower and encouraging the insects to emerge. In the study of this plant by Cammerloher (1933) the commonest visitors were flies of the family Sepsidae. The perianth lobe of *A. grandiflora* is about 12 cm wide and 20 cm long. The trap is U-shaped, and the hairs in it are all directed away from the entrance; as in *A. clematitis* these can bend inwards but not towards the entrance, so that the flies slip down the first part of the tube but are helped by them to climb up the second part into the brighter prison. The fly most commonly caught by *A. grandiflora* was a muscid 4-5 mm long that laid eggs in the prison (Cammerloher, 1923). Two more *Aristochia* species with U-shaped tubes are shown in Plate 5.

More recently some work has been done on *Aristolochia* in natural habitats in Brazil (Brantjes, 1980). On the whole it was found that different species of *Aristolochia* trapped flies of different genera or families, though occasionally with a small overlap. When the plants do share the same pollinators, they grow in different habitats. The size of the flower is not related to the size of the insect – the largest species of all, *A. cordifolia*, attracts the smallest flies. It is the width of the space in the prison around the sexual organs of the flower that is decisive: insects that do not fill the space do not pollinate the flower. The flowers tended to attract all flies but each species actually trapped only a limited range. Various factors were involved in this selection and some of the mechanical ones were not understood. All the *Aristolochia* species studied had unpleasant scents, but even these had a selective effect; for example, only one *Aristolochia* out of three trapped dung-frequenting Sepsidae. Although the various *Aristolochia* species trapped a range of different flies, the overall numbers of the sexes were similar and few eggs were laid in the flowers. Nevertheless, there must have been imitation of some feature of the brood-place; presumably male flies congregate on the materials where females are likely to come to lay eggs and where, in any case, both sexes may feed. Approximate equality of sexes has also been found among

Fig. 10.5 Sectioned flower of *Aristolochia sipho* (dutchman's pipe, Aristolochiaceae), back-lit to show 'window pane' round sexual organs; in the female stage with the cushion-like receptive stigmas shielding the anthers. This species is a liana.

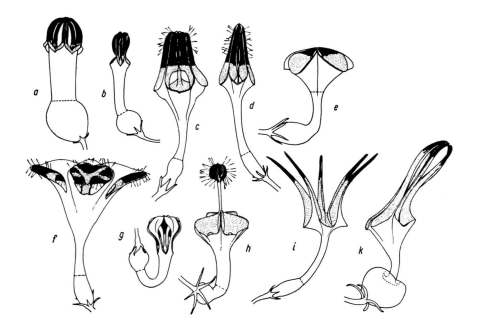

Fig. 10.6 Variation in the flowers of *Ceropegia* (Asclepiadaceae). The scent-producing areas are shown black, the slide-zones (as far as visible) stippled; shimmering hairs shown when present. Broken lines show lower limits of slide zones. **a**, *C. ampliata*; **b**, *C. woodii*, **c**, *C. sandersonii* × *C. nilotoca*; **d**, *C. radicans*; **e**, *C. elegans*; **f**, *C. sandersonii*; **g**, *C. euracme*; **h**, *C. haygarthii*; **i**, *C. stapeliiformis*; **k**, *C. robynsiana*. Not to scale; some slightly larger than life. From Vogel (1961).

the visitors to two species of *Arum* (see below).

Aristolochia species showing fungus-mimicry also occur. In the more-or-less trumpet-shaped entrance to the flower of *A. arborea*, there is what looks like a mushroom, complete with stalk. Although this plant is a shrub or small tree, its flowers are borne near the ground, breaking out of the woody stems. The underneath of the cap of the 'mushroom' is lamellate on the inner side where it stands over the entrance to the prison. The surfaces are very smooth and it is believed that the flies slip and fall. Once in the prison, they can ascend to a lighted area round the sexual organs of the flower (Vogel, 1978b, c).

In Asclepiadaceae imprisonment is practised by a single large genus, *Ceropegia*, occurring in Africa, Asia and Australia and comprising herbs, shrubs and climbers, often succulent. The lengthened corolla-tube in these plants forms a trap, similar to that of *Aristolochia clematitis*, in which flies are imprisoned for a time. The flowers are usually small compared with trap-flowers in other groups (about 10 cm or less long) and are often fascinatingly beautiful, the corolla most often being coloured in delicate shades of green, grey and brown, with elegantly shaped tubes and erect lobes which unite at their tips to give a lantern-like effect (Fig. 10.6). This lantern effect is found in some unrelated fly-trap flowers and appears to be a means of inducing insects to enter (Vo-

Fig. 10.7 Epidermal features of *Ceropegia* (Asclepiadaceae). **A**, hooked papillate cells inside entrance to corolla tube of *C. woodii*; **B**, blunt papillae from within upper part of tube of same; **C**, larger, sharper papillae from lower part of tube of same, in section; **D**, trap hair of *C. stapeliformis*, seen from above; **E**, section through papillate epidermal cells within tube of same, with base of a trap hair that is made from a single cell. A–C after L. Müller; D, E after Vogel (1961).

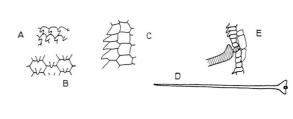

gel, 1954). A study of the tiny flowers of *Ceropegia woodii* (Müller, 1926), published in the same year as Knoll's investigation of *Arum* (see later), shows some remarkable parallels between these two unrelated plants, as well as with *Aristolochia*. As in these other plants, the interior of the flower tube is covered with lubricated papillae (Fig. 10.7), and as in *Aristolochia* it has hairs pointing away from the entrance. A day or two after the flower has opened, the tube becomes horizontal and the hairs inside shrivel so that the insects can escape. In cultivation in Vienna, this species trapped biting-midges (*Ceratopogon*), which were apparently attracted by a faint scent produced by the flower. Several species of *Ceropegia* were studied by Vogel (1961). In each he located the scent-producing area and mapped the area covered by the lubricated papillae (the slide-zone) (Fig. 10.6). He found that in cultivation in Germany five out of eight species attracted none of the available insects, while each of the others trapped female flies. *C. woodii* trapped biting-midges of the genus *Forcipomyia*, *C. stapeliiformis* trapped mainly *Madiza glabra* of the family Milichiidae, and a hybrid of *C. nilotica* trapped representatives of two other genera of this family. Thus the scents produced by the flowers are specific attractants to certain insects, and are presumably connected with egg-laying as in the open-flowered members of the family.

The flies approach with a typical scent-orientated flight and always alight on the scent-producing area; then they investigate the slide-zone, apparently being attracted by the dark interior of the flower, and soon slip into the tube, which is darkened by red colouring on the inner surface in some species. They then pass into the prison, which is usually partly darkened like the tube. The dark part frequently does not show on the outside (as in *Aristolochia lindneri*, p.299, and *A. fimbriata*, Fig 10.14, p.309). Again, as in fly-trapping flowers of other families, there is often a bright 'window pane', forming a ring round the sexual organs. The imprisoned insects, reaching the light end of the chamber, climb the pillar-like inner corona; here they drink from the nectarial cups formed by the outer corona, there being one opposite each groove on the column (Fig. 10.8). After drinking, the fly withdraws its head, and the throat membrane (in Milichiidae) or the base of the labellum (in *Forcipomyia*), catches in the groove, and receives the clip carrying the pollinia. If the insect already carries pollinia, one of these is caught lower down the groove and pulled off, coming to rest on the stigma. Thus the stimuli which the flower presents to the insects (and the needs to which their responses

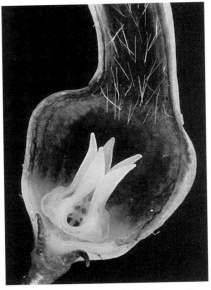

Fig. 10.8 Base of flower of *Ceropegia woodii* (Asclepiadaceae) cut open to show trap hairs, corona with upstanding lobes, a stamen visible between corona-lobes, and the 'window-pane' in the base of the prison.

are related) are successively: smell (egg-laying); dark cavity (egg-laying); bright light (escape from captivity); taste (nourishment). The duration of imprisonment varies from less than a day to four days, according to the species of *Ceropegia*. Arrangements for release are as in *Aristolochia*.

In some species, the hairs inside the tube are specially constructed with a narrow stalk and a wide asymmetric swelling just above (Fig. 10.7). These will thus bend downwards but not upwards or sideways. Each consists of a single cell, although it may be up to 5 mm long. They function in exactly the same way as the similarly-shaped but multicellular hairs of *Aristolochia* (Figs. 10.3, 10.4), presenting a striking case of evolutionary convergence.

Such remarkable fly-trapping plants occur in the European flora, where much the most widespread genus is *Arum*, which brings us back to the family Araceae. The curious inflorescence (Fig. 10.9) is borne on a stout stalk; its two main organs are the spadix, which is the axis of the inflorescence, and the leafy spathe. The female flowers form a zone at the base of the spadix and consist merely of ovaries topped by stigmas. Above them is a zone of male flowers consisting only of short-stalked stamens packed together. There are two groups of bristle-like appendages, considered to be sterile flowers. The club-shaped terminal part of the spadix forms a sterile appendix. Pollination is brought about by small insects that become trapped in the pollination chamber during its first, female, stage of development. In the subsequent male stage they are dusted with pollen, after which they are released.

The mechanisms involved in this process were studied by Fritz Knoll (1926), who worked mainly on the Mediterranean species *Arum nigrum*. The spathe opens overnight and during the following day the spadix produces a strong faecal smell. Insects, mostly dung-frequenting flies or beetles, are attracted in the morning. If they alight on the club of the spadix or inside the spathe they lose their grip and fall, because the surface is papillate, as described for *Aristolochia* and *Ceropegia*. Within the spathe, this type of surface is found from the top down to the upper part of the chamber. As they drop, the insects encounter the ring of bristles and if small enough they fall through into the chamber; large insects are arrested and can fly off. If the insects that are trapped have come from another *Arum* inflorescence, they may pollinate the female flowers (which have receptive stigmas during the first day), probably by climbing upon them in their attempts to escape. The pollen tubes grow quickly into the ovaries and the stigmas then wither, so that by the time the inflorescence sheds its own pollen, self-pollination is impossible. The pollen is shed in great quantity and thoroughly dusts the trapped insects. By the morning of the second day the surfaces of the bristles

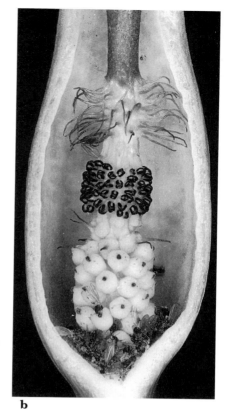

a   **b**

Fig. 10.9 Lords and ladies (*Arum maculatum*, Araceae). **a**, inflorescence at flowering time; **b**, inflorescence cut open on the second day of flowering, with darkened stigmas, dehisced anthers and numerous trapped owl-midges, *Psychoda*, sprinkled with pollen.

have become wrinkled, the papillae on the rest of the spadix have shrunk and scent-production has ceased. The change to the surfaces allows the insects to escape, and if they are then trapped by an inflorescence in its first day they can cause cross-pollination.

Knoll carried out experiments using imitation spathes of coloured glass. The inner surfaces of these were dusted with talcum powder, which made it impossible for insects to cling to them. The models, when provided with real *Arum nigrum* spadices, caught just the same kinds of insects as the real plants, though fewer of them, apparently because the smell of the detached spadices was weaker. These models also demonstrated the principle of capture by falling: it had earlier been thought that the insects entered the chamber voluntarily, to seek shelter and warmth. The models were also used to show that the attraction of the pollinating insects from a distance is purely by scent. Light coloured and dark coloured spathes gave identical results, and Knoll concluded that the colouring of very dark spathes (as in *Arum nigrum*) or very light ones (as in *A. italicum*) was only significant in so far as it made the spathes stand out from their surroundings and so induced insects to alight.

A feature of the spadix of *Arum*, which has long been known, is that it generates heat. This led to the theory that it was the warmth that attracted insects to enter the chamber. Knoll performed experiments with models having an artificial spadix which was electrically heated; this showed that the heat was no attraction. The rapid respiration which gives rise to the heating uses several grams of starch in the course of a few hours, which is out of all proportion to the metabolic effort required to generate the few milligrams of malodorous compounds (ammonia, amines, amino-acids, skatole and indole) that are produced (Fig. 10.13). Probably the main function of the heating is to help vaporise these compounds and intensify their dissemination (Meeuse, 1966). The smell itself is a purely deceptive attraction; the insects receive no food from the *Arum*, apart possibly from drops of a sweet secretion from the withered stigmas.

Arum maculatum (cuckoo pint, or lords and ladies), which occurs in Britain, has essentially the same mechanism as *A. nigrum*, but the heating of the spadix and scent production are at their height during the afternoon and evening. The pollinators are small flies of the genus *Psychoda* (family Psychodidae), of which large numbers may be found in the chamber (Fig. 10.9B) (Grensted, 1947; Knuth, 1906-1909; Müller, 1883; Prime, 1960). However, a single insect that has visited another compatible inflorescence of this self-incompatible plant may carry enough pollen to fertilise all the ovaries in one spathe (Lack & Diaz, 1991). Even so, in England fruit-setting may be limited by shortage of pollinators. Lack & Diaz also found that the stigmatic exudate of *A. maculatum* had a sugar concentration (sucrose-equivalent) only slightly higher than that of the exudate from cut stems, and that the flies did not drink it; indeed, it is reported that adult *Psychoda* do not feed.

Arum conophalloides from south-west Asia has a scent which attracts blood-sucking midges of the families Ceratopogonidae and Simuliidae. One spathe of this species which Knoll examined contained 600 Diptera, of which 461 were identified; these were females of three species, one of which was represented by 427 insects. Evidently the attraction of these plants is both effective and highly specific. Counts of insects trapped by *A. dioscoridis* and *A. orientale* revealed approximate equality of the sexes (Drummond & Hammond, 1991).

It was in other genera of this family that the function of 'window panes' was first discovered. In addition, in some tropical Araceae, insects are induced to enter the trap by the bright appearance of the interior of the spathe, viewed from the entrance. The tissue of the spathe seems to reflect and refract the light to produce this concentration of illumination, which is enhanced by dark surrounding colours. This effect was described in *Arisaema laminatum* by van der Pijl (1953) and is seen in Plate 5d. In the aquatic *Cryptocoryne griffithii*, both 'window panes' and a bright entry are found (Fig. 10.10). In another member of this family, *Amorphophallus titanum*, insects are trapped by being prevented from climbing to the top of the spadix by an overhanging ridge; this has such a sharp edge that the large beetles which are reported to pollinate the flowers fall off when they try to negotiate it (this plant also receives visits from *Trigona* bees). The species *Typhonium trilobatum*, on the other hand, is pollinated by minute beetles not more than half a millimetre long. They enter the spathe in the early morning and, after they have reached the female flowers at the base, the spathe becomes constricted just above them, making them captive. On the second day, pollen is shed and collects above the constriction; after this the constriction opens slightly and the insects crawl out through the mass of pollen. The trapping of fungus-gnats by *Arisaema* has already been mentioned, and there are degrees of fungus-imitation in this genus also. The way the hood of the spathe is illuminated may suggest the cap of a mushroom, while the striping in the tube may suggest gills. In more extreme cases, the inner surface is chambered or there is a visual imitation of a fungus by the spadix-appendage (species involved include *A. utile*, *A. griffithii*, already mentioned, and *A. costatum*) (Vogel, 1978b, c).

A one-way traffic system is found in this family in the taro (*Colocasia antiquorum*). Flies, attracted by an unpleasant smell, enter through a gap at the base of the spathe, and a constriction prevents them from going beyond the female flowers. Later the same day the smell fades, and the entrance closes up, imprisoning the insects. At night the flies are admitted to the upper part of the spathe where the pollen is shed, and on the second day this part opens, releasing the insects, now dusted with pollen (Cleghorn, 1913). Another one-way route for pollinators has been described in some Indian *Arisaema* species by Barnes (1934). In these, the spadix, unlike the spathe, is not slippery, and the insects climb down it, passing over the stamens and stigmas and out of a hole formed by unfurling of the spathe at the bottom. An example is *A. tortuosum*, which produces both male and hermaphrodite inflorescences. *A. leschenaultii* and some other species are dioecious, and only the males have the basal opening to the spathe. In the female inflorescences, the flowers are tightly packed and have stigmas that project out to the wall of the spathe. The insects – mainly tiny fungus-gnats (family Mycetophilidae) – push down between the stigmas until they are jammed, and die. These species of *Arisaema* have a great preponderance of male plants, which evidently reduces the risk of the female spathes being blocked by flies that are not carrying pollen. Larger flies are excluded by a ring of filamentous sterile flowers, as in *Arum*. For more on aroid pollination, see Meeuse & Morris (1984) and Bown (1988).

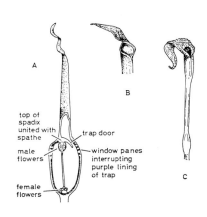

Fig. 10.10 Inflorescence of *Cryptocoryne* (Araceae). **A**, *C. griffithii*, whole inflorescence with entrance turned away from the viewer and spathe cut away below to show prison; **B**, view into brightly-lit entrance of same; **C**, inflorescence of *C. purpurea* in which the spathe is white outside, the throat yellow and the tip warty and purple inside. A after McCann (1943), B after Vogel (1963), C after W.H. Fitch in *Curtis's Botanical Magazine*, t. 7719.

Deception and manipulation by orchids

Deception by orchids ranges from a generalised floral mimicry on the part of unrewarding flowers, to deceptive sapromyiophily and sexual deceit (leading to pseudocopulation). The subject is mainly dealt with in Chapter 7, but here we give some further account of sapromyiophily in the family.

In order to get insects to operate the specialised pollination mechanisms found in orchids, they are often trapped for a short time and/or manhandled because in general only feeding insects can be made voluntarily to orientate themselves in a particular way on the flower. A good example (*Bulbophyllum macranthum*) is given on p.215. Others are *Cirrhopetalum* and *Megaclinium* (closely related to *Bulbophyllum*), and species

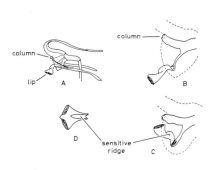

Fig. 10.11 Flower of the orchid *Masdevallia muscosa*. **A**, side view; **B**, side view of column and lip in open position; **C**, the same with lip in closed position; **D**, view of distal part of lip from above. Dotted lines show edge of chamber formed by sepals. After Oliver (1888).

in the unrelated genera *Anguloa, Masdevallia* (Dodson, 1962) and *Pleurothallis* (Chase, 1985). Of these, the last three are confined to the New World and the first two to the Old, while *Bulbophyllum* is found in both hemispheres.

Whereas some species of this type, especially those of *Cirrhopetalum*, attract flesh-flies by producing a bad smell and showing a greater or lesser resemblance to decaying flesh, others seem to be adapted to different tastes, for the flowers of *B. macranthum* smell of cloves and attract a single species of fly only, while *Pleurothallis endotrachys* attracts males of the fruit-fly *Drosophila* (Chase, 1985).

The large orchid genus *Paphiopedilum*, the tropical counterpart of the bee-pollinated slipper orchid, *Cypripedium*, is described on p.189. It displays most of the syndrome of sapromyiophily, combined with a one-way passage through the flower for the pollinators. The hoverfly visitors to the spectacular 30 cm-wide *P. rothschildianum*, in Sabah, Borneo, take about a minute get through the exit passage after laying eggs in the flower (Atwood, 1985).

Some orchids trap insects by a movement of the lip, induced physiologically when the pollinator touches a sensitive area. This effect is called irritability, and an example is provided by *Masdevallia muscosa* (Oliver, 1888). This New World orchid (Fig. 10.11) has been studied only in cultivation, but the likely pollinators are Diptera. The ridge on the distal, triangular part of the lip is the sensitive area, and touching it causes this part to rise up so that an insect settled on it will be carried into the funnel formed by the united bases of the sepals. The only escape passage now is between the lip and the column, where the pollinia and stigma are situated, and the lip remains in the trap position for about 30 minutes. A very similar arrangement is found in many species of the small terrestrial greenhood orchids (*Pterostylis*) in Australia, in which some of the perianth parts form a chamber with a hood above, and have their tips drawn out into antenna-like points. The flowers of these species thus look remarkably like miniatures of the fly-trap inflorescences of *Arisaema* (Plate 5d). This resemblance is heightened by dull green or reddish colouring with darker vertical striping, frequently by the bright-looking interior of the flowers (shown in coloured plates in Curtis's *Botanical Magazine* and in Nicholls, 1955, 1958) and by the tip of the narrow, upright lip showing itself at the entrance to the flower like a spadix. The sensitive area is at the base of the lip and often has the form of a filamentous appendage. The insect visitors are mosquitoes and other gnats or midges, and when they spring the lip they are thrown against the column with their backs towards it; if they are carrying pollen they pollinate the stigmas. The column has projections which function like those of *Bulbophyllum macranthum* (p.215), clasping the insect which backs against the column in its struggles and picks up fresh pollinia. There is evidence that in the course of repeated visits to the flowers the insects become intoxicated. A different species of insect appears to be responsible for the pollination of each species of *Pterostylis* (Coleman, 1934; Sargent, 1909, 1934). (Other species of *Pterostylis* combine the same mechanism with sexual deceit, p.212.)

There are also orchids that show features of the fungus-gnat syndrome. Vogel (1978b, c) has collected a series of examples of detailed fungus-mimicry in the South American genus *Masdevallia*, where it is the lip of the flower that is modified. It is semicircular or horseshoe-shaped and has radiating gill-like ridges on the side that faces downwards. Scent may be imperceptible or definitely present, and fungus-like or fishy. The fungus-mimics in this genus are now separated into the genus *Dracula*, but there are other species with a fungoid smell but no visual mimicry of fungi. In Australia, New Zealand and South Asia, tiny plants of the genus *Corybas* repeat the theme, while in Japan there is a *Cypripedium* (slipper-orchid) which bears much modified flowers drooping near the ground, in which the entrance to the pouched lip has the appearance of a small mushroom.

Further features of the syndrome of sapromyiophily

Floral motion as an attraction

The attractiveness to flies of some of the milkweed family is apparently enhanced by their possession of vibratile organs (Vogel, 1954). An example is supplied by the genus *Tavaresia* (allied to *Stapelia*), in which the corolla is tubular or bell-shaped and the base of the column is produced into ten long filaments, limp towards their tips and each terminated by a dark red knob (Fig. 10.12). These knobs hang down and constantly vibrate, apparently in response to air movements; the corolla is translucent so that the movement can be seen from all round the flower. Further, some *Ceropegia* species have large purple unicellular hairs, 3 mm long and with a constricted flexible base. These hairs are found on the borders of the corolla lobes, and hang down in still air; at the slightest breeze, however, they take on a rapid oscillation, producing a strange effect to the eye. Vibratile hairs, of similar size but multicellular in construction, are found in clusters on the tips of the petals of the tropical orchid *Cirrhopetalum ornatissimum*. In another tropical orchid, *Bulbophyllum medusae*, each perianth segment is drawn out into a

Fig. 10.12 Tavaresia grandiflora (Asclepiadaceae). Side view of the dangling knobs formed by lobes of the corona, with the outline of the corona added. Based on Jaeger (1957) and White & Sloane (1937).

thread many centimetres long, those from each cluster of flowers hanging down and forming a waving plume. The minute, pivoted lips of some species of *Megaclinium*, also a tropical orchid, likewise oscillate in the wind. (All three of these genera have been mentioned previously on p.305.)

Scents and scent sources

The molecular structures of some of the compounds used by sapromyiophilous flowers to attract pollinators are shown in Fig. 10.13. The tail-like structures that are widespread in fly-pollinated flowers may be formed by the perianth or, in the Araceae,

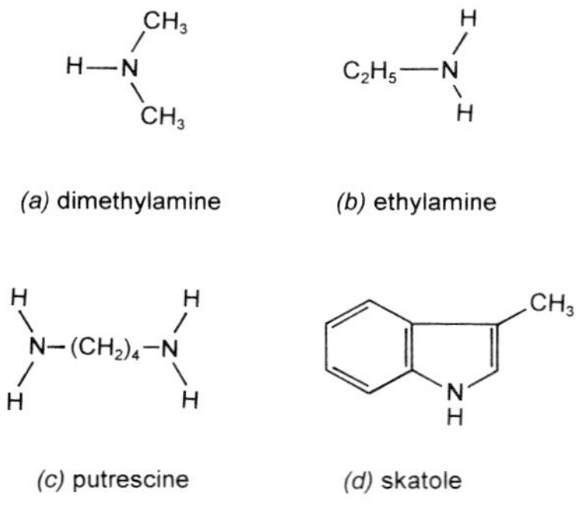

(a) dimethylamine

(b) ethylamine

(c) putrescine

(d) skatole

Fig. 10.13 Molecular structures of some compounds used by sapromyiophilous flowers to attract insects. **A**, dimethylamine; **B**, ethylamine; **C**, putrescine; **D**, skatole. All of these are found in *Arum*.

either the tip of the spathe or by the appendix of the spadix (Plate 5d). In considering the probable significance of 'tails' as alighting places for insects, van der Pijl (1953) drew attention to the propensity of flies for alighting on suspended objects such as fly-papers and electric light bulbs. There is, however, marked variation in the type of 'tail', some being erect, antenna-like rods, and others very long threads or ribbons which trail on the ground. An investigation by Vogel (1963) showed that most tail-like structures in fly-trapping or fly-deceiving flowers are scent-producing organs. He found that the scents are often highly specific in their effect on insects, and are responsible for the initial attraction of insects to the flowers – hence the prominent position of the scent-sources. These are called osmophores. Those which produce a very powerful scent during a short period contain big food reserves which are dissipated during the production of scent, and this is frequently accompanied by a pronounced rise in temperature caused by the rapid respiration in the organ. In addition to the classic example of *Arum* (p.304), this kind of heat-production is known in several other species of Araceae, as well as some species of *Ceropegia* and *Aristolochia*. There are, however, some species in each of these groups which produce their scent over a period of many days, without any abnormally rapid metabolism.

Sapromyiophily and food provision

All the special features of the flowers we have been describing contribute to the syndrome of sapromyiophily. Sapromyiophilous and other deceptive flowers operate by stimulating in the insect an instinctive response related to the fulfilment of some need which the flower does not satisfy. Even when food is provided, as in Asclepiadaceae and some *Aristolochia* species, the attraction of the insect is not normally by means of a food signal. Where male insects are attracted, the signal may indicate both a brood-site (as a place where females are to be found) and a source of food, but then there may be no food.

In this situation, co-evolution of plant and animal is precluded. It is therefore interesting to find that in sapromyiophilous flowers nectar has a high level of amino-acids. Baker & Baker (1975) scored the amino-acid content of many nectars and made a number of group-wise comparisons. They were able to draw general presumptive conclusions of a relationship between amino-acid content and the biology of the flower-visitors. The implication is that where insects have been selecting flowers for the amounts of amino-acid in their nectar, the plants have responded to this in the course of their evolution. The nectars of the group of carrion-fly flowers gave an amino-acid score that was higher than in any other group for which results were published, and about $2\frac{1}{2}$ times as high as in generalised fly-flowers. Further, Baker, Baker

& Opler (1973) observed that the stigmatic exudate that was supped by small flies trapped by two species of *Aristolochia* had an even higher concentration of amino-acids. As it is adaptively advantageous to insects to shun flowers that will trap them, higher amino-acid content cannot have evolved in response to insect 'preferences'. What is presumably happening here is that the plant enhances the chance of insects surviving until they reach the next flower.

Evolutionary convergence

Sapromyiophilous flowers and fungus-gnat flowers present adaptations to their special method of pollination in both coarse and fine details of the blossom, involving situation, shape, colour, pattern, hairs, surfaces, smells, heating, motile appendages and changes of posture. In these, the different families show most extraordinary parallels. Yet this repetitiveness is accompanied within families by a virtuoso display of variation in visual effects, as expressed in form, texture and colour (see, for Aristolochiaceae, Plate 5a, b & Figs. 10.3, 10.4 & 10.14, for Araceae, Plate 5d, e & Figs. 10.9, 10.10, and for Asclepiadaceae, Plate 5c & Fig. 10.6). The plant families involved are diverse, the Araceae and Orchidaceae being monocotyledonous and probably very distantly related, the Aristolochiaceae being primitive dicotyledons and the Asclepiadaceae advanced dicotyledons.

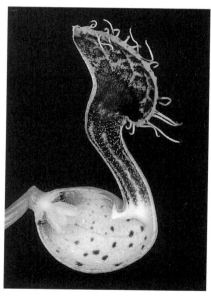

a   **b**

Fig. 10.14 Flower of *Aristolochia fimbriata* (Aristolochiaceae), a further example of the many types of bizarre decoration displayed by sapromyiophilous flowers. **a**, from the outside; **b**, with half the perianth removed: this flower has reached the male stage.

The pollination of water-lilies and philodendrons

Many water-lilies (family Nymphaeaceae) are night-flowering and attract beetles. The giant waterlily of the Amazon, *Victoria amazonica*, is such a plant. The flowers first open in the evening, and are then white and scented; they become heated to several degrees above ambient temperature by their own metabolic activity. Large beetles (*Cyclocephala* species, family Scarabaeidae, subfamily Dynastinae) (Prance & Arias, 1975) arrive, enter the flowers voluntarily and pollinate the stigmas. Overnight the outer floral parts close up and imprison the beetles, which then eat the starch-containing appendages of the carpels. Next day the flower gradually turns deep purple; the anthers dehisce and the beetles become completely covered with pollen. Adhesion of pollen to the visitors is promoted by the fact that when they eat the stigmatic appendages their bodies become sticky. The flowers re-open in the evening and the beetles fly off and head straight for the newly opened scented white flowers. The old flowers then close up and become submerged. Although the flower manages its pollinators by imprisoning them, it is not clear how far they can be said to be victims of deception because it is not known whether they have other food sources which the plant might be imitating.

An extraordinarily similar syndrome is shown by two giant species of the large tropical aroid genus, *Philodendron* (Gottsberger & Amaral, 1984). They have precisely timed opening and closing sequences, spells of warming in the spadix, bringing the temperature to as much as 24°C above ambient for up to an hour, and periods of scent emission. The lowest tenth of the spadix is female and the remainder is split into a lower half of wrinkled texture, formed by sterile male flowers, and an upper half of fertile male flowers. Dynastine beetles arrive in the evening and feed on exudates and on the sterile flowers. Next day the spadix enters the male phase, exuding pollen in sticky chains, while the spathe secretes a sticky resin on its inner surface and starts to close. The beetles are thus driven out, being pasted with adhesive and liberally covered with pollen as they go. These two species of *Philodendron* differ in many details and each attracts a different species of beetle, respectively 2.5 cm and 2 cm long, one being in the same genus as that which pollinates *Victoria amazonica*. There is clearly no trapping here.

11

A Home as a Reward: Brood-site Pollination

In the previous chapter we described the exploitation of insects by flowers that imitate the normal breeding sites of the insects but offer no food-base for the larvae. In the present chapter we turn to examples of symbiosis. In the more highly evolved cases, the plant provides a breeding site and the insect shows special pollinatory behaviour tending to ensure the survival of the host. But there are also simpler examples in which such behaviour is not apparent.

The simpler brood-site mutualisms

Here the pollinating activities of the insects are comparable to those of insect pollinators in general. In the aroid, *Alocasia pubera*, flies that may be pollinators complete their development in the inflorescence and pupate in the spathe (van der Pijl, 1953). Similarly, the biting midges that are pollinators of the cocoa tree (*Theobroma cacao*, family Sterculiaceae), breed in the decaying pods (Dessart, 1961). A parallel situation is said to be common in palms (Silberbauer-Gottsberger, 1991), but here the food-base for the insects is in the male inflorescences. Examples are the New World tropical palm *Orbignya phalerata*, the Old World oil palm, *Elaeis guineensis*, and probably *Nypa fruticans* in Papua New Guinea. The pollinator of the first is a beetle (*Mystrops*, family Nitidulidae) which lays eggs in the male flowers. These drop 48 hours after opening and the new generation of beetles emerges after 12-14 days. The beetles are only found in association with the palm (which is also partly wind-pollinated) (Anderson & Overal, 1988; Henderson, 1986). The pollinators of the oil palm are also beetles (see Chapter 13), whereas those of the *Nypa* are drosophilid flies (Essig, 1973). The breeding of pollinating flies also takes place in the male inflorescences of *Artocarpus heterophylla*, a tropical tree of the family Moraceae (van der Pijl, 1953) with edible fruits and nuts (jak or jack-fruit). The flowers are minute and are produced in large numbers on massive receptacles. The inflorescences have a smell of overripe fruit and are pollinated by small bees and by Diptera of two genera. After flowering the male flower-heads drop, and it is at this stage that the pollinating flies breed in them. In this way, the plants secure a population of pollinators constantly near to them. A similar pollination system is known in one of the cycads (see p.370).

The same end is achieved in a slightly different manner in some trees of Malaysia: *Shorea* section *Mutica* (family Dipterocarpaceae). The pollinators are thrips (Thysanoptera) and they begin to breed in the young flower-buds well before the flowers are ready to open. Within the time taken by the buds to develop into flowers they proceed through a number of generations, each lasting only eight days, destroying a proportion of the buds in the process. When mature, the undamaged buds open at dusk and the thrips then enter the flowers and feed from the petals and the pollen grains. They move readily from flower to flower, but as the trees are self-incompatible this is unlikely to lead to seed-set. By noon the following day, the propeller-shaped corollas have been shed and have whirled down to the forest floor. In the evening, when the new flowers are opening and producing their heavy scent, adult thrips that have remained in the fallen corollas fly up to the canopy and enter the new flowers. The descent from and return to the canopy may displace the insects, so that they reach a different tree. They can alight directionally and may also be carried about by wind movements. As the *Shorea* trees emerge from the canopy, the thrips have a chance of being carried further when they get above canopy level. The trees can produce millions of flowers at one brief flowering but occurrences of mass-flowering are occasional and sepa-rated by one or more much sparser flowerings. The thrips are generally available but their numbers are appropriately augmented by breeding in the buds that participate in a mass-flowering (Chan & Appanah, 1980; Appanah & Chan, 1981; see also p.392).

A temperate example of the same phenomenon is provided by the globe-flower (*Trollius europaeus*) (Pellmyr, 1989), which is also self-incompatible. The pollinators of these spherical flowers that never open are three species of *Chiastochaeta*, flies of the family Muscidae, the larvae of which live only in *Trollius* flowers. Adult flies of both sexes enter the flowers to mate and feed on pollen and nectar. The females then lay eggs (usually one per flower). The fly species differ in the stage of flowering at which they lay eggs, in the positioning of the eggs, and in the paths along which the larvae bore inside the carpels during development. Thus they largely avoid competition. The species that oviposits earliest in the 5-6-day life of the flower is the most effective pollinator. There is a fourth species that oviposits too late to cause any pollination (Pellmyr, 1992). Placement of pollen on the stigmas is incidental to the movements of the flies in the flowers. A flower produces nearly 400 ovules, but only a few of the young seeds are eaten by the fly larvae. Several other *Trollius* species are facultatively pollinated by *Chiastochaeta* species that breed in the flowers in Asia; here the flowers are flat and some pollination takes place in the absence of these flies (Pellmyr, 1992).

It seems that the insects pollinating *Alocasia pubera* (a fly-trap flower, p.311) and *Trollius europaeus* were pollinators before they evolved the habit of breeding in the plants, and the same is perhaps true of the insects breeding in palms. In the cases of the pollinators of *Artocarpus heterophylla* and *Theobroma cacao*, there seems to be no strong evidence either way.

In some of the relationships described above the insects live in tissue which the plant has discarded, though it is possible that it produces more of this than it other-wise would so as to feed the potential pollinators. However, in *Shorea* and *Trollius* what the insects eat is not waste tissue, and it clearly represents a trade-off on the part of the plant against pollinator service. All these relationships involve small or very small insects, so it looks as though the plants are minimising their costs in the deal.

The rest of this chapter is devoted to two relationships: that between figs and fig-wasps and that between yuccas and yucca-moths. As in some of the preceding exam-ples, the insects depend entirely on the plant and use the ovaries as the larval food-base but they display physical and behavioural adaptations for the pollination of their hosts.

Highly specialised brood-site mutualisms

Figs and fig-wasps

The figs (*Ficus*) belong to the same family, Moraceae, as *Artocarpus* and, like it, bear numerous tiny flowers on a massive receptacle. Here, however, the receptacle has be-come moulded into a hollow vessel with the unisexual flowers clothing its inner sur-face. This receptacle ('syconium') has a narrow opening ('ostiole') to the outside which is blocked by flexible scales; it becomes the fig 'fruit'. Female flowers contain only one ovule.

The best-known member of this very large genus is the edible fig (*Ficus carica*), which is probably native to south-west Asia. In the wild form of this species, three types of receptacle are produced at different times of the year. The type formed in winter contains many neuter (sterile female) flowers, and a smaller number of male flowers which are confined to the region of the entrance. This type of receptacle is invaded by tiny females of the gall-wasp species *Blastophaga psenes*, which lay eggs in the neuter flowers and then die. The offspring of the wasps complete their development in the ovaries of the flowers (one wasp to each flower). The male wasps hatch first and emerge from the ovary into the interior of the receptacle. They are highly modified, having reduced legs and eyes and no wings (Fig. 11.1A). They bore into the ovaries occupied by females, fertilise the females and then die. The females now emerge and leave the receptacle, receiving pollen at the entrance from the male flowers which have only just opened. It is now June, and the wasps find their way to the second type of receptacle. This type contains either a mixture of neuter and female flowers, or

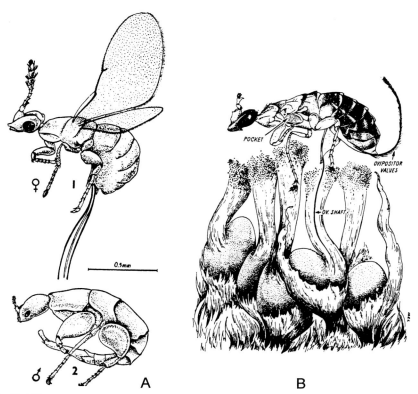

Fig. 11.1 Fig-wasps and fig flowers. **A**, a female and male of *Blastophaga quadraticeps*; **B**, female of *Ceratosolen arabicus* in the act of oviposition, with forelegs raised to extract pollen from the thoracic pockets. From Galil & Eisikovitch (1968b & 1969).

female flowers only. The wasps lay their eggs in the flowers, but only those laid in neuter flowers develop. The ovary of the neuter flower (a modified form of the female flower) is incapable of producing a seed, and the style is short with an open canal leading to the ovary. In the female flower, on the other hand, the solid style is too long to permit a wasp to reach the ovary with its ovipositor, and eggs not placed in the ovary fail to develop. The pollen which the wasps bring with them fertilises the female flowers, and so seed is set in these. The development of the wasps takes place as before, fertilised females emerging in autumn and going to the third type of receptacle which is smaller than the others. Here there are only neuter flowers, in which the insects develop that will emerge in winter and restart the cycle in the first type of receptacle (McLean & Cook, 1956; Grandi, 1961).

The entry of the female fig-wasps into a syconium is made difficult by the scales in the ostiole and they may lose wings and parts of their antennae in their struggles to get in (Grandi, 1961). The resistance offered is apparently an adaptation to prevent the entry of insects lacking the instinctive persistence of the female fig-wasp. When it lays an egg, the wasp injects a drop of a special secretion into the ovary of the neuter flower, and this stimulates the development of the the unfertilised ovule into a gall, which later provides the nourishment for the wasp larva. Thus the plant provides special flowers in which the pollinators breed, the winter syconia being devoted solely to this use.

Some cultivated forms of the edible fig are entirely female, but in these no pollination is necessary for the fruits to develop (McLean & Cook, 1956). All other sexually reproducing species of the mainly tropical genus *Ficus* provide neuter flowers for the pollinating wasps to breed in, but the life-cycle is normally less complicated than that of the edible fig. There are two main types (Wiebes, 1963). In one, exemplified by *F. fistulosa*, pollinated by *Ceratosolen hewittii* (Galil, 1973), there are two kinds of receptacle, one containing female flowers only and the other containing both neuter and male flowers. Fertilised female wasps emerging from the latter enter the next generation of either the same kind of receptacle or else the female kind. In the first case, the wasps lay their eggs, and in the second they pollinate the flowers. The two kinds of receptacle are normally on different plants (dioecy) (see also p.320). In the other type of life-cycle, all three kinds of flower occur in the same receptacle (monoecy). Fertilised female wasps pollinate the female flowers and breed in the neuter flowers, while their departing female offspring later acquire pollen from the male flowers. Usually the neuter flowers have short styles and the female ones longer styles, which makes oviposition in the latter difficult or impossible. In *Ficus sycomorus* (Fig. 11.1B) and *F. religiosa*, however, it has been found that long-styled and short-styled flowers are not physiologically different, for small percentages of the short-styled produce seeds, and of the long-styled produce galls (Galil & Eisikovitch, 1968a, b).

Another significant observation on *F. religiosa* is that female wasps newly emerged from the galls are attracted to the anthers, push their heads amongst them and even eat some of the pollen. The dependence of the wasp *Blastophaga quadraticeps* on this plant is emphasised by experiments on this species in which female wasps which had not been in contact with pollen were allowed to enter the receptacles. These dropped off without forming galls or seeds (Galil & Eisikovitch, 1968b).

During a study of the pollination of *Ficus sycomorus* in East Africa, Galil & Eisikovitch (1969) were puzzled at its effectiveness in view of the small amount of pollen carried on the body of the wasp, *Ceratosolen arabicus* – it always cleans itself carefully on emerging from the syconium. The accidental squashing of an insect led to the discovery of a pair of pouches on the underside of the thorax, each of which can probably carry 2,000-3,000 pollen grains. Because ovipositing females continued their activities despite the opening of the fig for observation, it was possible to see that after each egg had been laid, the forelegs were used to scratch some pollen out of the pouches and brush it on to the stigmas (Fig. 11.1B). In this way, a mixture of gall-flowers and seed-flowers develops in the area where the female has been working. When the receptacle has reached the male stage, male wasps emerge from their galls and after copulating with the females they cut off the anthers of the male flowers. The females emerge and with their fore-claws and a special comb on the fore-coxae load their pouches from the loose anthers. The males then collaborate in boring a hole to the outside and the females depart (Galil & Eisikovitch, 1968a, 1974). After the discovery of the pouches in C. *arabicus*, it was found that *Blastophaga quadraticeps* and some other fig-wasps had pollen pouches (Galil & Snitzer-Pasternak, 1970). Some of these are New World species, and the females of two of them, *B. tonduzi* and *B. esterae*, also search for and manipulate the anthers of the fig; they carry pollen in pouches sunk between the first and second thoracic segments on either side, and also in a recessed pollen basket (corbicula) on each of the fore-coxae (Ramirez, 1969; Galil, Ramirez & Eisikovitch, 1973). Thus the pouches are different in form and position from those of *Ceratosolen arabicus*. The claws of the forelegs can reach the pouches and the corbiculae; they appear to hook pollen out and are then knocked against each other as if to shake the pollen off on to the stigma. The ostiolar scales of the figs do not loosen but the male wasps bite a hole to the outside through which the fertilised female wasps carrying pollen depart. Ramirez found pouches in many other New World species of *Blastophaga* and in fig-wasps of other genera from various parts of the world.

Where pollination is 'deliberate' rather than random, it is of interest to know how the female wasps treat the sterile and fertile female flowers. In the dioecious *Ficus fistulosa*, wasps in the male syconia pollinate the gall-flowers, although these are not destined to produce seed, and in the female syconia they carry out oviposition movements (but probably do not lay eggs) before each pollination of a stigma, although no wasps will develop in the ovaries (Galil, 1973). Apparently the pollination of the neuter flowers stimulates growth of the endosperm of the ovule which is necessary for the nourishment of the wasp larva. This serves the same purpose as the injection of a growth-stimulating compound reported in *F. carica*.

A case of physiological control of the wasps' behaviour in *F. religiosa* was uncovered by Galil, Zeroni & Bar Shalom (1973). It was found that as the male wasps matured, their activity in emerging from their galls and mating with the females was stimulated by and dependent on a very high concentration of carbon dioxide in the receptacle. The males eventually bore holes through the fig wall and the atmosphere inside then equilibrates with that outside. In the lowered concentration of carbon dioxide the female wasps, previously inhibited, become active, emerge from their galls, load their pouches and exit through the borings. No such responses could be found in the wasps in *F. sycomorus*.

In non-seasonal tropical climates, where most figs grow, the plants can reproduce at any time of the year and there is usually a year-round availability of receptive syconia. This is achieved by occasional flowering of plants at irregular intervals, which is accompanied by synchronisation of flower and fruit development within the plant. Outbreeding will thus be enforced, and the pool of neighbours that are receptive will be changed at successive flowerings. However, when the wasps emerge from a particular tree, only a small proportion of the fig population is available to receive pollen, so that for pollination purposes the effective density of the population is greatly lowered and fig-wasps may have to travel relatively long distances to propagate themselves. It is inferred that receptive trees are located and identified by scent (Janzen, 1979; Addicott, Bronstein & Kjellberg, 1990).

Each species of fig-wasp (Hymenoptera-Parasitica, family Agaonidae) normally confines itself to one species of fig-tree (Wiebes, 1963; Hill, 1967). Exceptions are where one species of wasp pollinates two very closely related species of *Ficus* or two closely related wasps pollinate the same *Ficus* in different areas. This species-to-species relationship means that the rates of speciation of the fig-wasps and the fig-trees have been about equal, even though their generation times are enormously different. Some *Ficus* species are known to be interfertile, but no natural hybrids have been found, presumably because the wasps identify their particular host accurately.

Parasites in the fig system

Although there may be only one species of pollinating wasp in a species of fig, it is normal for parasites to be present. Thus in *F. sycomorus*, five species of wasp have been found in the syconia apart from *Ceratosolen arabicus* and none is known to cause pollination even though one belongs to this genus. The wasp *Sycophaga sycomori* enters the syconia and lays eggs in any female flower, since its ovipositor is longer than that of *C. arabicus*. However, when the latter is engaged in egg-laying, it 'stings' the styles of neighbouring long-styled flowers and bites their stigmas. This does not prevent the seed from developing but it does render them unsuitable for *Sycophaga*; thus different areas of the syconium come to be occupied separately by these two wasps. The other three wasps belong to family Torymidae; they have very long ovipositors with which they pierce the syconium from outside and lay eggs in galled flowers only, that is, those already occupied by *Ceratosolen* or *Sycophaga*. The larva of the corresponding torymid in *Ficus carica* kills the occupying larva. In the absence of *Ceratosolen*, oviposition by *Sycophaga* is effective in causing the syconium to complete its development, although it is seedless. This is the case in the East Mediterranean region where *F. sycomorus* is grown; the male wasps, like those of *Ceratosolen*, bore the hole to the exterior through which the females escape (Galil & Eisikovitch, 1968a, 1969; Galil, Dulberger & Rosen, 1970). Some torymids that live in figs appear to develop in unoccupied ovaries (Bronstein, 1991). This applies in *F. pertusa* where Bronstein found that Torymidae and Agaonidae did not seem to compete for food, and there were frequently more torymids in the syconia than agaonids (but the numbers were independent). The torymids depended on the agaonids to induce retention of the syconium on the tree and for the boring of the exit holes from the syconia.

Yuccas and yucca-moths

The relationship between the yucca plant and the moth that pollinates it is similar to that between fig trees and fig-wasps, and is equally famous. Pioneer studies on this relationship were carried out by Riley (1892) and Trelease (1893); later work by Busck was included in a taxonomic monograph on *Yucca* by McKelvey (1947). New information, summarised by Powell (1992), has greatly amplified the knowledge gained by these early investigators. *Yucca* (family Agavaceae) is a genus of North America and the West Indies. Typically, the plants produce showy inflorescences of numerous large creamy-white flowers (Fig. 11.2A). These are partially closed by the convergence of the tips of the perianth segments; they are scented and smell most strongly at night. Nectar is sometimes secreted at the base of the ovary but it is not drunk by the yucca-moths, which do not feed; it may, however, keep other insects, which are attracted to the flowers, away from the stigmas, though in some circumstances honeybees and bumblebees may cause pollination in yuccas (Powell, 1992).

The moths belong to the small suborder Monotrysia (p.86), family Incurvariidae, subfamily Prodoxinae (Davis, 1967); they are not unlike *Eriocrania*, a genus with primitive mouth-parts described in Chapter 4. Each of the maxillae of a yucca-moth comprises a galea similar to that of *Eriocrania* (Box. 4.2) and a palp, the latter with a special tentacle, prehensile and spinous, formed by the modification of the basal joint of the palp (Fig. 11.2C).

When ovipositing, yucca-moths have a stereotyped pattern of behaviour. The female moth, *Tegeticula yuccasella*, enters the flower, climbs along a stamen from the base and bends its head closely over the top of the anther. The tongue uncoils and reaches over the tip of the anther, apparently steadying the moth's head. All the pollen is then scraped into a lump under the head by the maxillary palps, and held fast by the maxillary tentacles and the trochanters of the forelegs (Fig. 11.2B). As many as four stamens may be climbed and the pollen collected in this way. As a rule, the moth then flies to a flower in another inflorescence, where it closely investigates the condition of the ovary, being able to tell if the flower is of the right age and whether eggs have already been laid in it. If the flower is suitable, the moth again climbs the stamens from the base but this time goes between them on to the ovary. It then reverses a little way back between the stamens and the ovary and lays an egg, boring into the ovary with its ovipositor. After this it at once climbs to the stigmas, which are united to form a tube, and thrusts some of its pollen down into the tube, working energetically with the galeae and tentacles. The most usual behaviour of the moth is to lay one egg in each of the three cells of the ovary, and to carry out pollination after laying each egg.

Since an unpollinated yucca flower soon dies, this behaviour of the moth ensures that there will be food for its larvae, which is provided by the abnormal growth of one or more of the ovules in the neighbourhood of each moth's egg. The remaining ovules, which are numerous, develop into seeds and, just as they are ripening, the moth larvae emerge and pupate underground. The adult moths always emerge in the flowering season of the yuccas in their area. The emergence of any one season's brood is spread over a period of some years after pupation, thus ensuring the continu-

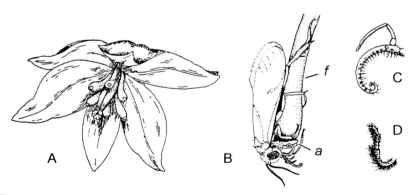

Fig. 11.2 Yucca and yucca-moth. **A**, flower of *Yucca aloifolia*, with the perianth segments parted to show six swollen stamens and the ovary with stigmatic lobes; **B**, female moth, *Tegeticula yuccasella*, gathering pollen from a stamen (*f*, filament of stamen; *a*, anther): **C**, maxillary palp and tentacle (coiled) of the moth; **D**, labial palp of same. After Riley (1892).

ance of the moth species even if, as occasionally happens, the yuccas fail to flower in a particular year. In this relationship the moth, like the fig-wasp, ensures the seed-pro-duction of its food-plants, while the plant provides food and shelter for the young of its pollinator. As the flowers are protogynous, the moth needs to move to another flower after gathering pollen in order to find an ovary in the receptive stage for polli-nation; but it does more – it goes to another *inflorescence*, thereby promoting outbreed-ing, which should enhance the genetic quality of the of the next generation of host-plants. (In the fig, the wasps that have gathered pollen are forced to find another syconium because at this time there are no receptive female flowers in the syconium from which the pollen was gathered.)

The currently recognised species of yucca-moth belong to two genera and are listed in Table 11.1, together with their host-plants. *Tegeticula yuccasella* is enormously widely distributed in the United States, mainly east of the Rockies but reaching the west coast in California. These small moths are active at night and spend the day at rest in the flowers, which they resemble in colour. *Parategeticula* also pollinates flowers at night; it occurs in two isolated areas and in each it pollinates a single species of *Yucca*. *T. maculata* pollinates only *Yucca whipplei*, which occurs in southern California, and has glutinous pollen massed into two pollinia in each anther; the moths carry out pollina-tion in the daytime. *T. synthetica*, pollinator of *Y. brevifolia* in the Mohave Desert of California, has a hard body and scaleless wings and appears to mimic sawflies of the genus *Dolerus*. Its host-plant has flowers with very firm perianth segments, scarcely parted at the tips, so that the moth has to force its way in as the fig-wasps do (Powell & Mackie, 1966; Powell, 1992).

Evidence has now been obtained that '*T. yuccasella*' is a complex of closely related species, each more or less specific to one *Yucca* species (Tyre & Addicott, 1993). The taxonomy and biology of these is not yet worked out.

Various statistics have been gathered in order to estimate the trade-off of seeds against pollination in yuccas. A study of eight species of *Yucca*, pollinated by *T. yuc-casella*, was carried out by Addicott (1986). Yucca ovaries contain many ovules, in these eight species ranging from 150 to 350 (lower numbers are found in berry-fruited yuc-cas) (Table 11.1). Seed production ranged from 90 to 200. Both these measures varied between species and between populations of the same species. The ratio of viable uneaten seeds to ovules in different species ranged from 0.36 to 0.60, but the variation was not statistically significant; however, differences between populations within spe-cies were significant. Fruit set also varied greatly between populations within species.

One yucca-moth larva may eat from 6 to 25 seeds. Oviposition and the develop-ment of moth larvae cause constrictions that distort the fruit, and some ovules addi-tional to those used as food may be damaged by these processes. Other failures could be due to inadequate pollination or limitation of resources available to the plant for seed-maturation. Individual yucca-moth larvae ate 18-43.6% of the ovules in a lo-cule; however, since not all ovules were potentially viable, the effective loss was only up to 13.6%. Adding damage by oviposition, the moths reduced seed production by 0.6-19.5%. The number of larvae per fruit was usually below ten but occasionally up to 24 (earlier, Addicott had found populations with 30-50 larvae per fruit). The high-est numbers of larvae probably occur as a result of yucca-moth visits to the young fruit stage, when additional eggs are laid. A high proportion of fruits contained no larvae, especially in berry-fruited species. Such fruits show signs of oviposition and it is thought that a high egg mortality accounts for the frequent lack of larvae.

Fruit-abortion rates are significant in maintaining a balance between seed-produc-tion and moth-production, as described below. Individual plants of *Yucca whipplei*

Table 11.1 Yucca-moths and the yuccas in which they live (based on Powell, 1992).

Moth species	Yucca section	Yucca species	Yucca fruit
Tegeticula yuccasella	Sect. *Sarcocarpa*	*arizonica, faxoniana, torreyi, treculeana, schidigera, baccata*	berry
Parategeticula pollenifera		*schottii, elephantipes*	berry
Tegeticula synthetica	Sect. *Clistocarpa*	*brevifolia*	berry
Tegeticula yuccasella	Sect. *Chaenocarpa*	nine species	capsule
Tegeticula yuccasella	Sect. *Heteroyucca*	*gloriosa*	capsule
Tegeticula maculata	Sect. *Hesperoyucca*	*whipplei*	capsule

aborted 29-72% of their fruits (Aker & Udovic, 1981), while in a further set of eight species fruit abortion rates ranged between 0-100% (Addicott, 1985) (but most of these species had lower abortion rates than *Y. whipplei*). Abortion of fertilised ovaries is attributable to resource-limitation. Fruit initiation was high in some species, but in others 23-33% of plants initiated no fruit at all. Low levels of fruit-set are presumed to be the result of pollinator-limitation.

Parasites in the yucca system

Just as the symbiotic system of *Ficus* carries its load of parasites, so does that of the yucca. Closely allied to the yucca-moth is the bogus yucca-moth (*Prodoxus*). Superficially it resembles *T. yuccasella*, but its maxillae have no tentacles. It breeds in the ovaries or peduncles of yucca flowers and, since it does not pollinate the flowers, it depends for its existence on the true yucca-moth. In addition, there are four species of non-pollinating 'advantage-takers' in the *T. yuccasella* complex: as species, these fly later than the mutualistic species and they too lack maxillary tentacles. Species of this kind have arisen at least twice (Addicott, Bronstein & Kjellberg, 1990).

Co-evolution of advanced brood-site mutualists

The yucca-moth and the fig-wasp both belong to groups in which feeding by the larvae on the internal parts of plants is frequent, so we may conclude that this process probably evolved first, the insects originally being at most incidental pollinators of the host-plant (Pellmyr & Thompson, 1992). It is thought that the yucca and moth interaction is probably quite recently evolved, since the form of most yucca flowers (large showy white bells) and the secretion of nectar by many of them, seem inappropriate to the pollination method. The fig and wasp relationship looks as if it is much older.

It is remarkable how the symbiotic pollinators have undergone structural modification and have developed very complex behaviour in relation to the flower in the course of the evolution of these partnerships. The plant-pollinator relationship is at the opposite extreme from that in deceit flowers, since mutual dependence is complete. Co-evolution has proceeded a long way, whereas in deceit flowers it is (by the definition of the term 'co-evolution') absent. However, it can be inferred that there is tension even in these relationships.

For climatic reasons, flowering in yuccas has to be strongly seasonal. Various factors promote synchrony of yucca-moth flight with flowering, but this is sometimes imperfect and in any case some yuccas have a long season (about 10 weeks in *Y. whipplei* [Aker & Udovic, 1981]). Those moths flying relatively late in the flowering season will consequently encounter young fruits as well as fresh flowers. If yucca pollination has been good, plants may abort some fruits because of shortage of resources. The fruits most likely to be aborted are the late ones in which less has been invested. The late moths' prospects are therefore best if they oviposit, not into the late fresh flowers, but into already fertilised and developing ovaries which have no need of pollination. The floral environment of relatively late moths thus favours 'cheating', and there is extensive evidence that this occurs (Addicott, Bronstein & Kjellberg, 1990). 'Cheating' has been observed in *Tegeticula maculata* on *Yucca whipplei* (Aker & Udovic, 1981) and on *Y. kanabensis*, where it has been shown that, although some moths oviposit when not carrying pollen, pollen-carrying moths may also 'cheat'. Here, however, non-pollination is strongly associated with oviposition in flowers already pollinated. 'Cheating' behaviour is not a characteristic of individual moths – it is facultative. Since 'cheating' is not here related to season but is observable on single oviposition bouts, it was concluded that it was related to an abundance of the moths, which could lay extra eggs in already-pollinated flowers and so leave more progeny (Tyre & Addicott, 1993).

In the case of fig-wasps, it can be surmised that if they grew longer ovipositors they could breed in more of the flowers in the syconium. The plants' best evolutionary response would at first sight seem to lie in growing longer styles. But because of the short generation time of the wasp compared with that of the fig-tree, the wasps might win the evolutionary race and in so doing prevent the plant from reproducing, so that both plant and wasp would become extinct. However, there is no evidence for disproportionate length in either ovipositors or styles. Murray (1985) has suggested how a balance might be achieved. It is supposed that the fig-plant has the power to abort young syconia and that abortion can be triggered by the level of infestation of the syconium by fig-wasps. Natural selection would then set the threshold level of infestation at which abortion began, so as to provide the best balance between seed-production and pollinator-production. If fig-wasp oviposition became too successful, fig-abortion would destroy the larvae, together with the genes for a long ovipositor that they carried.

When this system is in balance, the style-length variation places an upper limit on the number of ovaries in the syconium that contain fig-wasps. This means that abortions of fertilised syconia are rare and so one factor that at times makes 'cheating' advantageous to yucca-moths is absent in the fig. A very low availability of syconia, which might occur seasonally or locally, could make it advantageous for wasps to oviposit in developing pollinated syconia. However, there seems to be no facultative 'cheating' by fig-wasps, and the exploiting species *Sycophaga sycomori*, described by Galil & Eisokovitch (p.316), is apparently the only known case of such a species being closely related to the pollinator (Addicott, Bronstein & Kjellberg, 1990).

The relationship between dioecious figs and their pollinators presents an interesting problem: female wasps that enter female figs leave no progeny, so are the plants able to make the male and female syconia indistinguishable or are the wasps displaying 'altruism'? Various aspects of the fig/fig-wasp mutualism are discussed in detail by Addicott, Bronstein & Kjellberg (1990) and by Janzen (1979), who cite many other studies on these problems.

12

Breeding Systems: How Important is Cross-pollination?

In most animals, including nearly all the familiar ones, reproduction involves two separate animals, with male and female individuals coming together for the fertilisation of an egg. The majority of plants, by contrast, bear hermaphrodite flowers, with functional male and female parts in every flower. The earliest flowering plants almost certainly had both male and female parts on the same plant, probably as hermaphrodite flowers, and this is the basic structure of a flower described in Chapter 2. About 80% of all flowering plants have hermaphrodite flowers. In the remaining 20%, however, there are many variations and some species bear flowers without functional organs of both sexes; about 10% of the world's flora have two separate sexes, like animals. Even among the hermaphrodite majority there is a range of breeding systems, for instance some can successfully fertilise their own ovules; others require cross-fertilisation between different individuals, and there are several distinct mechanisms promoting this in different plants. Some plants have even dispensed with sexual reproduction altogether and reproduce without any fertilisation.

In this chapter we explore all these different forms of reproduction, how they work and how they are related to each other. We are concerned both with describing the different types of sexual system as we currently understand them, and with the immediate consequences of pollination as it leads to fertilisation. It is well to remind ourselves of the distinction between the two terms. Pollination is the transfer of pollen from an anther to a stigma, and this may or may not lead to fertilisation, the sexual union of one of the nuclei from the pollen grain with the egg nucleus to form an embryo, and an associated fusing of another pollen grain nucleus with the endosperm nucleus. An overall summary of the breeding systems of plants is presented in Box 12.1.

Why sexual reproduction evolved in the first place is a topic of great academic interest and debate, but is beyond the scope of this book (Maynard Smith, 1978). Plants, fundamentally, reproduce sexually and we will treat this as a starting point. All the adaptations, the intricate beauty of flowers and the details of all the interactions of pollination which are the substance of the whole subject, are really about maintaining at least the possibility of cross-fertilisation, but why is this necessary? In a hermaphrodite flower both pollen and ovules are present and it should be quite possible for a plant to fertilise itself. Sometimes this happens, and if self-pollination in a plant leads to successful self-fertilisation, none of these adaptations is required; the flower only needs to produce enough pollen to fertilise its own ovules, really with no other floral parts necessary.

For most plants, however, even if selfing is possible, at least some cross-fertilisation resulting from a transfer of pollen between individuals is favoured. Cross-fertilisation provides at least two vital advantages. Firstly it means that every new plant bears a new combination of genes, so variation is released and the population can potentially

Box 12.1 Summary of sexual forms and reproductive systems in flowering plants.

1. Hermaphrodite – one type of flower with both male and female parts. The single commonest form including about 80% of angiosperms. Flowers all morphologically alike except in heteromorphy (1C).
A. Gametophytic self-incompatibility: incompatibility by response to pollen tube in the style or stigma. Widespread.
B. Sporophytic self-incompatibility: incompatibility by response to pollen grain wall at stigma surface. Brassicaceae, Asteraceae, Betulaceae, Convolvulaceae, Caryophyllaceae, ?others.
C. Heteromorphic self-incompatibility: incompatibility from various interactions at stigma surface, style or ?ovule. Two or three types of flower, most with different lengths of style (heterostyly) or different pollen and stigma structures or (usually) both. Scattered occurrence in at least 24 families, e.g. two style forms in Primulaceae, Rubiaceae etc., three style forms in Lythraceae, Pontederiaceae, etc.
D. Late-acting self-incompatibility: incompatibility from response at embryo sac entrance or ovule abortion. A range of little known systems, perhaps connected with inbreeding depression. Probably widespread, particularly in tropics.
E. Self-compatibility: derived from any of the above self-incompatibility systems. Widespread, with some plants almost entirely selfing, particularly short-lived plants; others with mixed cross- and self-fertilisation.

2. Unisexual – more than one type of flower, some with only functional male parts, some with only functional female parts.
A. Monoecious: each plant with both male and female flowers. c.5% of flora but includes dominant temperate trees in Fagaceae, Betulaceae etc., *Carex* (Cyperaceae), *Euphorbia*.
B. Dioecious: two types of plant, one with only male flowers, the other with only female flowers. 5% British flora, 10% worldwide. Widespread; dominant in some families, e.g. Salicaceae, common in tropical trees and oceanic islands, also *Urtica*, *Ilex*, etc.
C. Gynodioecious: two types of plant, one with female flowers, the other with hermaphrodite flowers. Often in only some populations of a species. Widespread in Lamiaceae, also *Plantago*, *Saxifraga*, etc.
D. Androdioecious: two types of plants, one with male flowers, the other with hermaphrodite flowers (*Datisca*, *Phillyrea*) or with male and female flowers (*Mercurialis*). True androdioecy only known from three genera. Morphological androdioecy with sterile pollen in hermaphrodite flowers (i.e. functionally dioecious) more widespread.
E. Gynomonoecious: each plant with both female and hermaphrodite flowers. In specialised inflorescences, mainly Asteraceae, with female ray florets, hermaphrodite disc florets.
F. Andromonoecious: each plant with both male and hermaphrodite flowers. Widespread; particularly characteristic of Apiaceae.
(G. Sterile flowers: used for attraction only, always associated with fertile flowers. *Viburnum*, *Hydrangea*, Asteraceae, etc.)

3. Asexual – no sexual reproduction involved.
A. Vegetative spread: by rhizomes, stolons or budding. Widespread, always associated with flowers and sexual reproduction.
B. Bulbils: new plants produced in place of flowers in inflorescence; often associated with flowers and sexual reproduction. *Allium*, Poaceae, *Saxifraga*, etc.
C. Agamospermous: the production of seeds involving no fertilisation. Associated with sexual reproduction on same plant (e.g. *Citrus*) or separate plants (e.g. *Ranunculus*). A few species exclusively agamospermous in several families, particularly Asteraceae, Rosaceae.

adapt to a new or changing environment. This is important for colonising any new place since the conditions are likely to be at least a little different from those at the original site. Over the past million years, the world's climate has fluctuated quite radically too, with glaciers coming and going in the northern hemisphere, affecting the climate worldwide, and, more recently, man modifying most habitats, so adaptability has been important. Variation may also be essential for developing resistance to attack from insect herbivores or fungal diseases which are themselves constantly evolving.

The second, and more immediate, advantage is that each plant gets a set of chromosomes from each parent. When the chromosomes are in a cell, they normally make exact replicas of themselves, but there are occasionally errors in the replication known as mutations. Some of these mutations lead to a change in the function of the structure or substance that is coded for, and many are potentially damaging or lethal. Having two different parents means that a mutation on one chromosome is unlikely to be present on the corresponding chromosome from the other parent, so a lethal feature is either not expressed or is, at least, 'covered' by the presence of the normal form. In addition to this, if the function of, say, a substance vital for the healthy running of a cell, has been modified by the mutation so that it can work under slightly different conditions, then the offspring may have the benefit of two slightly different forms. This may allow it to live under a wider range of conditions or to grow faster because of its greater tolerance. For these reasons, cross-fertilisation in plants has important advantages over selfing, although self-fertilisation has some counterbalancing advantages which will be discussed later (p.330).

Self-incompatibility

The commonest way in which plants avoid self-fertilisation is by self-incompatibility, a physiological barrier making it difficult or impossible for a flower to fertilise itself even though it may be abundantly pollinated with its own pollen. There has probably been some form of self-incompatibility within the flowering plants since earliest times.

Self-incompatibility involves the ability of a plant to discriminate between its own pollen grains and those of another plant and only allow pollen from a different plant to grow and fertilise the ovules. It is unusual as a recognition system since most other systems (such as our own immune system) involve recognition and rejection of a foreign organism or protein, such as a disease organism or a tissue transplant. In a self-incompatibility system, it is the *same* type that is rejected and a different type leads to acceptance and fertilisation.

The presence or absence of a self-incompatibility system is often quite easy to detect in a plant, but the details of its operation are much harder to elucidate, requiring sophisticated techniques of microscopy and physiological study. The result is that only a few species have been studied in any detail, although the number is growing. Much of what has been written about self-incompatibility generally is an extrapolation from these few studies. From what we know at present it seems that there are, perhaps, four broad types of self-incompatibility system but each of these contains a number of variations and the relationship between them is not at all clear (de Nettancourt, 1977; Richards, 1986; Seavey & Bawa, 1986; Gibbs, 1986; Barrett, 1988).

Gametophytic self-incompatibility

This is a broad heading covering a number of systems, in all of which the growing pollen tube is recognised and rejected. The best-studied form involves the pollen grain germinating on the stigma and the tube growing down the style but being stopped before it reaches the ovule. The tube may be blocked or it may burst in the style in a way that is similar to what happens when it reaches the ovule in a successful fertilisation; it is as if the bursting is triggered too early. The nuclei produced by the pollen grain, one of which forms the pollen tube nucleus, are regarded as the 'gametophyte', i.e. that part of the plant that produces gametes or fertile cells, from its presumed origin in ancestral gymnosperms (Chapter 14). This is why the system is called gametophytic. It is known from a diverse range of plant families such as the legumes (Fabaceae), poppies (Papaveraceae), nightshades (Solanaceae) and lilies (Liliaceae), and thought or presumed to occur in more, although often the ground work of microscope study has not been done thoroughly enough to be sure.

A rather different gametophytic system is known from the grasses, although it is still the pollen tube that is recognised. In grasses, recognition takes place at the stigma surface and the pollen tube is blocked as it penetrates the stigma, making this a rapid and probably very efficient recognition and rejection system of incompatible pollen. In at least one other monocot family, the Commelinaceae (including *Tradescantia*, well known as a house plant) the recognition site is in the stigma papillae and in the evening primroses (*Oenothera*, Onagraceae) pollen tubes are blocked only just beneath the stigma.

A few of the species with gametophytic self-incompatibility have been studied genetically. In poppies (*Papaver*), clovers (*Trifolium*), evening primroses and probably many others, the incompatibility system appears to be controlled by a single gene locus that has many forms (alleles); the presence of an allele identical to either of the parents' alleles will stop the pollen tube growing. The corn poppy (*Papaver rhoeas*) is the best studied of these, and O'Donnell & Lawrence (1984) estimated that there were between 25 and 45 alleles in natural populations. They reported similar numbers from study on evening primroses and unconfirmed reports of higher numbers in clovers. They showed, in the poppy, that any one plant was fully capable of breeding with over 80% of the others in the population and totally incompatible with less than 5%, mainly close relatives.

In the grasses, the system is more complex and appears to be controlled by two interacting gene loci. In the buttercups (*Ranunculus* spp.) three gene loci may be involved, and in the beet (*Beta*) four loci, each of which probably has several alleles (Lundqvist, 1975; Osterbye, 1975). This adds to the number of compatible individuals, since the alleles at all the loci must be the same for the incompatibility reaction to occur. In so doing it does allow more fertilisation of close relatives than the single locus system (Bateman, 1952). The operation of these multi-locus systems is little known and is probably not the same as that described for a one-locus system.

Sporophytic self-incompatibility

In this form of self-incompatibility, the site of recognition is the stigma surface, as in the grasses, but it is the proteins in the outer coat of the pollen grain (Chapter 2) which are recognised, not the growing pollen tube. This pollen surface material derives from the parent plant, the sporophyte or spore-bearing plant, not from the pollen grain itself, hence the name of this system. The pollen grain is either inhibited from germinating or, as in the grasses, if it germinates, the pollen tube is blocked before it enters or just as it enters the stigma, stopping penetration. The system shares with that of the grasses the efficiency of early recognition and rejection of incompatible pollen grains. Sporophytic self-incompatibility is a feature of two of our largest and most important plant families, the crucifers (Brassicaceae) and the composites (Asteraceae). There are reports of similar systems from *Ipomoea* in the bindweed family (Convolvulaceae) and from hazel in the Betulaceae, but further confirmation is needed to see whether these have a similar action to that in the crucifers and composites (Gibbs, 1986, 1988). Lundqvist (1990) reported a similar system in the field mouse-ear (*Cerastium arvense*), in the pink family, Caryophyllaceae with, interestingly, a remnant of a gametrophytic interaction between the growing pollen tube and the style.

In the crucifers and composites and, probably, the other families, a single gene locus is involved (Gibbs, 1986; Stevens & Kay, 1988) and this may have many alleles. Ford & Kay (1985) estimated that there were at least 24 alleles in populations of the charlock (*Sinapis arvensis*), Sampson (1967) estimated 25-34 in the related wild radish (*Raphanus raphanistrum*), and Lundqvist (1990) 7-19 in the field mouse-ear, similar numbers to confirmed reports in the gametophytic system.

The self-incompatibility systems described so far are most efficient as breeding systems, since they allow any individual plant to breed successfully with most other members of the population but not itself or some of its nearest relatives. Since plants rely on external agents for pollination and so cannot choose their mates, this is a considerable advantage.

Heteromorphic self-incompatibility

This is a most distinctive form of self-incompatibility, since two or three recognisably different forms of flower occur, usually involving different lengths of style (when it is known as heterostyly). Any one plant produces flowers of only one of these forms. The most familiar example is the 'pin' and 'thrum' flower forms of the primrose, *Primula vulgaris*, and many other *Primula* species (Fig. 12.1). The styles are of two different lengths, with the stigma visible at the top of the corolla tube and the anthers half way down in the pin form, the anthers at the top of the tube and the stigma half way down in the thrum form. The pollen grains are of different sizes and the stigma of the thrum is much more papillose – almost mop-like – than that of the pin. In fact, a similar system with two different style lengths has been found in many other plants; in Britain it is found in the primrose, cowslip (*P. veris*), oxlip (*P. elatior*), birdseye primrose (*P. farinosa*) and water-violet (*Hottonia palustris*) (all in the primrose family, Primulaceae), the bogbean (*Menyanthes trifoliata*) and fringed water-lily, (*Nymphoides peltata*) (both Menyanthaceae), the lungworts (*Pulmonaria* spp., Boraginaceae), and the commonly cultivated forsythia (*Forsythia* spp.) and winter jasmine (*Jasminum nudiflorum*, Oleaceae) (Fig. 12.2). There are slight differences between the various groups (e.g. some show no differences in their pollen). Some members of four plant families, the wood-sorrel family, Oxalidaceae, the loosestrifes, Lythraceae, the water-hyacinth family, Pontederiaceae, and the daffodils, Amaryllidaceae, have gone a step further and have three forms of flowers rather than two, with three alternative style lengths and stamen positions. The one British representative is the

Fig. 12.1 Flowers of cowslip (*Primula veris*) with half the corolla cut away, showing a 'pin' (long-styled) flower on the left, and a 'thrum' (long-styled) flower on the right.

Fig. 12.2 Flowers of *Forsythia* cultivars (*Forsythia suspensa* and hybrids with *F. viridissima*): **a**, surface view of a short-styled flower; **b**, section of short-styled flower; **c**, section of long-styled flower. Each cultivar is a single clone, so will always have the same style form. In Britain, most cultivars appear to be short-styled with only about one in ten planted bushes being of a long-styled form.

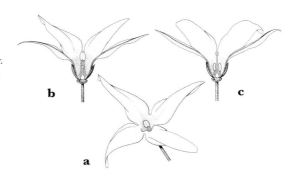

purple loosestrife, *Lythrum salicaria* (Fig. 12.3).

The thrift, *Armeria maritima*, and some of the sea lavenders (*Limonium* spp.), both in the Plumbaginaceae, have two flower types in which the styles are about the same length so they look the same to the naked eye, but the stigmas and pollen grains are strikingly different under the microscope. The stigmas have different sizes of projections (papillae) and the pollen grains have different sculpturing on the surface (Fig. 12.4) and it is possible that other plants with such a system remain to be discovered.

In all these heteromorphic plants, 'pin' pollen pollinates 'thrum' stigmas, and *vice versa*, and the various differences that are apparent between the types of flower are all related to this function. The pollen will be carried on different parts of the visiting insects, which will favour pollination between different forms and minimise stigma clogging by incompatible pollen (Wolfe & Barrett, 1989).

Heteromorphy is nearly always associated with a self-incompatibility system. Self-pollination or crossing between plants of the same type does not, in general, lead to fertilisation. The mechanism of recognition between stigma and pollen grain is complicated, and it differs in the different heteromorphic groups. In some of these plants it differs between the long-styled and the short-styled forms. It often involves a stoppage of penetration on the short style, and sometimes on the long style too, but in the long style pollen more commonly penetrates the stigma but bursts in the style as in the

Fig. 12.3 Three forms of purple loosestrife (*Lythrum salicaria*): **a**, short-styled; **b**, mid-styled; **c**, long-styled.

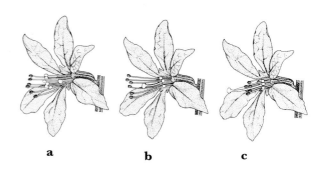

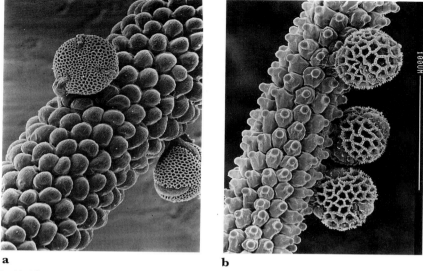

a   **b**

Fig. 12.4 Scanning electron micrographs of stigmas of thrift (*Armeria maritima*), with compatible pollen grains attached: **a**, 'cob' stigma with 'papillate' pollen; **b**, 'papillate' stigma with 'cob' pollen.

gametophytic incompatibility reaction. In some species, incompatible pollen tubes grow right down to the base of the style before growth ceases (de Nettancourt, 1977; Richards, 1986; Gibbs, 1986).

The genetic control of heteromorphy appears to be fairly simple. In species with two flower forms the evidence is consistent with there being a single gene with two alleles, both alleles being present in the thrum form, two copies of the same allele being present in the pin form. In those with three flower forms, two genes, each with two alleles, are involved (Ganders, 1979). The recognition is classed as sporophytic in that it is the parent genotype that is rejected (e.g. all thrum pollen is rejected on pin stigmas although half of it will be carrying the same allele as is present in the pin style), but the mechanism seems to bear little relationship to the sporophytic recognition described for the crucifers etc., and probably differs between the different heteromorphic groups (Richards, 1986; Gibbs, 1986).

Heteromorphic self-incompatibility seems to be a less efficient system than the other forms of self-incompatibility, firstly, because it means that each plant can only fertilise half the population (or two-thirds in those with the three style lengths) rather than the majority, and, secondly, because it seems that the system 'leaks' more often, in that more self-fertilisation happens, particularly on pin (long-styled) plants.

Heteromorphy appears in a diverse range of at least 24 families, including both monocots and dicots (Ganders, 1979). There is certainly no common relationship between the heteromorphy in these families, and all the families contain non-heteromorphic members. It does not occur among the groups which have the most primitive floral structure, such as the Magnoliidae, nor in the most advanced flowers such as orchids and composites and their relatives. It is characteristic of insect-pollinated flowers with few stamens, precise pollination, radial symmetry and in what we might regard as a middle evolutionary level (Endress, 1994).[1] Heteromorphy has clearly evolved many times and possibly several times even within one family such as the Rubiaceae (Barrett, 1988). It seems as if there must be a big advantage in having the two (or three) different forms, presumably for precision in pollination, when coupled with the simple two-allele system. In those with two forms we should expect approximately a 1:1 ratio of pin and thrum plants since either in the minority will have an advantage; this is usually found (Ganders, 1979).

Late-acting self-incompatibility

The types we know least about are those known collectively as late-acting self-incompatibility (Seavey & Bawa, 1986; Gibbs 1986, 1988). There are several potentially different systems coming under this broad category but, in all, the incompatibility acts at a late stage, after the pollen tube has grown right down the style and as it enters the embryo sac, or even after fertilisation. In these systems, the ovule (or the embryo) is aborted. Since late-acting self-incompatibility was discovered in cocoa (*Theobroma cacao*, Sterculiaceae) by Cope (1962), one form or another of it has been found in a number of plant families. It may be associated particularly with woody plants and is described from a number of tropical trees in the Bombacaceae and Bignoniaceae (Gibbs, 1988; Gibbs & Bianchi, 1993), but one of the earliest known examples is the birdsfoot trefoil (*Lotus corniculatus*), and it is found in other legumes, so it is certainly widespread. Godley & Smith (1981) found a type of late-acting system in *Pseudowintera* in the primitive winter's bark family, Winteraceae.

Because the ovules abort, this type of self-incompatibility might appear to be expensive for the plant in terms of lost reproductive potential. However, many plants always produce more flowers than can mature into fruits. Ovule abortion is common for many different reasons, particularly among woody plants, so it may not be too serious a loss for the plant. In many plants, self-fertilised ovules may abort because of the problems associated with self-fertilisation mentioned at the beginning of this chapter. This 'inbreeding depression' is discussed in more detail later, but it is possible that combinations of lethal genes arising from self-fertilisation lead to the ovule abortion that is the characteristic feature of late-acting systems. If this is so, a highly variable response to both self- and cross-fertilisation would be expected and this has been found in a few species (Krebs & Hancock, 1991; Manasse & Pinney, 1991; Seavey & Carter, 1994). In this interpretation there is no self-incompatibility system as such, just marked inbreeding depression. However, Waser & Price (1991) found almost total ovule abortion following selfing in the scarlet gilia (*Ipomopsis aggregata*) and suggested that a true self-incompatibility system must be involved. Late-acting self-incompatibility has been suggested for species in a wide range of families so a variety of mechanism is likely, but much more study is needed before the details are resolved.

Some general comments

Probably about a third of the British flora has one of these self-incompatibility systems, but amongst the grasses the proportion is greater, perhaps two thirds or more

---

[1] In two of the four tristylous families, Lythraceae and Pontederiaceae, the flowers show slight zygomorphy. It is possible that tristylous plants differ in this respect.

(Clapham *et al.*, 1987; Grime *et al.*, 1988). In most plants with one of these systems there is no self-fertilisation since, in an incompatible pollination the pollen tube either never reaches the ovule or the ovule is aborted. In addition to these precise systems, there is a 'grey area' of partial self-incompatibility. The borage family (Boraginaceae) probably has a late-acting type of system, but some self-fertilisation does occur (Crowe, 1971; Varapoulos, 1979); this is less likely if a plant is itself a product of a self-fertilisation. Several gene loci appear to be involved. One of the North American lupins, *Lupinus nanus*, is regarded as self-compatible, but, as in the Boraginaceae, this is partial: selfed flowers had lower self-fertility than crossed flowers (Karoly, 1994). In other members of the borage family, and in the willow-herb family, Onagraceae, Weller & Ornduff (1977) and Bowman (1987), have described what they term 'cryptic' self-incompatibility; selfing is possible but rarely happens since cross pollen grows faster in the style and most seeds are the result of cross-fertilisation. This is likely to occur widely and as in late-acting self-incompatibility is difficult to distinguish from the strong disadvantages of self-fertilisation which are shown by many plants.

In heteromorphic plants, the intermediate area is easier to study because of the morphological differences involved. There are some intermediate forms and the self-incompatibility has broken down in some plants. Casper (1985) demonstrated a breakdown of the self-incompatibility in *Cryptantha flava* (Boraginaceae), although in this species plants still have the two style forms. A breakdown in primroses has been accompanied by a disappearance of the heterostyly, and populations of these 'homo-style' primroses (pin-type long styles with thrum-type anthers) have been known from a few isolated places since the 1940s (Crosby, 1949). The homostyles are self-fertile and their relative advantages or disadvantages over the normal heterostyles are not clear, but they have persisted for many years (Piper *et al.*, 1984). Some *Primula* species are regularly homostyle, representing a permanent breakdown of the heterostyly. In the wood-sorrel family, Oxalidaceae, and in the water-hyacinth family, Ponted-eriaceae, both of which contain species with three style morphs and species with two morphs, the three-styled form has broken down to a two-styled form and sometimes to homostyly, probably more than once in each family (Weller & Denton, 1976; Barrett *et al.*, 1989). In many heterostylous plants it is not known whether or not there is an accompanying self-incompatibility system and in one of the European buglosses, *Anchusa officinalis*, there seems to be a self-incompatibility system which is not connected to the heterostyly (Philipp & Schou, 1981). In this species, most unusually, there are always more pins than thrums (from 2:1 to 28:1 in different populations) and there is some form of self-incompatibility but different pin (or thrum) plants may be fully interfertile.

Evolution of self-incompatibility

Self-incompatibility is a remarkable system in a number of ways: it is strange not only that a plant's own tissue should be rejected and foreign tissue accepted, but also that several quite different mechanisms are involved in different plant groups, with different sites of recognition and different types of response. How then are the various forms related? Are they all derived from the same system? Which is the most primitive and is there any particular advantage of one system over the others?

The evolutionary relationships between the systems are not at all clear and have been the subject of dispute. Fossil plants cannot tell us which are the primitive and which the advanced systems, so we have to make inferences from a study of living plants and their known or suspected relationships. Whitehouse (1950) and others have thought that self-incompatibility of some kind is ancestral in the flowering plants and may have contributed to their success. This idea has achieved wide support. Self-incompatibility of one kind or another is certainly widespread among angiosperms and includes some of the most primitive families. It is often considered that a form of gametophytic system is primitive, but most authors have not considered the late-acting systems such as have been shown in the Winteraceae. Gibbs (1986) has stressed the lack of evidence from most plants. Zavada & Taylor (1986) suggested that recognition at the stigma surface, probably sporophytic, is likely to be primitive, but this idea is not generally accepted. The efficient single-locus gametophytic system may well be derived from a multi-locus system such as that in buttercups, possibly more than once (Bateman, 1952; Lundqvist, 1975; Barrett, 1988). Sporophytic systems have probably evolved more than once since the families in which it has been found are not closely related. Heteromorphic incompatibility does seem to be a derived system, possibly from a self-fertile ancestor in each group in which it occurs. Speculation is rife here too and various authors have suggested that it originated from a gametophytic system or a sporophytic system (Muenchow, 1982; Gibbs, 1986). There is, clearly, a great deal still to learn about all these systems and, though the study can require sophisticated equipment and much patient and careful microscopy and dissection, it promises to be a rewarding field of research for many years to come.

Self-fertilisation

Many plants are capable of self-fertilisation, despite the enormous range of adaptations to attract pollinators and to disperse pollen onto the stigmas of other plants. Self-fertilisation is common in many species, and in some it is the norm. To start with, all the self-incompatibility systems sometimes 'leak'. The plant's own pollen can reach the ovules in some individuals, probably in all species with such a system. There is always the fact that cross-pollination relies on an external agent, often an animal, to be available to transfer pollen from one individual to another at the right time. This may not always be possible because pollinator populations, especially of insects, may fluctuate enormously from year to year, and in some years there may be inadequate numbers for full pollination of a species (Varley *et al.*, 1973; Taylor & Taylor, 1977). The ability to self-fertilise, to produce seeds without the aid of any pollinating agent, is then an advantage – producing any seed, even with some disadvantageous mutations, is better than producing no seed at all. In the majority of British plants, perhaps two-thirds of the species, all individuals are capable of self-fertilisation, and this is probably true of most temperate regions. The great majority of short-lived plants are capable of self-fertilisation but the proportion is much smaller in long-lived perennials and woody plants. In the tropics, particularly in the lowlands, the proportion may be smaller (Bawa *et al.*, 1985a).

Plants that can self-fertilise range from those which do so only occasionally through to those which very rarely cross-fertilise, so that full self-incompatibility and full self-compatibility should really be seen as the ends of a continuum of variation (Stephenson & Bertin, 1983). Having said this, there are a number of features associated with species which habitually self-fertilise and these will be considered first.

Habitual self-fertilisation

This is particularly common in ephemeral plants with short life spans of a few weeks and a single flowering episode followed by death (these plants are often, rather misleadingly, known as 'annuals' although most live for much less than a year). They occur in most open habitats and may form the majority of species in unpredictable environments, such as arable land or gardens and some coastal areas like parts of sand

a

Fig. 12.5 Flowers of cranesbills: **a**, self-incompatible flower of wood cranesbill (*Geranium sylvaticum*) soon after opening; one whorl of anthers shedding pollen, stigma not yet expanded; **b**, wood cranesbill flower at a later stage when the pollen has all been shed and the stigmas are receptive; **c**, self- compatible, smaller flower of dovesfoot cranesbill (*Geranium molle*) showing close proximity and simultaneous maturation of anthers and stigmas.

b

c

dunes. Habitual self-fertilisation is also characteristic of many plants living in environments with cold and wet weather during their flowering season, such as some northern coastal and montane environments, but most habitats have at least some habitual selfers.

Most habitually self-pollinating species have flowers which are smaller in all their parts than those requiring cross-pollination and they usually have fewer flowers with less or no nectar, fewer pollen grains and ovules, and a less well-defined colour and guide marks. Some genera demonstrate this well, for instance, in the bittercress genus, *Cardamine*, the flowers range from the large, showy and self-incompatible flowers of the lady's smock (*Cardamine pratensis*), which is visited by various insects, through the smaller-flowered large bitter-cress (*C. amara*), which attracts pollinating flies but is probably capable of self-fertilisation, through to the small-flowered hairy bittercress (*C. hirsuta*), common as a garden weed, which attracts only an occasional visiting insect and is automatically self-pollinated. A similar range is shown by the cranesbill genus, *Geranium*, with the large-flowered meadow cranesbill (*G. pratense*) and wood cransesbill (*G. sylvaticum*) being mainly or entirely cross-pollinated, chiefly by bees, and the small-flowered dovesfoot and cut-leaved cranesbills (*G. molle* and *G. dissectum*), automatically self-pollinated (Fig. 12.5); there are many other examples.

The number of pollen grains per ovule in a flower is less in habitual selfers than in cross-pollinating plants and, indeed, this pollen:ovule ratio can be quite a good guide to the occurrence of self-fertilisation in some groups. Flowers of habitual selfers usually have less than 100 pollen grains per ovule, while cross-fertilising flowers have 700 or more, although there are exceptions (Cruden, 1977).

Attraction of insects is not necessary for self-pollination, so the production of large or showy flowers will be selected against as a drain on energy supplies for no reproductive gain. There is another, probably more important, reason why so many ephemeral plants are habitual selfers, however, and that is connected with the fact that, for many ephemeral species, speed of development and production of seeds in the shortest possible time will be a great advantage. Many self-pollinating ephemeral plants show features in their attractive parts that are similar to juvenile flowers of outcrossing relatives, and it seems that development of the anthers is simply speeded up relative to maturation of the petals, so the flower opens at a younger developmental stage and earlier in the plant's life. This has two other consequences; firstly, the anthers and stigmas are more likely to be at the same height in the flower, making self-pollination more likely, and secondly, the self-incompatibility system can be circumvented, since normally it only develops when the flower opens; in fully self-incompatible plants self-pollination when still in bud leads to successful self-fertilisation (Richards, 1986). In this scenario, the main change is simply in the relative speed of development of the different floral organs (Guerrant, 1989; Diggle, 1992).

The change in flower form from outcrossing to habitual selfing can sometimes have a consequence in the way we classify plants, since so often the classification is based on the form of the flower. One American species of sandwort, *Arenaria alabamensis*, was separated from the widespread *A. uniflora* on the basis of smaller flowers and other characters associated with self-fertilisation, although these characters almost certainly arose independently from the parent species in two different places (Wyatt, 1988). Gottlieb (1973) and Crawford et al., (1985) give similar examples and there may well be others. Self-fertilisation over many generations will lead to large numbers of plants becoming nearly uniform genetically, since there will be no opportunity for new combinations to arise and certain genes will become fixed. Plants in one site may then differ slightly from those in another in a consistent way, different genes becoming sta-

Fig. 12.6 Silhouettes of
specimens of whitlow-grass
(*Erophila verna*), from three
different populations:
A, Oxford; **B**, Bishopsteignton,
Devon; **C**, Rattery, Devon.

bilised in the different places, or, within one site, there may be a small number of
slightly different morphological types. This is sometimes sufficiently noticeable that
they have been classified as a range of species in the past, for instance the tiny whitlow-grass, *Erophila verna*, common on thin soils in northern Europe in early spring, was
given about 200 different species names by the nineteenth-century French botanist,
Jordan (Fig. 12.6). Differences between the plants were quite consistent, and maintained in cultivation, but these 'species' simply represented different self-fertilising
lines. The various European species of eyebright (*Euphrasia*) show a suite of distinguishing characters but many of them also appear to represent a range of self-fertilising lines (Yeo, 1966).
    Not all habitual selfers have small and inconspicuous flowers and it cannot be used
as a character to indicate a predominantly selfing species. Indeed, some surprising
plants are predominantly self-fertilising, none more so than the bee orchid (Chapter
7). In this and other large-flowered species, habitual selfing may be a recently evolved
feature and, in other parts of their range, particularly in the centre of their geographical or ecological range, these plants may be mainly cross-pollinated (Schoen, 1982;
Wyatt, 1988).
    It does seem that even in those plants with very small flowers and automatic self-pol-
lination, cross-pollination can, and does, take place from occasional insect visits, leading to just a little cross-fertilisation. Marshall & Abbott (1982, 1984) showed that in
many populations of groundsel (*Senecio vulgaris*), over 99% of the seeds resulted from
selfing, but occasional cross-fertilisation occurs in all populations. This has led to hybridisation with the Oxford ragwort (*Senecio squalidus*), and subsequent incorporation
of ray florets into some populations. The eyebrights, mentioned in the previous paragraph, form fertile hybrid swarms in places. Stebbins's (1957) contention that no species is entirely self-fertilising throughout its range has not been disputed.

Partial or occasional self-fertilisation

Many species which do not automatically self-fertilise are, nevertheless, self-fertile. In some species self-fertilisation serves as a 'back-up' if cross-fertilisation fails. What may happen is that the pollen tubes deriving from a cross-pollination grow more quickly down the style, or otherwise outcompete pollen tubes from self-pollination (Stephenson & Bertin, 1983). If no cross pollen reaches the stigma, it will self-fertilise and produce seeds. Another sort of back-up system is shown by some of the bellflowers (*Campanula* spp.), in which the anthers mature first and the stigma pushes up through the anthers before opening. As they mature, the stigma lobes diverge, eventually bending right round so that they can come into contact with any of the flower's own pollen remaining on the style. Thus, self-pollination can take place if cross-pollination fails (Fig. 6.14) (Faegri & van der Pijl, 1979). In some species, rain may occasionally enhance self-pollination within a flower (Hagerup, 1950; Catling, 1980).

A few species even have two different sorts of flower, one largely or entirely cross-fertilising and the other entirely self-fertilising. The clearest examples in the British flora are the violet species (*Viola* spp.) and wood-sorrel (*Oxalis acetosella*), both of which produce showy flowers in the spring. These flowers are attractive to some insects around at that time of year, but these are rather sparse and often many flowers remain unvisited. They are incapable of automatic self-pollination (Plate 1c). Later in the year, small flowers are produced, mainly obscured by the leaves; these flowers never open and self-pollinate automatically. This way, if cross-pollination fails, some seed will still be set by the self-pollinating flowers (Beattie, 1969). A number of other species do this (particularly among short-lived species of legume and grass) with the self-pollinated flowers being produced under the ground. This establishes two simultaneous dispersal strategies, with genetic stability remaining at the parental site but more variability dispersed to other sites (Cheplik, 1987).

Evolution of self-fertilisation

The two stimuli for the evolution of self-fertilisation appear to have been shortage of suitable pollinators and/or speed of development in ephemerals (Jain, 1976; Charlesworth & Charlesworth, 1987; Wyatt, 1988; Diggle, 1992). The disadvantages of self-fertilisation, which arise mainly from the manifestation of deleterious genes, are likely to be great, at least initially (Charlesworth & Charlesworth, 1987), so we must assume that the advantages gained by self-fertilisation outweigh this in some circumstances. 'Hybrid vigour' has long been known from crop plant breeding, the 'hybrids' being crosses between two varieties, but it was Charles Darwin himself who first seriously studied the effects of cross- and self-fertilisation. Typically, the experiments were extensive and many of them remarkably complete (Darwin, 1876). He studied over 50 species in all and, although some provided rather equivocal results, mainly through small sample sizes, the great majority showed inbreeding depression. His main comparisons were with growth under greenhouse conditions of seedlings that were the result of cross- or self-pollination, but he also studied number of fruits set and weight of seeds. In many species, he found that the crossed flowers were superior, but there were different effects in different species. The disadvantages of self-fertilisation may be manifest at the stage of seed set, seed weight, germination, establishment or survival, and modern work has confirmed that different species show different responses. Some plants recorded as self-incompatible, particularly if that self-incompatibility is not complete, may, in fact, just show a very pronounced disadvantage of self-fertilisation. A few examples will suffice to give an idea of the variety of responses.

In chives (*Allium schoenoprasum*), Stevens & Bougourd (1988) showed that the biggest disadvantage of self-fertilisation over cross-fertilisation was in depressed seed set, whereas seed weight and germination were most important in some *Phlox* species (Levin, 1989). In the columbine (*Aquilegia caerulea*), self-fertilisations gave fewer and lighter seeds than crosses and selfed fruits had a 38% higher abortion rate (Montalvo, 1992). The mountain laurel (*Kalmia latifolia*) showed about 75% reduction in fruit set from selfs versus crosses (Rathcke & Real, 1993). Waller (1984) demonstrated that there was a small effect at almost all stages in the jewel weed (*Impatiens capensis*), leading to a distinct disadvantage overall. The effect can be quite subtle, and Kohn (1988) showed, in a wild cucumber species, *Cucurbita foetidissima*, that seedlings derived from cross-fertilisation were three times more likely to survive their first year in the field than those derived from selfing, although there were no detectable differences in seed set, seed size or germination. In *Hydrophyllum appendiculatum* no differences were apparent in the greenhouse but when the plants were grown in crowded, competitive conditions, the crossed progeny demonstrated their superiority (Wolfe, 1993). Finally, Snow & Spira (1993) emphasised the variety of response; some inbred plants in their study on *Hibiscus moscheutos* were vigorous, although overall, on average, there was some inbreeding depression. From these and other studies on the effects of inbreeding, we can conclude that inbreeding depression is a general phenomenon, and for many species it is marked, but that there is a great variety of response. It means that, for many self-compatible species, even if the overwhelming majority of pollinations are self-pollinations, this may not be reflected by the number of adult plants derived from selfing.

In some plants which habitually self-fertilise, inbreeding depression appears to be slight or apparently non-existent. Presumably in these plants there must have been a period of a few generations in which many plants died as a result of the manifestation of disadvantageous genes. In many, though, there is still inbreeding depression, often similar to that in outbreeders (Karron, 1989) and, in others, a plant resulting from a cross-fertilisation may show improved performance on the norm so cross-fertilisation is advantageous (Richards, 1986; Charlesworth & Charlesworth, 1987). New mutations will keep arising and will be the main source of variation, although most will be deleterious so there is likely to be a constant elimination from the population. One can imagine that cross-fertilisation in some habitual selfers could disrupt gene complexes that have proved themselves to be successful, so crossing would then be less favourable than selfing, but in no study on any species that we have come across have the seeds or seedlings derived from selfed flowers proved to be superior overall to those from crossed ones. Selfing is particularly common in polyploid plants (those in which the chromosome mumber has increased, usually doubled, and often following hybridisation). These may be less affected by deleterious mutations because of the extra genetic material present.

Selfing will allow rapid expansion of a population with a combination of genes that has proved successful. Assurance of seed set always seems to be the advantage of selfing. The fact that this can happen even when there is a single individual means that they have an advantage as colonisers. It is easy to imagine that, for ephemeral plants in weedy habitats which change rapidly but are fairly free from competition, the production of a large number of seeds in a short time will be much more significant than the production of a smaller number of stronger and more variable offspring. In these habitats, it is likely that populations will sometimes fall to very low numbers when the ability to recover even from a single individual will be at a considerable advantage. The production of fewer, but stronger and more variable offspring, from cross-fertilisation, is likely to be more suited to a stable habitat, where plants will be in competition with each other and can live longer. It is salutary to be reminded, however, that any generalisation has its exceptions and two of our most important short-lived cornfield weeds, the corn poppy (*Papaver rhoeas*) and the charlock (*Sinapis arvensis*), are both self-incompatible; different species have adapted to similar problems in many different ways[1]. Self-fertilisation may be a safe way to ensure that some seed is set, but the sacrifice, the reduction of variability on which the adaptability of the plant to a changing environment depends, is great.

Adaptations to limit self-pollination

Self-incompatibility stops self-fertilisation but, of course, it cannot in itself stop a plant's own pollen from landing on the stigma. This has long been thought of as a disadvantage since there is a limited amount of space on a stigma, and a plant's own pollen may clog the stigma or interfere with the germination and growth of pollen from another plant. In gametophytic systems, the self pollen germinates and in late-acting systems it leads to ovule abortion, so the interference is particularly marked. In *Polemonium viscosum*, a relative of the jacob's ladder with gametophytic self-incompatibility, self-pollination reduced the germination of compatible pollen from other plants by up to 32% and, in some years, reduced seed set by up to 40% (this was variable between years; Galen *et al.*, 1989). Similar reductions in seed set resulted from self-pollination (done either before crossing or simultaneously) in two milkweed species, *Asclepias exaltata* and *A. syriaca* (Broyles & Wyatt, 1993; Morse, 1994), and Ockendon & Currah (1977) showed similar interference by self pollen in cabbage (*Brassica oleracea*), with a sporophytic self-incompatibility system. A slightly different form of interference was recorded by Bertin & Sullivan (1988) on the trumpet vine (*Campsis radicans*). They showed that self pollen germinated and fertilised some of the ovules, but only when in the presence of cross pollen, so any self-incompatibility could be overriden. All these studies suggest that it is likely that self-pollination in self-incompatible plants does interfere with pollination and will, therefore, be selected against. The problems of inbreeding depression in most self-compatible plants mean that for many of these too there will be selection for enhanced cross-pollination. So, what can the plant do to limit the amount of its own pollen that lands on the stigma?

Separation of floral parts

One simple and common way to avoid too much self-pollination is to separate the anthers and the stigma, either in time or in space. A separation in timing of the anthers bursting and stigmas maturing is known as dichogamy (Chapter 2). The commoner of the two forms is 'protandry', anthers maturing first in any one flower and shedding their pollen before the stigmas become receptive (Fig. 12.5). The time interval is variable, but, typically, the stigmas mature one or a few days after the anthers. In 'protogyny', the stigmas become receptive first, before the anthers shed their pollen.

Lloyd & Webb (1986), and subsequently Bertin & Newman (1993), made wide-ranging surveys of the occurrence of dichogamy and came to some most interesting

---

[1] Both poppies and charlock have low chromosome numbers and outcrossing may be more important for such plants to maintain different alleles within the genome of each individual.

conclusions. Firstly, protandry is common generally among insect-pollinated plants and is characteristic of flowers with what are considered advanced evolutionary features, such as fused corollas, zygomorphy and specialised inflorescences. It is a feature of most species in some plant families such as the pinks (Caryophyllaceae), labiates (Lamiaceae) and composites (Asteraceae), and is frequently a feature of species with a spike-like inflorescence. Visiting insects will often visit more than one flower on each plant, but bees in particular normally move up a flower spike when foraging (Chapter 5). Familiar examples of flower spikes are the foxgloves (*Digitalis* spp.), in which the lower flowers open first, so the insect will meet first the flowers that have been open longest, perhaps in the female phase. If they are carrying pollen some may then be deposited, and as they move up the spike they reach the younger flowers, in the male phase, so collecting more pollen before moving to another inflorescence. Cross-pollination is then promoted. Many earlier workers assumed that protandry was associated with avoidance of self-fertilisation, but these surveys showed that it is common among plants with a physiological self-incompatibility system, in fact, if anything rather associated with such a system. This inevitably leads us to conclude that avoidance of interference between pollen and stigma – self-pollination without self-fertilisation – is the driving force of natural selection behind protandry. In protandry, pollen dissemination from the anthers and pollen receipt by the stigmas can be completely separated within a flower. This is not always so, and the degree of protandry differs between related species, and sometimes between different populations of a single species. Schoen (1982) demonstrated that populations of an annual member of the phlox family, *Gilia achilleifolia*, differed in their degree of protandry and those with the largest difference in timing were more cross-pollinated than those in which the anthers and stigma matured at around the same time.

Protogyny is not generally as common among insect-pollinated plants as protandry, although it is a feature of some wind-pollinated plants, many of which are monoecious rather than hermaphrodite (see p.343). It is associated with some primitive flowering plants, with certain families such as the buttercups (Ranunculaceae) and the crucifers (Brassicaceae), and with trap flowers such as *Aristolochia* and *Arum* (in *Arum* the female flowers maturing first) (Chapter 10). It is also characteristic of some spring-flowering plants in the temperate zones and becomes more frequent among Arctic plants. Many of these plants do share the fact that pollen transfer is rather imprecise, the insect-pollinated ones having an open generalist structure, and their average size is much smaller than those of protandrous flowers. In contrast with protandry, protogyny is, if anything, associated with self-compatibility. In protogyny, the stigma is receptive for a period before the anthers mature, but if not pollinated it may remain receptive when the anthers release their pollen, so the two functions are not completely separated. Self-pollination is possible should cross-pollination fail, and this may, in part, explain its association with flowers which have less reliable pollinators, such as those flowering early in the year.

Ecological associations of protandry or protogyny must be interpreted with caution, particularly since both features have strong links with certain plant families. Members of the plant family may be associated with the conditions mentioned and the dichogamy simply connected with the family. Some species, though, have both, for example in *Trochodendron aralioides* (the sole member of the east Asian family Trochodendraceae, related to the witch-hazels) individuals of any one population all flower synchronously, but some are protandrous and some protogynous, making it a kind of temporal equivalent of heterostyly (Endress, 1994). In the avocado pear all plants are protogynous, but there are two forms opening at different times of day, promoting outcrossing (Chapter 13, p.359).

Separation of anthers and stigmas in space, known as 'herkogamy', may also be effective to stop automatic self-pollination (Webb & Lloyd, 1986), and sometimes, as in the spring flowers of violets (*Viola* spp.), self-pollination is precluded by such a separation (Beattie, 1969). There is, potentially, the problem that the pollen, being placed away from a flower's own stigma, may then be on a part of a pollinator that does not touch the stigmas of another plant. This is avoided in species with a 'brush' type of inflorescence, such as the shrubby New Zealand *Hebe* species familiar in British gardens. Although the anthers and stigmas are separated, avoiding automatic self-pollination, each flower is small and a pollinator usually walks over much of the inflorescence. In some plants, the problem is solved by the anthers or stigmas moving as the flower matures. This is very often associated with a separation in timing (usually protandry) as described above. For instance, in the yellow mountain saxifrage (*Saxifraga aizoides*) just after the flower opens, the filaments curl one by one to place each anther in turn over the centre of the flower, where it dehisces. The filament then bends back to lie parallel with the petals and the anther drops off. The stigmas only mature after all the anthers have dehisced and are either well away from the centre of the flower or have dropped. In some cranesbills (*Geranium* spp.) and common St. Johns-wort (*Hypericum perforatum*), the stamens bend inwards after shedding their pollen, and in the common mallow (*Malva sylvestris*) they bend outwards; these movements push the stamens out of the way of the maturing stigmas. More dramatically, in the rockrose *Helianthemum nummularium*, the many anthers are sensitive to touch and become splayed after the visit of a pollinator (usually a bee) revealing the stigma in the centre (Fig. 12.7). This can be demonstrated easily by agitating the stamens with a pencil.

In some species with a bowl or disc-shaped flower, such as the greater celandine (*Chelidonium majus*) and, perhaps, the poppies (*Papaver* spp.) and buttercups (*Ranunculus* spp.), the stigmas are in the centre of the flower and the anthers are splayed apart. Visiting insects home in on the centre of the flower and land on the stigma, depositing pollen if they have some, and then move to the edge to take off, walking over the anthers and collecting pollen in the process (Webb & Lloyd, 1986). This seems an effective and simple method of promoting cross-pollination. Other mechanisms involving an insect moving the anthers or stigmas as it enters the flower have been described in Chapter 6.

Although automatic self-pollination in the absence of any pollinating agent is lim-

Fig. 12.7 Common rockrose (*Helianthemum nummularium*). **a**, an unvisited flower soon after opening; **b**, stamens splayed after agitation by a visiting insect (or a biologist).

a   b

ited or prevented by dichogamy or herkogamy, insect visitors usually visit more than one flower per plant and this can still lead to self-pollination. It is avoided only if the flowers open synchronously, which is certainly rare in temperate floras, but does occur in some umbellifers (Apiaceae), and in a few tropical plants (Cruden & Hermann-Parker, 1977; Webb & Lloyd, 1986).

Unisexual flowers

Some plants have flowers with only male or only female parts and, although they are much less common than hermaphrodites, they probably account for about 15% of all plants. Plants with only unisexual flowers have these either on separate plants, dioecy (adjective dioecious), the familiar form in higher animals, or on the same plant, monoecy. A number of other species, though, have unisexual flowers mixed with hermaphrodite flowers on the same or different plants.

Species with unisexual flowers belong to a wide range of plant families and unisexual flowers have clearly arisen many times in the evolution of flowering plants. As so often, though, there are taxonomic associations and nearly all members of one of the ten or so 'superorders' of flowering plants, the Hamamelidae, have unisexual flowers. This group includes many of our common trees, such as the birches and hazels, oaks and beeches, as well as the witch-hazels and some smaller families. Some other families or groups of families have exclusively unisexual flowers, such as the willows and poplars (Salicaceae), which are all dioecious, and the spurges and allies (Euphorbiaceae), which have monoecious and dioecious members. Other species with unisexual flowers belong to largely hermaphrodite families, or even genera, for instance the two sorrel species, *Rumex acetosa* and *R. acetosella* are dioecious, but the docks, such as *Rumex crispus* and *R. obtusifolius*, are hermaphrodite. The campions (*Silene* spp.) demonstrate a whole variety of sex expression: the red and white campions (*S. dioica* and *S. latifolia*), are dioecious; the Spanish catchfly (*S. otites*), is also dioecious but male plants may have some hermaphrodite flowers; the Nottingham catchfly (*S. nutans*) has male, female and hermaphrodite plants; the bladder campion (*S. vulgaris*) has male, female and hermaphrodite plants and plants with unisexual and hermaphrodite flowers, and the night-scented catchfly (*S. noctiflora*) is fully hermaphrodite. In salad burnet (*Sanguisorba minor*) the inflorescences may have all flowers of the same sex or a mixture of male and female (Fig. 12.8).

Fig. 12.8 Inflorescences of salad burnet (*Sanguisorba minor*), left with male flowers; right with female flowers.

Some plants can change sex and, frequently, a male plant will produce the occasional female flower and vice versa. So the expression of sex is somewhat flexible, unlike in most higher animals, and it may depend not only on a plant's genetic make-up but also on the environment in which it is growing (Schlessman, 1988). All sorts of different factors are known to affect the expression of sex in plants, ranging from day length and water or nutrient content of the soil, to levels of plant growth substances produced internally by the plant (Meagher, 1988). This flexibility is exemplified well by the bog myrtle (*Myrica gale*), which may be monoecious or dioecious under different conditions, and any one individual may change sex in different years. There is a well-known isolated plant of this species at Wicken Fen in Cambridgeshire, the sole survivor of a once more extensive population. This large plant is either male or female in any one year, but since it is all one individual it is only one sex at a time and, with no others in the vicinity, it remains sterile.

Dioecy

Dioecy occurs in about 10% of the world's flora. It comprises less than 5% of the British flora, but does include some very well-known plants, such as all the willows (Figs. 9.9, 9.10, pp.277-8), stinging nettles, creeping thistle, holly and mistletoe (most people realise, after collecting them for Christmas, that it is only certain holly bushes and mistletoe plants that ever produce berries). Dioecy is commoner in some parts of the world, with about 20% of Mediterranean shrubs being dioecious (Aronne & Wilcock, 1994), and up to a quarter of tropical rain forest trees (Schatz, 1990). It is especially common in the isolated floras of New Zealand and Hawaii, and the possible reasons for these distributions will be discussed later.

Most dioecious species are insect-pollinated, but they have smaller flowers than hermaphrodites, on average, frequently very small, and these are often greenish and rather inconspicuous with unspecialised pollinators. Some are wind-pollinated, such as the mercuries (*Mercurialis* spp.) (Fig. 12.9), stinging nettle (*Urtica dioica*), poplars and a few others and the willows may be both insect- and wind-pollinated. In many dioecious species, particularly those in largely hermaphrodite families, there are vestigial organs of the other sex present in the flower, and the sterile anthers in the females may be part of the attractant in female flowers; some provide sterile pollen. Normally, though, nectar is the attractant, since most female flowers do not produce pollen.

Fig. 12.9 Dog's mercury (*Mercurialis perennis*): left, a male shoot; right, a female shoot (stigmas arrowed).

Male flowers of dioecious species are very often larger and more showy than female flowers, produced in greater quantity and in larger inflorescences (Bawa, 1980; Givnish, 1982; Kay & Stevens, 1986). This will mean that male flowers are more likely to attract the attention of pollinating insects from a distance so that

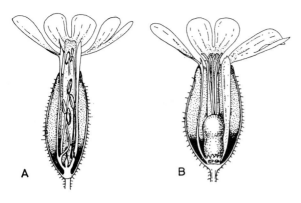

Fig. 12.10 Male (**A**) and female (**B**) flowers of red campion (*Silene dioica*). Note the larger corolla scales and rudimentary female parts in the male flower and the ring of rudimentary anthers in the female flower.

insects may visit them before the females, clearly essential for successful pollination. Indeed, for maximum male reproductive success, pollen must be disseminated as widely as possible, so many insect visits are needed, whereas female flowers may need just one successful visit for full seed set (Bell, 1985; and the consequences of this idea are followed up in Chapter 16). The development of petals and stamens is often connected, however, and it is possible that the frequently smaller size of female flowers may be a by-product of suppression of anther development, suppressing petal development at the same time.

The quantity or concentration of the nectar secreted may differ between male and female flowers and this may affect insect visits. More nectar may be secreted by males (e.g. in cloudberry [*Rubus chamaemorus*], Ågren *et al.*, 1986) or by females (e.g. in red campion [*Silene dioica*], Kay *et al.*, 1984), and in some circumstances this may lead to insects discriminating between males and females and visiting mainly one sex, although this will be most inefficient for the plant. The flowers of one sex appear to mimic those of the other in appearance in some species, e.g. staminodes or equivalent in female plants mimicking anthers (Ågren *et al.*, 1986) or, in red campion, corolla scales on the male perhaps mimicking stigmas (Fig. 12.10) (Kay *et al.*, 1984; Willson & Ågren, 1989). This implies that visiting insects are deceived into visiting one or other sex, though evidence for mimicry is mainly circumstantial (Chapter 10).

Some dioecious species show a chromosome difference between males and females, as in mammals and birds, though it is usually a minor change, not one so large and obvious as the XY system in mammals. This, too, is variable and in many dioecious species there is no distinct sex chromosome.

Dioecious animals mostly have an approximately 1:1 sex ratio and, indeed, selection should favour this (Maynard Smith, 1978). We perhaps expect dioecious plants to do the same, but in fact, the sex ratio varies considerably (Willson, 1983; Sutherland, 1986; Kay & Stevens, 1986). The ratio may be biassed in favour of males or females and, commonly, different populations of the same species differ in their sex ratios. The recording of the sex ratio may be confounded by a slight difference in the growth or ecology of males and females. Not only do males often produce more and larger flowers than females, but they may also start flowering at a smaller size or an earlier age (Kay & Stevens, 1986). Often, differential growth and survival may affect the sex ratio, particularly in a plant with vigorous vegetative spread. Males and females may even have slightly different distributions, either within a habitat or geographically. If this happens the male is normally tolerant of a wider range of habitat than the female, occurring on the thinner soil or the drier habitat, perhaps because the requirements for fruit and seed set are particularly exacting. The most striking example of this in Britain is the butterbur (*Petasites hybridus*), in which males are known from much of the country, spreading by vegetative means, but females are confined to a broad band in northern England and southern Scotland. Some other species show smaller differences in tolerance, but in many no differences have been recorded (Kay & Stevens, 1986; Iglesias & Bell, 1989). Again caution is needed here since, in the Canadian pondweed (*Elodea canadensis*), introduced into Britain and now very common in still and slow-moving water, almost all the plants are female. Males have been recorded, but are very rare. The vigorous vegetative spread of this plant has led to a highly biassed sex ratio, presumably by chance, and with most 'plants' really just fragments of a small number of extensive clones.

Dioecious plants have evolved many times from hermaphrodite plants, but, on the face of it, it is hard to see why, since it appears to involve the surrender of half of an individual's reproductive potential. The impossibility of self-fertilisation and its attendant problems was long thought to be its main advantage, but with the increased knowledge of self-incompatibility systems in hermaphrodite plants this is clearly too simple. Avoidance of stigma clogging and ovule wastage is likely to be a major stimulus, however, and the small size of the flowers may be significant here, since self-pollination is less easy to avoid in a small flower. One of the stimuli may be to do with the relative effectiveness of a flower as a male or as a female. A large floral display attracts insects from a distance, particularly unspecialised insects which are rather characteristic as pollinators of dioecious species. If they then move on to other smaller plants, the large plant may function mainly as a male, those with smaller floral displays being relatively more female; this could eventually lead to dioecy, though there is no direct evidence. Some dioecious species have heterostylous ancestors, usually pins becoming female and thrums becoming male. This has been recorded in several families and has almost certainly happened several times, sometimes associated with attraction of insects too small to reach the anthers of pins or the stigmas of thrums (Beach & Bawa, 1980; Muenchow & Grebus, 1989).

If we look a little further, we find that dioecy in plants is correlated with some other features not directly connected with the flowers. It is commoner generally in tropical floras than temperate and particularly among shrubs and small trees; there is an association with fleshy fruits and the presence of vegetative growth, and some oceanic islands have a large proportion of dioecious species (Bawa, 1980; Givnish, 1982; Thomson & Brunet, 1990; Aronne & Wilcock, 1994). Some of these associations do not apply to temperate regions, and island floras and fleshy fruits may be associated together and not directly with the dioecious habit, but these associations suggest that some other selective agents may be important. Fleshy fruits may be significant since they need more resources than most other types of fruits, and plant species generally allocate a similar proportion of their total resources to producing fruits, whatever their breeding system (Sutherland, 1986). Fewer large fleshy fruits can be produced than smaller ones, so female plants may save resources by dispensing with male parts and producing fewer, smaller flowers since fewer insect visits will be necessary.

It does seem that generalisations on the advantages of dioecy that apply to all dioecious species are most unlikely to be forthcoming, since dioecy has appeared in so many different and diverse groups; each individual species or group will have responded to a different set of selection pressures (Thomson & Brunet, 1990). Charlesworth (1993) turned the whole argument on its head by suggesting that dioecy (and monoecy) is simply less likely to evolve in specialist bee-pollinated systems since these flowers often use pollen as an attractant. When a species becomes free of this constraint, as in the development of wind-pollination, or pollination by undiscriminating insects, flowers become smaller and may become dioecious. In this argument, the pollination system comes first. A good example of this is the genus *Thalictrum*, the meadow-rues, which are among the only wind-pollinated members of the buttercup family, and a few of these wind-pollinated meadow-rues are the only dioecious members of the family. Charlesworth (1993) suggests a few other examples, but her argument is likely to apply mainly to those dioecious species closely related to hermaphrodites. It may partly explain the large number of dioecious species on New Zealand and other isolated islands where specialist bee pollinators were absent until introduced by European settlers.

As a postscript on the subject, dioecy and crop plants do not go well together as an external pollinating agent is essential and, under cultivation conditions, may not be sufficiently reliable. Dioecy is probably the ancestral condition in grapes and in hemp but it has been bred out of most cultivated varieties of these crops, emphasising the fact that breeding systems in plants are not fixed but can respond to changed conditions.

Monoecy

Monoecy occurs in a wide range of plants, including 5% or so of the British flora. Monoecious flowers are nearly all small or very small compared with hermaphrodite or even dioecious flowers, and monoecy is quite strongly associated with wind-pollination (Bertin & Newman, 1993). Most of the British woodland is dominated by monoecious, wind-pollinated species such as the oaks (*Quercus* spp.), beech (*Fagus sylvatica*), birches (*Betula* spp.), hornbeam (*Carpinus betulus*), hazel (*Corylus avellana*) and alder (*Alnus glutinosa*) (Fig. 12.11) and the same is true right across the temperate parts of the world. Since the majority of conifers are also monoecious, including our native Scots pine (*Pinus sylvestris*), it means that, despite the fairly small number of species involved, vast tracts of the world are dominated by monoecious species. It is more common in the temperate regions than the tropics, in contrast to dioecy, although the Fagaceae (the oak and beech family) all of which are monoecious, have their greatest diversity in the mountains of south-east Asia. In addition to the trees, all but one of the British sedges, *Carex* spp., are monoecious (*C. dioica* is

Fig. 12.11 Alder (*Alnus glutinosa*), showing the long male and much shorter female catkins (inflorescences) in early March. In the top right hand corner are the dried woody remains of last year's female catkins after shedding their seeds.

Fig. 12.12 Complex inflorescence of bur-reed (*Sparganium erectum*), with male inflorescences towards the tip and the larger female inflorescences near the base.

dioecious) and some water plants, e.g. hornworts (*Ceratophyllum* spp.), water milfoils (*Myriophyllum* spp.) and bur-reeds (*Spargamium* spp.) (Fig. 12.12), all of which are wind-pollinated (or water-pollinated in *Ceratophyllum*). There are a few insect-pollinated monoecious species in Britain, notably the spurges (*Euphorbia* spp.) and lords-and-ladies (*Arum maculatum*). Both of these have tiny individual flowers but they are organised into highly specialised inflorescences which, in many ways, resemble a single flower. Another highly organised inflorescence structure of tiny flowers is demonstrated by the figs, many of which are monoecious (Chapter 11).

In many monoecious species self-pollination is possible, although the hazel and at least some sedges have a self-incompatibility system, probably with sporophytic control (Thompson, 1979; Faulkner, 1973). In the majority of monoecious species, female flowers, often in catkins, mature earlier than male flowers on any one plant. This is a form of protogyny, and in monoecious species protogyny is particularly common and well marked, whereas protandry is almost absent. The main evolutionary pressure favouring protandry is the need for the removal of the anthers before the stigma matures; clearly this is quite irrelevant in monoecious flowers (Bertin & Newman, 1993). In the alder, and perhaps some other species, the whole tree is synchronised, with all female flowers opening first and, later, all the male flowers, further reducing the likelihood of self-pollination (Lloyd & Webb, 1986). In the walnut (*Juglans regia*) the trees occur in two forms, demonstrating a pattern analogous to the two dichogamous forms of *Trochodendron* and forming another parallel with heterostyly. On some individuals the female flowers mature first and in others the male flowers mature first (possibly the only protandrous monoecious plants). Each matures the flowers of the other sex about two weeks after the first flowers open. The plants occur in approximately a 1:1 ratio (Gleeson, 1982).

Growth substances produced internally by the plant exert control over which sex of flowers is produced in which part of the plant and this inevitably means that sex expression in many monoecious species, as usual, is not fixed. As in dioecy, in conditions that are favourable for the growth of any particular species, they will produce more female flowers, and in a more marginal environment, more male flowers, or even exclusively male flowers (Freeman *et al.*, 1980; Meagher, 1988). Many monoecious species are closely related to dioecious species and in the spurge family, for instance, which has many monoecious and dioecious members, diploid annual mercury (*Mercurialis annua*) is dioecious but its polyploid relatives of southern Europe (usually considered as varieties of the same species though sometimes referred to as *M. monoica*) are monoecious.

Monoecy, or at least the possession of unisexual flowers, seems to be the ancestral condition in the trees mentioned here and the sedges, spurges and arums, so its evolutionary origin is certainly ancient. Some of the evolutionary arguments associated with dioecy may apply equally well to monoecy, e.g. the greater likelihood of self-pollination in small flowers and the removal of constraints imposed by specialist animal pollination, but some correlations are quite different. For instance, monoecy is correlated with dry, often single seeded fruits, and no vegetative spread in the Mediterranean (Aronne & Wilcock, 1994), although with so few species this needs cautious interpretation. It does not seem to have arisen many times in evolution in the way that dioecy has and it is mainly confined to well recognised families or other taxonomic groupings which are mainly or entirely monoecious.

Gynodioecy

There are some further refinements in the breeding systems of plants which do not fall neatly into the categories so far considered, and one of the most important is 'gynodioecy', the production of female and hermaphrodite individuals. This usually comes about when some individuals produce flowers with sterile or aborted pollen, thus becoming functionally female. Plants without functional pollen have been reported in many species, usually as just the occasional individual (Stevens & Kay, 1991). In some species, however, it has become fixed in most, or all, populations, including our own wild thyme (*Thymus polytrichus*) (Fig. 12.13) and several other members of the labiate family, Lamiaceae (Kheyr-Pour, 1981; Belhassen *et al.*, 1989). In other species, such as ribwort plantain (*Plantago lanceolata*) and the meadow saxifrage (*Saxifraga granulata*), female plants occur in some populations but not others (van Damme & van Delden, 1984; Stevens 1988). Van Damme & van Delden (1982), Kheyr-Pour (1981) and others have shown that the inheritance and control of gynodioecy is complicated and, in some species at least, involves not just the ordinary chromosomes, but also some input from other cell parts, this all coming from the female parent.

The ratios of the two forms are often not 1:1 and, as in dioecious species, ratios vary between populations of a single species, e.g. Jane Cockram, working with AJL, found that it varied between approximately 1:1 and 4:1 ratios of hermaphrodite:female in wild thyme from different parts of one sand dune system, and Belhassen *et al.*, (1989) showed that the ratio in *Thymus vulgaris* (the Mediterranean common thyme) changed with time after a fire. This strongly suggests that, as in some dioecious species, the two

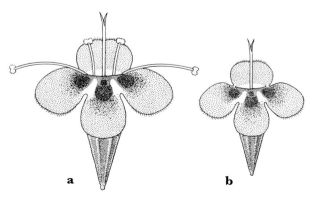

Fig. 12.13 Flowers of wild thyme (*Thymus polytrichus*): **a**, hermaphrodite flower; **b**, female flower to the same scale. Vestigial anthers are enclosed within the corolla tube in female flowers.

a   **b**

forms may have slightly different ecologies, shown most obviously in a species that spreads vegetatively, like wild thyme. Female flowers are nearly always smaller than hermaphrodite flowers of the same species, even when just the occasional individual has been recorded (Stevens & Kay, 1991).

Gynodioecy should only spread where there is a considerable advantage to the females in terms of seed set or vegetative growth (Charlesworth & Charlesworth, 1978; Stevens & van Damme, 1988). More seed is sometimes produced by the females, as predicted, but in other species the hermaphrodites produce more (Webb, 1981; Stevens, 1988). There may be advantages to the offspring of females in self-compatible plants, since they will all be the result of cross-fertilisation. This has been shown in several species (Stevens, 1988; Kohn, 1988) and, significantly, no such advantage was shown in the self-incompatible ribwort plantain (van Damme & van Delden, 1984). In general, much remains to be discovered about gynodioecy and how it is maintained in populations, but it seems that it is a condition that can turn up in almost any species. It has been regarded as one potential intermediate step in the evolution of dioecy by the Charlesworths (1978).

Androdioecy

The occurrence of males and hermaphrodites, androdioecy, is very rare, in contrast with gynodioecy, although some plants show a range of sex expression from male through hermaphrodite to pure female. There usually seems to be no advantage to androdioecy (Charlesworth, 1984). It has been proved to occur in the south-west American species, *Datisca glomerata* (Datiscaceae), which is related to dioecious species and derived from them (Liston *et al.*, 1990; Rieseberg *et al.*, 1992), and the Mediterranean shrubs *Phillyrea latifolia* and *P. angustifolia* in the olive family, Oleaceae (Traveset, 1994). In populations of *P. latifolia* studied by Aronne & Wilcock (1994) males were in the majority and produced denser flowers and three times as much pollen per flower as hermaphrodites. This superiority in male function may be sufficient to maintain males (Charlesworth, 1984).

Polyploid forms of the annual mercury (*Mercurialis annua*) in southern Europe are normally monoecious, but many populations have some male individuals, so this has a slightly different form of androdioecy, consisting of males and monoecious individuals. Males are always a minority of plants (up to 30%) but John Pannell (working partly with AJL) has demonstrated how the combination of males and monoecious plants can be maintained and spread in populations of this short-lived plant under conditions of repeated extinction and colonisation.

Most other suspected androdioecious plants have been shown to have sterile or near-sterile pollen in their hermaphrodite flowers, so they are functionally dioecious. In these species, the pollen may be important as an attraction for insect visitors to the females (Appanah, 1982; Kevan & Lack, 1985; Anderson & Symon, 1989).

Andromonoecy, gynomonoecy and sterile flowers

In some species male flowers or female flowers occur on individuals which also produce hermaphrodite flowers. This is quite common in plants with large or specialised inflorescences, such as some umbellifers (Apiaceae), composites (Asteraceae), and various other groups. In most of these, including the umbellifers, there are extra male flowers mixed in with hermaphrodites (usually males in the middle of the umbels) and, in andromonoecious umbellifers, only about a fifth of all flowers produced are hermaphrodite (Lovett Doust, 1980). These hermaphrodite flowers are usually markedly protandrous. Some of our commonest umbellifers have mainly male flowers, like

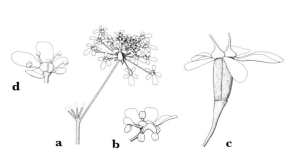

Fig. 12.14 Cow parsley (*Anthriscus sylvestris*): **a**, an umbel showing asymmetrical petals of outer florets; **b**, hermaphrodite flower in early, male stage; **c**, hermaphrodite flower in later, female stage after anthers have dropped; **d**, male flower. Male flowers are in the middle of the umbels and the proportion of male flowers increases as the season progresses.

the cow parsley (*Anthriscus sylvestris*) (Fig. 12.14) and the wild parsnip (*Pastinaca sativa*). Many umbellifers, even those producing only hermaphrodite flowers, mature rather few fruits relative to the number of flowers and it seems that producing some flowers with no carpels may be a further saving on resources while keeping the same floral display. The male flowers can still function as pollen donors, too, probably at fairly low extra cost to the plant, and similar arguments are presented for the evolution of andromonoecy in the nightshade genus, *Solanum* (Willson, 1983; Anderson & Symon, 1989). Some bat-pollinated species are andromonoecious (Chapter 8).

Gynomonoecy is mainly associated with composites (Asteraceae). In some composites, female flowers are produced around the edge of the highly specialised inflorescences and these usually have a different form from the hermaphrodite flowers (ray florets, as opposed to disc florets; Chapter 6, Fig. 6.52). Any advantage here seems to be derived from the increased outcrossing of female flowers, and these usually mature first (Willson, 1983; Ross & Abbott, 1987). In some other species the outer florets lose all their sexual function and become sterile, this being particularly common in composites again, but also occurring rather prominently in such plants as the guelder rose (*Viburnum opulus*) (Fig. 15.2) and *Hydrangea* species. They are there in these species purely to attract insect visitors and are often larger than the fertile florets. Garden forms of the guelder rose, the so-called snowball tree, and many hydrangeas have all their fertile flowers replaced by sterile ones.

Asexual reproduction

Vegetative spread has been mentioned already several times in this chapter. It is, really, horizontal growth – one can, fancifully, imagine an ever-growing oak tree lying on its side as an analogy for how a plant may spread vegetatively. It may be very significant for the spread of a species. Many plants produce side buds or runners, stolons or rhizomes, all horizontal stems which can root at a little distance from the original plant. This growth can sometimes cover a large area which is then, genetically, all one individual, and parts of it at any point may die, splitting up the 'clone'. It can allow one genetic individual to live to a great age; the oldest of all plants are probably clonal creosote bushes (Chapter 16). It is often impossible to tell where the boundaries of one clone are, though cross-fertility relationships in some self-incompatible species can give a good indication. Harberd (1961, 1962) studied clones of the red and sheep's fescue grasses (*Festuca rubra* and *F. ovina*), and concluded that some individual clones were growing over an area about 200 m long and were probably hundreds of years old. Many others are likely to be the same. Species spreading in this way include trees such as the elm species (*Ulmus* spp.), and aspen (*Populus tremula*), through a whole range of herbaceous plants (the all-too-well-known couch grass [*Elymus repens*] and ground elder [*Aegopodium podagraria*] are typical examples) to many water plants, such as the minute duckweeds (*Lemna* spp.).

In other species, pieces of stem or leaf may root, having been dispersed at least a little way from the parent plant, and bulbous plants can spread by budding of the bulbs. Probably all these species flower at least occasionally, and pollination remains a part of their life cycle. The formation of new genetic individuals from sexual fertilisation may be important for the species' reproduction only in the long term and when colonising new sites.

Somewhat more specialised are those plants which produce bulbils in place of some or all of their flowers, the so-called viviparous plants. The bulbils disperse from the parent plant as a clone. Several such plants occur in montane regions where pollination of any kind may be limited by cold, often wet, summers. Examples are the alpine bistort (*Persicaria vivipara*), drooping saxifrage (*Saxifraga cernua*), and some grasses such as the fescue, *Festuca ovina* var. *vivipara*. Some other plants are viviparous and it seems to be a characteristic feature of some of the onion tribe, including the common roadside weed crow garlic (*Allium vineale*). In a few other species (Mogie, 1992, lists 12 genera) a fertilised embryo, derived from normal pollination, may divide to produce two or more seeds, as in the production of identical twins.

Some plants produce flowers, but the seeds mature without any fertilisation from pollen. This is known as agamospermy. It happens in a variety of ways and, in some species, some seeds are produced by agamospermy and some by normal pollination and sexual reproduction. It appears that normal pollination of an ovule may stimulate the production of other unfertilised ones. This is quite common in some citrus trees and other fleshy-fruited tropical species and has been found in canopy rain forest trees in south east Asia (*Shorea* spp., Dipterocarpaceae) and in one genus, *Eriotheca* (Bombacaceae), in the Brazilian savanna (Kaur *et al.*, 1978; Richards, 1986; Oliveira *et al.*, 1992; Mogie, 1992). Among British species the meadow buttercup (*Ranunculus acris*) and common St. John's-wort (*Hypericum perforatum*) are sometimes agamospermous, although normally reproducing sexually (Grime *et al.*, 1988). It may be very difficult to detect agamospermy if normal pollination is required as an initial trigger or if it only happens in some individuals, so it may be much more widespread than generally assumed.

In some agamospermous plants, normal sexual reproduction rarely or never occurs. In the British flora, this condition is rather strongly associated with two families, the rose family and the composites. In the rose family, some of the cinquefoils (*Potentilla* spp.), the lady's mantles (*Alchemilla* spp.), parsley piert (*Aphanes arvensis*), some whitebeams (*Sorbus* spp.), and the brambles (*Rubus fruticosus* agg.), all have agamospermous species, and among the composites, dandelions (*Taraxacum* spp.), hawkweeds (*Hieracium* spp.), and hawksbeards (*Crepis* spp.). A few other families contain agamospermous members in Britain, such as some of the sea-lavenders (*Limonium* spp.), goldilocks (*Ranunculus auricomus*) and the meadow grass (*Poa pratensis*), and world-wide a number of other families and genera are involved (Mogie, 1992) Many of these plants have high numbers or odd numbers of chromosomes and this may inhibit sexual reproduction, although sometimes the plants produce viable pollen. Like self-fertilised species they will have an advantage in colonising since a single individual will be able to found a population.

All the seeds produced by agamospermy will usually be identical genetically to the parent plant (except for rare mutations) and they will, effectively, be a clone dispersed as seeds. Each clone, or set of closely related clones, shows a narrow, but often quite consistent, range of variation wherever it grows. Some of the best studied of the agamospermous groups have been divided into large numbers of 'microspecies' or 'biotypes', which have each been given species names in some floras, although they are, clearly, not equivalent to sexual species, and pose difficult problems for the taxonomist. The range of variation displayed by a collection of agamospermous microspecies is, in most groups, similar to that found in a small number (say, two to five) of sexual species. All the agamospermous plants are closely related to sexual species in the same genus. Some agamospermous plants can undergo chromosomal rearrangements in the process of producing seeds, so a little variation can be generated in them without sexual reproduction (Richards, 1986).

The viable pollen produced by agamospermous plants can fertilise related sexual species or occasionally other (or the same) agamospermous ones, and it is thought that this is common in some groups, although in others it is rare and intermittent. This will give rise to new agamospermous lines and is, presumably, how most of these groups have proliferated. It could also work against the invasion of a sexual species into an area dominated by agamospermous plants, because the sexual species may get swamped by agamospermous pollen, which is likely to be mainly at a different ploidy level from the sexual species, so the resultant seeds will be sterile or agamospermous. With the advantage in colonising ability that agamospermy brings, this may help to explain why, in widely agamospermous groups like the dandelions, the sexual species are restricted to the centre of the range of the genus as a whole (Mogie, 1992). A rather curious feature, and one that is probably a hangover from their evolutionary past, is that many agamospermous plants actually do need pollination for seed set despite the fact that the ovules require no fertilisation. Successful pollination and full pollen tube growth is required in some groups for fertilisation of the endosperm, that other half of the fertilisation process, unique to flowering plants, e.g. the brambles (Richards, 1986). Brambles are, in fact, very attractive to quite a wide range of insect visitors. All the apparatus for pollination remains in agamospermous plants, even though it is not necessary for seed set, probably because the plants are derived from one clone so are identical, or nearly so. There is very little potential for evolutionary change.

13

Plant Breeding and Crop Production: The Un-natural History of Pollination

Pollination, as well as being an object of study in its own right as a biological phenomenon, is an incidental but very important process in plant-breeding and in agriculture and horticulture. Plant-breeding requires controlled pollination on a small scale. It is usually a simple matter to transfer pollen from anthers to stigma with a suitable implement. The basic requirements for normal fertilisation are that the plants should be growing within a certain temperature range that suits them and that the pollen should be viable, which it normally is if transferred direct from flower to flower. However, some plants produce a daily flush of short-lived pollen, and then pollination must be carried out at a particular time of day, when the pollen is fresh. Plant breeders often need to prevent both self-fertilisation and chance pollination from unspecified plants. Prevention of self-pollination usually requires emasculation (removal of the anthers). This may unavoidably cause damage to the perianth, which in some plants unfortunately causes the flowers to drop or wither. However, in some of the cereal crops certain high or low temperatures can be used which destroy the pollen but do not injure any other parts of the flower. For lucerne (alfalfa: *Medicago sativa*) the explosive mechanism is first sprung ('tripped') by cutting off the keel petals. The pollen thus discharged is then either destroyed by immersing the flowers in 57% ethanol or removed by suction through a fine glass tube, the operator wearing a binocular magnifier. Chance pollination from unspecified plants can be prevented by enclosing the flowers in paper or muslin bags before they open or by growing the plants in an insect-proof greenhouse; with large flowers such as marrow (*Cucurbita pepo*), a rubber band placed over the bud prevents it from opening.

The development of pollen and the process of fertilisation are described on pp.28-34). The preservation of viable pollen may make crosses possible which would otherwise be prevented by differences of flowering time, and it enables the pollen to be sent from one part of the world to another without the trouble of transporting living plants and meeting the international plant health requirements. Summaries of research results on the storage of pollen are given by Hoekstra & Bruinsma (1975) and Shivanna & Johri (1985). Grass pollens have a very short life under natural conditions; some remain viable for only a few minutes, while others survive for up to five hours. They have a high water content (30-60%) and in some cases their lifetime can be extended to several days by storage at 50-100% relative humidity (RH). Other pollens, however, will last for several days, weeks, or even for many months, without any special treatment. These are released with a low water content (less than 20%) and the lifespan of those that are binucleate (with an undivided generative nucleus – see Chapter 2) can be further extended two- to ten-fold by storage at low RH (0-40%). Other measures to prolong the life of these pollens (especially binucleate ones) are storage at low temperature or under vacuum (preferably after freeze-drying), or storage in an atmosphere of nitrogen or one with a high concentration of carbon dioxide or in certain organic solvents. All operate by checking the consumption of nutrients by respiration in the pollen. Strong light (particularly ultra-violet) is harmful to the survival of pollen (p.272, Box 9.1).

Pollination normally leads to seed-production only if it is followed by fertilisation and by the adequate development of the embryo and the surrounding endosperm. Often there are no obstacles to these processes in crosses within species or in crosses between closely related species. Sometimes, however, growth of pollen tubes towards the ovary may be retarded or seed development may be imperfect. Sometimes embryos begin to develop but then abort; in such cases, they can sometimes be saved at an early stage by removal from the seed and transfer to a sterile culture medium. Slow growth of pollen tubes can often be countered simply by early pollination (as soon as the flower opens, or even before) provided pollen can be made to adhere to the stigma, which may not always be possible before the stigma is receptive (Lawrence, 1939). It can also be countered by surgical shortening of the style or even by intra-ovarian pollination, in which pollen is introduced directly into the ovary, the boring being sealed afterwards with petroleum jelly (Maheshwari & Kanta, 1964).

The same methods are used for the self-pollination of plants that are normally self-incompatible, for example, early pollination in breeding brassicas (Hayes, Immer & Smith, 1955) and truncation of the style in the hollyhock (*Alcea officinalis*). Various organic and inorganic substances are known to retard the withering of the style, or to increase the germination percentage of pollen or the rate and extent of pollen tube growth. The effect of the different substances varies from species to species, but some of them have been used to overcome self-incompatibility, and they include amino-acids, vitamins and hormones or other compounds with a hormone-like action.

Bee-keeping

Crop-production that depends on pollination by honeybees is accompanied by the production of honey. The behaviour of the honeybee (*Apis mellifera*) has been dealt with in detail in Chapter 5, and many of the discoveries in that field are of great importance economically. Honeybees, on the whole, behave in a manner which is for the good of the hive, and this means that they tend to seek out and exploit those crops from which they can most easily obtain large quantities of pollen and nectar. However, von Frisch (1954) found that, although the corollas of certain thistles were too long for honeybees to work easily, they could be artificially induced to visit these plants; they then obtained a worthwhile crop of nectar from them and so produced more honey. The method of inducement was to place sugar-water, in which thistle flowers had been soaked to give their scent to the liquid, near the hive. This procedure can be important when the pollination of an agricultural crop is the main concern and the flowers of the crop are relatively unattractive to the bees. It has been tried on many occasions (particularly in Russia) and is sometimes successful, but unfortunately it fails quite frequently. A whole series of other tricks has been tried or suggested in the attempt to modify honeybee behaviour, ranging from opening and closing hive entrances at the time of day when 'target' crops are in flower, to the use of bee pheromones. The density of hives and the way they are arranged are also important and the dispositions employed vary greatly according to crop, presence of competing crops, time of year and geographical location.

Occasionally the measures taken coincide with the interests of the bee colonies, but usually they are contrary to them, and this is evidently the reason why chemical tricks are so unreliable. Hive-closing and hive-positioning are things that the bees can do nothing about, and if these are contrary to the best interests of the colony, the colony suffers. All these matters and others are reviewed by Free (1993, Chapter 4), Jay (1986 [this paper covers much more than its title indicates]) and Robinson *et al.*, (1989a & b).

If bee colonies are weakened as a result of their use for crop pollination, they have to be restored after the crop has finished flowering by being re-sited at lower densities and/or in different districts where wild or cultivated flowers can be freely visited. Commercial apiarists in the USA hire out colonies for pollination twice or occasionally three times a year. Between rentals, they may migrate long distances with their hives (for example, from the north-eastern states to Florida). It is estimated that in California hiring out bees for pollination makes a net loss, but the apiarist makes a gain overall when honey yield for the year is taken into account. On average, the increase in the value of crops is estimated at 78 times the cost of hiring bees. Weakening of colonies during crop pollination mainly results from the high hive-densities used to saturate the crop and force the bees to forage at flowers they might not naturally choose (see later, under lucerne).

The use of artificially managed pollinators is necessitated by the transformation of the natural environment that is entailed by modern agriculture: huge areas of a single plant species in flower and little or no 'waste land' to provide a seasonal succession of wild flowers and living space. (See Westrich, 1990, Chapter 2, for the contrast between untidy and excessively tidy landscapes.) At present, the honeybee is by far the most important managed pollinator but, as we describe below, farmers have begun to use other species.

Self-pollinated crops

Crop plants that are readily self-pollinated are usually free of pollination problems. In fact, one possible method of dealing with such difficulties is to select self-fertilising cultivars. Several cultivated plants such as the sweet pea (*Lathyrus odoratus*) and the culinary pea (*Pisum sativum*) are regularly self-fertilised, although the structure of the flowers indicates that their ancestors were insect-pollinated. However, in some other cultivated plants the deterioration which normally occurs when an outbreeding species is forced to inbreed is so serious that self-fertilising cultivars are useless. This is true, for example, of lucerne (*Medicago sativa*) (Bohart, 1957). Oilseed rape (*Brassica napus*) benefits from insect pollination, despite being self-fertile, through an increase in yield resulting from a well-synchronised early seed-set.

Wind-pollinated seed crops

Cereals (members of the grass family, Poaceae) are wind-pollinated (if not self-pollinated). An example is maize or corn (*Zea mays*), which is peculiar in having separate male and female inflorescences on the same plant. The male inflorescences are panicles, referred to as tassels, produced at the top of the plant. The female inflorescences (ears or cobs) are produced lower down and are enclosed in sheaths, from each of which emerge many enormously long drooping stigmas, called silks. The tassels produce flowers for about two weeks, and the silks receive pollen which has been scattered by the wind. Maize is grown mainly for its seed, but also as green manure.

Contamination

When wind-pollinated crops are grown to provide seed for the following year, attention has to be paid to the possibility of contamination from different cultivars grown in the same area (as in maize, preceding section) or natural populations of the same species (as in beet, p.283). Experiments show that contamination may drop by 99% over a distance of 15 m from the contaminant crop of a wind-pollinated plant (Bateman, 1947b) (pp.271-2, 423). Such results provide a basis for deciding the adequate separation between maize crops which are liable to contaminate one another.

This problem also arises with outbreeding insect-pollinated plants that are grown as seed crops – many annual flowers and vegetables, for example. Consequently, appeals are regularly made to private gardeners not to let their brassica crops run to seed. In experiments with turnip (*Brassica napus*) and radish (*Raphanus sativus*) Bateman (1947a) found that contamination dropped by 99% over a distance of 50 m, while Lawrence (1939) stated that it was advisable to keep cultivars of nasturtium (*Tropaeolum majus*) very well separated. For both insect-pollinated and wind-pollinated crops, contamination may be countered either by growing a barrier of some other species round the field, or by discarding part of the crop from the edges of the field. Where a hybrid seed crop is being produced with male-sterile and male-fertile plants (see next section), the borders to be discarded can be planted with the male-fertile form. The aim is not necessarily to eliminate contamination absolutely, but to reduce it to a commercially acceptable level (Frankel & Galun, 1977).

Hybrid seed for crops and flowers

By avoiding contamination, growers are able to produce uniform seed, which is very desirable in crop plants because it ensures similarity of reaction to particular conditions and a uniform product. Better crops, however, can often be produced by the crossing of two different uniform cultivars, for by this means the phenomenon of hybrid vigour (or positive heterosis) manifests itself, without loss of uniformity. The progeny of these 'F1' hybrids are highly variable, so the the farmer cannot save his own seed for next year; the cultivar has to be produced anew by crossing the same two chosen parents. Maize was one of the first crops in which this method was used. Two chosen parent cultivars are interplanted and from one the tassels are removed so that when, later, its cobs are gathered, they will have been pollinated from tassels of the other. The female parent is made to outnumber the pollen source by up to four to one. By 1951, hybrid maize accounted for 81% of the total crop in the United States (Hayes, Immer & Smith, 1955). Later, the principle was applied to many insect-pollinated species. Insect-pollinated ornamentals, which have hermaphrodite flowers that have to be emasculated, are usually grown for seed production in greenhouses. In the case of the petunia, each 'female' plant remains in continuous production for five months and produces about 200 flowers (Bodger, 1960).

If male-sterile cultivars can be bred (as in carrot, onion and tomato) the laborious business of emasculation can be eliminated. Male-sterility is of two types, genetic and cytoplasmic (Frankel & Galun, 1977, Chapter 3.4). Only the second is commonly used in producing F1 hybrid seed; it depends on some factor in the cytoplasm (rather than in the nucleus) that is passed from a female parent to its offspring via the egg-cell.

Difficulties with insect-pollination of male-sterile members of a population arise either because they lack pollen as an insect-attractant, or because of the incidental development of other differences between the male-sterile and male-fertile flower types in the course of breeding. Thus the available rewards (not only of pollen, but also of nectar) and the 'signals' given to pollinators may differ, leading to inequality of visitation or even neglect of one or other form. Breeding programmes leading to male-sterility for the production of hybrid seed must, therefore, be carried out in the presence of the expected pollinator (normally the honeybee) to ensure that such unintended differences do not creep in (Erickson, 1983).

Bee-pollinated seed crops

Our examples of bee-pollinated seed crops belong to the family Fabaceae subfamily Faboideae. The very strong relationship between the flowers of this subfamily and bees, and between these plants and soil-condition and soil-fertility (through their nitrogen-fixing activity), has been emphasised by Leppik (1966). Leppik recognised the importance of these relationships both for agriculture and for the natural environment and in this he anticipated the concern of later biologists, to be mentioned at the end of this chapter.

Lucerne/alfalfa and red clover

The field crop lucerne (*Medicago sativa*) needs insect-pollination to produce a full yield of seed. There is some degree of genetic self-incompatibility, and pollen from another individual gives better progeny, while continual selfing leads rapidly to loss of vigour. When a flower of lucerne is visited, the stamens and style emerge explosively from the keel (see p.158); during the explosion, referred to as 'tripping', the surface of the stigma becomes abraded and, for the first time, receptive to pollen. Pollination normally takes place at this stage when the stigma forcibly strikes the underside of the visiting insect, on which there is usually pollen from a flower visited previously. After the flower is tripped the stigma presses against the standard, and it is then almost impossible for it to receive any further pollen. Lucerne is a native of the steppe regions of eastern Europe, but it is a very important forage crop in North America, where it is effectively pollinated by solitary bees, especially alkali bees (*Nomia melanderi*) and leafcutter bees (*Megachile* spp.). Unfortunately, the spread of agriculture causes the number of wild bees to decrease, so that when lucerne is grown in a newly cultivated area, yields of seed are very high at first but decrease after a number of years. Honeybees can then be imported, but they do not always do the job required of them because, except during a short period when they are unfamiliar with the crop, they trip less than 3% of the flowers they visit for nectar. By probing the keel from the side, they can get at the nectar without being buffeted on the underside by the tripping of every flower. They are much more effective when they are collecting pollen because then they have to trip the flowers, but they cannot be relied upon to collect pollen from lucerne, often preferring to get their supplies from other plants. Consequently, if the honeybee is used as a pollinator, very high densities have to be maintained and if the tripping rate falls below 1% it may be impossible to raise the density of honeybees sufficiently to produce a good seed-set.

A European bee that is strongly associated with lucerne is *Melitta leporina*. This has been found at the very low density of 168 bees per hectare in Denmark; the tripping rate was very high and it seemed that these bees were causing two or three times as much pollination in a given period as honeybees working the same crop at a density of 7,000 per hectare (Todd, Norris & Crawford, in Mittler, 1962).

An alternative to the use of honeybees is the restoration of the declining wild bee populations, and this has been done successfully in North America for lucerne pollination with native alkali bees and a species of leafcutter bee that was accidentally introduced from Europe and naturalised in the USA, both of which are non-social but gregarious.

Alkali bees nest in bare, damp, rather light soil, patches of which can be prepared for them near the crops, and this is particularly successful because of the bees' gregarious tendency (Frick, Potter & Weaver, 1960; Stephen, 1965). Bare, damp soil arises naturally where water (sometimes irrigation water) flows over an impervious layer just

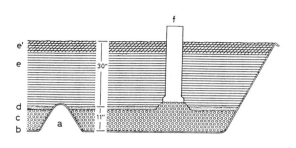

Fig. 13.1 Cross-section of part of artificial nesting bed for alkali bees. **a**, one of the soil ridges dividing the bed into segments so that a leak cannot ruin the whole bed; **b**, polythene lining; **c**, gravel layer forming water reservoir; **d**, coarse sand; **e**, nesting soil; e^1, upper 20 cm with common salt added; **f**, downspout for saturating the gravel layer (little water is needed after the reservoir is filled). From Stephen (1965).

underground, and percolates to the surface. Evaporation then causes an accumulation of salts, which keeps down the vegetation, reducing it to, at most, a thin growth of specialised salt-tolerant plants. Similar conditions can be created artificially in specially prepared beds. The area is excavated to a depth of a metre or more, lined with plastic sheeting and then filled as shown in Fig. 13.1. Carefully selected soil is imported to provide optimum conditions. The filling is spread and compacted by bulldozers, and common salt is mixed into to the top 12 cm. This greatly reduces summer evaporation, but may cause concretion of the soil after a year or two; calcium chloride or gypsum is then added to counteract this. The beds are stocked by importing, in winter or early spring, blocks of soil containing dormant prepupal bees from other areas; the density must be sufficient to prevent these gregarious insects wandering off to seek more thriving colonies. A bed of 10×17 m should supply pollinators for 12 to 16 hectares of lucerne, but far larger beds are sometimes made. Once a colony is established, the bees tend to fly in a stream from the beds to the fields, and their flight line may cross busy roads at a height of only a metre. Since each bee is reckoned to be able to set about 0.5 kg of seed, notices are sometimes put up requesting motorists to slow down for alkali bees.

The non-native leafcutter bee, *Megachile rotundata*, has been increased in lucerne-growing areas of the USA and Canada by the provision of artificial nest sites in various forms (Stephen, 1961; Bohart, 1962; Richards, 1984). Large-scale management involves extracting the cocoons, made up of bits of leaf, from the nests and discarding all defective cells. The healthy cocoons are then stored through the winter at 5°C until shortly before the lucerne begins to bloom, when they are transferred to incubation trays and kept at 30°C until about 75% of males and 20-50% of females have emerged. At this stage, the temperature can be lowered to 20°C if it becomes desirable to delay emergence somewhat. The trays have a gauze cover, except at the front where there is a wooden lid. Any parasites emerging during this period pass through the gauze and are disposed of. Then the trays are placed in shelters, which are spread evenly through the fields, and the lids removed. Each shelter is a box open at the sides and contains, in addition to space for trays, 10,000 to 50,000 nesting holes made by stacking boards, grooved on each side, one on top of the other, so that the grooves match to form tubes. The shelter may be mounted on posts or on a trailer. The bees leave the trays and begin to make nests in the tubes, which they line with pieces of lucerne leaf. Each female fills two or three holes 10-15 cm long. After the nesting season, the grooved boards are taken in and separated, and the cocoons are removed to storage. The 'hives' are cleaned and prepared for next year. Exact instructions for managing leafcutter bees for pollination are given by Richards (1984). This publication covers all the equipment required, including a storage/incubation room that will store 8 million cocoons or accommodate half this number for incubation. The room has to be fitted out with racks for incubation trays, humidifier, cooler, heater, fans, an ultraviolet insect-trap for parasites, and thermostats, as well as alarms that respond to departures from the programmed conditions. The bees themselves have become a commodity that is traded on the market (in the form of cocoons).

The alkali bee and the leafcutter bee provide the most successful cases of management of non-social bees as pollinators, though the use of the former has declined as the technique of managing the latter has been refined (but see later, under apples). Even these bees are not used throughout the climatic range in which lucerne is grown for seed (Robinson *et al.*, 1989a), though in western Canada the threshold temperature at which *Megachile rotundata* starts foraging dropped from 21°C to 18°C between 1962 and 1984 (presumably as a result of selection acting on the genetic constitution of the

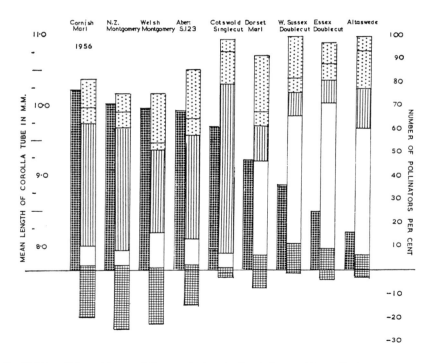

Fig. 13.2 Bumblebee visitors and flower-tube length of red clover cultivars. Black squares (left-hand columns) show tube length, right-hand columns show visit-frequency; vertical hatching: *Bombus hortorum* group, crosses: *B. pascuorum* group, white: *B. terrestris* and *B. lapidarius* groups. Negative values represent biting or feeding through holes by *B. terrestris* group. Based on figures for 1956 in Hawkins (1965).

bees). Tests in north-west Canada show, however, that *Megachile rotundata* can pollinate and increase its numbers on several other leguminous crops (Fairey & Lefkovitch, 1991; Richards, 1991).

Another field crop requiring insect-pollination is red clover (*Trifolium pratense*) which is highly self-incompatible. Its natural pollinators are bumblebees of various species, whose importance seems to vary with the tube-length of the clover cultivar being grown; for example, Hawkins (in Mittler, 1962) found in England that the *Bombus lapidarius* group predominated on short-tubed cultivars, the *B. hortorum* group on long-tubed cultivars in the same trial, while the *B. pascuorum* group was equally frequent on all cultivars tested (Fig. 13.2). The workers of the short-tongued *B. terrestris* group are, unfortunately, persistent corolla-biters, though less troublesome on the shorter-tubed red clover flowers than on the longer. However, in Sweden, where the *B. terrestris* and *B. lapidarius* groups are usually the commonest bumblebees on red clover, the former do not rob at the beginning of the season, so that they may be useful pollinators in years when the clover is early (Mittler, 1962, Part I). As with lucerne, pollination is dependent on the availability of nesting sites for the bees near the crop. For example, Hawkins found that one late-flowering cultivar of red clover, largely dependent on *B. pascuorum* for pollination, yielded 50 kg of seed per hectare when growing in open country, whereas a few kilometres away a field of the same cultivar, growing near a forest, yielded 135 kg per hectare, apparently because the forest provided a suitable nesting habitat for the bees. Hawkins pointed out the need for a full study of bumble-bee ecology, and he indicated the value of white dead-nettle (*Lamium album*) and also crops of the winter and spring cultivars of field bean (*Vicia faba*), which all provide nourishment for these bees before the flowering of the clover. In general, wild flowers and crops attractive to bumblebees are valuable outside the clover season.

Free & Butler (1959) discussed the reduction of bumblebee populations caused by intensive agriculture, which involves the elimination both of hedges and of patches of neglected land. The resulting shortage of nesting places may be offset either by the restoration of suitable ground for nesting sites (in ways suggested by these authors) or by the provision of artificial nest boxes. The lack of a seasonal succession of flowers at which the bees can forage can also be remedied by leaving patches of waste ground. In North America, artificial nest boxes were set out in wild places and, when the bumblebee queens had established colonies in them, they were moved to the crops. It was found necessary to capture any workers not in the box at the time of removal, and also to use variously coloured boxes with the entrances facing in different directions to avoid confusion of neighbouring nests by the queens (Hobbs, in Mittler, 1962). In Denmark, nests from the wild were placed initially in greenhouses; all the young queens raised by each colony then hibernated and in the following spring founded many more new colonies (Holm, in Mittler, 1962, and in Åkerberg & Crane, 1966). Very good percentages of survival and nest-establishment were achieved, using sub-stantial numbers of bees, but there were serious losses when the established nests were moved from the greenhouses to the open fields. Similar work in New Zealand resulted in a higher degree of success in this direction than previously achieved in Europe (Griffin *et al.*, 1991). As it is estimated that in New Zealand red clover needs 2,000-4,000 long-tongued bumblebees per hectare for maximum seed-yield, and natural lev-els of these introduced bees are 300-500 per hectare, any appreciable artificial increase in density can produce profits. (For another use of bumblebees, see later, under tomato.)

For the shorter-tubed cultivars of red clover, honeybees are used as pollinators in some countries. They are more effective on the second flowering of the crop than on the first, possibly because the later flowers tend to be shorter-tubed or because the nectar is both sweeter and more abundant in warm dry weather (Bohart, 1957). In England, Free (1965a) found that honeybees visiting red clover caused the same percentage of florets visited to set seed as did bumblebees; in both cases, the set by pollen-gatherers was much higher than by nectar-gatherers.

Introduction of pollinating bees

The introduction of new species of wild bees from foreign countries has been urged by Bohart (in Mittler, 1962) and discussed more recently by Torchio (1991). Not only would this increase the number of species available in the receiving countries, but the introduced bees, if brought in free from disease and parasites, would probably flourish better than the native bees. The success of the introduction of bumblebees to New Zealand to fertilise the introduced red clover is well known. Moreover, several other species of wild bees, introduced into various countries by chance, have, like *Megachile rotundata*, become useful pollinators in their new areas. Both the alkali bee and *M. rotundata* have been introduced into Canada, but both are limited to certain areas with sufficiently high summer temperatures, the alkali bee being the more severely restricted (Arnason, in Åkerberg & Crane, 1966). Some authorities who favour the use of wild bees are against introduction of species because of the unpredictable effect on the existing ecosystem.

Other insect-pollinated seed crops

A largely self-incompatible field crop, which is pollinated mainly by short-tongued insects (bees, wasps and flies), is the carrot (*Daucus carota*, family Apiaceae). In an investigation carried out at Logan, Utah, all the insects that could be found visiting carrot flowers over a period of four years were identified, and a total of 334 species, belonging to 37 families, was recorded (Hawthorn, Bohart & Toole, 1956). The abundance of particular species varied from year to year and from one part of the flowering season to another. Among the genera commonest, on average, the following were judged, from their behaviour at the flowers, to be the most efficient: *Apis* and *Halictus* and/or *Lasioglossum* (bees), *Tachytes* (a sphecid wasp), *Eristalis* and *Syritta* (hoverflies), and *Stratiomys* (a soldier-fly). It was suggested that growers might improve the pollination of carrots by maintaining a good supply of honeybees, avoiding the presence of competing bloom, and providing decaying vegetation nearby for Diptera to breed in (Bohart & Nye, 1960).

The pollination of the self-incompatible cabbage and brussels sprout for commercial seed-production is carried out in isolation-greenhouses in some areas, owing to the great danger of contamination (see p.352). Insect pollinators are introduced into the houses and, while honeybees are the easiest to procure, bumblebees and blow-flies (*Calliphora*) give a greater seed yield. The blowfly pupae can be obtained from factories that produce maggots for anglers (Faulkner, 1962).

Fruit crops

Tomato

The tomato (*Lycopersicon esculentum*) is naturally buzz-pollinated (pp.125, 179). Therefore, although it is self-fertile, in cultivation it usually requires some disturbance to the flowers for full pollination and maximum fruit-yield, particularly with long-styled cultivars. Under glass, traditional growers would tap the stakes to which the plants were tied or, more recently, use a hand-held electrical truss-vibrator. An alternative now available is to use artificially prepared colonies of the bumblebee, *Bombus terrestris* (Eijnde *et al.*, 1991). To start a colony, a hibernated queen bumblebee is placed in a small dark nesting box with three or four newly-emerged honeybee workers which help to calm her (Ptácek, 1991) (for the life-cycle of *Bombus*, see Chapter 5). Sugar solution and pollen collected by honeybees are supplied and the queen uses these foods to start her brood. When the first bumblebee workers emerge, the colony is transferred to a much larger box with two compartments. The same food supply is continued. When 80 workers have emerged, the colony is sent to the grower. Some colonies are kept by the producer in order to raise queens and drones. Each new queen is allowed to mate with a drone from a different colony, after which she would normally hibernate. However, she is now subjected to two bouts of CO_2 narcosis, which prevents hibernation, so that the normal annual cycle is broken. This makes it possible to supply bee colonies to growers from January to September, a much longer span than is covered by the bees' natural period of activity.

Long-styled tomato flowers are characteristic of the wild species in its native South America. Short-styled flowers occur in some cultivars, apparently representing another example of inadvertent change brought about by selection for fertility (see peas, p.352). Rick (1950) found that whereas the most active pollinator of the tomato in an area of California was a bee of the genus *Anthophora*, a bee of the family Halictidae pollinated the same cultivar twice as efficiently in Peru.

Avocado pear

The sub-tropical fruiting tree, avocado pear (*Persea americana* [*P. gratissima*], family Lauraceae), is also self-fertile, but requires insect pollinators because its timing mechanism prevents self-pollination. Avocado trees are normally grown commercially as vegetatively propagated clones of two distinct types, which may be called A and B. The flowers each have a male and a female stage, separated by a period of closure (Fig. 13.3). According to early reports, such as that of Robinson & Savage (1926), the timing of their activity completely restricts pollination to cross-pollination between trees of different type. For each tree there is usually a short interval of less than an hour between the closing of the morning flowers and the opening of the afternoon ones. (The same arrangement is found in other New World Lauraceae [Kubitzki & Kurz, 1984] and in the shrub *Zizyphus spina-christi*, family Rhamnaceae [Galil & Zeroni, 1967].) More recent studies, such as that of Ish-Am & Eisikovitch (1991), using named cultivars, have shown overlaps of flowering between different cohorts of flowers on a single plant or clone, allowing the possibility of fertility in an orchard

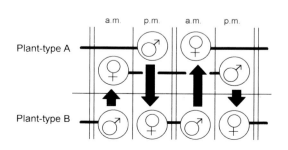

Fig. 13.3 Diagram to show when the flowers of the two types of avocado plant are open over two successive days. Bold horizontal lines connect the male and female stages (circles) of a single flower. Arrows show possibilities for pollination.

planted with a single cultivar. In plantations in Florida, California and Israel honey-bees are supplied to pollinate the avocado, though the flowers are apparently adapted to fly-pollination (Robinson *et al.*, 1989a; Ish-Am & Eisikowitch, 1993).

Vanilla

This tropical American orchid (*Vanilla planifolia*) is particularly successfully cultivated in Madagascar. However, its pollinating bee, *Melipona*, is not available there and grow-ers have to pollinate the flowers artificially to ensure the ripening of the pods, which are the source of the flavouring. A small stick is used to remove the anther cover and tuck the rostellum (see Chapter 7) out of the way; the anther and stigma are then pressed together, causing self-pollination.

Date palm

The dioecious date palm (*Phoenix dactylifera*) is also propagated vegetatively, the clones being either male or female. The trees are wind-pollinated, and it is only necessary to tie part of an inflorescence from a male tree into each female tree to secure pollina-tion (see p.12). For this purpose, it is sufficient to grow one male tree for every hundred females, and the male inflorescences may be cut and tied up a little before their flow-ers open (Cobley, 1956). When, by careful experiment, the same female date palm is pollinated by male trees of several different clones, the resulting dates differ according to the source of pollen. The pollinations which produce larger seeds yield dates with more flesh (maternal tissue), the pollen having influenced the flesh indirectly ('metaxenia') through its direct genetical effect ('xenia') on the seed (Nixon, 1928). By selecting suitable male clones, the size of dates can thus be increased. The larger fruits, however, ripen more slowly, and although in some climates this is an advantage, in others rapid ripening of smaller fruits may give the best yield.

Oil palm and a multi-million dollar weevil

The oil palm (*Elaeis guineensis*), a monoecious species native to West Africa, is a very important source of oil for making margarine and soap. Pollination is not a problem in West Africa, but this palm has also been cultivated in south-east Asia since the early twentieth century, and here yields were inferior until the matter was scientifically in-vestigated. Delay in dealing with the problem arose from a belief that pollination was by wind. Investigation in West Africa showed that wind-borne pollen was unlikely to be important in fruit set, especially during the rainy season. On the other hand, both male and female inflorescences were found to be visited by thousands of insects, mainly beetles and flies, and substantial numbers of viable oil palm pollen grains were found on beetles of the genus *Elaeidobius* (belonging to the weevil family, Curculioni-dae, subfamily Derelominae).

In south-east Asia, it appeared that a nocturnal moth was pollinating the oil palm in Sabah, and thrips were pollinating it in the Malay Peninsula. However, in the for-mer area fruit set was poor, and in the latter the presence of thrips and relatively good yields of fruit were found only in old plantations, young ones being apparently too breezy for these minute insects (Syed, 1979). (For thrips-pollination of dipterocarp trees in Malaysia, see p.311.)

In Cameroon, six species of *Elaeidobius* were identified on the oil palm flowers. They breed in the spent male inflorescences (p.311), each species occupying distinct struc-tures. The beetles apparently locate the male flowers by their aniseed-like scent. The female flowers emit the same scent in pulses, thus periodically attracting the beetles, which arrive with pollen on their bodies and then return to male inflorescences (a nice example of deceit pollination, the female flowers being rewardless – see Chapter 10).

The researchers decided to introduce *Elaeidobius kamerunicus* to south-east Asia. Insects were bred in captivity to free them from mites and nematodes that infest them. They were released at two sites in the Malay Peninsula in 1981, and within a year this species had become abundant throughout the territory. It was then taken to Sabah, Papua New Guinea, the Solomon Islands and Thailand. Yield levels of the oil palm rose to those prevailing in West Africa. The chemical attraction of the beetles by the flowers is highly specific. Thus the introductions are thought unlikely to upset the ecosystem while the plant/pollinator relationship is likely to remain stable. The gain to the planters in the Peninsula was estimated as US$115,000,000 annually (Greathead, 1983; Hussein *et al.*, 1991).

Temperate orchard fruits

The pollination of apples, pears, plums, cherries etc., has been reviewed by Free (1962, 1970b & 1993) and by Luckwill, Way & Duggan (1962). These vegetatively propagated fruit trees, although having bisexual flowers, are usually genetically self-incompatible. The degree of self-incompatibility varies and there are some clonal cultivars that are partially or wholly self-compatible, although this property may change somewhat from year to year. However, in many cases, particularly among apples, self-incompatibility is so great that it is essential to grow two different suitable cultivars together. Only a few cultivars of apples are really well-known to the general public, and the cultivars with which they have to be planted in the orchards are called pollinisers, being grown mainly for their pollen rather than for their fruit. Indeed, crab-apples may be used (Mayer, 1983). Crane & Lawrence (1952) took the view that since it is only necessary for a small proportion of the flowers to set fruit to produce a good crop, a moderate degree of self-compatibility will often suffice. A more modern approach adopted in America is to go for maximum fruit set and then thin the young fruit manually or chemically (Robinson *et al.*, 1989a).

Bees are the most important pollinators in orchards, but some other insects that probably help to pollinate the blossom from time to time are, among Diptera, midges and fungus-gnats, St Mark's flies and fever flies, some representatives of the house-fly and blowfly families and, among beetles, small flower-beetles of the family Nitidulidae.

Among solitary bees, the commoner species of the genus *Andrena*, together with *Osmia rufa*, are the most important fruit-tree pollinators in Britain. They can only be significant, however, when local conditions enable good numbers to nest in or near the orchards. Even then, these bees are much affected by the weather, requiring higher temperatures for their activities than honeybees, and occurring in reduced numbers if the previous season has been unfavourable to them. It has been found that *Andrena* visit half as many flowers per minute as honeybees but, in partial compensation, pollen adheres to *Andrena* more loosely and to a greater area of the body. In Nova Scotia, pollination of fruit trees is almost entirely dependent on *Andrena* and *Halictus*, or the related *Lasioglossum*, while in Norway solitary bees and bumblebees are effective together, honeybees being almost absent in these countries. Bumblebees are subject to the same disadvantages of scarcity and variation from year to year as solitary bees, but they operate at a lower temperature than honeybees and are more likely to move from tree to tree. Moreover, in certain apple and plum cultivars with long styles projecting up beyond the stamens, only bumblebees are obliged, by their size, to straddle the stigmas.

Honeybees have, of course, been carefully studied on fruit-blossom and are its most important pollinators when foraging for pollen. They are not very effective when collecting nectar because then, besides visiting fewer flowers per minute, they often stand on the petals and probe between the stamens from the side, without touching the stigmas, particularly if the stamens are long and stiff. In order to increase the proportion of pollen-gatherers, sugar-syrup is sometimes offered near the hives. As the bees keep an approximate balance in the intake of the two foods, their requirement for pollen increases if they can rapidly collect a large quantity of syrup from an artificial supply. It has been pointed out that attempts to direct bees to a crop by feeding scented sugar-water (p.351) may produce better pollination simply by increasing the collection of pollen (Free, 1965).

Honeybees work more on the sunny side than the shady side of large fruit trees and, when leaving one tree for another, they fly to its nearest neighbour. The bees can easily collect a full load of pollen or nectar from a single tree, but in fact they visit, on average, two trees per trip, though young scout bees may visit considerably more. The bees move more often from tree to tree if small fruit trees are used. Because bees respond to the productivity of the flowers, it is desirable that the attractiveness to them of the crop cultivar and the polliniser should be similar, so that neither monopolises their attention (Free, 1966; Free & Spencer-Booth, 1964) (compare production of hybrid seed using a male-sterile seed-parent, p.353). Some studies have shown that inter-plant movement is inadequate to account for the levels of fruit set in apples due to honeybees. It was found that bees in the hive carry pollen from trees they have not visited (identifiable by scanning electron microscopy). This pollen, acquired by contact with their sisters in the hive, is viable and plays a part in the fruit set of flowers on which the bees subsequently forage (DeGrandi-Hoffman et al., 1984, 1986).

A single honeybee colony may forage over an area of about 48 hectares, but in poor weather only those bees with established foraging areas near the hive venture out. Bee-hives are best brought into the orchard after flowering has begun, as the bees will then concentrate for a time entirely on the fruit blossom, and only later give a share of their time to flowers outside the orchard. In orchards that are grassed down, the flowering of dandelion (*Taraxacum officinale*) is prevented, as it produces nectar at a lower temperature than the fruit blossom and may be constantly preferred by the bees (Free, 1968).

The density of hives in orchards is traditionally one to the acre (approximately two to the hectare). They are placed in groups for convenience, and it has been found that four hives grouped in every two hectares give an even density of bees throughout the orchard, even in poor weather. However, modern apple orchards are usually based on dwarfing rootstocks, producing far more apples per hectare than older orchards. Consequently, four or more colonies per hectare may be recommended, though growers usually try to make do with less than the recommended density (Robinson et al., 1989a). Sometimes, at flowering time, bouquets of blossom may be cut from the pollinisers and tied into the cropping trees in order to improve the fruit set. The proportion of polliniser trees and their positioning has in the past been considered an important matter. The proportion chosen is usually a compromise between what gives maximum inter-plant movement and the commercial value of the fruit, bearing in mind that the polliniser is normally less valuable than the main crop. If bees from different parts of the foraging area exchange pollen in the hive, the positioning of the polliniser is much less constrained than was previously thought. (In the blackberry [*Rubus*], a polliniser has been specially bred in America to have the same cultural and fruit qualities as the original cropping cultivar [Shoemaker, 1962].)

In the United States, hand-pollination has been used in orchards in areas where pollinating insects are scarce. Although laborious, it has the advantage of giving control over the distribution of fruit on the tree, and it may also eliminate the need for thinning the young fruit. In Japan, pollen has sometimes been mixed with lycopodium powder to make it go further in hand-pollination, and much effort is being put into devising economic methods for pollinating apples mechanically (Sadamori *et al.*, 1958; Ohno, 1963). The problem here is that insecticides and clean agricultural practices have virtually eliminated natural pollinators. However, use of the wild non-social bee *Osmia cornifrons* has proved more economical than hand-pollination. Measures are taken to increase the bee outside areas of intensive agriculture and then bring a proportion of the nests into the orchards. Management of the related bee *O. lignaria propinqua* in North America is now a commercial operation and, more recently, progress has been made in Europe in increasing the populations of *O. cornuta* for the same purpose (Torchio, 1991).

It has been found in Canada that some virus and fungal diseases of fruit trees may be spread in pollen or by pollinating insects. One such disease is fire-blight of apple and pear, and here an attempt has been made to turn the tables, so to speak, by supplying streptomycin to beehives (in pollen-inserts), so getting the bees to distribute the fungicide (Arnason, in Åkerberg & Crane, 1966).

Economic importance and impact on nature conservation

The expenditure on studies of bee behaviour described in Chapter 5 and of crop pollination described here underline the economic importance of adequate pollination. Similar indications are provided by the scale of operations involved in preparing alkali bee beds and managing leafcutter bees for pollination. The vast scale of honeybee culture in North America has already been indicated; an estimate of its economic significance by Robinson *et al.*, (1989a & b) was that it contributed about 9,300 million US$-worth of crop value in 1985. However, spread of mite infestations and of Africanised honeybees is reducing the availability of honeybee pollination there. Currently, a need has been recognised for much more research into the pollinatory environment as a whole. Extensive monitoring could give us a better understanding of the relationship of crop pollination and apiculture with the wild insect-pollinated flora and wild bees (Corbet *et al.*, 1991; Osborne *et al.*, 1991; Williams *et al.*, 1991). This kind of research may bring increased yields, but it may also prevent decline of yields by staving off ecological disaster.

The extent to which crop-pollination currently depends on wild bees remains unknown, but there is a widespread belief that inadequate pollination of almost any crop can be remedied by bringing in hives of honeybees. In so far as crop pollination can be performed by the honeybee, there is nothing to stop the spread on suitable land of 'total agriculture', involving the elimination of natural and semi-natural habitats. There is a positive incentive for this when wild flowers are considered to be competing with the crops for the attention of honeybees; then they may be mown when the crop is in flower, or even destroyed. Independently of any efforts to increase the effectiveness of honeybees, the value of unploughable terrain to potential natural pollinators can be negated by dropping fertilisers from the air, so increasing the pasture value of the land and drastically reducing the diversity of its plant and animal life. Against this, there are some alleviating factors. The destruction of wild insects by insecticides is suspended at certain times of the day or year, for the sake of the honeybees. Where bumblebees and other wild bees are recognised as useful pollinators, or where parasitic ichneumon wasps are important for pest-control (see p.101), habitats of wild flowers may be left to give them food when the crops are not in flower, and small wild areas may be preserved for the bees to nest in. These areas should be allowed to de-velop seminatural vegetation, undisturbed by ploughing and uncontaminated by fer-tilisers, herbicides or insecticides; they should not be allowed to develop into woodland. The ineffectiveness of honeybees on many crops (see below) and the threats to apiculture already mentioned, mean that it may be unwise to allow further decrease in wild bee populations.

There is, in fact, currently some partisanship over honeybees versus wild bees. On behalf of honeybees, some American observers consider that the honeybee is benefi-cial to wildlife (by which they seem to mean only birds and mammals) because their pollinating activities promote plant reproduction, so improving the availability of seeds and fruits as food for vertebrates (Barclay & Moffett, 1984), but this only applies to a landscape that includes some wilderness areas. On the other hand, it is contended that honeybees individually are poor pollinators of specialised flowers and injurious to the more diverse natural bee fauna through competition for food (Westerkamp, 1991).

Aside from this, however, some scientists are anxious about the consequences of the destruction of natural habitats and of the extreme dependence of agriculture on the honeybee, and are advocating management of other bees for pollination and a 'more environmentally sensitive human exploitation of the world,' (Kevan, 1991; see also Corbet et al., 1991; Torchio, 1991). In addition, there are factors in the current situ-ation that suggest that there may be a move away from dependence on the honeybee. One is a pressure for crop diversification away from cereals, which may lead to in-creased demand for pollination services. The arrival of 'set-aside' policies to reduce overproduction of food also offers hope. However, Fussell & Corbet (1992) have shown that, generally, flowers of perennial plants are far more important for bees than those of annuals. Therefore, any set-aside policy that, like the British one introduced in the early 1990s, requires the land to be annually ploughed or otherwise treated, may not help this problem.

So perhaps even now, the wild bees and the natural habitats that support them will be deliberately conserved. This could benefit both crops themselves and the world-wide biosphere of which they are part, by relieving the pressure on wild pollinators. The gathering strength of the conservation movement is also a help here, and fortu-nately some of the people who understand and value the natural heritage are farmers.

A case where exploitation was halted to save a pollinator comes from Malaya. If the quarrying of a limestone outcrop near Kuala Lumpur containing the Batu Caves had not stopped, the bats that pollinate the durian fruit (p.246-7) over a wide area would have lost their home (Pye, 1983).

14

Pollination Through Geological Time

When we turn to the subject of the how pollination has evolved through geological time we immediately find ourselves inescapably caught up with the evolution of the flowering plants generally, and so this chapter must embrace the whole of that story, at least in outline. The early evolution of flowering plants has been the subject of a great deal of speculation over more than a hundred years, with many theories developing, some of them mutually contradictory. Charles Darwin famously regarded the origin of flowering plants as an 'abominable mystery', but since about 1980 there has been a considerable increase in the number of relevant fossil finds and something of a consensus is emerging on the origin of flowering plants and aspects of their early evolution. Interpretations of these new fossils and a reappraisal of others have overturned many of our earlier ideas on what the most primitive of flowering plants looked like and their relationships with other living and extinct seed-plant groups, and, with this, our ideas of what pollinated them and how pollination has shaped flowering plant evolution. Frequently, the fossils are such that we must reconstruct, say from a pollen grain or a partially preserved seed or leaf, what we think the whole plant may have looked like, so there is certainly some speculation involved, but speculation can be informed. Examination of particularly well preserved fossils and a knowledge of living seed plants give many clues, and it is on these interpretations that we can base conclusions on how the plants are likely to have evolved.

Early land plants

Plants colonised the land some 420 million years ago, in the Silurian era (Edwards, 1993). These early land plants were derived from green algae and their success on land depended on the presence of some kind of internal 'vascular' or conducting system for the transport of water and mineral nutrients between the roots and the parts above the ground (Raven, 1993).[1] They were up to a metre or so high and reproduced using spores, dry, resistant and dust-like, which dispersed in the wind. Gradually, through the eras, different and more complex plants evolved from these. A huge diversity was reached in the Carboniferous, the time of the coal measures about 300 million years ago, and, by then, there were many trees. The ferns and the first seed-plants, known collectively as seed-ferns or pteridosperms, dominated the forests, and the horsetails and clubmosses, groups that survive today only as modest herbaceous plants, were also trees. Today, the ferns are still a group of major importance in the world's flora, although most modern ferns differ in a number of ways, particularly in

---

[1] The only successful non-vascular land plants are the mosses and liverworts; without a vascular system, they are restricted in size to a few centimetres or so. A rudimentary vascular system, which has arisen independently of the true vascular plants, is seen in a few moss genera, e.g. *Polytrichum*, allowing height up to c. 50 cm.

their spore dispersal, from these ancient plants. The seed-ferns were highly heterogeneous and do not constitute one natural group but a number of groups of seed-plants.
Their descendants diversified for a further 300 million years, throughout the age of
the dinosaurs, the Mesozoic, showing a wide range of variation in details of their
wood, vegetative and reproductive structures. It is possible that some were visited, and
even pollinated, by primitive insects.

The flowering plants do not appear in the fossil record until long after the first
appearance of these other groups, late in the Jurassic period or early in the Cretaceous. At first they were a minor, perhaps marginal, part of the flora, but they diversified during the Cretaceous, the last period of the Mesozoic which saw the final
adaptive radiation of the dinosaurs, alongside some other seed plants. All but five of
these Mesozoic seed-plant groups are now extinct, some succumbing with the dinosaurs at the end of the Cretaceous. Flowering plants not only survived but had a
further, explosive, adaptive radiation in the early Tertiary period, a mere 60 million
years ago, which led to the great diversity of flowering plants that we see today and
their dominance of the world's flora. They now far outnumber all the other vascular
plant groups put together.

In order to understand how pollination in the flowering plants evolved – indeed,
how seeds and pollen evolved at all – it is necessary to look at the reproduction of the
spore-bearing plants that preceded them in the evolutionary time scale. The living
ferns, horsetails and clubmosses give us a picture of possible stages of evolution in the
history of seed plants, and by examining their reproduction we can see how seeds
must have evolved.

Alternation of generations

Fundamental to the reproduction of all the land plants, and most algae, is an alterna-
tion of two generations, one of which has just a single set of chromosomes (haploid),
the other having the two sets that we consider to be the normal state in most organisms (diploid). The life-cycle of the great majority of modern ferns and one of the

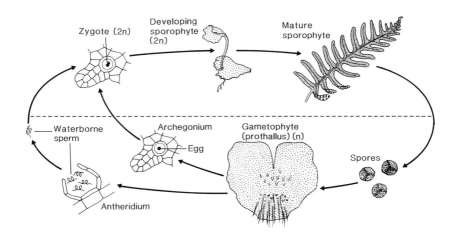

Fig. 14.1a Life-cycle of a typical homosporous fern or clubmoss, e.g. polypody (*Polypodium vulgare*).

three groups of modern clubmosses, the Lycodiales, is set out in Fig. 14.1a and is, in all basic respects, the same as that in many algae and in mosses and liverworts. In the ferns and clubmosses, the main plant is diploid and reproduces by spores, so is known as the sporophyte. The spores are the result of a reduction division typical of sexual reproduction and are haploid. In the non-flowering vascular plants they are generally dispersed widely in the wind before germinating, without any fertilisation, to form the other generation, a prothallus. This prothallus is haploid, like the spores, and forms sex organs: antheridia producing sperms and archegonia producing eggs. It is known as the gametophyte generation. The sperms can swim to the eggs in water, as in many

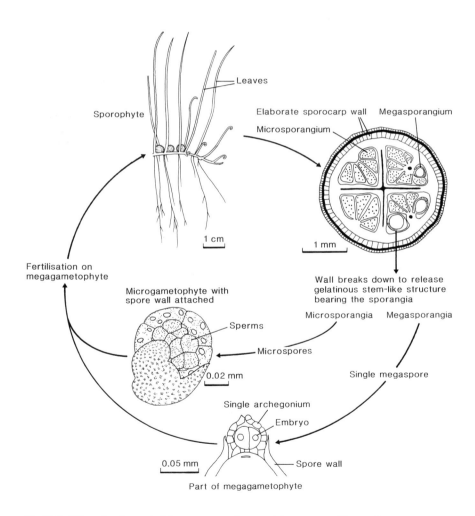

Fig 14.1b Life-cycle of a typical heterosporous fern or clubmoss, e.g. pillwort (*Pilularia globulifera*). Based mainly on Bell (1992).

animals, and fertilisation takes place in the archegonia. The fertilised egg cell grows into a new diploid plant, without any further dispersal.

In the horsetails, a few ferns and the two other clubmoss groups there is a partial or complete division of the sexes in the gametophytes. In the horsetails the spores are identical but each prothallus produces either antheridia or archegonia, a labile character depending, in part, on the environment in which it is growing, and the female prothalli produce antheridia as they age (Duckett & Pang, 1984). A further stage is exemplified by the fern *Platyzoma microphyllum*, in which there are two different types of spore, so it is known as heterosporous. These spores give rise to free-living prothalli, the smaller microspores producing gametophytes with antheridia only, the megaspores producing gametophytes normally with archegonia only. As in the horsetails, the sex of the prothalli can be modified by manipulating growth conditions (Duckett & Pang, 1984). One further stage is demonstrated by two groups of aquatic ferns, the Marsileales, including the rare British native pillwort, *Pilularia globulifera* (Fig. 14.1b) and Salviniales (including the water-fern, *Azolla*), and two groups of clubmosses, the Selaginellales and Isoetales (quillworts). They are similar to *Platyzoma* just described, but the prothalli are much reduced in size and the megaspores are, in some, retained on the parent plant; the microspores are spread by the water in the fern groups and Isoetales, and by the wind in the Selaginellales. In both types of spore the whole gametophyte is contained within the spore wall, except where it is split by the sex organs. Fertilisation then takes place and the new fertilised egg develops to form a new sporophyte. Heterospory has arisen independently in these five modern groups and almost certainly arose repeatedly in earlier, fossil groups. These heterosporous plants are not directly related to the seed plants, but give us a glimpse of the probable way in which seeds may have originated.

Seed plants have a life-cycle similar to the heterosporous ferns and clubmosses, with a further reduction in size of the gametophyte and a development of the megasporangium, now always retained on the adult plant, into a hard-coated dispersal unit, the seed. The microspores are now called pollen grains and the microsporangia are known as stamens. In the flowering plants, the size and organisation of the gametophyte has been so reduced that, without knowledge of the other groups, we should not think of it as a separate generation at all. The male gametophyte of flowering plants consists of just three nuclei enclosed within the male spore, the pollen grain, and the female gametophyte consists of the embryo sac, normally with eight nuclei, one of which will be fertilised to form the new embryo. One of the defining features of the flowering plants is that this developing seed is enclosed in a carpel. Uniquely in the flowering plants, two other nuclei of the female gametophyte fuse and this 'diploid fusion nucleus' then fuses with a second pollen grain nucleus; subsequent divisions of this cell form the endosperm, the food store for the developing seed. The endosperm then has three sets of chromosomes, two from the seed parent and one from the pollen parent (see section on the Gnetales for possible fore-runners of this endosperm). More details of the life-cycle in flowering plants are given in Chapter 2, and descriptions of the various life-cycles outlined here are given in many biological text books, e.g. Purves, Orians & Heller (1995), Gifford & Foster (1989).

There is no doubt that the ancestors of the flowering plants are to be sought among the diversity of seed-plants that appeared during the Mesozoic. These seed-plants, which may represent more than one line of evolution from a fern-like ancestor, are loosely termed gymnosperms, referring to all seed plants except the true flowering plants. It is among the living and fossil gymnosperms that we must look to find the origins of flowering plants. A great diversity of gymnosperms dominated the land for over 200 million years; they are represented by numerous fossils and form a major constituent of coal. Today, in many parts of the world, their place has been taken by the flowering plants, but those that remain are divided into four rather distinct orders. We will first examine these living orders, before looking at the fossils.

Reproduction in living gymnosperms

Much the most diverse, widespread and important order of gymnosperms today is the conifers. They still dominate the greatest of all the world's forests across the northern cool temperate regions of Europe, Asia and North America, and occur throughout the world. Most conifers have their reproductive parts aggregated into unisexual cones, and cones of the two sexes may be borne on the same tree (monoecious) or, more rarely, on different trees (dioecious). All the conifers are wind-pollinated, often producing copious amounts of pollen which, in a pine wood, can carpet the ground with a yellow dust during late May or June. In the conifers and in other gymnosperms, the pollen is caught by the ovule in a small drop of fluid at the micropyle. Details of their pollination are given in Chapter 9. The adaptive radiation of the conifers partly coincided with that of the flowering plants, but they are derived from a different gymnospermous ancestor and are not closely related to the flowering plants.

A second order, the cycads, are a tropical and subtropical group of rather palm-like small trees. They are all dioecious, so far as is known, and bear cone-like structures, very large in some species, either laterally or at the tips of the stems. The microsporangia (pollen sacs) are borne on the underside of thick, scale-like leaves. The ovules are borne in pairs, or as a group of six or eight, on the sides of leaves which may be scale-like, or pinnate and resemble a small foliage leaf (Fig. 14.2). The male gametophyte is represented by a pollen tube which penetrates the micropyle, similar to the conifers and the flowering plants. One of the most interesting features of the cycads, linking them with the spore-bearing plants, is that, despite the presence of a pollen tube, the male gamete has a mass of flagella and swims through a fluid-filled chamber to fertilise the egg. This gamete is huge relative to most sperms, being about a thousand times the volume of a fern sperm and ranging in diameter from 80 to 400 µm (Norstog, 1987). The whole process of fertilisation in cycads takes up to five months, and seems to be rather cumbersome.

Cycads have long been thought to be wind-pollinated, like the conifers, but detailed

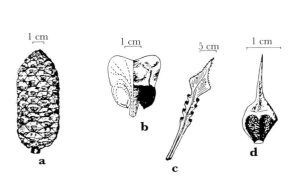

Fig. 14.2 Reproductive structures of cycads: **a**, female cone of *Zamia floridana*; **b**, female sporophyll of *Zamia skinneri*; **c**, female sporophyll of *Cycas circinalis*; **d**, male sporophyll of *Macrozamia spiralis*. Redrawn from Sporne (1965).

observation of cultivated specimens and study in wind tunnels led Norstog (1987) to believe that some, at least, may involve insect pollination, confirming some earlier reports (Willemstein, 1987). His first observation was that many cycads, when cultivated outside their natural range, remain sterile despite the presence of both sexes, and that wind was not transferring pollen effectively. He then demonstrated that a minute weevil, *Rhopalotria mollis*, was essential for the pollination of the Mexican cycad, *Zamia furfuracea*. Larvae of this weevil feed on parts of the developing male cones but not the developing pollen. Adult weevils are produced at the time of pollen release, chew their way through the pollen sacs and gather a dusting of pollen as they leave. They are attracted strongly to the female cones, probably by the fluid secreted by the micropyle which has sugar and amino acids in it. This micropylar fluid, present in all gymnosperms, usually functions as a trap for wind-borne pollen; in cycads it may have the dual function of trapping pollen and attracting insects. In most cycads the ovules are not exposed to the air, but in some species, unlike *Zamia furfuracea*, wind may be the initial agent of pollen transfer to the edges of the cone scales. Following this initial pollen dispersal, transport of pollen into the concealed ovules may well require insects, and, if so, they may be pollinated by a combination of wind and insects. Very few observations have been made on cycad pollination, particularly in their natural environments, and there is, undoubtedly, a great deal still to find out about these enigmatic plants.

A third gymnosperm group is represented today by a single species, the maidenhair tree (*Ginkgo biloba*), a native of China but widely planted in temperate regions, including Britain. It is highly ornamental as a tree, with its elegant architecture, curious leaf shape and spectacular yellow autumn colour, and it is fairly resistant to pollution and insect attack so thrives in towns. It is a truly remarkable survivor, in that leaves which are almost identical to the living species have been found in Jurassic deposits in many different parts of the world. These fossils can be regarded as the same genus and Seward remarked that the tree is '...an emblem of changelessness, a heritage from worlds of an age too remote for our human intelligence to grasp...' (Sporne, 1965).

Ginkgo biloba is dioecious and has male parts in catkin-like 'cones' which hang loosely, each scale with two pendent pollen sacs; pollen is dispersed by the wind (Fig. 14.3). The ovules are borne in pairs at the end of a short stalk in the axils of the leaves and mature into seeds about 3 cm across, with a fleshy coat. This seed is edible, and cultivation of the tree in China and Japan has probably contributed to its survival, since the tree is now unknown in the wild. The mature seed coat smells somewhat unpleasant, rather like rancid butter, and, for ornamental purposes, often only the male trees are planted. *Ginkgo*, like the cycads, has large, multicellular female gametophytes and motile male gametes, although these male gametes are a little smaller than those of the cycads and they have flagella in bands at one end. In *Ginkgo*, cells of the female gametophyte are full of chlorophyll although they are surrounded by a thick sporangial wall, seemingly a throw-back to its ancestry from free-living gametophytes.

The fourth living gymnosperm group is the Gnetales, which consists of three rather disparate genera. These are a tropical genus of small trees and vines, *Gnetum*, which strongly resemble dicotyledonous angiosperms except in their reproductive structures; the rather broom-like *Ephedra*, from Mediterranean climate regions and semi-desert in Eurasia and North and South America; and the extraordinary *Welwitschia mirabilis* (sometimes known as *Welwitschia bainesii*), confined to the Namib Desert. They are dioecious, or less often monoecious, and bear their reproductive organs in compound cones, like the other groups of gymnosperms (Fig. 14.3). The pollen is borne in sacs on short shoots and the ovules, often only two or even one per cone, in the axils of

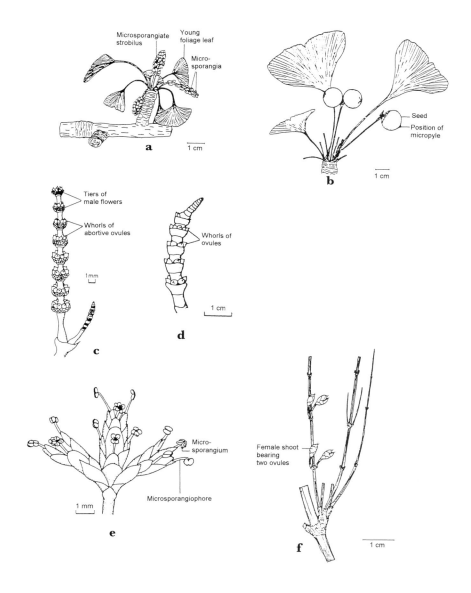

Fig. 14.3 Reproductive structures of *Ginkgo biloba* and Gnetales: **a**, catkin-like male cones of *Ginkgo*; **b**, 'fruits' of *Ginkgo*; **c**, male cones of *Gnetum*; **d**, female cones of *Gnetum*; **e**, male cones of *Ephedra*; **f**, female shoot of *Ephedra*. From Bell (1992), with permission.

bracts. The pollen is caught in a pollination drop secreted by the ovules, similar to that in other gymnosperms. Some species are probably wind-pollinated but a few *Ephedra* species can be successfully pollinated by unspecialised insects (Kato & Inoue, 1994), *Welwitschia* may be occasionally pollinated by insects and, in *Gnetum*, insect pollination may be the rule. *Gnetum* is a little-known genus, but most species produce their 'cones' in the understorey of the rain forest where wind-pollination is rare. Kato & Inoue (1994) observed moths (Pyralidae and Geometridae) feeding at the pollination drop of *Gnetum gnemon* and transferring pollen; they suggested that it was a specialist moth-pollinated plant.

We can see in the cycads and, particularly, in the Gnetales, how double fertilisation may have evolved. In cycads usually two male gametes are produced by a germinating pollen grain, as in the flowering plants, and these can both fertilise eggs in the female gametophyte (Norstog, 1987). Normally, only one of these fertilised eggs matures into a seed but the process in cycads seems to be fairly flexible. The embryo is nourished in its development on the parent plant by the female gametophyte which consists of many hundreds of cells. The process may be similar in some other gymnosperms and two or more fertilisations in one female gametophyte is probably a common occurrence. In the Gnetales, *Ephedra* has been studied in some detail, and it seems that the fertilisation process has been refined, since the nucleus of the egg cell divides to form a central nucleus in the egg cell and one peripheral one. The sperm nucleus from the pollen grain also divides and one fuses with the central egg nucleus, the other with the peripheral one. Both fertilised nuclei then divide, but the peripheral one aborts while still small and takes no further part in seed formation (Friedman, 1990, 1992). Again nourishment for the developing embryo is provided by the female gametophyte, but the process of double fertilisation bears a strong resemblance to that in the flowering plants.

In angiosperms the process is refined further, particularly with the huge reduction in size of the female gametophyte, and the diploid fusion nucleus leading to the endosperm, but the fertilisation process in *Ephedra* is so similar that it may well be the direct fore-runner, the double fertilisation being homologous in the two groups. Angiosperms have a much more efficient process, since not only is the female gametophyte reduced in size and much cheaper to produce, but its production takes only a matter of days rather than months. In addition, the food store for the seed only grows after fertilisation, so no resources are wasted if there is no fertilisation. This efficiency and speed of the whole reproductive process must have allowed much greater flexibility in lifestyle generally and may well have contributed to the current dominance of angiosperms (Stebbins, 1974; Norstog, 1987; Friedman, 1992).

The Gnetales are considered, on morphological grounds, to be the gymnosperm group most closely related to the flowering plants and the discovery of the details of double fertilisation suggests this even more strongly (Doyle & Donoghue, 1987). As a modern group they are restricted in diversity and range, but there is fossil evidence of greater diversity in the Cretaceous at the time of angiosperm origins. At that time, there were various other gymnosperm groups that are now extinct which also give some clues to the origin of angiosperms.

Fossil gymnosperms

Some of the Mesozoic seed-plant groups had large, pinnate fern-like leaves, others had palmate leaves and still others bore leaves like the needles of conifers. Their wood anatomy differed in a number of details of pitting in the conducting cells and the details of xylem and phloem formation, variously resembling different living groups

(Gifford & Foster, 1989; Doyle & Donoghue, 1987). In their reproductive structures some, such as the Caytoniales and Corystospermales, bore their ovules on leaf-like structures, pinnate in the corystosperms, whereas others, such as the Bennettitales and the Gnetales, bore them directly on stems. The symmetry of the seeds varied too, some bearing radially symmetrical seeds and others with bilateral symmetry. Some, such as the Bennettitales, bore their reproductive structures on fertile stems surrounded by modified leaves, giving them a close resemblance to flowers.

Only some groups in the great diversity of fossil gymnosperms are mentioned in the previous paragraph; many other fossils have been found, variously assigned to genera,

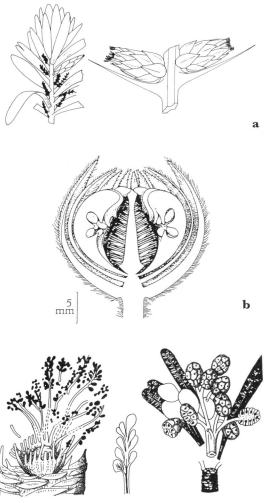

Fig. 14.4 Reproductive structures of fossil gymnosperms:
a, *Cordaianthus concinnus* (Cordaitales), male shoot (female shoot looks similar, with ovules at the end) (from Bell, 1992, with permission);
b, hermaphrodite 'flower' of *Williamsoniella* (Benettitales) (from Crepet & Friis, 1987, with permission);
c, *Pentoxylon sahnii* (Pentoxylales), male shoot, male sporophyll and female cones. From Sporne (1965).

families or orders, with many of the affinities uncertain, and there is still uncertainty as to which of these groups, if any, is the direct ancestor of the flowering plants. The most likely ancestors, along with the Gnetales, are the Caytoniales (sometimes included among the pteridosperms), the Bennettitales or the Pentoxylales, but some others have also been suggested. These groups were widespread from the mid-Triassic to the end of the Cretaceous period, and many may have been visited by insects.

The Caytoniales had aggregations of ovules, each in a small outgrowth ('cupule') on the side of a leaf-like structure (Fig. 14.4). The pollen was borne on anthers on separate inflorescences which were quite unlike the anthers of any other group. They were flattened, branching structures, with opposite pinnae, each pinna being branched and bearing an anther on the end of each branch. The Bennettitales had remarkably flower-like reproductive structures, rather like large buttercups. There were thick broad scales around the stem which bore the stamens on their inner faces, and inside them, at the tip of the stem, ovules, each one surrounded by a cupule of five or six scales (Fig. 14.4). Both of these gymnosperm groups had similarities with the flowering plants, although the flower-like appearance of the Bennettitales does not itself indicate a direct relationship with the angiosperm flower. The Pentoxylales superficially resembled the Bennettitales in their reproductive structures (but not in their wood anatomy). They had whorls of staminal shoots at the tips of short stems, each bearing a spiral arrangement of stamens, and ovules borne in aggregations of about 20 on short branches from separate shoots. The mature 'fruits' must have resembled stalked mulberries (Fig. 14.4) (Sporne, 1965; Doyle & Donoghue, 1987; Crane & Lidgard, 1990).

It is highly likely that some of these gymnosperms were insect-pollinated. Scott & Taylor (1983) suggested that even some of the earliest gymnosperms in the Carboniferous may have had spores dispersed by arthropods. The flower-like reproductive structures of the Bennettitales suggest at least unspecialised insect pollination and some of the fossil Gnetales were almost certainly insect-pollinated, like the living members of the group. Add to this the importance of insect pollination in the cycads and it suggests that the association of reproductive structures with insects goes back much further than the origin of the flowering plants. Insect pollination in the angiosperms was probably a co-option of an insect/plant relationship existing in contemporary Cretaceous gymnosperms (Crepet & Friis, 1987; Crane *et al.*, 1995).

Evolution of angiosperms from gymnosperms

The angiosperms are defined as a group by their reproductive structures. They are much the most successful of modern land plants in terms of number of species, number of individuals and dominance of today's vegetation, so the precise differences between angiosperms and other seed plants are worth looking at in detail. The two most obvious morphological features that distinguish the angiosperms from the gymnosperms are the endosperm, which has already been discussed, and the fact that the seed is totally enclosed by a carpel. The evolutionary origin of the angiosperm carpel is still obscure. It may be derived from a modified leaf or from a cupule, as in the Caytoniales, or some other structure (Crane *et al.*, 1995), but, whatever its origin, it surrounds the sporangium, now called an ovule, and it will develop into the fruit. The term 'angiosperm' refers to this, meaning hidden seed; 'gymnosperm' means naked seed. The anthers of angiosperms are simpler than those of Bennettitales or Gnetales. The reasons that the carpel evolved are likely to be connected with early insect pollination, and the evolution of the flowering plants generally is intimately bound up with insect pollination right from the start. It is likely that pollination by unspecialised insects is the primitive condition in the angiosperms. Wind-pollination undoubtedly appeared early, but this, and all other forms, are secondarily derived from insect pollination within the flowering plants.

The ovules of a plant form the seeds and they are rich in nutrients. The pollen, too, is rich in protein and the result is that flowers provide a potentially important food source for insects. Today, many insects use flowers as a vital source of food, with pollen the main attraction for insects in many modern flowers. Pollen is collected by most modern flower-visiting groups. Some insects, such as many of the flower-visiting beetles, eat both pollen and ovules and are responsible for considerable losses of potential fruits and seeds. The beetles as a group date back to the Carboniferous, long before the angiosperms appeared, and they may have been the main pollinators of some of the insect-pollinated gymnosperms in the Mesozoic. The reproductive parts of plants must always have been attractive to insects and one can easily imagine that enclosure of the ovules by the angiosperm carpel gave increased protection to these most precious and vulnerable structures within the flower.

Another feature associated with the appearance of the carpel is the presence of a style and stigma, which the pollen tube must pass through in order to fertilise the ovules. This is almost a by-product of ovule protection, but raises the possibility of an effective system of self-incompatibility, the inhibition of a plant's own pollen from fertilising the ovules (Chapter 12). The pollen recognition mechanism for most self-incompatibility systems occurs in the style or on the stigma surface. Such a system ensures that all successful fertilisation derives from pollen from another plant and that there is, as a result, a constant supply of new gene combinations and increased potential for rapid evolution (Whitehouse, 1950). Even without complete self-incompatibility, the occurrence of a style allows the potential for pollen tubes to compete with each other on their way to the fertilisation of the ovules. It may also have brought about increased selection of genetic differences between pollen grains that may have been beneficial for the success of the flowering plants (Mulcahy, 1979; Mulcahy & Mulcahy, 1987).

The earliest flowers

Various rather fragmentary fossils, which may be angiosperms, have been found in late Jurassic and early Cretaceous deposits about 130-140 million years old. These are mainly fruits and seeds and give us little idea of what the flowers (if they had true flowers) looked like (Friis & Crepet, 1987). The earliest undoubted angiosperm fossils to have been found are some pollen grains about 110 million years old. These pollen grains, which have been given the name *Clavatipollenites*, resemble closely pollen of the genus *Ascarina*, a living member of the Chloranthaceae found in Australia and New Zealand (Fig. 14.5). Other Lower Cretaceous pollen grains show striking resemblances to the pollen of two other living genera in the same family, so there is good evidence that it is one of the most ancient of all living plant families (Walker & Walker, 1984; Crane *et al.*, 1989, 1995). Preserved fruits in these deposits are rather different from those in living Chloranthaceae, so, clearly, these early fossils were not the same as living genera, but they do show numerous similarities. The flowers of members of this family, living and fossil, are very simple indeed (Fig. 14.6). They are bisexual in two of the four living genera, *Chloranthus* and *Sarcandra*, and unisexual in the other two genera, *Ascarina* and *Hedyosmum*. Each flower contains one or three stamens and/or one carpel and some have a small bract associated with the flowers. These tiny flowers (up to about 3 mm long) are borne in catkin-like inflorescences (Endress, 1987a). Fossil flowers from the Lower Cretaceous deposits (whole flowers are preserved in slightly

Fig. 14.5 SEM of pollen of *Ascarina rubricaulis* from New Caledonia, closely resembling the fossil pollen-type *Clavatipollenites* from Cretaceous deposits.

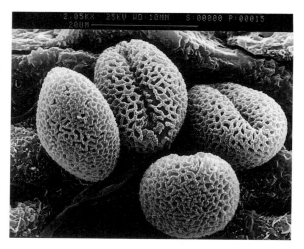

later deposits than the earliest pollen) include very similar bisexual and unisexual forms. Living bisexual members of the Chloranthaceae are pollinated by unspecialised beetles and short-tongued flies (and possibly other unspecialised insects), whereas the unisexual ones are thought to be wind-pollinated, though this requires confirmation (Endress, 1987a).

Although the fossils of Chloranthaceae-like pollen are the earliest which are unquestionably angiosperms, in only slightly younger Cretaceous rocks there is evidence for another living family, the winter's bark family (Winteraceae), and by the mid-Cretaceous, about 95 million years ago, fossils bearing strong resemblances to present-day magnolias (Magnoliaceae) and allied families have been found, along with planes (Platanaceae), saxifrages (Saxifragaceae) or related families, lilies (Liliaceae) or their allies and, probably, hornworts (Ceratophyllaceae) (Walker *et al.*,1983; Walker & Walker, 1984; Dilcher & Crane, 1984; Crane & Dilcher, 1984; Crane *et al.*, 1986; Friis & Crepet, 1987; Doyle *et al.*, 1990a, 1990b; Crane *et al.*, 1995). The magnolias, in total contrast to the Chloranthaceae, have very large showy flowers, up to 25 cm across, with numerous stamens and carpels, surrounded by an indefinite number of pink or white bracts or perianth segments (these two types of floral organ are not distinct in

Fig. 14.6 Living Chloranthaceae: **a**, inflorescence of *Sarcandra chloranthoides*; **b**, flower of *Sarcandra chloranthoides* showing single stamen and carpel with subtending bract; **c**, male inflorescence of *Hedyosum mexicanum* containing many unistaminate flowers. From Endress (1994), with permission.

a **b** **c**

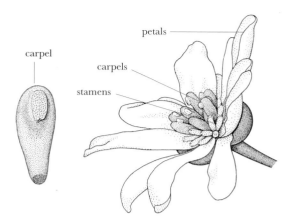

carpel

petals

carpels

stamens

Fig. 14.7 Flower of *Drimys winteri* showing indefinite numbers of all floral parts and separate carpels. Each carpel has a tiny slit by the stigma.

these plants). The appearance of this variety of flower form by the mid-Cretaceous suggests that the very first angiosperms had appeared some time before this, probably in the Jurassic, or even the late Triassic, but in these deposits the features that distinguish angiosperms from other seed-plants were not clear cut and the flora consisted of a mixture of seed plants, the majority of them clearly not angiosperms (Cornet, 1989; Crane & Lidgard, 1990).

The evidence leads to the conclusion that, once the unique features of flowering plants had been firmly established together, there must have been rapid evolution and diversification of flowering plants during the early to mid-Cretaceous, with the appearance of a great variety of flower form. Perhaps this is to be expected. The innovation of enclosed ovules, along with, perhaps, other associated characters in the flowering plants, was successful and may have led to a new range of evolutionary possibilities.[1]

These flowers did not have a rigidly set structure; they had variable numbers of carpels and stamens which, in some, were surrounded by modified leaves forming bracts (or a perianth). This open organisation can still be seen today in the Winteraceae (Fig. 14.7). Most species in this family have variable numbers of floral organs, the most extreme example being *Drimys* (*Tasmannia*) *piperita* which has 0-15 petals, 7-109 stamens and 1-15 carpels per flower (Vink, 1970; Endress, 1987b)[2]. *Ascarina*, in the Chloranthaceae, is also variable and may have unisexual or bisexual flowers and one or more stamens per flower (Endress, 1987a, 1990).

[1] There is an interesting parallel with the range of multicellular animals that appeared in the geologically short space of time in the Cambrian era. This is thought to have occurred as a result of another successful innovation, multicellular life, which led to all sorts of previously unknown possibilities and evolved very rapidly (Gould, 1990).

[2] The Winteraceae has some other apparently primitive features, notably that there is a tiny slit in the carpel by the stigma so the ovules are not completely enclosed and, in some species, there are no vessels in the xylem. The primitive nature of these features is questionable, however, since the mignonettes, Resedaceae, also have a (larger) slit in the carpel, and vessel-less wood may be derived within the Winteraceae (Doyle *et al.*, 1990b).

The flower fossils mentioned so far were highly variable and diverse in form and they appeared within a geologically short space of time. As a consequence they give clues only as to which characters are genuinely the most primitive, but they do share certain features. For instance, each carpel and each stamen is separate from the others and they are all inserted above the insertion of the perianth/bracts, if these are present. The contrast in size and complexity in the extreme forms, and the variation in structure, have given rise to several alternative ideas about the early evolution of angiosperms. One suggestion is that the large flowers of, for example, a magnolia may be derived from an inflorescence of the tiny Chloranthaceae-type flowers, each reduced to a single stamen or a single carpel (some have little more than this, anyway). These are aggregated into a monoecious inflorescence, with the female flowers at the tip, becoming the centre of the inflorescence surrounded by male flowers, and these are surrounded by modified leaves which have become the attractive perianth. This, then, is one magnolia 'flower' (Burger, 1977; Willemstein, 1987). Most flowers, according to this theory, are, in fact, inflorescences of the most primitive flowers, but the simple primitive flowers are retained in the Chloranthaceae and related families and in catkin-bearing trees such as the plane and oak. Attractive as this idea is, however, there were many flowers of intermediate size around at the same time, and there is little evidence to support it (Endress, 1987b).

Another theory is that angiosperms are not a single group at all, but the combination of features that define them has appeared more than once in evolution. In this theory, the group including the Chloranthaceae and catkin-bearing trees are derived from one gymnospermous ancestor, probably among the Gnetales, the magnolias and relatives from another, and perhaps the buttercups or the monocots from other groups (Krassilov, 1977, 1991). Again, the weight of evidence is probably against this theory, but it demonstrates that there is still much speculation and few certainties in our views on early angiosperm evolution (Crane et al., 1995).

During this early period of flower evolution, the petals differentiated. In the Chloranthaceae and related families there are no petals, but in some species the stamens are coloured and attractive to insects. In the Winteraceae there may be several whorls of stamens, but some stamens may be replaced by petals in the outer whorls (Vink, 1970). In some other plants with a primitive type of flower, such as water lilies (*Nymphaea* spp.), there are structures that seem to be intermediate between the stamens and the petals. Also, in many cultivated 'double' flowers, most clearly in roses, some of the stamens are replaced by petals. This evidence suggests that petals are modified sterile stamens which, in most plants, have taken over the function of attracting insects. The sepals, by contrast, have probably been derived from another whorl of leaves or bracts that have become intimately associated with the flowers. Subsequently, in some plants the petals and sepals have become alike, particularly where both are involved in attracting insects (e.g. in many lilies).

Putting all the evidence together, we can speculate on what the first angiosperm flowers looked like when they appeared, perhaps 150 million years ago. They were, almost certainly, small or very small in size, with small but variable numbers of stamens and carpels. The variation in numbers of fertile organs may have allowed them sometimes to have both sexes in the same flower and sometimes just one; unisexual flowers certainly appeared very early. These flowers may have been associated with bracts (modified leaves) but they had no clearly differentiated petals or sepals. They were pollinated by a range of unspecialised insects, with beetles prominent among them, but also short-tongued wasps and flies (Endress, 1987b; Willemstein, 1987). Amongst living flowering plants, the Magnoliidae has long been regarded as the most primitive class of angiosperms, and a range of plant families and a diversity of flower form, including the Chloranthaceae, Winteraceae and Magnoliaceae, are classified within it. It is perhaps significant that pollination by beetles and unspecialised flies predominates among the whole group, and only a very few are pollinated by butter-flies, moths, birds, bats or the more advanced groups of bees, all of which evolved in the Tertiary (Endress, 1990).

It has long been thought that the first angiosperms were trees or shrubs, but Taylor & Hickey (1992) have suggested that they were scrambling rhizomatous herbs, like the modern Chloranthaceae and the related Piperaceae. It does seem likely that they were fairly small plants anyway (no more than, say, 2 m high), of early successional habitats at low latitudes (Upchurch & Wolfe, 1987; Crane & Lidgard, 1990).

Early in the evolution of flowering plants came two distinct lines, with different specialisations in their pollination. One was a specialisation for pollination by beetles, excluding other less specialised insects. Beetles are the main pollinators of some living Magnoliaceae and there are very similar fossil flowers; this involved a great increase in size and complexity of the flower, with numerous floral parts and an attractive perianth. The second was a specialisation to wind-pollination, seen in some present-day Chloranthaceae (almost certainly in *Hedyosmum*) and in the great majority of cat-kin-bearing trees. Pollen similar to that of living wind-pollinated trees, particularly plane pollen, is well represented as fossils. This led to a retention of small, simple flowers with small numbers of parts but, for many, a fixed separation of the sexes into different flowers. The climate throughout the world during much of the Cretaceous was drier than today, giving good conditions for wind-pollination to evolve and spread (Whitehead, 1969; Upchurch & Wolfe,1987). Early diversification of flowers appears to have involved, initially, a stabilisation of flower structure and an increase in com-plexity, and thereafter both increases and decreases in complexity in different lines.

One anomaly has arisen in the position of the hornworts (Ceratophyllaceae), a small but widespread family of aquatic plants. They have a simple vegetative structure with no roots and are monoecious with unisexual flowers of either 10-20 stamens or a single carpel, each surrounded by a perianth. They are water-pollinated (Chapter 9). Some analyses of relationships within modern flowering plants, using molecular techniques, suggest that this family may be quite separate from all others, branching off in evolution from the ancestral stock in the early Cretaceous (Chase *et al.*, 1993; Qiu *et al.*, 1993). Fruits similar to those of living *Ceratophyllum* are among the oldest fossils of potential angiosperm affinity, occurring in rocks about 120 million years old, so this is quite possible, but further evidence is needed (Crane *et al.*, 1995). If it is so, this one instance of water pollination, which is so rare among flowering plants gener-ally, may have arisen very early.

Later floral evolution

Plants continued to evolve through the Cretaceous era, and by the late Cretaceous open, radially symmetrical flowers appear as fossils. Unlike earlier plants, in some of these groups the number of floral parts has become fixed, many of them in whorls of five, and there are fossils of flowers bearing resemblance to present-day pinks (Caryo-phyllaceae) and heathers (Ericaeae) (Friis, 1985, 1990). By this stage, the early rapid evolution of angiosperms seems to have slowed and there was a period of relative stability of floral form (Lidgard & Crane, 1988). At the very end of the Cretaceous, however, and through the beginning of the Tertiary, from about 75 to 50 million years ago, there was a second great period of adaptive radiation in the angiosperms. By the end of this period the majority of living plant families had appeared in the fossil record. The massive extinction of animals at the end of the Cretaceous, which saw the final demise of the dinosaurs, appears to have left flowering plants almost completely unaffected. This second adaptive radiation is even more closely linked with the evolution of insects than the earlier one.

Specialist insect pollinators, today's most important flower-visiting animals, appear in the fossil record at the same time as the increasing diversity of flowering plants. Butterflies and moths, long-tongued flies and bees all show adaptive radiation in parallel with that of flowering plants. They provided the major stimulus for the evolution of the huge variety of flower shapes and colours seen today, and the insects may have been significant in fuelling this second period of rapid evolution in the angiosperms generally.

This adaptive radiation was not just to do with flowers and pollination; several other, vegetative, features undoubtedly contributed to the overall success of the angiosperms and helped in their rise to being the dominant land plants. One important feature was that the climate of the world became much wetter allowing, among other vegetation types, the development of ever-wet tropical rain forest. The range of vegetation types across the world that are regarded as tropical rain forest contains more than half of all the living flowering plant species and, clearly, the wet early Tertiary climate acted as a catalyst of angiosperm evolution. Some specialised insect groups have an intimate relationship with flowering plants, and the butterflies and moths, which together comprise one of the most abundant of all insect orders, feed on flowers as adults but feed on leaves and other vegetative parts as larvae. This larval feeding has probably been a major stimulus for angiosperm diversification (Ehrlich & Raven, 1965).

There were all sorts of different evolutionary trends in flower form during this time, the most important of which are summarised in Box 14.1. The first flowers with fused petals had appeared by the late Cretaceous and this period saw the particularly significant advent of bilateral symmetry in flowers (zygomorphy). Both of these traits are associated with precise pollination by specialist insect visitors (Chapter 6). The trends mentioned in Box 14.1 have been followed independently by several different plant families or groups. One of the latest significant developments, in the early to mid-Tertiary, was the advent of pollination by specialist birds and bats, derived from pollination by the specialist insect groups and sometimes from each other. Bird and bat pollination occur in a wide range of different plant families, most of which have many insect-pollinated members too, though evolution in certain genera or families seems to have been stimulated by vertebrate pollination (Chapter 8). A few plant families have gone down the evolutionary road of extreme specialisation to a particular insect genus or species, most obviously the orchids (Chapter 7), figs and yuccas (Chapter 11), but these are the exceptions. Aggregation of small flowers into inflorescences is a feature seen in some of the earliest flowering plants and is another repeated trend, with its culmination in the highly successful daisy family (Asteraceae) (Chapter 6); a few members of the family having apparently gone through the process twice, having inflorescences of inflorescences, e.g. hemp agrimony (*Eupatorium cannabinum*).

Since this great adaptive radiation, many of the trends mentioned can be seen to have gone in the reverse direction within one family or genus. This includes the frequent advent of less specialised flowers available to a range of insect visitors, particularly in temperate environments where pollinating insects vary greatly in abundance. Other trends towards less specialisation include the evolution of self-pollination which may be associated with small flowers or rapid development, and wind-pollination. Wind pollination nearly always involves reduction in the number and size of floral

---

Box 14.1 Some major evolutionary trends in flowers.

These trends have happened across families and orders. Most of the trends can be seen in the second major angiosperm radiation in the late Cretaceous-early Tertiary. Since this radiation occurred, within one plant family or genus, many of the trends can be seen to have gone in the reverse direction. (Based mainly on Crepet, 1983; Willemstein, 1987; and Friis & Endress, 1990.)

Open organisation with indefinite numbers of floral parts $\longrightarrow$ fixed numbers of floral parts
Undifferentiated bracts/perianth $\longrightarrow$ separate petals, sepals and/or bracts
Flowers with no symmetry $\longrightarrow$ flat radial symmetry $\longrightarrow$ three-dimensional radial symmetry $\longrightarrow$
    bilateral symmetry
Separate petals $\longrightarrow$ fused petals
Small flowers $\longrightarrow$ 1. large flowers
       $\longrightarrow$ 2. small flowers in inflorescences
Inflorescences of small flowers $\longrightarrow$ specialised inflorescences with differential functions of
    constituent flowers (Apiaceae, Asteraceae)

Ovaries superior $\longrightarrow$ ovaries inferior
Flowers with pollen as food reward $\longrightarrow$ flowers with nectar as food reward
Hermaphrodite flowers $\longrightarrow$ monoecy, dioecy and related conditions
Multi-allelic self-incompatibility $\longrightarrow$ self-compatibility
Self-compatibility $\longrightarrow$ heterostyly

Pollination by short-tongued unspecialised insects (beetles, wasps, flies) $\longrightarrow$
    pollination by   1. beetles
                    2. long-tongued flies and wasps
                    3. wind
Pollination by long-tongued flies and wasps $\longrightarrow$
    pollination by   1. long-tongued bees
                    2. butterflies
                    3. moths
Pollination by bees/butterflies $\longrightarrow$ pollination by birds
Pollination by moths $\longrightarrow$ pollination by bats
Pollination by birds $\longrightarrow$ pollination by bats

---

parts, seen most obviously in the grasses, sedges and related families, but wind-pollination occurs in some species belonging to a wide range of families that are mainly insect-pollinated. Waser *et al.*, (1995) have emphasised that generalisation of pollination systems will be favoured for plants under many conditions and that many of the most advanced flower structures are generalist in their pollinator attraction. This has sometimes been obscured by emphasis on pollination 'syndromes' of floral characters associated with a particular pollinator group. Frequently flowers show greater flexibility than this suggests.

Lines of evolution in pollinating animals

Flower-visiting insects had a major effect on the evolution of flowers, but the effect of the flowers on the evolution of pollinating insects has, by and large, been more generalised, although there are exceptions. The unspecialised flower-visiting beetles, flies and wasps show, by definition, no adaptations specifically for exploiting flowers but have various organs, mainly associated with their mouth-parts, which can be used for flower feeding or for feeding at other food sources. The long-tongued groups, though, do show a number of characters associated specifically with feeding from flowers and show a wide adaptive radiation in association with the flowers that they visit (Chapters 3-5). These insect groups all appeared, probably, during the Cretaceous but their main radiation occurred with the second radiation of flowering plants in the late Cretaceous and early Tertiary (Crepet & Friis, 1987; Michener, 1979; Michener & Grimaldi, 1988; Scott & Wright, 1990).

In the flies, the most important group, the hoverflies, have an elongated and attenuated proboscis which is clearly associated with flower feeding, but most feed on fairly open flowers and show no precise adaptations to specific groups. A few, e.g. *Rhingia* spp., feed on flowers with long corolla tubes.

The Lepidoptera have a much attenuated proboscis. Some Lepidoptera have developed the most extreme lengths of proboscis, and Nilsson (1988) has elegantly demonstrated how a plant and a pollinator can become locked into an evolutionary road leading to ever-increasing spur length in the plant and tongue length in the lepidopteran pollinator. In this scenario, the most successful pollinations will be by insects that have to probe deepest, i.e. in those flowers with the longest spurs which will stimulate the evolution of a longer proboscis in the moth. This reaches its extreme form in the seemingly absurd length of the spur in some Madagascan orchids. *Angraecum sesquipedale* is the best known of these, with a spur 28-32 cm long, although even this is exceeded by *A. eburneum*, which can reach an almost unbelievable 40 cm (L.A. Nilsson, pers. comm.). Darwin knew of no potential pollinator of *A. sesquipedale*, but at the turn of the century a hawkmoth, *Xanthopan morgani* spp. *praedicta*, with a proboscis up to 25 cm long, was discovered, although it has still not been seen to visit these extraordinary flowers. Such apparently precise adaptation is rare, and when specialisation has reached this sort of extreme, the plant and its pollinator must be at risk of extinction if conditions change. Nilsson *et al.*, (1985) emphasised that the relationship between hawkmoths and long-spurred orchids in Madagascar not only involves a primitive group of hawkmoths and a long-standing relationship in an isolated, stable environment, but is also probably more flexible than previously thought (Chapter 7, pp.214-15; Chapter 15, pp.408-9).

It is in the Hymenoptera that we see the greatest variety of specialist adaptations, perhaps predictably since they are the group most intimately and completely associated with flowers (Michener,1979; Roubik, 1989). Indeed, the evolution of sociality in bees may have been stimulated particularly by flower feeding. Flowers provide some food that is protein-rich, some that is full of energy, and both are readily stored, and they are resources that are constantly shifting in position and quality. Sociality would be a most advantageous strategy to exploit this type of resource. Various other features associated specifically with flower feeding and pollination in Hymenoptera are listed in Box 14.2.

Precise structural adaptation is shown by bees of several families in relation to certain plant groups, e.g. a brush tongue type with hooked hairs and a rake on the fore-metatarsi is found in certain species in several families and appears to be associated specifically with visiting flowers of the borage family (Boraginaceae). In a detailed study on *Rediviva* bees, which collect oil as a food source from *Diascia* species in South Africa, Steiner & Whitehead (1990) showed that the length of the spur in which the oil was secreted by *Diascia* varied between species and between races of one species, and that the length of the forelegs of the bees, used to collect the oil, varied precisely with it. Some very long proboscides, associated with particular flowers, are recorded from bees, as in the moths mentioned above (Roubik, 1989).

Among birds, three large families, the sunbirds of the Old World, the honeyeaters

---

Box 14.2 Evolutionary trends in bees in relation to flower visiting.
(Based on Michener, 1979; Michener & Grimaldi, 1988; and Roubik, 1989.)

Insect prey → pollen for larvae

Solitary life style → primitive social → advanced social ('eusocial')

Short-tongued → long-tongued

Simple labial palps in proboscis → sheath-like palps

Complete wing venation → reduced wing venation

Grooming as a cleaning operation → grooming as pollen gathering

Pollen in crop → dry collection in scopa → wet collection in corbiculae

---

of Australasia and, most particularly, the hummingbirds of the New World, along with some other smaller groups, e.g. the honeycreepers (part of the tanager family), and the flowerpeckers (part of the mistletoe-bird family), have radiated in direct association with flower visiting. These birds are mainly small (including the smallest of all birds among the hummingbirds) and brightly coloured, but represent several different avian groups. The hummingbirds are a unique group of uncertain affinity, possibly related to the swifts, whereas the sunbirds and honeyeaters are both passerine families which may or may not be closely related to each other. Flower feeding is found in many other bird families and probably evolved from feeding on insects found in flowers. Modifications include features of the tongue shape, such as the appearance of a brush-type tip, and, in hummingbirds, a striking adaptation of the wing which allows them to hover, with their bodies motionless, by a flower. The problem of ever-increasing length of corolla tube and, in birds, the beak, appears to have evolved in hummingbirds in a similar way to the insect groups described. The swordbill (*Ensifera ensifera*) of South America has a bizarre bill about 10 cm long, longer than the rest of its body and longer in proportion than the bill of any other bird; the flowers it visits, e.g. *Datura* species (thorn-apples) and *Tacsonia* (related to the passion-flowers), have a similarly long corolla. It is related to the hermit hummingbirds of the rain forest understorey and, as with the hawkmoths of Madagascar, this is a primitive group of hummingbirds in a stable environment; the more advanced trochiline hummingbirds are generalists (McDade, 1992).

The flower-visiting bats almost certainly evolved from those that feed on fruit. They show some parallel adaptations to those of birds, in particular a brush-ended tongue. The flower-visiting habit has evolved quite independently in the Old and New World bats and, probably, independently in south-east Asia and in Africa. The differences between flower-visiting bats and other bats are less striking than those among the insects and birds, and many of them feed on fruit as well as nectar.

15

Pollination, Community and Environment

The science of ecology arose in the early years of the twentieth century directly from the great compendium of natural history observations that had grown with such vigour during the nineteenth century. Initially, plant ecology revolved around descriptive aspects of vegetation, and this and theories of succession dominated the subject for the first half of the century. Genetics and cytology were growing rapidly too, giving the whole discipline of biology a firmer foundation than before. Pollination studies had a low profile during this time, but after the Second World War there was an increasing interest in the ecology of populations and the ways in which populations of plants and animals interact. This led to a new look at pollination, and 'pollination ecology' or 'anthecology' began to build on the massive turn-of-the-century compilation of Knuth (1906-9). Pollination, after all, is not only an essential step in maintaining a reproducing plant population, but it also maintains a 'service industry' – populations of insects or other visitors, many of which depend on pollen or nectar as a food source. Pollination research in an ecological context began to gather momentum around 1950, and has steadily increased in prominence ever since. In this chapter, our aim is to put pollination into its ecological context. How far do communities show structure in their pollination relationships? How do pollination patterns differ and the communities of pollinating animals vary in different parts of the world and between different habitats?

Life on earth depends on solar energy, fixed initially by photosynthesis in green plants in the form of carbohydrates. In temperate grasslands or forests, somewhere in the range of a few hundred grams up to two kilograms or more dry weight of carbohydrate is fixed per square metre each year (up to 20 tonnes per hectare) (Whittaker, 1970). This, with the mineral nutrients that the plants extract from the soil, provides the base of the food webs of animal communities. Typically herbivores consume around a tenth of it; the rest eventually decomposes. Plants commit a proportion of their photosynthesis and mineral uptake to flowers and floral rewards. At a community scale, this resource is substantial. In plants of the milkweed *Asclepias syriaca* in Michigan, up to 37% of daily photosynthesis during flowering went to nectar, and nectar production accounted for almost 3% of the plant's total photosynthesis over the year; an alfalfa crop produced nearly two tonnes of nectar per hectare (equivalent to 13 gigajoules/ha of energy), representing about 15% of above ground production (Southwick, 1984). It is probably common for nectar to account for a few percent of total primary production in insect-pollinated plants.

Nectar is a substantial source of energy, providing also some proteins and other substances (Chapter 2). How many pollinators can this energy support? A herb-rich grassland may produce about 25 g sugar per square metre over a year. Of course, this will only be produced in the summer months and insect visits will take place during the day, but making these assumptions and knowing the approximate energy consumption of some pollinators for flight we can calculate that this is about enough to support on average one honeybee per square metre throughout the summer. One bumblebee would need 3-4 m². The alfalfa field, at the height of production, could support up to 50 honeybees per square metre, but only for a very short season (calculations based on Heinrich, 1975a). Some pollinators, notably the vertebrates, require much more energy and equivalent calculations could be made. Pollen production is substantial too. Annual deposition of wind-borne pollen may locally approach 100 kg/ha (Solomon, 1979). Production of pollen by insect-pollinated flowers is more difficult to estimate and is likely to be less, but on a community scale may often be in the region of a tenth to a fifth of nectar production, in terms of energy and a substantial amount of protein too.

At the scale of a community, these resources are provided by many plant species in the course of a season, and this is essential to sustain populations of most pollinators. But plant communities are highly variable in species composition and in overall productivity, depending on latitude, temperature, soil type, etc. In some, the majority of plants are wind-pollinated, so do not provide any resources for pollinators. The relative importance of the numerous ways in which pollination may be achieved – insects, vertebrates, wind, water or selfing within a flower – differs in different regions and in different habitats. In general, there is a greater diversity in all forms of life in the tropics than in higher latitudes and a greater diversity of flower morphology and of agents of pollination, but there are many exceptions and the precise relationships differ markedly in different parts of the world. Enough work has now been done on the pollination relationships in different plant communities for us to survey the communities of the world and suggest tentative conclusions on the relative importance of different pollinators.

The distribution of pollination types in plant communities of the world

Temperate habitats

In the temperate parts of Europe, and at equivalent latitudes in North America and across Asia, a striking feature of many communities of plants is that the dominant plants are wind-pollinated. This is true of the majority of temperate trees such as the oaks, beeches, birches and all the conifers, and of the grasses and sedges that dominate so much open land (Chapter 9). In the temperate rain forests of the world, e.g. in Chile, New Zealand and the Pacific north-west of North America, wind-pollination is

Fig. 15.1 Southern English temperate woodland in May; wind-pollinated oak (*Quercus robur*), ash (*Fraxinus excelsior*) and hazel (*Corylus avellana*) are the dominant woody plants, with insect-pollinated bluebells (*Hyacinthoides non-scripta*) and primroses (*Primula vulgaris*) on the ground. Dorset.

predominant despite the very wet climate. Wind-pollination will be particularly advantageous where a plant is common and gregarious and living in an exposed habitat. Some plants with flowers that look as if they are insect-pollinated may, in such circumstances, be at least partially wind-pollinated. Good examples are heather, *Calluna vulgaris*, growing as a dominant on so much moorland in Britain, and the Öland rockrose (*Helianthemum oelandicum*), and perhaps other rockroses (Widén, 1982) (see pp.284-5).

Wind-pollination may be characteristic of many of the dominant species, but there are insect-pollinated flowers in these communities too. In some habitats, insect-pollinated plants are abundant and can be among the dominants, for instance in the herb layer of a woodland (Fig. 15.1) and in some meadow areas and disturbed ground, where they live alongside the wind-pollinated grasses. Throughout the floras of temperate regions, however, many plant species, including probably a majority of animal-pollinated plants, are capable of self-pollination. Insect populations are notoriously variable from year to year and may be unreliable. For many plants cross-pollination is favoured, but they are capable of selfing, although the ability to self-pollinate and the amount of self-pollination differs greatly between species (Chapter 12). Some plants are predominantly selfed, and in places with particularly harsh conditions, for instance in the cool, wet and miserable summers on the Faroes (or in the exceedingly *in*temperate hot dry climate of Timbuktu!), self-pollination may be the major pollination type (Hagerup, 1932, 1951).

Most shrub species, in contrast to the trees, are insect-pollinated and most flower slightly later than the trees, in May and June. They are characterised by lack of specialisation in their flowers, with unspecialised flowers available to a large range of insects. Some specialist pollinators are flying then, but their numbers are still low at this time of year and plants such as the elder (*Sambucus nigra*), hawthorn (*Crataegus monogyna*), dog roses (*Rosa* spp.), and guelder rose (*Viburnum opulus*) (Fig. 15.2) are visited and can be pollinated by numerous different insects (Knuth, 1906-9; Yeboah Gyan & Woodell, 1987). Many of these shrubs are plants of woodland edge and clearings, and lack of specialisation might be expected in such opportunistic plants. Generalist pollinators may be numerous at this time of year, and many woodland herbs flower then, too, including those that are bee-pollinated.

In Rhode Island, U.S.A., Rathcke (1988) demonstrated a similar flowering pattern with quite different shrubs belonging to such genera as *Ilex*, *Vaccinium* and *Gaylussacia*. Rathcke could not suggest good reasons why they flowered so early in the year, and

Fig. 15.2 Guelder-rose (*Viburnum opulus*), showing small white fertile flowers, accessible to generalist insect pollinators, typical of temperate shrubs. In this species, unusually, the fertile flowers are surrounded by a ring of sterile flowers adding to the floral display; in some cultivated forms the fertile flowers are replaced by sterile flowers (the 'snowball tree').

she showed that the specialist pollinators in her study area reached their peak of abundance later in the year than this. There may be other reasons for early flowering not directly connected with pollination. Many of these species have fleshy fruits dispersed by birds and the timing of fruiting may be crucial (Snow & Snow, 1988); time of fruit maturation may constrain the flowering time, or it may be constrained by their phylogenetic relationships, a point discussed later in this chapter (Kochmer & Handel, 1986).

On the floor of a temperate woodland, there is an abundance of spring flowers opening before the leaf canopy is complete, and many woodland herbs have finished flowering before the end of May. There are a few plants flowering in mid-summer or early autumn but these are mainly associated with clearings and semi-shade. Light on a woodland floor is limited in mid-summer and there is little energy for plant growth after the trees leaf out. Some of the most characteristic of the plants that flower under the canopy belong to specialized families such as the orchids, which rely on fungi associated with their roots (mycorrhiza) to provide energy and nutrients for flowering.

Among the early-flowering woodland herbs in Britain, there is a mixture of open flowers, such as wood anemone (*Anemone nemorosa*), celandine (*Ranunculus ficaria*) and wild garlic (*Allium ursinum*), with pollen or nectar available to many insects, and more specialised tubular flowers, such as bugle (*Ajuga reptans*), and other labiates, violets (*Viola* spp.), primrose (*Primula vulgaris*) and bluebells (*Hyacinthoides non-scripta*). The pollinators consist of unspecialised flies, beetles and small Hymenoptera visiting the open flowers, and solitary bees, some more specialised flies like the beeflies (Bombyliidae) and bumblebees on the tubular flowers. Bumblebees in general are regarded as uncommon in woodland (Williams, 1988), but they may be common in clearings and rides, and can be important as pollinators of the open flowers as well as the tubular ones (Saville, 1993). This is particularly true of *Bombus pascuorum* and, to some extent, the early flying *B. pratorum* and *B. hortorum*. Saville (1993) found, in a Cambridgeshire woodland that tubular bee flowers were commoner than the open flowers and that they occurred through the season and in the most shady parts. Some were freely visited by bumblebees.

In two areas of North American woodland, in Illinois and North Carolina, there appeared to be a higher proportion of open bowl-shaped flowers with unspecialised pollinators than there is in Britain, and a smaller number of specialist bee flowers (Schemske *et al.*, 1978; Motten, 1986). These authors suggested that, in the spring flowering season in many woodland areas, there is a limited number of pollinators, favouring the unspecialised flowers. The flowers of many of the woodland herbs in these studies had a long period of receptivity and most were capable of self-pollination. Reduced seed set as a result of limited pollination occurred in a few species, mainly the bee-pollinated ones, and would probably have been more widespread if fewer species were capable of self-pollination.

Among the insect-pollinated herbs of non-wooded communities throughout temperate regions, probably a majority rely, at least in part, on bees for successful pollination[1]. In Europe, except for parts of the Mediterranean and the Arctic, and equivalent

---

[1] The umbellifers (Apiaceae) are a notable exception to this. They are almost all unspecialised in their pollination, attracting mainly flies and unspecialised Hymenoptera (p.173).

latitudes in North America and eastern Asia, bumblebees (Bombinae) are the most important pollinators. Pollination by butterflies, moths, flies and other insects is widespread, and vital for some species, but even many of the flowers pollinated by these groups can be successfully pollinated by bees, and bee-pollination frequently dominates the flowering ecology of temperate communities. This dominance by social bees extends, in Europe, south at least to northern Spain, particularly in the mountains (Obeso, 1992).

In most open, grassy communities there is a preponderance of flowers in June and July and they exemplify the dominance of bee-pollination (Pojar, 1974; Heinrich, 1976a, b; Lack, 1982d). They include specialist bee flowers, such as the legumes (Fabaceae) and labiates (Lamiaceae), and generalist flowers pollinated mainly by bees but visited by many insect groups, such as the thistles and knapweeds (*Cirsium, Carduus* and *Centaurea* spp.) and scabiouses (Dipsacaceae). These last two groups produce flowers in capitula and flower in mid to late summer; they require many visits per capitulum for full pollination and flower at the time of peak bee numbers. These plants are extremely attractive to bees, dominating their attentions (Lack, 1982d; Saville, 1993). Many of the other plants, some apparently attractive to and pollinated by other insects, are also pollinated by bees. The morphology and behaviour of bees often means that, even where they are in a minority as visitors, they may often be the most important pollinators (Chapter 16).

In open communities in temperate North America, there is a distinct group of plants pollinated by hummingbirds. Most of these are red and tubular and they appear to have converged in appearance and often, in any one area, in flowering time too, both traits probably a response to migrating pollinators (Chapter 8). They come from a number of different plant families and all are related to insect-pollinated species. Many appear to be recently evolved from insect-pollinated species and only one genus, *Castilleja* (Scrophulariaceae), has had an evolutionary radiation specifically of bird-pollinated flowers (Brown & Kodric-Brown, 1979; Stiles, 1981; Grant, 1994). Several are capable of being pollinated by insects, again mainly bees (Waser & Real, 1979). The hummingbirds colonised North America from South America, probably arriving in the Eocene and colonising mainly in the mid-Tertiary, after the major plant groups were established (Grant, 1994).

Moving north in Eurasia and North America, bee-pollination remains the commonest type of animal pollination into the boreal zone (Ranta *et al.*, 1981). Even further north, two species of bumblebee were present and important pollinators for a few species, such as louseworts, *Pedicularis* spp., at 81°N on Ellesmere Island in the Canadian Arctic (Hocking, 1968; Kevan, 1972). The predominant types of flowers in these regions are open and cup-shaped, however, and unspecialised in their pollination. The most important pollinators generally are dipteran flies (Kevan, 1972; Philipp *et al.*, 1990; Pont, 1993), and, as visitors to the Arctic will know, flies such as mosquitos thrive and become extremely numerous in the short Arctic summers.

Mediterranean-climate regions

Southern Europe and regions of California, Chile, southern Australia and South Africa have a 'Mediterranean' climate of cool wet winters and hot dry summers. These areas, with the possible exception of Chile, are all centres of plant evolution, and many genera, mainly of small shrubs and herbs, appear to have evolved and radiated in each of these regions. It is mainly from these areas that plants have spread to adjoining temperate regions. Many plants in the Mediterranean-climate areas have restricted distributions and there is a high degree of endemism. Except for South Africa,

Fig. 15.3 Diverse shrubby vegetation in Mediterranean France, dominated by bee-pollinated plants, e.g. the sun-rose *Cistus albidus* in the foreground.

there is a high diversity of bees in these regions, and in Europe and California there is the highest diversity of bees in the world, most of these being solitary species (Roubik, 1989).

In all these areas except South Africa, bee-pollinated plants dominate the shrub and ground flora, as in the north temperate regions, and there appears to be have been much mutual evolution between the plants and the bee pollinators (Fig. 15.3). The dominant shrubs are mainly pollinated by a variety of solitary bees, with the addition of the honeybee in Europe (Herrera, 1988). Many of the herbs in these regions are specialised for pollination by one or a few species of solitary bee, which divide the flora between them (Moldenke, 1976; Simpson, 1977; Armstrong, 1979; Petanidou & Vokou, 1990). Social bees occur here, but are relatively less important than they are in other parts of the world. It seems that the evolutionary radiation of small shrubs and herbs has combined with a high degree of specialisation in their pollination compared with other regions. The climate in these regions is relatively predictable, giving benign conditions for flowering and pollinator activity in the spring, mainly during April and May in the northern hemisphere, followed by intense insolation and summer drought with little flowering in mid-summer. It has, perhaps, not changed so radically as further north over Pleistocene and recent times. This climate and flowering regime may suit solitary bees more than the long-lived colonies of social bees. Those plants that have spread northwards from the Mediterranean-climate regions into the more seasonally unpredictable temperate areas are most likely to be the adaptable generalists, and this may explain their greater dependence on social bees. Social bees are better equipped to respond to an unpredictable resource than are solitary bees, which have greater potential for specialisation (Roubik, 1989; and see discussion of tropical communities).

In Chile and California, up to 10% of the flora is bird-pollinated, but this does not include the dominant plants (le Maitre & Midgley, 1992). In western Australia and South Africa there is a similar proportion of bird-pollinated species, but this does include many of the most abundant plants, particularly the shrubs, and one of the most striking features of these two Mediterranean-climate areas of the southern hemisphere, compared with the northern hemisphere, is the abundance of bird-pollination by sunbirds and sugarbirds in South Africa, and by honeyeaters in Australia (Johnson, 1992; Keeley, 1992). There are bees as well, though the dominant bees in Australia belong mainly to different families from those in the northern hemisphere;

the rather primitive family Colletidae which occurs throughout the world is well represented (Armstrong, 1979; Roubik, 1989).

In South Africa, the pollination community is different from the other regions. The Cape region of South Africa has, on a regional basis, the most diverse flora in the world but a relative poverty of pollinating insects, and of bees in particular in both diversity and numbers. Bees, mainly Anthophoridae, do occur and account for some pollination (Johnson, 1992; McCall & Primack, 1992), but of a smaller proportion of plants than in the other Mediterranean-climate regions. Specialist long-tongued flies are diverse and important pollinators in South Africa. These belong to the beefly family, Bombyliidae, and to two families not so well known as flower visitors, the Tabanidae and Nemestriniidae. These, along with the birds, take the bees' place, to some extent. There is, in addition, a high proportion of wind-pollination in South Africa compared with the other regions, despite the floral diversity, and a range of unusual pollination types normally regarded as less efficient than bee-pollination, such as pollination by beetles and non-flying mammals. This may, at least in part, be a response to the poverty of the pollinating insect fauna. With its staggering floral diversity, however, there is, as one would expect, a very wide range of pollination types with some specialist systems such as that of oil-collecting *Rediviva* bees and *Diascia* flowers (Scrophulariaceae) (pp.43, 115, 382), and some orchids such as the sexually-deceptive *Disa* species and their wasp pollinators (Chapter 7, p.205).

South Africa and Australia are characterised by poor soils and particularly intense solar radiation and Johnson (1992) suggested that this favours bird-pollination since, unlike bees, birds require mainly energy from flowers, obtaining their other nutrients from insects (Chapter 8). In South Africa some other conditions are different; for instance it appears that there are more fires than in the northern hemisphere and the plants flower over a longer season. There is, no doubt, a variety of interacting ecological reasons for the differences between South Africa and the other Mediterranean-climate regions, as well as biogeographical differences in the fauna and flora (Johnson, 1992; Keeley, 1992).

Some diverse regions

Arroyo *et al.* (1982) studied the high Andes of central Chile and found bee-pollination to be dominant there, as in so many places, along with fly-pollination which became most important just beneath the snow line. Inouye & Pyke (1988) studied pollination in the Australian mountains where fly-pollination was the most important type, although over 30% of the flora was bee-pollinated.

New Zealand has a unique, isolated flora and fauna and many plant and insect groups occurring in most parts of the world do not occur naturally in New Zealand. There is a preponderance of white flowers, mostly pollinated by flies or beetles, with just a few bee- and bird-pollinated flowers (Godley, 1979). The pollinator fauna has been supplemented by numerous introductions from Europe, particularly bees, so some of the natural patterns have been severely disrupted, and, as elsewhere, bees have proved to be important as pollinators of introduced and, now, some native plants.

Pollination communities in the tropics

The tropical rain forests form the greatest contrast to temperate communities with their almost non-seasonal warm and wet climate and their fabled diversity of plants. Unlike the floral diversity of the Mediterranean-climate regions, here there is a diversity not just of species but of life form. Canopy trees of many species dominate the

Table 15.1 Frequencies of different pollination systems in lowland tropical rain forest in Costa Rica (from Bawa, 1990).

Pollination type	% of tree species (number)	
	canopy (total 52 spp)	subcanopy and understorey (total 220 spp)
Medium-sized to large bee	44.2 (23)	21.8 (48)
Small diverse insect	23.1 (12)	7.7 (17)
Small bee	7.7 (4)	16.8 (37)
Hummingbird	1.9 (1)	17.7 (39)
Moth	13.5 (7)	7.3 (16)
Beetle	0	15.5 (34)
Butterfly	1.9 (1)	4.5 (10)
Bat	3.8 (2)	3.6 (8)
Wasp	3.8 (2)	1.8 (4)
Wind	0	3.2 (7)

environment and a quite different set of species occupy a subcanopy; further species occur in the understorey along with saplings of the larger trees and there is a profusion of species of herbs, vines, lianas, stranglers and epiphytes. With such a diversity of species, it is not surprising that there is a wide range of ways in which pollination is effected.

In tropical rain forests the majority of species are incapable of self-fertilisation, either through a physiological incompatibility mechanism or through a breeding system precluding selfing, usually dioecy. In total contrast to the forests of temperate regions, wind-pollination is rare and confined to just a few canopy and subcanopy trees, particularly the occasional conifer such as *Araucaria*, and possibly a few rain forest grasses on the forest floor. The vast majority of species depend on animals for their pollination (Bawa *et al.*, 1985b; Bawa, 1990; Endress, 1994).

In the canopy of rain forests throughout the world, medium-sized to large bees are the most numerous pollinators (Table 15.1). In Costa Rica, Bawa *et al.* (1985b) recorded 44% of canopy tree species pollinated by medium-sized to large bees, and there is probably a similar proportion in other rain forests of the world, since these insects are abundant in most rain forests. Many of the other canopy trees had small, often greenish, inconspicuous flowers, pollinated by a range of generalist insects including flies, unspecialised bees, beetles and others (Table 15.1); many of these trees were dioecious. The subcanopy had more tree species and more diversity in pollination, but those pollinated by medium-sized to large bees still formed the largest single group. Pollination by hummingbirds, by beetles and specialist pollination by small bees, mainly Apidae, Halictidae and Megachilidae, were important in the subcanopy, although hardly represented in the canopy. Moth-pollination was important throughout, although the large, hovering hawkmoths occurred mainly in the subcanopy, and bat-pollination occurred in a few plants throughout.

These studies have concentrated on the trees, but many of the lianas, epiphytes, etc. are also pollinated by medium-sized to large bees or have unspecialised flowers, so the overall proportions would not change very much if these were included. Humminbgird-pollination is more prominent among non-woody plants because in neotropical forests it is strongly associated with four common non-woody plant families: the bromeliads (Bromeliaceae), the most numerous of all neotropical epiphytes; the african-violet family (Gesneriaceae), a family of herbs and epiphytes; the large herb

Heliconia spp., related to the bananas (Fig. 8.3, p.230); and the climbing passion-flowers (Passifloraceae). Some of these are associated with clearings and forest gaps, and hummingbird-pollination is frequently concentrated in gaps.

Other pollinators have particular associations with plant families, e.g. moths and butterflies (and, to some extent, hummingbirds) with the bedstraw family (Rubiaceae) which, in the tropics, is an enormous family of understorey and subcanopy trees and shrubs. Beetles are associated with the primitive custard-apple family (Annonaceae) and the true laurels (Lauraceae), both of which are mainly understorey trees, and with the huge and diverse arum family (Araceae), which is often dominant among the herbaceous climbers. Bat-pollination and hawkmoth-pollination are associated with the baobab and silk-cotton family (Bombacaceae). Where these families are well represented, these pollination types are well represented too. One of the most numerous families in neotropical forests, mainly as trees and lianas, is the pea family (Fabaceae:Faboideae), and many members of this family are exclusively pollinated by medium-sized or large bees, so this one family makes a large contribution to the dominance of these bees as pollinators.

On the rain forest floor, there are often rather few flowering herbs in the deep shade, but there may be ferns and clubmosses. In places there are orchids, reliant as they are on mycorrhiza for much of their energy; many terrestrial orchids have small and inconspicuous flowers and their pollination is little known. There are a few grasses in the rain forest understorey, and even these have evolved pollination by insects, emphasising the ineffectiveness of wind-pollination in rain forests (Soderstrom & Calderon, 1971).

Hummingbirds are confined to the New World, making a striking contrast with the rain forests of the Old World in which pollination by birds generally appears to be less common and almost confined to the canopy, in total contrast to the hummingbird-pollinated plants (Pettet, 1977; Appanah, 1981). The passerine flower visitors involved differ from hummingbirds not only because they normally perch while feeding but also because they often forage in flocks. These two traits appear to make them less suitable for pollinating understorey plants and more suitable for the large forest trees (Stiles, 1981). One family of plants that is strongly associated with sunbirds is the semi-parasitic mistletoes (Loranthaceae), most of which are bird-pollinated.

In the tropical rain forests of south-east Asia, the climate is the most non-seasonal in the world and the forests are dominated by a single plant family, the Dipterocarpaceae, contributing the great majority of canopy species and containing over 400 species in all. Many of these trees do not flower every year or produce very few flowers, but then flower abundantly, sometimes after a decade or more, the cue being a drop in temperature (Ashton, 1988). This phenomenon, known as mast-flowering, occurs elsewhere among certain species (e.g. the bamboos) but nowhere else is the whole community so influenced by it. The reasons why it has arisen may involve a physiological constraint developed in a more seasonal habitat where a trigger for flowering like a drought or temperature drop is more reliable. Once developed, the irregular fruiting can be advantageous since seed predators may not be able to maintain numbers through lean, non-fruiting years, so they are swamped in the mast years (Janzen, 1974; Ashton, 1988). Medium-sized to large bees, mainly social bees, dominate the pollinator fauna in south-east Asian rain forests for most of the time, but even these opportunistic insects cannot pollinate this vast number of flowers after what may be a long gap (one tree can produce four million flowers in two weeks). Many mast-flowering dipterocarps are pollinated by thrips, which breed in the flowers and can produce four generations, up to 4,000 individuals, from an original pair in the space of two to three weeks (Chapter 11), and are probably one of the only insect groups capable of reaching large enough numbers in such a flowering regime. A few other plants are pollinated by thrips (Chapter 3) but nowhere else are they so significant as pollinators at the community level.

There are some other regional differences, for instance there appears to be a particularly high diversity of beetle-pollinated plants in Australian rain forests, associated with a high diversity of Annonaceae and other typical beetle-pollinated families (Bawa, 1990). In Madagascar there are many hawkmoth-pollinated plants and some that are pollinated by lemurs; this may be connected with the relative scarcity of flower-visiting bats compared with other tropical regions (Sussman & Raven, 1978; Nilsson *et al.*, 1985, 1987).

It is in tropical rain forests that we find many of the most specialised of pollination relationships, with such one-to-one relationships as figs and fig wasps (Chapter 11), and orchids and their specialised bees, seen most obviously in the New World euglossine bees and scent-producing orchids (Chapter 7). Figs and orchids occur throughout the tropics and are often abundant. The figs are one of the most significant single plant groups since not only are they the most numerous, and often the only, group of strangling plants (germinating on branches as epiphytes but growing roots to the ground and eventually smothering the host tree), but also their fruits are produced throughout the year and many birds and some primates depend on figs for their survival, particularly at lean times of year (Bawa, 1990; Mabberley, 1991). They are integral to the ecology of the whole community, and their specialist pollination system can be seen as a vital link in its ecology.

Other extreme specialisations such as the hawkmoth/long-spurred orchid relationships in Madagascar, and the specialist hummingbirds and their associated plants, occur in rain forests (Chapter 12), but all these specialist systems together account for only a small part of the vegetation of the rain forest. Many species may be involved, so they make a contribution to the diversity of plants and pollination types disproportionate to their importance in the flora.

With an increase in altitude on tropical mountains, the frequency of bird-pollination increases (Cruden, 1972; Linhart *et al.*, 1987). Temperatures on mountains are lower and cloud much more frequent than at lower altitude and these two features probably limit insect activity (Fig. 15.4). More of the plants are capable of self-pollination too, again a reflection of poor conditions for pollination (Sobrevilla & Arroyo,

Fig. 15.4 Tropical mountain scrub forest, Dominica, West Indies, with characteristic foggy climate and prominent bird-pollinated bromeliads, *Guzmania plumieri*. A.J.L.

1982). In places, hummingbird-pollination dominates the community on neotropical mountains (Linhart *et al.*, 1987) and, at high altitude in Africa, many of the dominant genera of shrubs such as *Leonotis* (Lamiaceae), *Protea* (Proteaceae), *Erica* and *Lobelia* are primarily pollinated by sunbirds (Dowsett-Lemaire, 1989; Matthew Evans, pers. comm.).

Outside the rain forests, in the more seasonal parts of the tropics, there are a number of obvious differences from the rain forests. For one thing, wind-pollination is frequently important, particularly where grasses form a herb layer, as in nearly all savannas. Flowering of the community is more predictable and, for the trees, is concentrated in the dry season, when some trees are leafless. Many herbs flower in the wet season, and shrubs throughout the year, so there is a shift in flower availability, and, frequently, a seasonal shortage of flowers in the late wet season (Janzen,1967; Frankie, 1975; Frankie *et al.*, 1983). There is a wide range of pollination types in these habitats as in the rain forests.

In dry forest areas of Costa Rica, medium-sized to large bees were abundant, as in the rain forests, and many of the trees in Frankie's (1975) study were pollinated by them, with many others having open and unspecialised flowers. It appears that, among the medium-sized to large bees, there is usually a smaller proportion of the extreme generalist social species than in the rain forests (Roubik, 1989). One of the most important plant families throughout the seasonal tropics is the legume family, including such genera as *Acacia*, dominating vast areas; this family is particularly associated with bees throughout its range. Moth-pollination was common in Frankie's (1975) study too, and, in Costa Rica and elsewhere, was particularly associated with trees and lianas flowering in the wet season. There was a group of hawkmoth-pollinated plants, particularly trees but including other plant life forms, comprising about 10% of the flora of the Costa Rican dry forest, most of which flowered in the early wet season (Haber & Frankie, 1989).

There is certainly some variation. In Australia, bird-pollination is particularly important in the savannas and other areas of seasonally dry vegetation (Fig. 15.5). A large proportion (up to half) of the *Eucalyptus* species that dominate these areas are pollinated by birds and bird-pollination occurs in several other genera of Myrtaceae (Plate 7e) and, prominently, in *Banksia* and *Grevillea* in the Proteaceae. The main pollinators are honeyeaters and brush-tongued parrots. In contrast, hummingbirds in tropical America are mainly a rain forest group, and in the savannas and seasonal parts of central and South America, bird-pollination is less common than in the rain forest but honeycreepers and icterids are important pollinators for some species (Endress, 1994). Percival (1974) found that the plants in her dry

Fig. 15.5 Savanna vegetation near Darwin, northern Australia, with bird-pollinated shrub, *Grevillea* sp. (Proteaceae). A.J.L.

scrub study site in Jamaica were pollinated predominantly by butterflies. Of lesser importance was pollination by birds, flies and solitary bees.

In deserts, growing and flowering seasons are unpredictable and do not occur annually. Many desert plants are generalist in their pollination, presenting open flowers or capitula (the daisy family is well represented in deserts), perhaps not surprisingly in view of the habitat's seasonal unpredictability. Wind-pollination is scarce in true deserts, though it is not altogether clear why; it is much commoner in the semi-deserts where wind-pollinated plants may form a substantial part of the flora (Regal, 1982).

Ecological importance of pollination within plant communities

The plants that coexist in a community are likely to interact in some way. At a general level, members of the same species can compete for a position in the plant community; one species can form part of a successional stage on a sand dune; one species can facilitate the presence of another, as with trees providing shade for a ground flora, and legumes, with their nitrogen-fixing root nodules, paving the way for other plants. Pollination is, inevitably, another way in which plants interact. Viable populations of most plants are maintained ultimately by replacement from seed set, and to set seed requires pollination. For a few species, seeds and pollination may only be necessary at long and rare intervals, but even long-lived or clonal perennials usually require seeds for any long distance colonisation (Chapter 16).

Effects of pollination on coexisting plant species

There is one point that comes out strongly in most critical studies of pollination in plant communities, namely the great inherent flexibility of the relationships. The presence of a plant species in a community is likely to be influenced mainly by the physical conditions at a site, competitive interactions for space or nutrients etc., diseases and herbivory. The particular pollination interactions of a plant species that can be observed are likely to be only a minor influence on its presence or abundance, though a few specialised species may be limited by their pollinators.

One problem when dealing with ecological processes in relation to pollination is the great variability in so many of the pollinator populations. Many insect populations fluctuate enormously from year to year in any one place (Taylor & Taylor, 1977); in Britain, there are 'butterfly years' when certain species are particularly common, and researchers on bumblebees have recorded at least 100-fold differences in numbers between years (C. O'Toole, pers. comm.; AJL). Even in the tropics, abundances may vary widely (Roubik, 1989). These fluctuations will mean that different pollinators are likely to have differential importance in different years and only broad generalisations on community structure can be made. With such fluctuations, we would also expect the plant/pollinator relationships to be flexible, with both plants and insects keeping many options open. Many temperate plants can be successfully pollinated by a range of insects, and most insect visitors can visit a wide range of plant species. Bee-pollinated plants, even specialist ones, are frequently capable of being pollinated by flies or butterflies; bird-pollinated plants may be pollinated by bees, and vice versa. Even insect- and wind-pollination, although they appear to involve such different floral adaptations, are interchangeable in some plant groups, e.g. plantains (*Plantago*) (Chapter 9). Fundamentally, the pollination relationship is not tightly constrained and it seems that the species composition of any plant community is only influenced in a minor way by the pollination requirements of its constituents. Over a long time span one can imagine that total, or even a major, reliance by a plant on one species or one group of insect pollinators is vulnerable. Specialisation has occurred and we see a number of

specialised interactions today, mainly in environments which are thought to have remained fairly stable over a long period. Specialised interactions may dominate among
rain forest epiphytes and, to some extent, in the Mediterranean-climate regions, but
in most plant communities, specialisation is the exception. In the long term, it is this
fundamental flexibility that has led to the evolution of different pollination types and
the great range of floral adaptation seen, often within one plant family.

The pollination requirements may have affected the flowering behaviour or breeding systems of the constituent plant species. The community of plants and their pollinators will be responding all the time to a range of evolutionary forces.
Fundamentally, despite their mutual dependence, the requirements of the plants and
those of their visitors are different. In general, the plants must provide sufficient reward to make visits by any pollinators worth while, (deception, to succeed, must be
uncommon relative to real sources of food; see Chapters 7 & 10). For successful cross-
pollination, a plant species requires visiting insects to move frequently between flowers, and between plants. The visitors, for maximum efficiency, will visit as few flowers
as possible, with least expenditure of energy, favouring those with greatest rewards,
taking into account the density of the flowers and proximity to a nest site, if they have
one. There will be a balance, in any one plant species, between providing sufficient
nectar and/or pollen to attract and maintain visitors, and providing so much that
visitors are satiated with one or very few visits to flowers.

If coexisting plant species overlap in flowering period and attract the same pollinators, they may compete for pollinators or, particularly if they are not growing densely,
actually facilitate each others' pollination by attracting more visitors overall (Rathcke
& Lacey, 1985). This will be a relationship that is sensitive to small shifts in flower or
pollinator abundance which so frequently happen in different seasons, and plants may
facilitate each other in one season but compete in another. The outcomes of these two
opposing forces will be different and, by considering each in turn, we may be able to
understand whether either or both is important in any one community.

Interspecific competition for pollination can take two forms: either one species can
simply be more attractive than another, so the second one loses out, or, if there are
many flights between species, much pollen can be wasted or the stigmas of a species
can get clogged with pollen grains from another species.[1] Mosquin (1971) showed that
one or two species can be so attractive to pollinators that anything else flowering at the
same time may be very rarely visited. If either of these types of competition limits the
numbers of seeds set, and this is important for an individual's reproductive success,
then selection will favour a shift in some aspect of the flowering biology of one or
both species. Over time, if individuals differ in their floral ecology, there are several
possible evolutionary outcomes:

1. Those plants with flowers of a slightly different shape, e.g. with a longer corolla
   tube, may attract different visitor species and so avoid direct competition.
2. If a plant is less favoured by pollinators, individuals that produce more food reward
   and attract more visitors may be at an advantage.
3. Plants with a different flowering time, overlapping less with a competitor, will be

---

[1] Stigma clogging as a general topic is discussed in Chapter 12, mainly in the context of a plant's own pollen
on a self-incompatible species taking up space and inhibiting fertilisation from another individual.

favoured.

4. Individuals capable of self-fertilisation may be at an advantage, so avoiding competition and potential shortage of pollinators.

5. Those plants which position anthers and stigma in a slightly different place may avoid contact with the pollen or stigma of different species through being carried on a different part of a visitor's body. This would avoid pollen wastage and stigma clogging (Chapter 6, pp.163, 166-7).

If plants facilitate each others' pollination we would expect a convergence of flowering time and form of the attractive parts. Problems of stigma clogging and competition between pollen grains will still arise, perhaps more so, and differences in pollen stigma placement, etc., may be particularly favoured.

Constraints on the responses of plants

In any study on the outcome of natural selection it is difficult to know the likely ancestral form, because this will be affected by the evolutionary background of the species and past ecological conditions which may have been different from those it encounters. A few generalisations can be made now. Firstly, fundamental features of floral morphology, like numbers and arrangement of floral parts, are generally regarded as conservative in evolutionary terms, and plant classification has floral morphology as its basis. It is true, though, that even small changes in some aspects of floral morphology, such as length of corolla tube or spur, can lead to large differences in pollination relationships (Grant & Grant, 1965) (Fig. 15.6). Fruiting characters, which may have a strong indirect effect on flowering are also, often, characteristic of particular plant families or genera. Secondly, closely related species are derived, by definition, from a common ancestor usually in the recent past. When related species occur together in a community they are the species most likely to interact in terms of pollination, suggesting that the most likely starting point is that they have similar flowering characteristics.

One feature of particular importance is flowering time. This has usually been regarded as flexible, but each plant species, and certain plant genera and families, have a characteristic flowering period (Kochmer & Handel, 1986). For instance, many plant families occur in the north temperate areas of western Europe, North America and Japan, but the species, and sometimes the genera, differ. Throughout these areas, though, violets (*Viola* spp.) all flower in the spring, heathers (*Calluna* and *Erica* spp.) mostly flower in late summer, and even large families like the pea family (Fabaceae: Faboideae) nearly all flower in early summer and the daisies (Asteraceae) in late summer. There are exceptions in any large grouping, but the numerous broad associations between a plant group and its flowering period do suggest that this is a conservative character which, for most species, has not responded closely to prevailing ecological conditions. There may be developmental constraints which can explain some of these patterns (Grime & Mowforth, 1982), but it does mean that where these plant groups are particularly common or dominant, there will be more flowers available at certain times of year than others. These constraints may, in some places, have affected which plants or which pollinators can colonise the habitats.

How important the constraints are and how important ecological conditions are, is not understood, but both features may work together, e.g. members of the daisy family, such as some thistles, require many insect visits for successful pollination since each floret produces a single seed and the florets in one capitulum open sequentially over several days. In most open habitats where these species grow there are more pollinating insects in late summer than earlier in the year (Lack, 1982d).

There is evidence of a shift in flowering time in some environments, showing that,

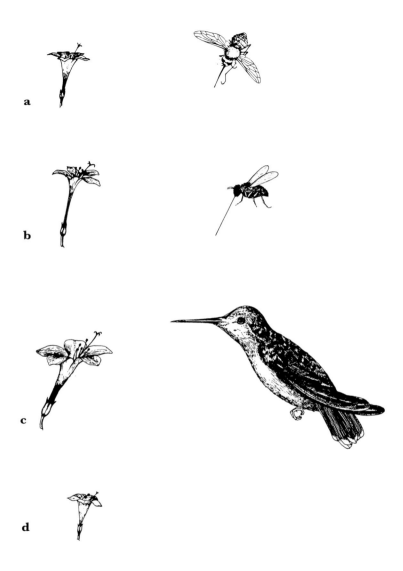

Fig. 15.6 Four subspecies of *Gilia splendens* (Polemoniaceae), which occurs in mountain ranges in California, and their principal pollinators: **a**, widespread subspecies and characteristic bee-fly, *Bombylius lancifer*; **b**, San Gabriel mountain race and the fly *Eulonchus smaragdinus* (Cyrtidae); **c**, San Bernadino mountain race and hummingbird *Stellula calliope*; **d**, desert race, largely autogamous. The bee-flies may pollinate any of the subspecies and some other insects, e.g. solitary bees, may sometimes be effective. From Grant & Grant (1965).

Fig. 15.7 Traditional hay meadow in June, Oxford, showing diversity of insect-pollinated flowers such as ox-eye daisy (*Leucanthemum vulgare*), dropwort (*Filipendula vulgaris*), birdsfoot trefoil (*Lotus corniculatus*), greater burnet (*Sanguisorba officinalis*) common knapweed, (*Centaurea nigra*) etc. A.J.L.

if there are constraints, at least some can be overcome by strong selection. For instance, one habitat of great interest in Britain is traditional hay meadows which have been cut during July every year for centuries (Fig. 15.7). Many plant species have colonised these meadows and, with strong selection to flower and set seed before the hay cut, there is a massive peak of flowering in June, including some species, like the knapweed (*Centaurea nigra*) and devil's-bit scabious (*Succisa pratensis*) which, in other habitats, flower in mid July, August or even later (Lack, 1982a; AJL).

Studies on pollination of coexisting plant species

The evidence suggests that there are serious constraints on any response to ecological conditions, but that certain responses are more likely than others. Exclusion of a plant species from a site on the basis of competition for pollinators has not been demonstrated and would be hard to prove. A shift to self-fertilisation is one of the more likely outcomes of long term competition for pollinators and may have occurred frequently, despite its potentially severe genetic disadvantages (Levin, 1972) (Chapter 12). Usually we will see only the results of such competition, but two examples provide convincing evidence that this has occurred. The North American mountain laurel (*Kalmia latifolia*) was automatically self-fertilising in the presence of *Vaccinium erythrocarpum* but outbreeding when *V. erythrocarpum* was not present, despite similar levels of inbreeding depression in both places (Rathcke & Real, 1993). Both species are pollinated by bees but the bees almost completely ignored the *Kalmia* when *V. erythrocarpum* was present.In the south-western United States, populations of the sandwort *Arenaria uniflora* had outcrossed flowers except when growing with the showier *Arenaria glabra*, when they had smaller self-pollinating flowers. This may have been due mainly to greater numbers of pollen grains of the other species reaching the stigmas, as well as competition (Wyatt, 1983, 1986). Another likely outcome that may be widespread is a change in quantity or quality of floral reward. The potential is there, since there can be much variation within a species in nectar production (Kauffeld & Sorensen, 1971; Lack, 1982b), and Brown & Kodric-Brown (1979) found a nectarless population of *Lobelia cardinalis*, which they interpreted as floral mimicry (see p.402). Although this is not directly to do with competition, it is likely that examples of a response in food reward could be found.

Table 15.2 Insect visitors to three legumes of chalk grassland in southern England (from Lack, 1982d).

Species	Peak flowering date 1978	Bombus lapidarius	% of visits by Bombus hortorum	Apis mellifera	others
Hippocrepis comosa	13 June	30	1	57	12
Onobrychis viciifolia	27 June	82	4	<1	13
Anthyllis vulneraria	4 July	9	83	0	7

Most study has centred on small differences in floral morphology and flowering time and differences in the species of pollinator visiting or the way in which they visit. The following examples of flowering and visitation patterns reveal that different communities and groups of plants appear to have responded to these various pressures and constraints in different ways.

Three species of peaflower flowering in early summer on English chalk grassland, sainfoin (*Onobrychis viciifolia*), kidney vetch (*Anthyllis vulneraria*) and horseshoe vetch (*Hippocrepis comosa*), were all pollinated by bumblebees. Different species of bumblebee were the most important pollinators of each one, and in this way they may have avoided direct competition (Table 15.2) (Lack, 1982d). In this example, the community is heavily influenced by human activity and is unlikely to have existed in its present form for more than a few centuries, so competition in the one community is unlikely to have been the main influence, but perhaps historical competition in parts of the ranges of these and other related plants has led to the evolution of a diversity of flower form. Coexistence may now be helped by the differences in pollinators that visit them. In arctic and alpine regions of North America, Macior (1973, 1975) studied a group of closely related species of lousewort (*Pedicularis*) (Chapter 6). He showed that they share flowering time and pollinators to a large extent, perhaps constrained by the short summer period, but differ in corolla length and details of floral colour reflectance, particularly in the ultra-violet. Some flower constancy may be promoted by the floral differences and there are differences in positioning of the pollen, some species placing their pollen on the back of the pollinators and some on the front thus avoiding pollen wastage or stigma clogging with inconstant pollinators.

There are several examples of related plant species flowering in succession when pollinated by the same insects; competition for pollinators may well have been one of the most important influences on many of them, but for most it is inferred rather than proved (Rathcke & Lacey, 1985). One of the most convincing examples of flowering time displacement is provided by the scarlet gilia (*Ipomopsis aggregata*) in association with *Penstemon barbatus* in different parts of the central and southern United States (Waser, 1983). In some places, *I. aggregata* flowers earlier than *P. barbatus*, and in others later, suggesting strongly that the difference in flowering time is a response to competition, not a natural constraint of either species.

In the tropics it is likely that at least some of the flowering time constraints will be more relaxed because of the relative uniformity of the climate and continuous growing conditions. It is often found that related species do flower at different times of year (Bawa, 1990). Ashton *et al.* (1988) demonstrated a succession of flowering in canopy species of *Shorea* (Dipterocarpaceae) in Malaysia, although they all fruited at the same time. They suggested that selection for thrips-pollination, probably involving selection against interspecific pollen transfer, was important. One of the most detailed studies has been that of Armbruster (1986, 1992) on the large genus of climbers *Dalechampia*, in the spurge family, Euphorbiaceae, in tropical America. He concluded that reproductive interaction between coexisting species had resulted in shifts in the food reward offered (some offered nectar, others resin), and that there was less overlap in flowering time and species of visitor than would be expected if there was no pollination interaction between the species. Other aspects of the distribution and ecology of the species led him to think that the ecological effect of interactions in pollination had been selection for character shifts in coexisting species, and that pollination interactions had not been important in determining which species were capable of coexisting. One of the particular strengths of his study was his consideration of several potential ancestral positions for the interaction, demonstrating that it genuinely was likely to be pollination that selected for the differences observed between species.

Gentry (1974) studied the flowering and pollination of members of one of the largest families of woody plants in South America, the Bignoniaceae. He revealed four different broad types of flowering pattern which, he suggested, have arisen from selection on the floral characters. Gentry called these types 'cornucopia', a provision of substantial numbers of flowers and food reward over a period of two weeks or more; 'steady-state', few flowers at a time produced over several months; 'big bang', a synchronous mass of flowers lasting only a few days; and 'multiple bang', like big bang but on several separate occasions and involving flowers with no food reward (Fig. 15.8). He proposed that these strategies represented different solutions to the problem of pollination in a diverse plant community and contributed to the coexistence of the species. The cornucopia strategy was, in his opinion, primitive and similar to nearly all temperate plants; the steady-state plants exploited the 'trap-lining' learning behaviour of social bees, birds and bats which visited a set round of foraging sites (Chapters 5 & 12) and the other two strategies relied on short-term disruption of regular visiting patterns by massive short-lived displays. How important these strategies are in other

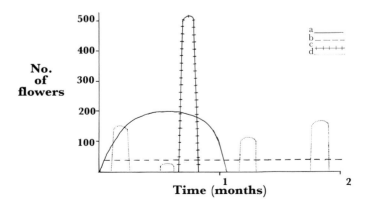

Fig. 15.8 Flowering strategies of Bignoniaceae in tropical South America: **a**, 'cornucopia'; **b**, 'steady state'; **c**, 'big bang'; **d**, 'multiple bang'. Based on Gentry (1974).

plant groups and how they contribute to the overall structuring of a flowering community are not known in detail.

Facilitation of the pollination of one species by the presence of another may lead to a convergence of flowering time. Rathcke (1988) suggested that the shrubs in her study site on Rhode Island had converged in this way, although there was no clear evidence of facilitation in pollination. Similarly, a convergence of floral form may arise to facilitate pollination of two or more species, a phenomenon known as Mullerian mimicry. Brown & Kodric-Brown (1979) demonstrated that the bird-pollinated flowers of Arizona were all red and tubular, and many flowered together, suggesting that all species were favoured by this convergence.

In Brown & Kodric-Brown's (1979) study, most species supplied nectar for the birds, including, in most places, *Lobelia cardinalis*, but one population of *L. cardinalis* looked similar but had no food reward, so was reliant on fooling the birds. This provides an example of the other well known form of mimicry, Batesian mimicry. This form of mimicry by plants depends for its success on the fact that other, usually more common, members of the community are attracting visitors and providing incentive for them to stay. Pollinators will often spend most of their time on known rewarding species, but are constantly investigating other members of the community as potential food sources and some plants with no food reward do not mimic others but simply produce showy flowers, relying on these exploratory visits. The deceivers can be successfully pollinated by such occasional visits by pollinators and save the cost of providing a reward. The orchids are, perhaps, the deceivers *par excellence*, with their precision pollination requiring so few visits and the fact that many orchid species are rare in the communities that they occupy; examples of deception and mimicry in this family are numerous (Chapter 7).

Often it may not be realistic to distinguish between the two forms of mimicry; for instance, a rare species may mimic a commoner one but still supply a food reward. Dafni (1984) showed that about 50 examples of floral mimicry had been demonstrated, but that it is probably more widespread. As so often in ecological and evolutionary study, the outcome, here enhanced pollination in one or both species, has usually been inferred rather than demonstrated. One possible example in Britain is seen by the presence of ray florets in the knapweed, *Centaurea nigra* ssp. *nemoralis*, in parts of southern England where it grows with the greater knapweed (*Centaurea scabiosa*). *C. nigra* only possesses ray florets in some parts of its range (Fig. 15.9), whereas *C. scabiosa* always has ray florets and is more attractive to bees. Lack (1976,

Fig. 15.9 Common knapweed (*Centaurea nigra*): **a**, capitulum without ray florets; **b**, capitulum with sterile ray florets.

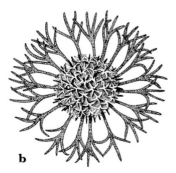

a   **b**

1982c) suggested, with some evidence, that *C. nigra* is mimicking the more attractive *C. scabiosa*, and demonstrated that, when ray florets were removed from *C. nigra* capitula, they received fewer insect visitors than intact inflorescences.

How important this type of selection on pollination attributes is for the structure of the plant communities is only partially known. For many habitats it is likely that the plants have not grown together as a community for very long, particularly considering the enormous influence of human utilisation and management over the past few centuries or, at most, two to three thousand years. We might expect more mature habitats to show the results of such selection most clearly. This has been addressed by Parrish & Bazzaz (1979), who studied three different successional stages in grassland communities of Illinois, and by Fussell & Corbet (1992) and Saville (1993) in farmland and woodland in Britain. These studies agreed that short-lived ephemeral species were generalist in their pollination, usually presenting open flowers, and were frequently self-pollinating; many attracted very few pollinator visits. Mid-successional plants, particularly the biennials and short-lived perennials like the thistles and other Asteraceae, appeared to attract the most pollinators, some of these plants producing abundant nectar over quite a long flowering season, but most were generalist in their attraction. They provided some of the most important food sources for bees and could maintain large populations of many pollinators. The latest successional stages, tall-grass prairie and mature woodland glades, had the most specialised pollination types and the greatest differentiation between plant species in the most important pollinators. They concluded that, in the mature community, interactions between the plants and insects and, in particular, competition for pollination had played a role in structuring the community but at the earlier successional stages it was of minor or no importance.

Other interactions

Plant species may interact in their pollination even without flowering at the same time. One example is a relative of the kapok tree, *Ceiba acuminata*, in thorn scrub of western Mexico, which flowers late in the dry season when little else is flowering. Baker *et al.*, (1971) showed that these flowers were pollinated mainly by bats, but that the flowers were providing floral food for a number of other pollinators and maintaining populations of important pollinators during an otherwise lean period, so providing a 'service' for plant species flowering at other times. A similar 'mutualism' was invoked by Waser & Real (1979) for *Delphinium nelsonii* and *Ipomopsis aggregata* in the Rocky Mountains. These two herb species overlap to some extent in flowering time and may compete for pollinators, but maintain the pollinator fauna for each other by flowering sequentially.

Pollination may be just one of a number of selection pressures on flowering. Insects which eat flowers or seeds, the time taken to mature fruits from a fertilised flower, ecological pressures on the timing of fruit production, or the impact of land management by humans, can all have a large effect on flowering attributes. Augspurger (1981), working on the shrub *Hybanthus prunifolius* in central American rain forests, found higher seed predation on fruits from plants which were artificially induced to flower away from their normal season. This species was highly synchronous in its flowering and she regarded saturation of seed predators at the peak of its fruiting as one of the main factors favouring this synchrony. De Jong & Klinkhamer (1991), working on dune systems in the Netherlands, found reduced seed set on plants of the hound's-tongue (*Cynoglossum officinale*), which were induced to flower before and after the population mean date, although here the reason was not so clear but was unlikely to be connected with limitation of pollination.

Conclusions on the role of pollination in plant community organisation

If we take all the evidence together, we can draw some conclusions about how pollination interactions are affecting plant communities and how the community dictates the pattern of pollination. Other pollination biologists may dispute the details of these conclusions and they are not definitive, but they fit with the evidence as it stands and are not likely to be changed in essence.

1. The direct ecological effects of pollination are normally not manifest in terms of presence or absence of plant species; other environmental factors are likely to override any pollination effects, particularly in the temperate zone. Selection on pollination-related characters may well have an effect on coexisting species, particularly in mature habitats, but it is unlikely to have an effect on which species can coexist.

2. The breeding system of plants is flexible, especially the balance between self- and cross-pollination. Self-pollination often appears spontaneously in normally self-incompatible plants and the degree of self-pollination is likely to be affected by pollination interactions. In the highly seasonal communities of the world where flowering is restricted by seasonality, this may be the most likely feature to respond.

3. Attraction of different pollinators by related plant species, or differential placement of pollen, has been selected for, mainly in association with avoidance of hybridisation or stigma clogging.

4. Flowering phenology of coexisting species may be responsive to selection on pollination, but this remains contentious. In highly seasonal environments, there appear to be phylogenetic constraints (Kochmer & Handel, 1986), and selection on flowering may involve mainly factors other than pollination (Ollerton & Lack, 1992); in the wet tropics, where seasons may be less marked, flowering phenology can probably vary more and does seem to have responded to selection on pollination.

5. Most plants are capable of being pollinated by a range of visitors, even if they are specialised and pollinated most efficiently by one or a small related group of visitors.

6. Other attributes associated with reproduction, such as fruiting or developmental phenology, may affect the pollination relationships profoundly.

The longer term population and genetic effects of the reproductive system are explored in Chapter 16.

Community structure of pollinating animals

Most pollinating animals visit a wide range of different plant species. If they are active over more than a week or two they must be somewhat opportunistic since most plants do not flower throughout the active period of the animals concerned. The distributions of pollinating insects are not normally limited by the distributions of the plants they visit and most pollinators have adaptations which allow them a wide range of flower choice. The tight relationships such as figs and fig wasps, yuccas and yucca moths and some orchid/bee and *Diascia*/bee relationships (Chapters 7, 11 and 14) are exceptions. Most of the flower-visiting flies, beetles and wasps, and many of the birds and bats, do not even depend on flowers and feed on other food such as fruit or insects as well.

Where there are many plant species usually there are many animal visitors (e.g. Heithaus, 1974; Moldenke, 1976), but the relationship is a loose one and the relative numbers of species are determined by numerous other factors of ecological importance.

Structure of temperate bee communities

Many bee species do depend on floral food at all stages of their life cycle, unlike most pollinating animals, and bees predominate as pollinators in many parts of the world. They are, therefore, the group most likely to be affected by pollination interactions, particularly competition, between the species. Such competition may lead to structured communities of bees with coexisting species differing in aspects of their ecology as flower visitors. The enormous fluctuations in abundance between years may reduce some of this but it is worth examining bee communities in detail. Bumblebees are the best studied group, although solitary bees are also important, particularly in spring.

To start with temperate Europe, in southern England there are six common species of bumblebee, demonstrating a range of tongue length, flying season and foraging preferences (Brian, 1957; Prys-Jones & Corbet, 1987; Fussell & Corbet, 1992). These species differ ecologically and each visits a different suite of flowers (Table 15.3). There are inherent differences between the bumblebee species in abundance and food requirements related to the colony cycle, in addition to those mentioned in Table 15.3. For instance, *Bombus hortorum*, the species with the longest tongue, flies mainly in early summer, is never particularly numerous and visits horizontal, mainly single, flowers such as the dead-nettles (*Lamium* spp.) and other members of that family (Lamiaceae). The short-tongued *Bombus lapidarius* has a later flying period, may be abundant and prefers open flat-topped flowers or inflorescences such as thistles (*Cirsium* and *Carduus* spp.). *Bombus pascuorum* has a medium tongue length and a long flying season, with a wide tolerance of conditions and plant species. Some species, e.g. *Bombus pratorum* and *B. terrestris*, hardly overlap seasonally since *B. pratorum* colonies are almost finished by mid-July when *B. terrestris* colonies are only beginning to build up numbers. By contrast, on chalk grassland in southern England, the two most common bumblebee species, *Bombus lapidarius* and *B. terrestris*, have a similar tongue length and were most abundant at the same season in late summer (Lack, 1982d). Worker bumblebees within one colony may differ in size and tongue length (mainly related to their nutrition as larvae), so there can be differences in flowers visited even within one colony.

In eastern North America Heinrich (1979) and Pleasants (1983) proposed that there were usually four species of bumblebee in any one habitat as a result of interspecific

Table 15.3 Differences in ecology between six common species of bumblebee, *Bombus*, in lowland England (from Prys-Jones & Corbet, 1987).

% of visits for:	lapidarius	lucorum	Bombus terrestris	pratorum	pascuorum	hortorum
nectar only	75	71	80	23	60	38
pollen only	12	28	11	10	7	5
nectar and pollen	13	1	9	67	33	57
% of visits according to orientation of flower entrance						
down	0	1	7	39	13	16
horizontal	35	38	40	28	61	71
up	65	61	53	33	26	13
Average corolla depth visited for nectar (mm)	5.1	5.1	6.3	7.4	8.3	8.7
Average tongue length (mm)	8.1	7.2	8.2	7.1	8.6	13.5
Max. no. of workers observed	August	August	July-Aug	June	August	June-July

competition. These included one long-tongued bumblebee, one with a medium-length tongue and one short-tongued at any one stage in the season, the fourth one fitting into another part of the season. This is undoubtedly too simple, and there are many exceptions, but it does suggest a degree of ordered community structure in North American bumblebees. Some North American bumblebee species have shorter tongues than any European species, perhaps owing to the absence of native honey-bees from North America, which have a similar tongue length to these bumblebees (Heinrich, 1979).

Inoue & Kato (1992) studied many aspects of the ecology of Japanese bumblebees and found that these bees differed in a range of ecologically important characters, such as tongue length, head shape and size of corbicula. These attributes were impor-tant influences on which plants they visited, again suggesting that selection is acting on community organisation.

Curiously, as one moves north in Europe the diversity of bumblebees does not di-minish and, if anything, actually increases to the sub-Arctic. Ranta & Vepsäläinen (1981) and Ranta et al., (1981) argued that, since bee colonies are static and have a limited foraging range, the instability and unpredictability of the habitat leads to hardship at different times of year for different species or different colonies of each species and it was this that limited numbers and allowed coexistence. In their study site at Abisko, in northern Sweden (lat. 68° 22′ N), they found that, despite differences in tongue length and other aspects of their ecology, the bee species overlapped exten-sively in the flowers that they visited, and, statistically, there were no differences be-tween them in their flower visiting. There was a longer mean tongue length overall at Abisko, probably related to greater general body size in the cooler climate.

The general conclusion from all of these studies of bumblebee communities in the north temperate zone is that they are structured in their pollination relationships to some extent, but less so as one moves north. In all parts there was much overlap in flowers visited by the different species all of which showed a high degree of flexibility.

Bumblebees occur in large numbers in many temperate habitats, and many species are widely distributed. They often dominate the bee fauna numerically. Solitary bees, though less numerous, usually have more species with each one more restricted in its food sources and its distribution. In the eastern United States, there are nearly 800 species of solitary bee and only 16 of bumblebees, but in many areas the bumblebees outnumber the solitaries and are more important as pollinators (Heinrich, 1979).

The great diversity of bees in Mediterranean climate regions of the northern hemi-sphere has not been adequately explained. There is an abundance of solitary bees in parts of California, including many species that are specialist flower feeders on one or a group of related plant species (Moldenke, 1976), and this is probably true of the European Mediterranean regions too (Petanidou & Vokou, 1990). There is much bare ground suitable for the nesting of these bees in California (Moldenke, 1976) and per-haps this, coupled with a greater predictability of flowering than in most parts of the world, may have led to greater specialisation of flower visitation and a greater diver-sity of solitary bees. The seasonal shortage of floral food in the summer months may also contribute to the paucity of social bees, such as bumblebees, some of which re-quire a longer period of food availability to sustain their colonies.

One factor which may have disrupted the natural communities of bees is the wide-spread introduction of the honeybee from southern Europe into more temperate re-gions of Europe and North America. Its impact on the populations of other bees is not well understood, although it may well have partially outcompeted some short-tongued bumblebees, particularly in North America. There is no evidence that any bee has become extinct as a result of honeybee invasion.

Bee communities in the tropics

In the tropics there is generally a greater diversity and a more complex structure to the plant and animal communities. There has been less study than in temperate regions, but a few patterns are emerging. One striking feature is that there are actually fewer species of bees in wet tropical areas than there are in the Mediterranean-climate regions, although they still dominate the pollinator fauna (Table 15.4). Differences between the different continents are clear too, with south-east Asia having only about half the number of species occurring in the neotropics (Africa is intermediate). Almost half of the tropical bee species in each area are social bees, a higher proportion than in temperate communities, and they do encompass the whole range of morphology shown by temperate bee species (Roubik, 1989, 1992). This is a well studied group and we can have confidence in the published figures. Sociality in these bees reaches great complexity and tropical social bees are regarded as the most evolutionarily advanced of the Hymenoptera. They form large and often abundant long-lived colonies and are extreme generalists, visiting a wide range of flowers. This may at least partly explain the small number of species. In an environment with such a diversity of tree species, any one of which may have widely scattered individuals flowering over just a short season, flexibility in flower visiting will be favoured. This is particularly true in the south-east Asian rain forests, where some of the canopy tree species flower so intermittently that specialisation on these by any pollinators will be almost impossible. The bee fauna of south-east Asia is dominated by a few species of abundant social bees, particularly four species of the honeybee genus, *Apis*, which have a wide flight range and visit many different plant species. They may make up in numbers for the many central American species. One other factor is that many tropical social bees nest in trees and there may be unlimited availability of nest sites in tropical forests (Roubik, 1989).

There is a similar proportion of social bees in different tropical regions (Table 15.4) and apparent convergent evolution between different bee groups in morphology and in sociality in south-east Asia and the neotropics (Roubik, 1992). Unrelated genera and families have come to resemble each other in morphology and in their ecology on the different continents, suggesting that competition, and perhaps other interactions between species, are influential in determining community structure of the bees. There is at least one apparent seasonal trend in neotropical bee communities: an increase in the number of long-tongued and more specialist social bees towards the end

Table 15.4 Species richness of bees in lowland tropical and temperate areas (from Roubik, 1992).

Area	No. of species	Percent social
Central Sumatra	110	50
Belem, Brazil	250	50
French Guiana	245	50
W. Costa Rica	200	25
S. California	500	~15
S.W. France	500	~15
C. Japan	170	~15
Illinois, U.S.A.	300	10

of wet seasons, associated with a lower number of species flowering, but many of those flowering in the steady-state pattern of Gentry (1974), with few large flowers produced over a long period.

Despite the evidence of community structure in tropical forests, the bees are so adaptable and flexible in their flower visiting that it is likely that constraints on community structure are loose. There has been an extraordinary 'natural experiment' exemplifying this over the last three decades or so with the invasion of south and central America by 'Africanized' honeybees from colonies introduced to Brazil in the 1950s (Roubik, 1989). These bees are now abundant over a vast area and visit up to one fifth of all the tree species. In some habitats the native bee populations do appear to have decreased following this invasion, but quantifying the effects on the native bee populations is extremely difficult, given the limited knowledge of their ecology and the natural fluctuations in numbers which are a feature of insect populations. In general, they do not appear to have excluded the native bees. Some ecological changes may be subtle and those involving the plants will take a long time, so it is really too early to assess the overall effects of this invasion. It does clearly illustrate the vulnerability of tropical pollinator communities and emphasises the fact that both plants and pollinators are flexible in their interactions with each other.

In the more seasonal tropics, particularly in dry regions, the importance of solitary bees increases, as in the Mediterranean, and probably for similar reasons of availability of nest sites, predictability of flowering and seasonal flower shortage inhibiting colonisation by some social bees (Roubik, 1989; Percival, 1974).

Community structure of other flower-visiting insects

Most flower-visiting insects except bees can feed on food other than that provided by flowers. In butterflies most of the feeding occurs in the larval caterpillar stage and the adults mainly require energy. Where there is a greater diversity of nectar sources there does normally appear to be a greater diversity of butterfly species, and different butterfly species have been shown to visit different flowers preferentially. Competition for floral resources is unlikely to be involved, however, and, indeed, it is doubtful if it is anything to do with their feeding as adults. Porter *et al.* (1992) considered that, in Britain, the butterfly community structure was largely determined by habitat and larval food plant interactions. A variety of nectar sources will normally be associated with plant diversity, and therefore a diversity of potential larval food, too. Factors other than pollination are likely to be dominating influences in determining species distribution and population sizes, which fluctuate widely, as in so many insects, in all parts of the world. Many adult butterflies feed on fruit, dung and other food sources as well as flowers.

Hawkmoths and their flowers form one of the most mutually dependent associations, and in parts of the tropics there are numerous species of hawkmoth, in which the proboscis lengths of the different species vary hugely. Some do not feed as adults. Study on the group is limited, owing to the obvious problems of observing their highly active nocturnal behaviour. Haber & Frankie (1989) studied dry forest hawkmoths in Costa Rica and found that numbers of hawkmoths were at their peak in the first half of the wet season, apparently responding to the rains. A number of plants were specifically adapted to the hawkmoths, but the moths often visited other flowers with short corolla tubes and seemed to be fairly unselective. In Madagascar, there is a group of long-tongued hawkmoth species and there are differences in corolla tube length of orchid species which correspond with the tongue length variation among the hawkmoths. However, Nilsson *et al.* (1985, 1987) found that several species of long-spurred orchids are pollinated by just one hawkmoth species, *Panogena lingens*, despite the presence of other hawkmoths, and the insects again seemed to be generalist foragers. The relationship between tongue length and flower visiting may affect coexistence, but it is clearly not directly one-to-one and we do not know whether the variety of orchids or other long-spurred plants plays a part in maintaining the diversity of long-tongued hawkmoths (Chapter 7, pp.213-15).

Opportunism in flower visiting is greater still in most flower-visiting flies and beetles and, similarly, community structure is likely to be based on factors other than the pollination relationships of these insects. The hoverflies (Syrphidae) are among the most specialised flower visitors and, within the group, there are pollen specialists and mixed-diet feeders, but Gilbert & Owen (1990) concluded that hoverfly communities in Britain are 'coincidences of species in space and time', and demonstrated very little in the way of community structure in their pollination relationships. Specialist species appeared to fluctuate together in response to ecological factors other than flower visiting. In late summer in temperate Eurasia, when hoverflies are especially abundant, there may be competition for the limited floral resources and dominance hierarchies of the species build up (Kikuchi 1962). Indeed, on some popular late-flowering species in Britain, such as devil's-bit scabious (*Succisa pratensis*), the number of visitors around each inflorescence may be so large that the scene resembles aircraft stacked in the air waiting to land at a too-busy airport (AJL). Even so, coexistence of the different species of hoverfly or their relative abundance is unlikely to be connected with the flowers that they visit.

Structure of pollinating bird communities

Hummingbird communities in central and south America appear to demonstrate some community structure with coexisting groups of long-, medium- and short-beaked species and a variety of foraging and migratory patterns (Stiles, 1975; Feinsinger, 1976; Feinsinger & Colwell, 1978). Stiles (1975) suggested that, in the mature lowland rain forest of Costa Rica, the community of hummingbirds was strongly affected by interspecific competition for floral resources. These hummingbirds were closely tied to their main nectar sources, species of *Heliconia*. The nine species of hummingbird in his study included four specialist trap-lining hermit hummingbirds visiting three *Heliconia*s with sequential flowering peaks, three short-beaked territorial species visiting primarily three short-tubed *Heliconia* species which occurred in large clumps in different habitats, and two more generalist hummingbird species.

Feinsinger & Colwell (1978) considered the hummingbirds on islands and suggested that there were usually two or three species of hummingbird on an island, each with a different set of characteristics associated with its flower visiting. There would be a territorial short-beaked species defending nectar sources, a long-beaked non-territorial species visiting specialised flowers with long corollas in a 'trap-line', and a short-beaked non-territorial generalist, or perhaps trap-liner, visiting less specialised flowers. In the more diverse and complex continental communities, several species occupied each of these types and there were some further types like 'marauders' and 'filchers', using the same flowers as the territory holders but not holding territories, and some particularly long-beaked trap-liners specialising on large-flowered understorey plants offering a high reward (see also Chapter 8). Feinsinger *et al.* (1985) compared the hummingbirds of the large, diverse, sub-continental island of Trinidad with that on the nearby smaller Tobago. Eleven species of hummingbird occur on Trinidad and a subset of five of these on Tobago. The flora of the two islands is similar and the flowers visited by the five species did not differ on the two islands within one habitat. What they found, though, was that, in Trinidad, some species moved out of a habitat to forage elsewhere at times of flower shortage and that these were the ones that were absent from Tobago. The overall diversity of habitats on the island was contributing to the diversity in any one habitat, a conclusion likely to hold true for diversity in general. Similarly, in Costa Rica, Feinsinger (1976) found that a seasonal increase in *numbers* of flowers available in any one habitat for hummingbirds led to a temporary increase in hummingbird diversity there, owing to movement between habitats; an increase in *diversity* of flowers in a habitat did not have this effect.

The studies on hummingbirds demonstrated, as with bees, that for many species, though not all, there is considerable flexibility in the flowers that they visit and that, although the flower community is exerting an influence on which species can coexist, much of the community organisation is based on other aspects of the ecology of the species, such as insect food, migration and movement patterns.

The other flower-visiting birds, including all those of the Old World such as sunbirds and honeyeaters, are less dependent on floral food. In Australia, although there is a large diversity of species of honeyeater, Carpenter (1978b) suggested that their community structure was more related to the insect resources than nectar. Hawaiian drepanidids showed dominance hierarchies at floral food sources, and two of the most important floral food sources appeared to diverge in flowering time owing to competition for pollinators, but again there was little evidence of structuring within the bird community. The only group of birds other than hummingbirds which may show evidence of structure in their floral resource utilisation is group of flower-visiting birds in South Africa made up of the sunbirds (Nectariniidae) and sugarbirds (*Promerops*). There are many coexisting species and they differ in morphology, notably in beak length, and in degree of territoriality and aggression. It seems likely that some of this diversity is a result of competition and other interactions at the floral food sources, although this is less studied than the hummingbird communities of central America. With a greater ability to use insect food as an alternative and a highly mobile bird community, the interactions are likely to be less precise than those involving hummingbirds.

The community of bat pollinators is also likely to be structured largely by factors other than their floral food sources. The most detailed study has been that of Heithaus *et al.* (1975) on the flower-visiting bats of a seasonal forest in Costa Rica. They found that only one of the seven bat species visited flowers all the time and that the others switched to fruits in times of flower shortage. The species overlapped considerably in their floral resource utilisation, but much less in which fruits they ate, so the authors concluded that the bat community may be partly structured by fruit resources, but only in a very minor way by flowers.

There are some other communities of pollinators that have been little studied, for instance that of flower-visiting lemurs in Madagascar or other non-flying mammals in South Africa and Australia, the importance of which is only beginning to be understood.

The rather loose-knit nature of the animal communities studied and, in general, the relative lack of precise community structure means that it is difficult to see, among the animal flower visitors, what are the effects of competition for visits to flowers. The effects are likely to be more diffuse than the effects on the plants except, perhaps, among the bees, because of the utilisation of a variety of food sources other than flowers. Most of the animal pollinators have short generation times compared with the plants, so, even more than in the plants, we are likely to see the results of competition or other interactions because response time can be short.

Conclusion on structure of animal-pollinator communities

In every part of the world, most of the common pollinators are opportunistic to a considerable extent, and will feed on whatever flowers are available. Only if there are no flowers at all suitable will there be no pollinators present. Differences between years in flowering patterns and weather patterns, and the large population fluctuations of so many pollinators, mean that flexibility by both plants and pollinating animals will be selected for in most situations. There is undoubtedly some structuring to the animal communities, particularly among the bees and hummingbirds, and the interactions of pollination can have quite profound ecological consequences in community organisation, but a high degree of flexibility in pollination is the rule in most communities of the world. This is further emphasised by the fact that many of the most successful flower visitors, and usually the most abundant, are the generalists, with the honeybee as a prime example.

16

Flowers, Genes and Plant Populations

We have examined in this book the ways in which plants achieve successful pollination and the various properties of the pollinators themselves. What we have not examined so far is how significant pollination is in the lives of plants. A plant's life-cycle goes from the germination of a seed, to its initial establishment as a seedling, its growth into a vegetative and then a flowering adult, through pollination leading to the development and dispersal of seeds and fruits. Each step is vital for the survival of a plant population and the first aim of this chapter is to see whether pollination is a potentially limiting step in this life-cycle. Dispersal is vital for all organisms, but most plants cannot move once they are growing and rely on pollen and seeds and/or fruits. The whole individual disperses as a seed, but pollination is the one way in which a plant maintains contact with other members of its population, and the stage at which the genes can be reassorted for a new generation. It is genetic variation that is the starting point for evolutionary change. Different methods of pollination and seed dispersal have a profound impact on how widely the genes of plants spread, and the second aim of this chapter is to examine this gene flow in plant populations. We can then draw some conclusions about how important pollination has been for plant evolution.

Pollination in the life-cycles of plants

We are accustomed to the idea that animals live in populations, the members of which interact with each other, both competitively and for breeding. This is also true for plants, although there is such variety in their life-cycles, mode of growth and breeding systems that a plant population can be very hard to define, and sometimes interaction between individual plants can be minimal. Successful pollination, however, is the main mechanism, and usually the only mechanism, by which seeds are formed, so it is a vital link in the persistence of most plant populations.

There is great variation in how long plants live. The shortest-lived plants are garden weeds such as the bittercress (*Cardamine hirsuta*) or groundsel (*Senecio vulgaris*), or some of the ephemeral desert plants which appear after rains. These germinate from seed, grow, flower and die leaving the next generation of seeds – all in as little as three weeks, in prime growing conditions. For such a short-lived plant, pollination is vital during each flowering season. Some longer-lived plants flower just once or a few times in their lives, and, again, pollination is essential. Many others, though, grow continuously and retain vegetative shoots throughout their lives, and so there is potential for any one individual to reach a great age. The oldest known plants are the creosote bushes (*Larrea tridentata*, Zygophyllaceae) in the Mojave desert of California, which are thought to have been continuously growing from seeds which germinated over 11,000 years ago when the environment first became suited to them (Silvertown & Lovett Doust, 1993).[1] Many plants may live for centuries. Some plants spread vegetatively and fragments can split off to form new plants. Individual plants become almost impossible to define, as a single clone can cover large areas, as in the common reed (*Phragmites australis*), one clone of which can cover a whole marsh, and can spread to new detached areas. For such a long-lived or clonally-reproducing species, pollination may be quite incidental in the short term, except, in some, for colonising new sites and on the rare occasions when an adult dies. Breeding system, considered in Chapter 12, also has a profound effect on the way a population functions.

Male and female function of flowers

A hermaphrodite flower fulfils two functions: firstly, pollen is brought in to the stigma to fertilise the ovules, and secondly, the plant's own pollen is dispersed to other stigmas. Both functions contribute genes to the next generation, but, within one flower, the setting of seed and fruit requires a greater investment of energy and nutrients than producing pollen, and the two functions have rather different consequences and generate different selection pressures, so they need to be considered separately. A predominantly outbreeding hermaphrodite plant species pollinated by insects is taken as the norm here; the justification for this is that it is the single commonest type, and it is thought to be the primitive type from which other breeding systems evolved (Chapter 12).

The importance of pollination and other factors in seed production

The female function of the flower, the successful setting of seed, is easier to measure than pollen dissemination since the seeds are set on the maternal parent. As a consequence, we have more information about what is important in determining the effectiveness of a plant as a seed parent. When we start to dissect out the major influences on seed production, we see quickly that there are several stages which may influence the successful setting of seed and, therefore, the contribution to the next generation. Pollination is just one factor. Developing fruits and seeds require nutrients and energy after the flower is fertilised. A shortage of resources for these developing fruits could limit the number produced. Fruits and seeds vary enormously in size, depending on mode of dispersal and the ecology of the seeds, so the amount of resources required will be quite different in different species; imagine here the contrast between the microscopic pollen-like seeds of orchids and large, nutritious coconuts, or the chaffy fruits of grasses compared with a water melon.

Another factor is simply space. For large-fruited plants it may be physically impossible for more than a fraction of the flowers to produce fruits, because a mature fruit is so large relative to the size of a flower. The apple and other fruit trees are familiar examples. When large fruits are produced on dense inflorescences the effect is even more marked, e.g. in the Australian shrub, *Banksia spinulosa*, inflorescences contain hundreds of flowers, but only about 10% can mature into fruits (Vaughton, 1991). Inevitably, whether fertilised or not, some flowers must abort.

---

[1] Newspaper reports in February 1995 suggested that a Tasmanian huon pine (*Dacrydium franklinii*), a conifer in the mainly southern-hemisphere family Podocarpaceae, may exceed this, one individual possibly being 30,000-40,000 years old and covering more than a hectare of montane forest.

We may ask, then, whether pollination limits fruit or seed set in plants. A number of research workers have experimentally pollinated flowers to see whether they do obtain an increased fruit set over normal open pollination and, frequently, the answer has been that they do (Zimmerman, 1988; Nilsson, 1992b; Burd, 1994). If they do, then this means that pollinator activity is limiting fruit set in that flowering season, or in those particular flowers. The problem is that this can interact with a limitation of resources in various different ways – it is well known, for instance in fruit trees, that poor fruiting years regularly follow good years, suggesting a limitation of resources that the plant can supply over two years at least. Orchids have received more study than most other plants and, in many species, it has been shown that there may be reduced growth, or a plant may even die, the next season if it sets many fruits in one season (Zimmerman & Aide, 1989; Nilsson, 1992b). In the pink lady's slipper orchid (*Cypripedium acaule*) the major costs of setting fruit (from manipulated flowers) were not seen for three to four years (Primack & Hall, 1990). Paige & Whitham (1987) showed that some plants of the scarlet gilia (*Ipomopsis aggregata*), which usually dies after flowering once, produced a side rosette and flowered the following season if pollinators limited fruit set below 40% of the maximum. Another possibility is that plants may be able to shift resources from one part to another, so higher seed set on one branch might lead to lower set elsewhere. Really what needs to be studied is the overall lifetime reproductive success of one individual compared with another, usually a time-consuming and difficult feature to study. This gets further complicated by the fact that delaying reproduction may carry a cost in terms of contribution to the next generation – a plant which has set fruit in one season may contribute new plants in the next or subsequent seasons increasing the impact of its own genes on the population generally. In any study, the resources available to different individuals are likely to differ substantially and this can have a huge bearing on fruiting success; a plant growing in a favourable place may be larger, with more flowers, may attract more pollinators and have more resources available for fruit set than one in a poor place (see also pp.223-4).

Any study which sets out to analyse the subject needs to take all these considerations into account and, not surprisingly, they all have limitations. Different species clearly have different requirements and the relative importance of the interacting factors will differ widely. Stephenson & Bertin (1983) and Lee (1988) reviewed evidence on what limits seed production, and concluded that nutrient or energy resources, or both, were likely to be important limiting factors in fruit production in many species. In dioecious plants, female flowers are often smaller than the male and there are fewer of them per plant (Chapter 12), features which, in themselves, suggest that the resources required for successful fruit or seed production may be limiting and that male flowers are cheaper to produce. A male-biased sex ratio in a dioecious plant, which is often found, may be the result of increased mortality of females because of more resource use (Allen & Antos, 1993).

Despite the evidence for limited resources affecting seed set, Burd (1994) analysed published work on a total of 258 species and showed that pollination was a limitation in at least 159 of them (62%) at some stage and that, even if resources do limit them in another season, overall pollination is still important. One of the advances that mathematical models and computer simulations have brought is that they can express, given quite limited information, what is likely to happen under various conditions. In this context, Calvo & Horvitz (1990) developed a model which showed that, despite the resource costs, shortage of pollination was likely to be limiting fruit set, particularly in orchids, but in other species too. They included costs, pollination level and, crudely, inherent fecundity differences in their model but not timing of first reproduction, so even here there are difficulties in interpretation.

If pollinator abundance is severely limiting, it is possible that selection will lead to a modification in some aspect of a plant's ecology. One possible consequence is self-fertility, or at least reduced self-incompatibility, though this is normally inferred, not proven. Many short-lived plants, and those living in environments where pollinators are scarce, are self-fertile and largely self-pollinating (Chapter 12).

There is little doubt that the relative importance of all the factors considered differs among different species, between years, and between geographical areas in one species. In a few populations it may be clear that a single factor is limiting fruit set. Bierzychudek's (1982) study on the jack-in-the-pulpit (*Arisaema triphyllum*) showed convincingly that pollination was the overriding factor limiting seed set, but for most species all the factors are interacting and all may, to some extent, be limiting at once. The search for one limiting factor may be flawed from the start, since changing one will immediately alter the status of the others, as we have seen in some of the studies where nutrients became more limiting after artificially high numbers of cross-pollinations. This can be likened to a road system with intersections – speed up the flow of traffic at one intersection and the next one becomes more congested.

Most studies on limitation of fruit set have stopped at the maturation of seeds on a parent plant. This does not necessarily give a true picture of how successful that plant has been as a mother; what matters is how many of those seeds germinate, establish and produce flowers for the next generation. In many plant populations, there is especially high mortality between seed germination and establishment and this is regarded as a critical stage (Harper, 1977; Grubb, 1977). We have already seen (Chapter 12) that self-fertilisation and cross-fertilisation can result in seeds with markedly different properties and these may be crucial at this stage. Seedlings resulting from crossed flowers are usually more vigorous than those from selfed flowers at some stage of their subsequent development, but this must just represent the extremes. Many plants are highly variable and this includes differences in properties of seeds and seedlings. This subject is considered more fully in the section below on mate choice.

The dispersal of pollen – the flower as a male

The second function of a flower is to disperse pollen so that it can reach the stigmas of other individuals. This makes a genetic contribution equal to that of the seed parent in any one seed, but only needs the resources to make the pollen itself and its attendant attractants for pollinators. Because resources so often have a large limiting role in seed set, a greater floral display – indeed, most aspects of attractiveness of a flower – may generally be concerned primarily with the functioning as a male, a pollen donor, rather than as a female (Stanton *et al.*, 1986), although this is certainly not true of all plants (de Jong & Klinkhamer, 1989; Nilsson, 1992a). The spread of pollen has been studied in a number of different ways and is closely tied in with the behaviour of pollinators.

Pollinator effectiveness

Different groups of flower visitors carry different amounts of pollen and with different degrees of effectiveness as pollinators (Schemske & Horvitz, 1984). Intuitively, we would expect the furry bodies of bumblebees, bats and other mammals, and birds' feathers to be effective in carrying many pollen grains, whereas the small and relatively smooth bodies of flower-visiting beetles and some flies will carry many fewer. Butterflies may be less effective than, say, bees, on generalist flowers since pollen will sometimes touch only a small surface area of their thin proboscis which they probe into the flowers (Wiklund *et al.*, 1979). In more specialist butterfly flowers, the anthers and stigmas may be exserted far enough for pollen to be carried on the body or wings, depending on the species of plant and of butterfly (Cruden & Hermann-Parker, 1979; Courtney *et al.*, 1981). Coupled with this inherent ability to transport pollen grains, is the behaviour of pollinators: an animal may be totally ineffective as a pollinator if it stays in one flower or only takes pollen to other flowers on the same plant, however many it can carry. Some bees, particularly bumblebees and honeybees, groom themselves regularly and remove pollen into their pollen baskets where it is not available for pollination. Herrera (1987) studied the effectiveness of a range of bees, flies and butterflies on the Mediterranean lavender (*Lavandula latifolia*). He found that bees were generally more effective than flies or butterflies, which did not differ from each other, but the species within each insect group differed markedly. Some properties of the plants can affect pollinator behaviour, e.g. timing of nectar production may differ between individual plants and this may be critical in making pollinators move between plants and so enhance the pollination of all of them (Frankie & Haber, 1983).

Pollinators will forage in a way that is most efficient for themselves in their gathering of food, not necessarily in the way that is the most efficient for pollinating the plant. This may mean visiting many flowers on each plant, perhaps crawling rather than flying to new flowers if they are close, and, when flying off, moving to an adjacent or nearby plant. In the short term, some pollinators such as bees can become familiar with the pattern of available flowers and visit these repeatedly, although they are always ready to examine new possibilities (Heinrich, 1979, 1983; Roubik, 1989). It is difficult to establish accurately what the true foraging patterns of bees are. Researchers find it much easier to observe visits to adjacent flowers and when a pollinator flies off, perhaps to a new foraging area, observations usually cease. Observations will therefore be biased towards short movements. Short flights and visits to nearby flowers may disperse a lot of pollen in a restricted area, but, in a self-incompatible plant, much of this will be ineffective. Even if pollen lands on a neighbouring plant, this plant may be an offspring and incompatible or partially incompatible with the pollen donor. Long flights, although few compared with short flights, may be disproportionately important in terms of effective pollen flow. There are a number of features of flowering behaviour which do encourage pollinators to travel long distances, and there is mounting evidence that some bees and some birds and bats may travel long distances between flower visits.

Long flights may be inferred from indirect observations, like transporting marked bees from a known nest site (Roubik, 1989). Marked euglossine bees in Central America returned to the nest from 23 km distance in Janzen's (1971) study. These bees often forage using 'trap-lining' behaviour (Chapter 5), and the plants produce only a few flowers at a time, spread out over a long season, using this behaviour for efficient cross-pollination. The potential for long-distance pollen flow from such long flights and this behaviour pattern is great. Some bee species do not travel so far and some are territorial, such as *Centris* species (Anthophoridae), and if their territories are held around just one flowering tree they may be totally ineffective as pollinators themselves. Their territorial behaviour, however, may lead to aggressive encounters with other flower visitors, driving them off the defended flowers, perhaps to forage on another individual, taking some pollen with them, so even they may indirectly promote cross-pollination (Frankie, 1976; Frankie & Haber, 1983; Roubik, 1989).

The differences in pollen-carrying capacity and foraging behaviour of different pollinator species mean that careful observation may be needed to establish which visitors are the most effective as pollinators. Another consideration is the schedule of

Fig. 16.1 Bladder campion (*Silene vulgaris*); the flowers can be pollinated by a range of moth species with varying degrees of effectiveness.

flower opening and insect visits on any one plant species. Just after anthesis, a flower-constant bee that takes long flights may be responsible for much successful pollination but the same insect later in a flower's life, or an insect which is less flower-constant, may be less efficient. Some visitors could even be detrimental to successful pollination if they remove pollen before more effective pollinators can reach the flowers (Thomson & Thomson, 1992; Stanton *et al.*, 1992). Honeybees may well fall into the last category on some plants, since they forage in a highly efficient way but have smooth bodies compared with bumblebees and carry most of the pollen in their corbiculae, where it may be unavailable for deposition on stigmas. On other plants they may be as effective as bumblebees (Cresswell *et al.*, 1995).

Most butterflies do not forage so systematically as bees but, as always, species differ and many take long flights between flowers and may carry pollen for a long time (Courtney *et al.*, 1981; Herrera, 1987) so, in this respect, they may be effective as dispersers of pollen. Moths, especially hawkmoths (Sphingidae), can be strong fliers and effective pollen dispersers. Again, different species may differ and Pettersson (1991), in his study on pollination of bladder campion (*Silene vulgaris*) (Fig. 16.1), showed that the small elephant hawkmoth (*Deilephila porcellus*) was particularly effective as a pollinator while the silver-Y moth (*Autographa gamma*) was much less so, despite its greater abundance.

Of the other insect groups that are important pollinators, flies are usually opportunistic feeders (Olesen & Warncke, 1989), so can be responsible for long distance dispersal, but many have rather smooth bodies and they may visit several species on a foraging flight, so much pollen may be wasted. Beetles and thrips generally wander rather less between flowers, although they are, clearly, effective pollinators in certain situations (Chapter 3).

Birds are similar to bees in that some are territorial and others cover large distances (Linhart, 1973; Waser, 1983). Feathers can be effective pollen carriers and the active nature of nectar-seeking birds when foraging, particularly hummingbirds, may mean that they are especially effective as pollinators. Flower-visiting bats are hard to study, but from the evidence gathered so far, some, at least, are strong fliers and may travel long distances (Heithaus *et al.*, 1975; Start & Marshall, 1976). Although most appear to be somewhat opportunistic when foraging, the fur can carry a lot of pollen and bats are probably efficient pollinators. Many of the flowers which are pollinated by bats are large and heavy, requiring considerable resources from the plant: an indication

Fig. 16.2 The hoverfly *Syrphus vitripennis* on a white flower of wild radish (*Raphanus raphanistrum*). Pollinators can discriminate between the white and yellow-flowered forms of this species in mixed populations. Q.O.N. Kay.

that bats are efficient as pollinators and worth the investment.

There are two further aspects of pollinator behaviour which may affect pollen dispersal. The first is directionality of movement; insects, notably bees, like to maintain one direction when flying between forage plants. They do this even when they have visited flowers on different sides of a single inflorescence, and its most obvious advantage is that it will stop them visiting flowers that they have just visited. Bees frequently forage with short flights into the wind, taking longer flights downwind (Levin *et al.*,1971; Pyke & Cartar, 1992). The overall effect may be greater pollen flow than would be expected just by observation of individual distances of insect movement. The direction of pollen flow is likely to even out over a season. The second is the ability of some insects to discriminate between different flower colours or flower forms in a mixed population, and visit one morph during a foraging bout (e.g. Lack & Kay, 1987). This may contribute to the maintenance of two colour forms in one population, e.g. the yellow and white colour forms of the wild radish (*Raphanus raphanistrum*) (Fig. 16.2) (Kay, 1982), but it depends on the conditions in which the plants are growing and its importance in the maintenance of colour variation is disputed (Stanton *et al.*, 1989). Pollinators can sometimes discriminate between the sexes in a dioecious plant, only visiting males or females which is, clearly, detrimental to the pollination of the species (Kay *et al.*, 1984). In some orchids, it is the pollinators' ability to discriminate between different lip patterns or fragrance profiles that is the main mechanism of isolation between two species that are otherwise interfertile (Chapter 7), and a similar isolating mechanism is important in the interfertile columbines (*Aquilegia*).

Measures of pollen dispersal

Most pollen is deposited near its source and the quantity deposited falls off rapidly with distance (Levin & Kerster, 1974). This is true for wind- and insect-pollinated species, although it is most obvious for wind-pollinated ones, with their indiscriminate dispersal (Chapter 9). This does not give a proper impression of effective pollen flow, however, since in many wind-pollinated species, such as *Plantago lanceolata*, the pollen is dispersed in clumps. This means that effective pollen flow is greater, since the larger clumps are deposited nearer the source but each clump only pollinates one or two stigmas; smaller clumps travel further and single grains further still, so effective pollen flow is greater than the number of grains deposited suggests (Tonsor, 1985).

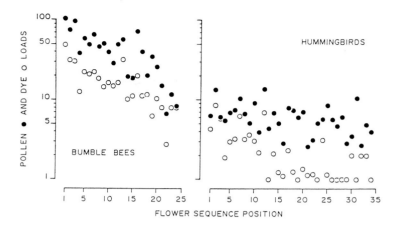

Fig. 16.3 The amount of pollen deposited by bee pollinators on stigmas of successive flowers visited in the larkspur *Delphinium nelsonii*. From Waser (1988).

An assumption of a number of early studies on pollen dispersal by insects, and the basis of some mathematical interpretations of pollen flow, was that all pollen is deposited on the next flower visited, although all were aware that this was an oversimplification. Waser (1988) removed the anthers from flowers of *Delphinium nelsonii*, a larkspur of the Rocky Mountains pollinated by bees and hummingbirds, leaving one source flower, and then examined pollen deposition on the sequence of flowers visited following a visit to the pollen source. He showed that there was a gradual decrease of pollen deposited, but that some may be carried over for 20 flowers or more (Fig. 16.3). Other studies have demonstrated pollen carryover and it can be extensive (e.g. Thomson & Plowright, 1980; Pettersson, 1991; Stanton *et al.*, 1992). What was clear from Pettersson's (1991) study was that the amount of pollen deposited by a flower visitor is highly variable and dependent not just on the amount being carried, but also on the length of the visit, the condition of the individual flower and the behaviour of the insect at that visit.

Another research technique is to use fluorescent dust as a pollen analogue. If the dust is placed on the anthers of a flower, any dispersal can be traced using an ultraviolet light after dark. As expected, these studies have shown that the great majority of the dust, and presumably the pollen, is deposited on nearby flowers, but some goes further and provides more evidence for pollen carryover. Waser (1988) demonstrated that the dispersal of dust may be less than that of pollen, but the patterns are much the same.

Pollination, population genetics and natural selection

So far we have considered the two sexual functions of a flower mainly in terms of the number of seeds produced or the quantity of pollen dispersed. We have not considered the *quality* of that reproductive effort, which may be vitally important for the effective contribution to the next generation. Individual plants may differ inherently in their effectiveness as pollen or seed parents or there may be an interaction between particular individual pollen or seed parents which may lead to a differential contribution to the effective seed pool. Such individual variation is the raw material for natural selection to work on and, as such, provides the basis for evolution. In pollination, it has been referred to as mate choice, or sometimes sexual selection, although it is not true sexual selection in the sense that Darwin originally coined the phrase (Grant, 1995).

Selection on reproductive characters in plants can act at all the different stages, from flower production to seed production. There can be competition between individual plants to attract pollinators, both to spread pollen and to receive pollen; competition between pollen grains at the stigma to germinate and grow down the style; 'selection' by females of particular pollen tubes, either in the style or at fertilisation, or selective abortion of particular fruits or seeds by the female. All of these will affect the genetic contribution to the next generation.

In animals the production of extravagant ornaments, displays and songs, particularly by male birds, has long attracted the attention of biologists. This is selection for secondary sexual characteristics which allow more effective reproduction, either through establishing the best territories, usually as a result of aggressive interactions between males, or through attracting the opposite sex, usually females, most effectively. Transmitting genes to the next generation is the result, so those most successful at it will be favoured by natural selection. There will be a balance between sexual success and survival – adding extra 'eyes' to a peacock's tail may attract more females and lead to more successful matings, but if it compromises the male's ability to escape predation he may not survive to mate again next year. In plants, sexual reproduction is achieved through an intermediary agent, a pollinator, but there are some analogies between sexual selection in animals and selection on floral attributes in plants.

Competition between males and selection by females

In floral display, there does seem to be an analogy with a peacock's tail (though Grant [1995] disputes this) – the larger the display, or the greater the attraction of the reward in a flower, the more likely it is to be visited and the greater is its likely success as a pollen donor. Display may also affect the female function since, even if resources are a limiting factor in fruit set, with the attraction of more pollinators there may be a greater genetic range of pollen landing on the stigma and more potential for other levels of selection. Constraints on size of the display will come mainly from a physical limitation of space and the amount of resources required to produce the flowers affecting survival, again an analogy with the displays of birds.

Once pollen grains are on the stigma, there are a number of factors that can affect which ones actually fertilise the ovules. One of the most significant factors here is which ones arrive first and where on the stigma they are deposited, e.g. in crevices or on the papillae. Relative investment in floral display can be a most important determinant of which flowers are visited first, but chance must have a role here too (Stephenson et al., 1992). The environment in which the pollen grain was formed can be important. Pollen grains produced by plants in low nutrient environments or at lower temperature may be less vigorous than those grown under more favourable conditions (Young & Stanton, 1990; Stephenson et al., 1992). All pollen grains may be stimulated by a greater density of pollen grains on the stigma, up to a certain point (Mulcahy et al., 1992), and there is some evidence from the trumpet vine (*Campsis radicans*) that many outcross pollen grains can stimulate the germination of self grains, allowing at least a partial self-compatibility to occur under these conditions (Bertin & Sullivan, 1988) (see Chapter 12).

There may be differential germination or growth of the pollen tubes, depending on their genetic constitution. Pollen tubes may differ inherently, presumably because of their genetic make-up, in their vigour of growth (Mulcahy *et al.*, 1992), and there is some evidence from the North American shrub *Hibiscus moscheutos* that this results in differential fertilisation (Snow & Spira, 1991, 1993). This stronger growth of pollen tubes may occur in pollen from certain individuals in all styles (which leaves one wondering how those with weak growth have survived) but vigour of growth can, and usually will, depend on an interaction between one individual's pollen and the style of another (Marshall & Folsom, 1991; Stanton, 1994). This is most clearly seen in the self-incompatibility systems described in Chapter 12, but can extend to partial incompatibility between relatives, with self or closely-related pollen usually growing more slowly than outcross pollen.

Mulcahy & Mulcahy (1975) suggested, with a little evidence, that the most vigorous pollen tubes produced seeds which were also likely to be vigorous. This evidence is disputed by Snow (1990), but it is a very tricky area of research, since a flower must usually be destroyed to study the vigour of the pollen tubes, making it impossible to assess the vigour of the resultant seeds. Stephenson *et al.*, (1992) found that herbicide resistance in pollen resulted in herbicide-resistant seeds, so there may be a connection.

The female part of the flower can exercise 'choice' once pollen grains have landed on the stigma. The interaction between particular pollen grains and styles can mean that certain pollen grains are selectively blocked from germinating, or the tubes can be blocked on their way down the style or at the point of fertilisation (Marshall & Folsom, 1991). The mechanisms for selectivity are poorly known and may involve precise chemical recognition systems, closely connected with self-incompatibility systems (Chapter 12). The result will be differential inhibition of pollen tube growth and non-random fertilisation. The distinction between female choice and male-male competition in these circumstances is a subtle one and usually it will not be possible to tell them apart (Stanton, 1994).

At the next level, after fertilisation, fruits can be aborted selectively. In a partially self-compatible species, for instance, fruits resulting from self-pollination could be aborted if there were other fruits resulting from cross-pollination. There is some evidence for this and, even among cross-pollinations, abortions may be selective (Lee, 1988; Marshall & Ellstrand, 1988; Bertin & Peters, 1992; Rigney, 1995). Stephenson & Winsor (1986) manipulated fruit set in the self-incompatible species, birdsfoot trefoil (*Lotus corniculatus*) (Fig. 16.4). They determined the natural level of fruit abortion and

Fig. 16.4 Birdsfoot trefoil (*Lotus corniculatus*), with flowers and developing fruits (the 'birds' feet' that give the plant its name). Not all flowers mature into fruits, and there may be selective abortion.

then thinned flowers themselves to a similar level, while leaving other plants to abort their own fruits. The result was that the seeds from natural fruit set germinated better and established more strongly than those from plants that had been manipulated.

As in the studies on pollen tube vigour and inhibition in the style, it appears that some pollen donors may be favoured on many female parents, or certain male/female combinations may be particularly successful; perhaps certain combinations of alleles work particularly well (Lee, 1988). One little-studied area is that of inbreeding depression from crossing between close relatives. This is the type of inbreeding we think of and know to be detrimental in animals (including ourselves), in which the sexes are separate as they are in dioecious plants. In plants, a clear effect was demonstrated by Heywood (1993) in the composite *Gaillardia pulchella* when he artificially crossed known siblings producing abnormalities and slower growth.

Genetic consequences of mate choice

Since greater vigour is often shown by progeny from outcrossing, the type of competition between pollen grains and pollen tubes and the choice being exercised by the female part of the flower will lead to a greater mixing of the genes within a population. This will mean that there will be a greater number of heterozygotes, individuals with a different allele from each parent, in the resultant seeds. If fruit production is stimulated when pollen grains from several different parents are present on the stigma, this will also lead to greater genetic mixing. Some self-fertilisation may well be advantageous in certain species, however, and there may be a balance between selfing and crossing, perhaps involving differences in the precise recognition systems as the flower ages or under different conditions (Holsinger, 1992). In a mixed selfing and crossing system, the responses of plants can be particularly subtle and the resultant seeds can have different properties. In the thistles *Cirsium vulgare* and *C. palustre*, selfing resulted in many fewer seeds in a capitulum, but these were larger and showed reduced dormancy compared with crossed seeds (van Leeuwen, 1981). The direct effects of selfing were not separated from the effects of small numbers of seeds in one flower head, and they are often connected, but it suggests that there may be two different strategies being employed here. It is possible that the self-pollinated seeds do not disperse far before germinating, perhaps consolidating an existing population, whereas the crossed seeds can be dispersed further to found new populations. The evidence for this kind of difference is very limited, but as biologists become more aware of mixed mating systems, it seems likely that more such subtleties will be discovered (Holsinger, 1992).

Despite the potential for these elaborate mechanisms, some plants have abandoned nearly all of them and are >95% self-fertilising. Plants with a high level of selfing will rapidly reach a situation where there is little variation in the progeny of any one individual and almost no 'choice' is being exerted by the female. Further evolution in such a plant will depend on mutations and occasional outcrossing and will generally be much slower, although high levels of selfing are particularly associated with short-lived plants, and short generation time will tend to compensate for this.

Genetic evidence of pollen flow

A few workers have looked at the genetic constitution of seeds to infer information about effective pollen flow. This requires an ability to detect some genetic marker in the potential pollen parent which is not present in the seed parents. Griffiths (1950) studied this on perennial rye grass (*Lolium perenne*) using a variety with red-based shoots as a marker pollen parent. Like nearly all grasses, rye grass is wind-pollinated, and he found that pollen had fertilised seeds at least 30 m from the source (Chapter 9). This shows that pollen can potentially travel long distances, although the great majority will be deposited near the source plant. It emphasises the importance of distinguishing between measurable pollen flow from a plant and *effective* pollen flow, i.e. that which sires seeds, and these two things may be very different. Among insect-pollinated plants, Ellstrand & Marshall (1985) compared foraging behaviour of pollinators, dispersal of fluorescent pollen analogue and actual paternity of seeds in a radish species, *Raphanus sativus*, and showed that effective pollen flow was more than twice as far as that shown by either of the other methods, demonstrating conclusively the limitations of many studies. They went on to show that selective abortion is one influential factor, and that pollen can travel between populations at least 150 m apart (Marshall & Ellstrand, 1988; Devlin & Ellstrand, 1990).

There is a little evidence that, in an outcrossing plant, the degree of relatedness of two plants can affect the success of seed set. Outcrossing across a great distance may break up combinations of genes which are well adapted to the environment in which a population or sub-population is growing. Waser & Price (1983, 1989, 1994) showed, in the outcrossing larkspur (*Delphinium nelsonii*) and scarlet gilia (*Ipomopsis aggregata*) that, optimally, a plant will mate with another between about 10 m and 100 m away; further than that and vigour of the seedlings decreased again. In the willowherb *Epilobium obcordatum*, crosses between populations 1-3 km apart were more effective than crosses within a population (Seavey & Carter, 1994). There is another interesting parallel with animals here: Bateson (1982) studied mate selection in quails, and when males were presented with a range of females from siblings to those distantly related, they chose first cousins selectively – birds that were similar, but not too similar.

Seed dispersal

To complete the picture, only a part of gene flow takes place through dispersal of pollen. Seed dispersal will affect the overall distribution of the genes within a population, though it will not affect the genetic constitution of an individual. Like pollen flow, seed dispersal has been seriously underestimated in the past, owing to a concentration of study on initial seed dispersal (Levin & Kerster, 1974). Further dispersal, often occurring by chance while a seed lies dormant, may be of overriding importance in affecting how a population is organised genetically, and its potential for colonising new sites. There are numerous specialisations of seeds for dispersal, such as a feathery sail or an edible fleshy fruit (van der Pijl, 1969), although many seeds have no obvious mechanism. It is likely that seed dormancy, coupled with the vagaries of winds, floods and moving animals, are the most significant means by which many seeds are dispersed beyond the immediate vicinity of the parent plant.

The genetic constitution of plant populations

The final step in a study of this kind is to see which seeds, which combinations of genes, are actually effective in forming the next generation. The most comprehensive study on this has been that by Meagher (1986, 1991) and Meagher & Thompson (1987) on the lily *Chamaelirium luteum*. They analysed the potential parentage of all the plants in a population based on an extensive study of biochemical variation, and showed that genetic mixing was limited, that a plant was most likely to sire seeds on a nearby female, and that most seeds were dispersed only short distances. There has long been the suggestion that heterozygotes are favoured in populations because of their vigour (Brown, 1979). Much of the evidence is circumstantial and mainly based on the observation that there is frequently a greater number of heterozygotes in a population than is expected, given what is known of potential genetic mixing. Schaal & Levin (1976) demonstrated this directly in field populations of the North American herb *Liatris cylindracea* (Asteraceae). A greater proportion of adults in the population was heterozygous than among the seedlings, i.e. the heterozygotes survived better to adulthood. In experimental situations, it has been shown that outcrossed progeny frequently do better than selfed, particularly when in competition at high density, and outcrossed plants will be more heterozygous on average than selfed plants (Darwin, 1876; Schmitt & Ehrhardt, 1990).

There have been many studies on the genetics of adult plant populations and these have shown that different pollination modes result in differences in the way the genes of a population are distributed (Loveless & Hamrick, 1984; Hamrick & Godt, 1990). Plants that are mainly self-pollinating have the most restricted pollen flow and, as expected, the genes are clumped within populations and different populations tend to differ genetically, so there is less mixing of genes generally than in outcrossing plants. Wind-pollinated plants showed greater mixing of genes than those pollinated by animals, so although every population varied, each population contained a similar range of variation. Many sources of variation have been discussed in this chapter, and these may influence the populaton genetics of any one species as profoundly as pollination type, but numerous species have been studied and there do seem to be real average differences in genetic structure between the different pollination categories.

The impact of pollination on the evolution of plants

The conclusion from all these studies of pollen flow, female choice, effective gene flow and germination, is that there is a great deal of variation thrown up in seed formation and that, in many species, there is the potential for strong selection of favoured genotypes to make up the next generation. Pollen tube growth, seed maturation, seedling germination and establishment are all crucial steps at which weaklings may be eliminated. Some of the genetic variation may be eliminated at the various stages, depending on the conditions, but there will still be considerable variation left in the population, partly maintained by the advantages of heterozygotes. It must be remembered, however, that there have only been a few detailed studies on a small number of species. These conclusions may not apply to all species. As with nearly all aspects of biology, a variability of response and different solutions to the various problems of life are the rule.

There is another aspect of a plant's ecology which may limit the importance of genetics. Ultimately, both pollen and seeds are dependent on external forces for their final positions, and there is a large element of chance involved as to where they will end up. A successful seed is often a lucky one, as much as it is an adapted one (van der Meijden, 1989). A particular freak seed-dispersal event may even found a new population, so chance must play a large part. Coupled with this, there is no doubt that plants can respond to particular environmental conditions by modifying their growth or life history, and that genotypes which are capable of such a plastic response have clearly been favoured. The relative importance of the chance factors, plasticity on the part of the growing plant and the plant's own genotype, continue to arouse much attention and debate from biologists around the world. Most natural populations fluctuate in number, often enormously, between generations, and plants do this in a similar way to animals. Plant ecologists have, understandably, concentrated detailed population studies on short-lived plants and most have not considered the differences in quality of seeds in their studies. Over the time scale of a few generations, differences in quality may have a minor, and often undetectable, effect on any properties of the population. In addition to this, many short-lived plants have a narrower range of genetic variation within each population than long-lived plants, owing to higher incidence of selfing. In the short term, then, over a few generations, the role of chance, plasticity of response to the environment, and immediate consequences of local ecological conditions are likely to be the predominant influences on the dynamics and the genetics of plant populations.

If we take a longer term view, the picture is different. Some long-lived species may germinate and have a chance of establishment only at quite long intervals, the most obvious examples being the germination of tree seeds after a gap is created by a hurricane, fire or other environmental catastrophe. We saw this in Britain most strikingly after the storms of October 1987. Dense swathes of tree seedlings came up in the woodland gaps, and competition between them must have been intense. Here, although chance still undoubtedly plays a part, speed of growth, vigour of the seedling and other differences between individuals must affect survival. The success of the individuals at producing vigorous and successful seeds will be severely tested under these circumstances and pollination mechanisms and features of mate choice will affect which seedlings will succeed. In the short-lived plants, over a large number of generations, similar tests of vigour are likely to come, either in particular seasons or just in small differences accumulating over the generations. It is over this longer time scale that we should expect to see the real importance of pollination, although studying this directly is extremely difficult.

If we step back and look at the sheer range of flowers, which did so much to inspire the study of pollination, we can only marvel at the variety of mechanisms, utilising so many different pollinators, that can bring about successful pollination. Despite the variety, however, each plant family usually has a uniform fundamental floral pattern with many features in common. Any major disruptions to this pattern are likely to be disadvantageous for the plant's reproduction and it will not survive; perhaps it is not surprising that, with many parts working together, basic flower form is so strongly conserved. The chance appearance of new mutations and gene combinations provide the raw material for evolution and, in certain species, we often see minor variants of flower form, some of which may not affect the success of pollination at all. But the basic patterns and the precise intricate adaptations of flowers for pollination must owe only their initial appearance to chance; chance cannot be a major driving force behind the evolution of pollination. Richard Dawkins (1986) graphically described such a possibility as like hurling scrap metal around and creating a jumbo jet.

Pollination is a vital link in flowering plant reproduction, mediates the genetic constitution of the next generation and affects which ones will survive. In so doing it is a major influence on the evolution of plants. Cross-pollination, particularly by insects, has very likely been central to the success of the flowering plants. The wealth of species of potential pollinators and the wealth of variation within the flowering plants have stimulated each others' evolution, leading to the remarkable diversity of flower visitors and, within a basic structure, the seemingly limitless variation and often beautiful adaptation of flowers.

References

Pages on which references are cited in this book are shown in square brackets.

Ackerman, J.D. 1981 Pollination biology of *Calypso bulbosa* var. *occidentalis* (Orchidaceae): a food-deception system. *Madroño* 28: 101–110. [222]

Ackerman, J.D. & Mesler, M.R. 1979 Pollination biology of *Listera cordata* (Orchidaceae). *American Journal of Botany* 66: 820–824. [102, 195]

Adams, R.A. & Goss, G.J. 1976 The reproductive biology of the epiphytic orchids of Florida III. - *Epidendrum anceps* Jacquin. *American Orchid Society Bulletin* 45: 488–492. [81]

Adams, W.T., Birkes, D.S. & Erickson, V.J. 1992 Using genetic markers to measure gene flow and pollen dispersal in forest tree seed orchards. *Ecology and Evolution of Plant Reproduction* (ed. R. Wyatt), pp. 37–61. Chapman & Hall, New York. [271]

Addicott, J.F. 1985 Competition in mutualistic systems. *The Biology of Mutualism* (ed. D. Boucher), pp. 217–247. Croom Helm, London. [319]

Addicott, J.F. 1986 Variation in the costs and benefits of mutualism: the interaction between yuccas and yucca moths. *Oecologia (Berlin)* 70: 486–494. [318]

Addicott, J.F., Bronstein, J. & Kjellberg, F. 1990 Evolution of mutualistic life-cycles: yucca moths and fig wasps. *Genetics, Evolution and Coordination of Insect Life Cycles* (ed. F. Gilbert), pp. 143–161. Springer, London. [316, 319, 320]

Ågren, J., Elmqvist, T. & Tunlid, A. 1986 Pollination by deceit, floral sex ratios and seed set in dioecious *Rubus chamaemorus* L. *Oecologia* 70: 332–338. [341]

Ågren, J. & Schemske, D.W. 1991 Pollination by deceit in a neotropical monoecious herb, *Begonia involucrata*. *Biotropica* 23: 235–241. [295]

Aker, C.L. & Udovic, D. 1981 Oviposition and pollination behavior of the yucca moth, *Tegeticula maculata* (Lepidoptera: Prodoxidae), and its relation to the reproductive biology of *Yucca whipplei* (Agavaceae). *Oecologia (Berlin)* 49: 96–101. [319, 320]

Åkerberg, E. & Crane, E. (eds.) 1966 *Proceedings of the Second International Symposium on Pollination (Bee World 47, Supplement).* [357, 358, 363]

Allen, G.A. & Antos, J.A. 1993 Sex ratio variation in the dioecious shrub *Oemleria cerasiformis*. *American Naturalist* 141: 537–553. [414]

Anderson, A.B. & Overal, W.L. 1988 Pollination ecology of a forest-dominant palm (*Orbignya phalerata* Mart.) in northern Brazil. *Biotropica* 20: 192–205. [311]

Anderson, G.J. 1976 The pollination biology of *Tilia*. *American Journal of Botany* 63: 1203–1212. [285]

Anderson, G.J. & Symon, D.E. 1988 Insect foragers on *Solanum* flowers in Australia. *Annals of the Missouri Botanical Garden* 75: 842–852. [180]

Anderson, G.J. & Symon, D.E. 1989 Functional dioecy and andromonoecy in *Solanum*. *Evolution* 43: 204–219. [346, 347]

Andrews, H.W. 1953 Flies at parsley blossom. *Entomologist's Record and Journal of Variation* 65: 58–59. [73]

Appanah, S. 1981 Pollination in Malaysian primary forests. *Malaysian Forester* 44: 37–42. [392]

Appanah, S. 1982 Pollination of the androdioecious *Xerospermum intermedium* Radlk. (Sapindaceae) in a rain forest. *Biological Journal of the Linnean Society* 18: 11–34. [346]

Appanah, S. & Chan, H.T. 1981 Thrips: the pollinators of some dipterocarps. *Malaysian Forester* 44: 234–252. [312]

Arber, A. 1920 *Water Plants.* Cambridge University Press, Cambridge. [285]

Arizmendi, M. del Coro & Ornelas, J.F. 1990 Hummingbirds and their floral resources in a tropical dry forest in Mexico. *Biotropica* 22: 172–180. [231, 295]

Armbruster, W.S. 1984 The role of resin in angiosperm pollination: biological and chemical considerations. *American Journal of Botany* 71: 1149–1160. [43, 115]

Armbruster, W.S. 1986 Reproductive interaction between sympatric *Dalechampia* species: are natural assemblages "random" or organized? *Ecology* 67: 522–533. [401]

Armbruster, W.S. 1992 Phylogeny and the evolution of plant-animal interactions. *BioScience* 42: 12–20. [401]

Armbruster, W.S., Keller, S., Matsuki, M. & Clausen, T. P. 1989 Pollination of *Dalechampia magnoliifolia* (Euphorbiaceae) by male euglossine bees. *American Journal of Botany* 76: 1279–1285. [115]

Armbruster, W.S., Edwards, M.E. & Debevec, A.M. 1994 Floral character displacement generates assemblage structure of western Australian triggerplants (*Stylidium*). *Ecology* 75: 315–329. [184]

Armbruster, W.S. & Webster, G.L. 1979 Pollination of two species of *Dalechampia* (Euphorbiaceae) in Mexico by euglossine bees. *Biotropica* 11: 278–283. [115]

Armstrong, J.A. 1979 Biotic pollination mechanisms in the Australian flora - a review. *New Zealand Journal of Botany* 17: 467–508. [247, 251, 262, 389, 390]

Aronne, G. & Wilcock, C.C. 1994 Reproductive characteristics and breeding system of shrubs of the Mediterranean region. *Functional Ecology* 8: 69–76. [340, 342, 345, 346]

Arroyo, M.T.K., Primack, R. & Armesto, J. 1982 Community studies in pollination ecology in the high temperate Andes of central Chile. I. Pollination mechanisms and altitudinal variation. *American Journal of Botany* 69: 82–97. [390]

Ashton, P.S. 1988 Dipterocarp biology as a window to the understanding of tropical forest structure. *Annual Review of Ecology and Systematics* 19: 347–370. [392]

Ashton, P.S., Givnish, T.J. & Appanah, S. 1988 Staggered flowering in the Dipterocarpaceae: new insights into floral induction and the evolution of mast fruiting in the aseasonal tropics. *American Naturalist* 132: 44–66. [400]

Atwood, J.T. 1985 Pollination of *Paphiopedilum rothschildianum*: brood-site deception. *National Geographic Research* 1: 247–254. [189, 306]

Augspurger, C.K. 1981 Reproductive synchrony of a tropical shrub: experimental studies on effects of pollinators and seed predators on *Hybanthus prunifolius* (Violaceae). *Ecology* 62: 775–788. [403]

Baker, H.G. 1957 Plant notes: *Calystegia*. *Proceedings of the Botanical Society of the British Isles* 2: 241–243. [69, 150]

Baker, H.G. 1961 The adaptation of flowering plants to nocturnal and crepuscular pollinators. *Quarterly Review of Biology* 36: 64–73. [85, 150]

Baker, H.G. 1970 Evolution in the tropics. *Biotropica* 2: 101–111. [135, 254]

Baker, H.G. 1975 Sugar concentrations in nectars from hummingbird flowers. *Biotropica* 7: 37–41. [243]

Baker, H.G. 1976 "Mistake" pollination as a reproductive system with special reference to the Caricaceae. *Tropical Trees: Variation, Breeding and Conservation* (ed. J. Burley & B.T. Styles), pp. 161–169. Academic Press, London. [295]

Baker, H.G. & Baker, I. 1973 Some anthecological aspects of the evolution of nectar-producing flowers, particularly amino acid production in nectar. *Taxonomy and Ecology* (ed. V.H. Heywood), pp. 243–264. Academic Press, London. [42]

Baker, H.G. & Baker, I. 1975 Studies of nectar-constitution and pollinator-plant coevolution. *Coevolution of Animals and Plants* (ed. L.E. Gilbert & P.H. Raven), pp. 100–140. University of Texas, Austin. [308]

Baker, H.G. & Baker, I. 1983a Floral nectar sugar constituents in relation to pollinator type. *Handbook of Experimental Pollination Biology* (ed. C.E. Jones & R.J. Little), pp. 117–141. Van Nostrand Reinhold, New York. [40, 41, 245, 255]

Baker, H.G. & Baker, I. 1983b A brief historical review of the chemistry of floral nectar. *The Biology of Nectaries* (ed. B. Bentley & T. Elias), pp. 126–152. Columbia University Press, New York. [40, 41, 245, 255]

Baker, H.G. & Baker, I. 1986 The occurrence and significance of amino acids in floral nectar. *Plant Systematics and Evolution* 151: 175–186. [42]

Baker, H.G. & Baker, I. 1990 The predictive value of nectar chemistry to the recognition of pollinator types. *Israel Journal of Botany* 39: 157–166. [242]

Baker, H.G., Baker, I. & Opler, P.A. 1973 Stigmatic exudates and pollination. *Pollination and Dispersal* (ed. N.B.M. Brantjes & H.F. Linskens), pp. 47–60. Dept. Botany, Univ. Nijmegen, Nijmegen. [308]

Baker, H.G., Cruden, R.W. & Baker, I. 1971 Minor parasitism in pollination biology and its community function: the case of *Ceiba acuminata*. *BioScience* 21: 1127–1129. [403]

Baker, H.G. & Harris, B.J. 1957 The pollination of *Parkia* by bats and its attendant evolutionary problems. *Evolution* 11: 449–460. [244, 246, 247]

Baker, H.G. & Harris, B.J. 1959 Bat-pollination of the Silk-Cotton Tree, *Ceiba pentandra* (L.) Gaertn. (sensu lato), in Ghana. *Journal of the West African Science Association* 5: 1–9. [247]

Baker, H.G. & Hurd, P.D. 1968 Intrafloral ecology. *Annual Review of Entomology* 13: 385–414. [178]

Barclay, J.S. & Moffett, J.O. 1984 The pollination value of honey bees to wildlife. *American Bee Journal* 124: 497, 498–561. [364]

Barnes, E. 1934 Some observations on the genus *Arisaema* on the Nilgiri Hills, South India. *Journal of the Bombay Natural History Society* 37: 630–639. [305, 306]

Barrett, S.C.H. 1988 The evolution, maintenance, and loss of self-incompatibility systems. *Plant Reproductive Ecology* (ed. J. Lovett Doust & L. Lovett Doust), pp. 98–124. Oxford University Press, Oxford. [323, 328, 330]

Barrett, S.C.H., Morgan, M.T. & Husband, B.C. 1989 The dissolution of a complex genetic polymorphism: the evolution of self-fertilization in tristylous *Eichhornia paniculata* (Pontederiaceae). *Evolution* 43: 1398–1416. [329]

Barth, F.G. 1982 *Biologie einer Begegnung: Die Partnerschaft der Insekten und Blumen.* Deutsche Verlags-Anstalt, Stuttgart. [50, 125, 128, 131]

Barth, F.G. 1985 *Insects and Flowers: The Biology of a Partnership.* Transl. M. A. Biedermann-Thorson. George Allen & Unwin/Princeton, London/Princeton. [50, 96, 125, 128, 131]

Bateman, A.J. 1947a Contamination of seed crops. I. Insect pollination. *Journal of Genetics* 48: 257–275. [353]

Bateman, A.J. 1947b Contamination of seed crops. II. Wind pollination. *Heredity* 1: 235–246. [271, 353]

Bateman, A.J. 1947c Contamination in seed crops. III. Relation with isolation distance. *Heredity* 1: 303–336. [271]

Bateman, A.J. 1952 Self-incompatibility systems in angiosperms. I. Theory. *Heredity* 6: 285–310. [324, 330]

Bateson, P. 1982 Preferences for cousins in Japanese quail. *Nature* 295: 236–237. [423]

Baumann, H. & Künkele, S. 1982 *Die wildwachsenden Orchideen Europas.* Kosmos, Stuttgart. [208]

Bawa, K.S. 1980 Evolution of dioecy in flowering plants. *Annual Review of Ecology and Systematics* 11: 15–40. [340, 342]

Bawa, K.S. 1990 Plant-pollinator interactions in tropical rain forests. *Annual Review of Ecology and Systematics* 21: 399–422. [391, 393, 400]

Bawa, K.S., Perry, D. & Beach, J.H. 1985a Reproductive biology of tropical lowland rain forest trees. I. Sexual systems and incompatibility mechanisms. *American Journal of Botany* 72: 331–345. [330, 391]

Bawa, K.S., Bullock, S.H., Perry, D.R., Coville, R.E. & Grayson, M.H. 1985b Reproductive biology of tropical lowland rain forest trees. II. Pollination systems. *American Journal of Botany* 72: 346–351. [391]

Beach, J.H. & Bawa, K.S. 1980 Role of pollinators in the evolution of dioecy from heterostyly. *Evolution* 34: 1138–1142. [342]

Beardsell, D.V. & Bernhardt, P. 1983 Pollination biology of Australian terrestrial orchids. *Pollination '82* (ed. E.G. Williams, R.B. Knox, J.H. Gilbert & P. Bernhardt), pp. 166–183. University of Melbourne Press, Parkville, Victoria. [213, 222]

Beardsell, D.V., O'Brien, S.P., Williams,

E.G., Knox, R.B. & Calder, D.M. 1993 Reproductive biology of Australian Myrtaceae. *Australian Journal of Botany* 41: 511–526. [251]

Beattie, A.J. 1969 Studies in the pollination ecology of *Viola.* 1. The pollen content of stigmatic cavities. *Watsonia* 7: 142–156. [170, 334, 338]

Beattie, A.J. 1972 Studies in the pollination ecology of *Viola.* 2. Pollen loads of insect visitors. *Watsonia* 9: 13–25. [62, 170]

Becher, E. 1882 Zur Kenntnis der Mundtheile der Dipteren. *Denkschrift der Kaiserlichen Akademie der Wissenschaften, Mathematisch-Naturwissenschaftliche Klasse [Wien]* 45: 123–162. [62, 65]

Belhassen, E., Trabaud, L., Couvet, D. & Gouyon, P.H. 1989 An example of nonequilibrium processes - gynodioecy of *Thymus vulgaris* L. in burned habitats. *Evolution* 43: 657–661. [345]

Bell, G. 1985 On the function of flowers. *Proceedings of the Royal Society of London, B* 224: 223–265. [341]

Bell, P.R. 1992 *Green Plants: Their Origin and Diversity.* Cambridge University Press, Cambridge. [371, 373]

Bené, F. 1941 Experiments on the colour preference of Black-chinned Hummingbirds. *Condor* 43: 237–242. [228]

Bené, F. 1946 The feeding and related behaviour of hummingbirds, with special reference to the Black-chin, *Archilochus alexandri* (Bourcier & Mulsant). *Memoirs of the Boston Society for Natural History* 9: 395–480. [227, 228, 229]

Benham, B.R. 1969 Insect visitors to *Chamaenerion angustifolium,* and their behaviour in relation to pollination. *Entomologist* 102: 221–228. [138]

Bennett, A.W. 1883 On the constancy of insects in their visits to flowers. *Journal of the Linnean Society (Zoology)* 17: 175–185. [69, 85, 95]

Benson, R.B. 1950 An introduction to the natural history of British sawflies (Hymenoptera Symphyta). *Transactions of the Society for British Entomology* 10: 45–142. [99]

Benson, R.B. 1959 Sawflies (Hym., Symphyta) of Sutherland and Wester Ross. *Entomologist's Monthly Magazine* 95: 101–104. [99]

Bergström, G. 1978 Role of volatile chemicals in *Ophrys*-pollinator interactions. *Biochemical Aspects of Plant-Animal Co-evolution* (ed. J.B. Harborne), pp. 207–232. Academic Press, London. [37, 221]

Bergström, G. 1991 Chemical ecology of terpenoid and other fragrances of angiosperms. *Ecological Chemistry and Biochemistry of Plant Terpenoids* (ed. J.B. Harborne & F.A. Tomas-Barberan), pp. 287–296 (Proceedings of the Phytochemical Society of Society of Europe 31). Clarendon Press, Oxford. [38]

Bergström, G., Birgersson, G., Groth, L. & Nilsson, L.A. 1992 Floral fragrance disparity between three taxa of Lady's slipper *Cypripedium calceolus* (Orchidaceae). *Phytochemistry* 31: 2315–2319. [189]

Bergström. G., Dobson, H.E.M. & Groth, I. 1995 Spatial fragrance patterns within the flowers of *Ranunculus acris* (Ranunculaceae). *Plant Systematics and Evolution* 195: 221–242. [38]

Bertin, R.I. & Newman, C.M. 1993 Dichogamy in angiosperms. *Botanical Review* 59: 112–152. [336, 343, 344]

Bertin, R.I. & Peters, P.J. 1992 Paternal effects on offspring quality in *Campsis radicans*. *American Naturalist* 140: 166–178. [421]

Bertin, R.I. & Sullivan, M. 1988 Pollen interference and cryptic self-fertility in *Campsis radicans*. *American Journal of Botany* 75: 1140–1147. [336, 420]

Biedinger, N. & Barthlott, W. 1993 Untersuchungen zur Ultraviolettreflexion von Angiospermenblüten I. Monocotyledoneae. *Tropische und subtropische Pflanzenwelt* 86: 1–122. [133]

Bierzychudek, P. 1982 The demography of Jack-in-the-pulpit, a forest perennial that changes sex. *Ecological Monographs* 52: 335–351. [415]

Bino, R.J., Dafni, A. & Meeuse, A.J.D. 1982 The pollination ecology of *Orchis galilea* (Bornm. et Schulze) Schltr. (Orchidaceae). *New Phytologist* 90: 315–319. [199]

Bischoff, H. 1927 *Biologie der Hymenopteren.* Springer, Berlin. [100, 133]

Blair, Patrick 1720 *Botanick Essays.* London. [16]

Blair, Patrick 1721 Observations upon the generation of plants. *Philosphical Transactions of the Royal Society* 369: 216–221. [15]

Block, J.M. 1962 Insect damage to flowers. *Gardeners' Chronicle* 152: 375. [106]

Bodger, H.S. 1960 Buckets of pollen and tons of seed. *Proceedings of the 15th American Horticultural Congress etc.*, pp. 18–19. American Horticultural Society, Washington, D.C. [353]

Bohart, G.E. 1957 Pollination of Alfalfa and Red Clover. *Annual Review of Entomology* 2: 355–380. [352, 358]

Bohart, G.E. 1962 *How to manage the Alfalfa Leaf-cutting Bee* (Megachile rotundata *Fabr.) for Alfalfa pollination* (Agricultural Experiment Station, Utah State University, Logan, Circular no. 144). [355]

Bohart, G.E. & Nye, W.P. 1960 *Insect pollinators of Carrots in Utah* (Agricultural Experiment Station, Utah State University, Logan, Publ. no. 419). [358]

Bolten, A.B. & Feinsinger, P. 1978 Why do hummingbird flowers secrete dilute nectar?

Biotropica 10: 307–309. [243]

Bolwig, N. 1954 The role of scent as a nectar guide for honeybees on flowers and an observation on the effect of colour on recruits. *British Journal of Animal Behaviour* 2: 81–83. [126]

Borg-Karlson, A.-K. 1990 Chemical and ethological studies of pollination in the genus *Ophrys* (Orchidaceae). *Phytochemistry* 29: 1359–1387. [37, 210]

Borg-Karlson, A.-K., Valterová, I. & Nilsson, L.A. 1993 Volatile compounds from flowers of six species in the family Apiaceae: bouquets for different pollinators? *Phytochemistry* 35: 111–119. [173]

Bowman, R.N. 1987 Cryptic self-incompatibility and the breeding system of *Clarkia unguiculata* (Onagraceae). *American Journal of Botany* 74: 471–476. [329]

Bown, D. 1988 *Aroids: Plants of the Arum Family.* Century, London. [305]

Boyden, T.C. 1982 The pollination biology of *Calypso bulbosa* var. *americana* (Orchidaceae): initial deception of bumblebee visitors. *Oecologia* 55: 178–184. [222]

Bradbury, J.W. & Emmons, L.H. 1974 Social organization of some Trinidad bats. I. Emballonuridae. *Zeitschrift für Tierpsychologie* 36: 137–183. [254]

Bradley, Richard 1717 *New Improvements in Planting and Gardening.* London. [14]

Brantjes, N.B.M. 1973 Sphingophilous flowers, function of their scent. *Pollination and Dispersal* (ed. N.B.M. Brantjes & H.F. Linskens), pp. 26–46. Dept. Botany, Univ. Nijmegen, Nijmegen. [91, 92]

Brantjes, N.B.M. 1976 Senses involved in the visiting of flowers by *Cucullia umbratica* (Noctuidae, Lepidoptera). *Entomologia Experimentalis et Applicata* 20: 1–7. [89]

Brantjes, N.B.M. 1980 Flower morphology of *Aristolochia* species and the consequences for pollination. *Acta Botanica Neerlandica* 29: 212–213. [299]

Brantjes, N.B.M. & Leemans, J.A.A.M. 1976 *Silene otites* (Caryophyllaceae) pollinated by nocturnal Lepidoptera and mosquitoes. *Acta Botanica Neerlandica* 25: 281–295. [58, 60]

Braue, A. 1913 Die Pollensammelapparate der beinsammelnden Bienen. *Jena Zeitschrift für Naturwissenschaften* 50: 1–96. [117]

Brian, A.D. 1954 The foraging behaviour of bumble-bees. *Bee World* 35: 61–67. [141]

Brian, A.D. 1957 Differences in the flowers visited by four species of bumblebees and their causes. *Journal of Animal Ecology* 26: 71–98. [113, 140, 141, 171, 405]

Brian, M.V. & Brian, A.D. 1952 The wasp *Vespula sylvestris* Scopoli: feeding, foraging and colony development. *Transactions of the Royal*

Entomological Society of London 103: 1–26. [106]

Briggs, D. & Walters, S.M. 1969 *Plant Variation and Evolution.* Weidenfeld & Nicolson, London. [21]

Briggs, D. & Walters, S.M. 1984 *Plant Variation and Evolution,* ed. 2. Cambridge University Press, Cambridge. [22]

Bronstein, J.L. 1991 The nonpollinating wasp fauna of *Ficus pertusa*: exploitation of a mutualism? *Oikos* 61: 175–186. [316]

Brown, A.H.D. 1979 Enzyme polymorphism in plant populations. *Theoretical Population Biology* 15: 1–42. [423]

Brown, J.H. & Kodric-Brown, A. 1979 Convergence, competition and mimicry in a temperate community of hummingbird-pollinated flowers. *Ecology* 60: 1022–1035. [388, 399, 402]

Broyles, S.B., Schnabel, A. & Wyatt, R. 1994 Evidence for long-distance pollen dispersal in milkweeds (*Asclepias exaltata*). *Evolution* 48: 1032–1040. [186]

Broyles, S.B. & Wyatt, R. 1991 Effective pollen dispersal in a natural population of *Asclepias exaltata*: the influence of pollinator behaviour, genetic similarity and mating success. *American Naturalist* 138: 1241–1249. [186]

Broyles, S.B. & Wyatt, R. 1993 The consequences of self-pollination in *Asclepias exaltata,* a self-incompatible milkweed. *American Journal of Botany* 80: 41–44. [336]

Buchmann, S.L. 1983 Buzz pollination in angiosperms. *Handbook of Experimental Pollination Biology* (ed. C.E. Jones & R.J. Little), pp. 73–113. Van Nostrand Reinhold, New York. [40, 70, 125, 179]

Buchmann, S.L. 1987 The ecology of oil flowers and their bees. *Annual Review of Ecology and Systematics* 18: 343–369. [43]

Buisseret, P.T. 1982 Allergy. *Scientific American* 247(2): 82–91. [30]

Burd, M. 1994 Bateman's principle and plant reproduction: the role of pollen limitation in fruit and seed set. *Botanical Review* 60: 89–137. [414]

Burger, W.C. 1977 The Piperales and the monocots: alternative hypotheses for the origin of monocotyledonous flowers. *Botanical Review* 43: 345–393. [378]

Burkhardt, D. 1964 Colour discrimination in insects. *Advances in Insect Physiology* 2 (ed. J.W.L. Beament, J.E. Treherne & V.B. Wigglesworth), pp. 131–173. Academic Press, London. [49, 130]

Burkill, I.H. 1897 Fertilization of spring flowers on the Yorkshire coast. *Journal of Botany [London]* 35: 92–99, 138–145, 184–189. [50]

Búrquez, A. 1989 Blue tits, *Parus caeruleus,* as pollinators of the crown imperial, *Fritillaria imperialis,* in Britain. *Oikos* 55: 335–340. [226]

Burr, B. & Barthlott, W. 1993 Untersuchungen zur Ultraviolettreflexion von Angiospermenblten II. Magnoliidae, Ranunculidae, Hamamelididae, Rosidae. *Tropische und subtropische Pflanzenwelt* 87: 1–193. [133]

Burr, B., Rosen, D. & Barthlott, W. 1995 Untersuchungen zur Ultraviolettreflexion von Angiospermenblüten III. Dilleniidae und Asteridae, s.l. *Tropische und subtropische Pflanzenwelt* 93: 1–185. [133]

Burtt, B.L. 1961 The Compositae and the study of functional evolution. *Transactions of the Botanical Society of Edinburgh* 39: 216–232. [174]

Butler, C.G. 1951 The importance of perfume in the discovery of food by the worker honeybee (*Apis mellifera* L.). *Proceedings of the Royal Society of London* 138: 403–413. [127, 141]

Butler, C.G. 1954 *The World of the Honeybee.* Collins, London. [41, 125, 133, 136, 137]

Cady, L. 1965 The flying duck orchids. *Australian Plants* 3: 174–177. [212]

Caldwell, J. & Wallace, T.J. 1955 Biological Flora of the British Isles. *Narcissus pseudonarcissus. Journal of Ecology* 43: 331–341. [152]

Calvo, R.N. 1990 Inflorescence size and fruit distribution among individuals in three orchid species. *American Journal of Botany* 77: 1378–1381. [224]

Calvo, R.N. 1993 Evolutionary demography of orchids: intensity and frequency of pollination and the cost of fruiting. *Ecology* 74: 1033–1042. [223]

Calvo, R.N. & Horvitz, C.C. 1990 Pollinator limitation, cost of reproduction, and fitness in plants: a transition matrix demographic approach. *American Naturalist* 136: 499–516. [223, 414]

Camerarius, R.J. 1694 *Academiae Caesareo Leopold. N. C. Hectoris II. Rudolphi Jacobi Camerarii, Professoris Tubingensis, ad Thessalum, D. Mich. Bernardum Valentini, Professorem Giessensem excellentissimum, de sexu plantarum epistola.* Tübingen. [13]

Cameron, S.A. 1981 Chemical signals in bumble bee foraging. *Behavioral Ecology and Sociobiology* 9: 257–260. [139]

Cammerloher, H. 1923 Zur Biologie der Blüte von *Aristolochia grandiflora* Swartz. *Österreichische Botanische Zeitschrift* 72: 180–198. [298, 299]

Cammerloher, H. 1933 Die Bestäubungseinrichtungen der Blüten von *Aristolochia lindneri* Berger. *Planta* 19: 351–365. [298, 299]

Campbell, B. & Lack, E. (eds.) 1985 *A Dictionary of Birds.* Poyser, for British Ornithologists' Union, Calton. [232, 235]

Campbell, C.S., Quinn, J.A., Cheplick, G.P. & Bell, T.J. 1983 Cleistogamy in grasses. *Annual Review of Ecology and Systematics* 14: 411–441. [280]

Carpenter, F.L. 1978a Hooks for mammal

pollination? *Oecologia (Berlin)* 35: 123–132. [257, 261, 262]

Carpenter, F.L. 1978b A spectrum of nectar-eater communities. *American Zoologist* 18: 809–819. [410]

Carpenter, F.L. 1983 Pollination energetics in avian communities: simple concepts and complex realities. *Handbook of Experimental Pollination Biology* (ed. C.E. Jones & R.J. Little), pp. 215–234. Van Nostrand Reinhold, New York. [244]

Carthew, S.M. 1993 An assessment of pollinator visitation to *Banksia spinulosa*. *Australian Journal of Ecology* 18: 257–268. [258]

Carthew, S.M. 1994 Foraging behaviour of marsupial pollinators in a population of *Banksia spinulosa*. *Oikos (Copenhagen)* 69: 133–139. [258]

Casey, T.M. 1989 Oxygen consumption during flight. *Insect Flight* (ed. G. Goldsworthy & C. Wheeler), pp. 257–272. CRC Press, Boca Raton. [93]

Casper, B.B. 1985 Self-compatibility in distylous *Cryptantha flava* (Boraginaceae). *New Phytologist* 99: 149–154. [329]

Catling, P.M. 1980 Rain-assisted autogamy in *Liparis loeselii* (L.) L.C. Rich (Orchidaceae). *Bulletin of the Torrey Botanical Club* 107: 525–529. [334]

Catling, P.M. 1983 Pollination of northeastern North American *Spiranthes* (Orchidaceae). *Canadian Journal of Botany* 61: 1080–1093. [196]

Chamberlain, A.C. & Little, P. 1981 Transport and capture of particles by vegetation. *Plants and their Atmospheric Environment* (ed. J. Grace, E.D. Ford & P.G. Jarvis), pp. 147–173. Blackwell Scientific Publications, Oxford. [272]

Chambers, V.H. 1945 British bees and wind-borne pollen. *Nature* 155: 145. [113]

Chambers, V.H. 1946 An examination of the pollen loads of *Andrena*: the species which visit fruit trees. *Journal of Animal Ecology* 15: 9–21. [113, 134]

Chambers, V.H. 1947 A list of sawflies (Hym., Symphyta) from Bedfordshire. *Entomologist's Monthly Magazine* 83: 91–95. [99]

Chambers, V.H. 1949 The Hymenoptera Aculeata of Bedfordshire. *Transactions of the Society for British Entomology* 9: 197–252. [103, 106]

Chan, H.T. & Appanah, S. 1980 Reproductive biology of some Malaysian dipterocarps, I: flowering biology. *Malaysian Forester* 43: 132–143. [312]

Charlesworth, B. & Charlesworth, D. 1978 A model for the evolution of dioecy and gynodioecy. *American Naturalist* 112: 975–997. [346]

Charlesworth, D. 1984 Androdioecy and the evolution of dioecy. *Biological Journal of the Linnean Society* 23: 333–348. [346]

Charlesworth, D. 1993 Why are unisexual flowers associated with wind pollination and unspecialized pollinators? *American Naturalist* 141: 481–490. [343]

Charlesworth, D. & Charlesworth, B. 1987 Inbreeding depression and its evolutionary consequences. *Annual Review of Ecology and Systematics* 18: 237–268. [334, 335]

Chase, M.W. 1985 Pollination of *Pleurothallis endotrachys*. *American Orchid Society Bulletin* 54: 431–434. [306]

Chase, M.W., Soltis, D.E., Olmstead, R.G. et mult. al. 1993 Phylogenetics of seed plants: an analysis of nucleotide sequences from the plastid gene *rbc*L. *Annals of the Missouri Botanical Garden* 80: 528–580. [379]

Cheplick, G.P. 1987 The ecology of amphicarpic plants. *Trends in Ecology and Evolution* 2: 97–101. [334]

Christy, R.M. 1922 The pollination of the British primulas. *Journal of the Linnean Society (Botany)* 46: 105–139. [63]

Church, A.H. 1908 *Types of Floral Mechanism*. Clarendon Press, Oxford. [21]

Clapham, A.R., Tutin, T.G. & Moore, D.M. 1987 *Flora of the British Isles, ed. 3*. Cambridge University Press, Cambridge. [329]

Clavaud, A. 1878 Sur la véritable mode de fécondation du *Zostera marina*. *Actes de la Société Linnéenne de Bordeaux* 32: 109–115. [293]

Cleghorn, M.L. 1913 Notes on the pollination of *Colocasia antiquorum*. *Journal and Proceedings of the Asiatic Society of Bengal* 9: 313–315. [305]

Clements F.E. & Long, F.L. 1923 *Experimental Pollination*. Washington. [23]

Cobley, L.S. 1956 *An Introduction to the Botany of Tropical Crops*. Longmans, Green & Co., London. [360]

Coe, M.J. & Isaac, F.M. 1965 Pollination of the baobab (*Adansonia digitata* L.) by the lesser bush baby (*Galago crassicaudatus* E. Geoffroy). *East African Wildlife Journal* 3: 123–124. [256]

Cohen, D. & Shmida, A. 1993 The evolution of flower display and reward. *Evolutionary Biology* 27: 197–243. [222]

Coleman, E. 1927 Pollination of the orchid *Cryptostylis leptochila*. *Victorian Naturalist* 44: 20–22. [205, 210]

Coleman, E. 1928a Pollination of *Cryptostylis leptochila* F.v.M. *Victorian Naturalist* 44: 333–340. [205, 210]

Coleman, E. 1928b Pollination of an Australian orchid by the male ichneumonid *Lissopimpla semipunctata* Kirby. *Transactions of the Entomological Society of London* 76: 533–539. [205, 210]

Coleman, E. 1929a Pollination of an Australian orchid, *Cryptostylis leptochila* F. Muell. (with a note by Col. M.J. Godfery). *Journal of Botany [London]* 67: 97–99. [205, 210]

Coleman, E. 1929b Pollination of *Cryptostylis subulata* (Labille) Reichb. *Victorian Naturalist* 46: 62–66. [211]

Coleman, E. 1930 Pollination of *Cryptostylis erecta* R. Br. *Victorian Naturalist* 46: 236–238. [211]

Coleman, E. 1931 Mrs Edith Coleman's further observations on the fertilisation of Australian orchids by the male ichneumonid *Lissopimpla semipunctata*, Kirby. *Proceedings of the Entomological Society of London* 6: 22–24. [211]

Coleman, E. 1934 Pollination of *Pterostylis acuminata* R. Br. and *Pterostylis falcata* Rogers. *Victorian Naturalist* 50: 248–252. [213, 307]

Coleman, E. 1938 Further observations on the pseudocopulation of the male *Lissopimpla semipunctata* Kirby (Hymenoptera Parasitica) with the Australian orchid *Cryptostylis leptochila* F.v.M. *Proceedings of the Royal Entomological Society of London, A* 13: 82–83. [211]

Coles, S.M. 1971 The *Ranunculus acris* L. complex in Europe. *Watsonia* 8: 237–261. [45]

Collins, B.G., Newland, C. & Briffa, P. 1984 Nectar utilization and pollination by Australian honeyeaters and insects visiting *Calothamnus quadrifidus* (Myrtaceae). *Australian Journal of Ecology* 9: 353–365. [236]

Collins, B.G. & Rebelo, T. [A.G.] 1987 Pollination biology of the Proteaceae in Australia and southern Africa. *Australian Journal of Ecology* 12: 387–421. [262, 263]

Colyer, C.N. & Hammond, C.O. 1951 *Flies of the British Isles*. Warne, London. [73, 74]

Cook, C.D.K. 1988 Wind pollination in aquatic angiosperms. *Annals of the Missouri Botanical Garden* 75: 768–777. [285]

Cooke, B. 1749 On the effects of the mixture of the farina of Apple-trees; and of Mayze or Indian Corn . . . *Philosophical Transactions of the Royal Society* 493: 205. [16]

Coombe, D.E. 1956 Biological Flora of the British Isles. *Impatiens parviflora* DC. *Journal of Ecology* 44: 701–713. [70]

Cope, F. 1962 The mechanism of pollen incompatibility in *Theobroma* L. *Heredity* 17: 157–182. [328]

Corbet, S.A. 1970 Insects on Hogweed flowers: a suggestion for a student project. *Journal of Biological Education* 4: 133–143. [50, 58, 60]

Corbet, S.A., Beament, J. & Eisikowitch, D. 1982 Are electrostatic forces involved in pollen transfer? *Plant, Cell and Environment* 5: 125–129. [272]

Corbet, S.A., Kerslake, C.J.C., Brown, D. & Morland, N.E. 1984 Can bees select nectar-rich flowers in a patch? *Journal of Apicultural Research* 32: 234–242. [139]

Corbet, S.A., Chapman, H. & Saville, N. 1988 Vibratory pollen collection and flower form: bumble-bees on *Actinidia, Symphytum, Borago*

and *Polygonatum. Functional Ecology* 2: 147–155. [125, 180]

Corbet, S.A., Williams, I.H. & Osborne, J.L. 1991 Bees and the pollination of crops and wild flowers in the European Community. *Bee World* 72: 47–59. [363, 364]

Cornet, B. 1989 The reproductive morphology and biology of *Sanmiguelia lewisii*, and its bearing on angiosperm evolution in the Late Triassic. *Evolutionary Trends in Plants* 3: 25–51. [377]

Correvon, H. & Pouyanne, A. 1916 Un curieux cas de mimétisme chez les Ophrydées. *Journal de la Société Nationale d'Horticulture de France* 1916: 29–31, 41–47, 84. [205]

Correvon, H. & Pouyanne, A. 1923 Nouvelles observations sur le mimétisme et la fécondation chez les *Ophrys speculum* et *lutea. Journal de la Société Nationale d'Horticulture de France* 1923: 372–377. [207]

Courtney, S.P., Hill, C.J. & Westerman, A. 1981 Pollen carried for long periods by butterflies. *Oikos* 38: 260–263. [416, 417]

Cox, P.A. 1982 Vertebrate pollination and the maintenance of dioecism in *Freycinetia. American Naturalist* 120: 65–80. [245]

Cox, P.A. 1983a Extinction of the Hawaiian avifauna resulted in a change of pollinators for the ieie, *Freycinetia arborea. Oikos* 41: 195–199. [241, 244]

Cox, P.A. 1983b Search theory, random motion, and the convergent evolution of pollen and spore morphologies in aquatic plants. *American Naturalist* 121: 9–31. [285]

Cox, P.A. 1984 Chiropterophily and ornithophily in *Freycinetia* (Pandanaceae) in Samoa. *Plant Systematics and Evolution* 144: 277–290. [241, 251]

Cox, P.A. 1988 Hydrophilous pollination. *Annual Review of Ecology and Systematics* 19: 261–280. [285, 292]

Cox, P.A. 1991 Abiotic pollination: an evolutionary escape for animal-pollinated angiosperms. *Philosophical Transactions of the Royal Society of London, B* 333: 217–224. [265]

Cox, P.A. 1993 Water-pollinated plants. *Scientific American* 269: 50–56. [285, 293]

Cox, P.A. & Knox, R.B. 1988 Pollination postulates and two-dimensional pollination in hydrophilous monocotyledons. *Annals of the Missouri Botanical Garden* 75: 811–818. [289]

Cox, P.A. & Knox, R.B. 1989 Two-dimensional pollination in hydrophilous plants: convergent evolution in the genera *Halodule* (Cymodoceaceae), *Halophila* (Hydrocharitaceae), *Ruppia* (Ruppiaceae), and *Lepilaena* (Zannichelliaceae). *American Journal of Botany* 76: 164–175. [289, 292]

Cox, P.A., Laushmann, R.H. & Ruckelshaus, M.H. 1992 Surface and submarine pollination in the seagrass *Zostera marina* L. *Botanical Journal of the Linnean Society* 109: 281–291. [293]

Cox, P.A. & Tomlinson, P.B. 1988 Pollination ecology of a seagrass, *Thalassia testudinum* (Hydrocharitaceae), in St. Croix. *American Journal of Botany* 75: 958–965. [293]

Cox, P.A., Tomlinson, P.B. & Nieznanski, K. 1992 Hydrophilous pollination and reproductive morphology in the seagrass *Phyllospadix scouleri* (Zosteraceae). *Plant Systematics and Evolution* 180: 65–75. [293]

Crane, M.B. & Lawrence, W.J.C. 1952 *The Genetics of Garden Plants, ed. 4.* Macmillan, London. [361]

Crane, P.R. & Dilcher, D.L. 1984 *Lesqueria*: an early angiosperm fruiting axis from the Mid-Cretaceous. *Annals of the Missouri Botanical Garden* 71: 384–402. [376]

Crane, P.R., Friis, E.M. & Pedersen, K.R. 1986 Angiosperm flowers from the Lower Cretaceous: fossil evidence on the early radiation of the dicotyledons. *Science* 232: 852–854. [376]

Crane, P.R., Friis, E.M. & Pedersen, K.R. 1989 Reproductive structure and function in Cretaceous Chloranthaceae. *Plant Systematics and Evolution* 165: 211–226. [375]

Crane, P.R., Friis, E.M. & Pedersen, K.R. 1995 The origin and early diversification of angiosperms. *Nature* 374: 27–33. [374, 376, 378, 379]

Crane, P.R. & Lidgard, S. 1990 Angiosperm radiation and patterns of Cretaceous palynological diversity. *Major Evolutionary Radiations* (ed. P.D. Taylor & G.P. Larwood), pp. 377–407. Clarendon Press, Oxford. [374, 377, 379]

Crawford, D.J., Ornduff, R. & Vasey, M.C. 1985 Allozyme variation within and between *Lasthenia minor* and its derivative species, *L. maritima* (Asteraceae). *American Journal of Botany* 72: 1177–1184. [332]

Crepet, W.L. 1983 The role of insect pollination in the evolution of the angiosperms. *Pollination Biology* (ed. L. Real), pp. 29–50. Academic Press, Orlando. [381]

Crepet, W.L. & Friis, E.M. 1987 The evolution of insect pollination in angiosperms. *The Origins of Angiosperms and Their Biological Consequences* (ed. E.M. Friis, W.G. Chaloner & P.R. Crane), pp. 181–201. Cambridge University Press, Cambridge. [373, 374, 382]

Cresswell, J.E., Bassom, A.P., Bell, S.A., Collins, S.J. & Kelly, T.B. [1995] Predicted pollen dispersal by honeybees and three species of bumblebees foraging on oil-seed rape: a comparison of three models. *Functional Ecology* (in press). [417]

Crosby, J.L. 1949 Selection of an unfavourable gene complex. *Evolution* 3: 212–230. [329]

Crowe, L.K. 1971 The polygenic control of outbreeding in *Borago officinalis*. *Heredity* 27: 111–118. [329]

Crowson, R.A. 1956 *Coleoptera; introduction and keys to families. Handbooks for the Identification of British Insects* 4(1). Royal Entomological Society, London. [53]

Cruden, R.W. 1972 Pollinators in high-elevation ecosystems: relative effectiveness of birds and bees. *Science* 176: 1439–1440. [393]

Cruden, R.W. 1977 Pollen-ovule ratios: a conservative indicator of breeding systems in flowering plants. *Evolution* 31: 32–46. [332]

Cruden, R.W. & Hermann-Parker, S.M. 1977 Temporal dioecism: an alternative to dioecism? *Evolution* 31: 863–866. [339]

Cruden, R.W. & Hermann-Parker, S.M. 1979 Butterfly pollination of *Caesalpinia pulcherrima*, with observations on a psychophilous syndrome. *Journal of Ecology* 67: 155–168. [416]

Cruden, R.W. & Toledo, V.M. 1977 Oriole pollination of *Erythrina breviflora* (Leguminosae): evidence for a polytypic view of ornithophily. *Plant Systematics and Evolution* 126: 393–403. [240]

Crüger, H. 1865 A few notes on the fecundation of orchids and their morphology. *Journal of the Linnean Society (Botany)* 8: 127–135. [216]

Dafni, A. 1984 Mimicry and deception in pollination. *Annual Review of Ecology and Systematics* 15: 259–278. [197, 221, 402]

Dafni, A. 1990 Pollination in *Orchis* and related genera: evolution from reward to deception. *Orchid Biology: Reviews and Perspectives* 4: 79–104. [222]

Dafni, A. 1992 *Pollination Ecology: A Practical Approach.* IRL/OUP, Oxford. [22]

Dafni, A. & Bernhardt, P. 1990 Pollination of terrestrial orchids of southern Australia and the Mediterranean region. *Evolutionary Biology* 24: 193–252. [210]

Dafni, A., Bernhardt, P., Shmida, A., Ivri, Y., Greenbaum, S., O'Toole, Ch. & Losita, L. 1990 Red bowl-shaped flowers: convergence for beetle pollination in the Mediterranean region. *Israel Journal of Botany* 39: 81–92. [56, 179]

Dafni, A. & Ivri, Y. 1979 Pollination ecology of, and hybridization between *Orchis coriophora* L. and *O. collina* Sol. ex Russ. (Orchidaceae) in Israel. *New Phytologist* 83: 181–187. [200]

Dafni, A. & Ivri, Y. 1981a Floral mimicry between *Orchis israelitica* Baumann and Dafni (Orchidaceae) and *Bellevalia flexuosa* Boiss. (Liliaceae). *Oecologia* 49: 229–232. [199]

Dafni, A. & Ivri, Y. 1981b The flower biology of *Cephalanthera longifolia* (Orchidaceae) - pollen imitation and facultative floral mimicry. *Plant Systematics and Evolution* 137: 229–240. [192]

Dafni, A. & Woodell, S.R.J. 1986 Stigmatic exudate and the pollination of *Dactylorhiza fuchsii* (Druce) Soo. *Flora [Jena]* 178: 343–350. [199]

Darwin, C. 1858 On the agency of bees in the fertilisation of papilionaceous flowers. *Gardeners' Chronicle* 1858: 824–844. [19]

Darwin, C. 1859 *On the Origin of Species by Natural Selection.* John Murray, London. [19]

Darwin, C. 1862a On the two forms, or dimorphic condition, in the species of *Primula* and on their remarkable sexual relations. *Journal of the Linnean Society (Botany)* 6: 77–96. [19]

Darwin, C. 1862b *The Various Contrivances by which Orchids are Fertilised.* John Murray, London. [19, 187, 213]

Darwin, C. 1876 *The Effects of Cross and Self Fertilisation in the Vegetable Kingdom.* John Murray, London. [19, 334, 424]

Darwin, C. 1877 *The Various Contrivances by which Orchids are Fertilised, ed. 2.* John Murray, London. [191, 196, 203, 216, 218, 219, 220]

Darwin, F. 1875 On the structure of the proboscis of *Ophideres fullonica*, an orange-sucking moth. *Quarterly Journal of Microscopic Science, New Series* 15: 385–389. [81]

Daumann, E. 1932 Über die 'Scheinnektarien' von *Parnassia palustris* und anderer Blütenarten. *Jahrbücher für wissenschaftliche Botanik* 77: 104–149. [78]

Daumann, E. 1935 Über die Bestäubungsökologie der *Parnassia*-Blüte II. *Jahrbücher für wissenschaftliche Botanik* 81: 707–717. [78]

Daumann, E. 1941 Die anbohrbaren Gewebe und rudimentären Nektarien in der Blütenregion. *Beihefte zum Botanischen Centralblatt* 61A: 1282. [197]

Daumann, E. 1963 Zur Frage des Ursprung der Hydrogamie. Zugleich ein Beitrag zur Blütenökologie von *Potamogeton*. *Preslia* 35: 23–30. [288]

Daumann, E. 1967 Zur Blütenmorphologie und Bestäubungsökologie von *Veratrum album* subsp. *lobelianum* (Bernh.) Rchb. *Österreichische Botanische Zeitschrift* 114: 134–148. [73]

Daumann, E. 1968 Zur bestäubungsökologie von *Cypripedium calceolus* L. *Österreichische Botanische Zeitschrift* 115: 434–446. [188]

Daumer, K. 1956 Reizmetrische Untersuchungen des Farbensehens der Bienen. *Zeitschrift für vergleichende Physiologie* 38: 413–478. [130, 131]

Daumer, K. 1958 Blumenfarben, wie sie die Bienen sehen. *Zeitschrift für vergleichende Physiologie* 41: 49–110. [131, 132, 179]

Davidar, P. 1983 Similarity between flowers and fruits in some flowerpecker pollinated mistletoes. *Biotropica* 15: 32–37. [236]

Davis, D.R. 1967 A revision of the moths of the subfamily Prodoxinae (Lepidoptera: Incurvariidae). *United States National Museum Bulletin* 255: 1–170. [317]

Davis, P.H. & Heywood, V.H. 1963 *Principles of Angiosperm Taxonomy.* Oliver & Boyd, Edinburgh. [22]

Davis, R.W. 1986 The pollination biology of *Cypripedium acaule* (Orchidaceae). *Rhodora* 88: 445–450. [223]

Dawkins, R. 1986 *The Blind Watchmaker.* Longman/Norton, London/New York. [425]

De Cock, A.W.A.M. 1980 Flowering, pollination, and fruiting in *Zostera marina* L. *Aquatic Botany* 9: 202–220. [293]

de Jong, T.J. & Klinkhamer, P.G.L. 1989 Size-dependency of sex-allocation in hermaphroditic, monocarpic plants. *Functional Ecology* 3: 201–206. [415]

de Jong, T.J. & Klinkhamer, P.G.L. 1991 Early flowering in *Cynoglossum officinale* L. Constraint or adaptation? *Functional Ecology* 5: 750–756. [403]

De Nettancourt, D. 1977 *Incompatibility in Angiosperms.* Springer-Verlag, Berlin. [323, 327]

De Vos, O.C. 1983 *Scrophularia nodosa*, adapted to wasp pollination? *Acta Botanica Neerlandica* 32: 345. [107]

DeGrandi-Hoffman, G., Hoopingarner, R.A. & Baker, K.K. 1984 Identification and distribution of cross-pollinating honey bees (Hymenoptera: Apidae) in apple orchards. *Environmental Entomology* 13: 757–764. [362]

DeGrandi-Hoffman, G., Hoopingarner, R. & Klomparens, K. 1986 Influence of honey bee (Hymenoptera: Apidae) in-hive pollen transfer on cross pollination and fruit set in apple. *Environmental Entomology* 15: 723–725. [362]

Delpino, F. 1868 *Ulteriori osservazioni e considerazioni sulla dicogamia nel regno vegetale.* Milano. [20]

Delpino, F. 1871 *Annotations to H. Müller, Application of the Darwinian theory to flowers and the insects which visit them.* Transl. R.L. Packard. Naturalists' Agency, Salem. [309]

Delpino, F. 1874 Ulteriori osservazioni e considerazioni sulla dicogamia nel regno vegetale. 2(IV) Delle piante zoidifile. *Atti della Società Italiana di Scienze naturali* 16: 151–349. [20]

Demoll, R. 1908 Die Mundteile der solitären Apiden. *Zeitschrift für wissenschaftliche Zoologie* 91: 1–51. [118, 152–4, 159]

Dessart, P. 1961 Contribution à l'étude des Ceratopogonidae (Diptera: Les *Forcipomyia* pollinisateurs du cacaoyer). *Bulletin Agricole du Congo Belge* 52: 525–540. [311]

Devlin, B. & Ellstrand, N.C. 1990 The development and application of a refined method for estimating gene flow from angiosperm paternity analysis. *Evolution* 46:

1030–1042. [423]

Dexter, J.S. 1913 Mosquitoes pollinating orchids. *Science* 37: 867. [60]

Diggle, P.K. 1992 Development and the evolution of plant reproductive characters. *Ecology and Evolution of Plant Reproduction* (ed. R. Wyatt), pp. 326–355. Chapman & Hall, London. [332, 334]

Dilcher, D.L. 1995 Plant reproductive strategies: using the fossil record to unravel current issues in plant reproduction. *Experimental and Molecular Approaches to Plant Biosystematics* (ed. P.C. Hoch & A.G. Stephenson), pp. 187–198. Missouri Botanical Garden, St Louis. [290]

Dilcher, D.L. & Crane, P.R. 1984 *Archaeanthus*: an early angiosperm from the Cenomanian of the Western Interior of North America. *Annals of the Missouri Botanical Garden* 71: 351–383. [376]

Dimmock, G. 1881 *The Anatomy of the Mouthparts and the Sucking Apparatus of some Diptera (dissertation)*. Boston. [65]

Disney, R.H.L. 1980 Records of flower visiting by scuttle flies (Diptera) in the British Isles. *Naturalist* 105: 45–50. [61]

Dixon, K.W., Pate, J.S. & Kuo, J. 1990 The Western Australian fully subterranean orchid *Rhizanthella gardneri*. *Orchid Biology: Reviews and Perspectives* 5: 37–62. [221]

Dobat, K. & Peikert-Holle, T. 1985 *Blüten und Fledermäuse: Bestäubung durch Fledermäuse und Flughunde (Chiropterophilie)*. Waldemar Kramer, Frankfurt am Main. [244, 250, 251, 252, 253, 254, 255] .

Dobbs, A. 1750 Concerning bees and their method of gathering wax and honey. *Philosophical Transactions of the Royal Society* 46: 536–549. [16]

Dobson, H., Bergström, J., Bergström, G. & Groth, I. 1987 Pollen and flower volatiles in two *Rosa* species. *Phytochemistry* 26: 3171–3173. [38]

Dobzhansky, T. 1937 *Genetics and the Origin of Species*. New York. [21]

Docters van Leeuwen, W.M. 1931 Vogelbesuch an den Blüten von einigen *Erythrina*-Arten auf Java. *Annales du Jardin Botanique de Buitenzorg* 42: 57–96. [239]

Docters van Leeuwen, W.M. 1938 Observations about the biology of tropical flowers. *Annales du Jardin Botanique de Buitenzorg* 48: 27–68. [246]

Dodson, C.H. 1962 The importance of pollination in the evolution of the orchids of Tropical America. *Bulletin of the American Orchid Society* 31: 525–534, 641 - 649, 731 - 735. [215, 306]

Dodson, C.H. 1965 Studies in orchid pollination: the genus *Coryanthes*. *Bulletin of the American Orchid Society* 34: 680–687. [217]

Dodson, C.H. & Frymire, G.P. 1961a Preliminary studies in the genus *Stanhopea* (Orchidaceae). *Annals of the Missouri Botanical Garden* 48: 137–172. [217]

Dodson, C.H. & Frymire, G.P. 1961b Natural pollination of orchids. *Missouri Botanical Garden Bulletin* 49: 135–152. [215]

Dowsett-Lemaire, F. 1989 Food plants and the annual cycle in a montane community of sunbirds (*Nectarinia* spp.) in northern Malawi. *Tauraco* 1: 167–185. [394]

Doyle, J. 1945 Developmental lines in pollination mechanisms in the Coniferales. *Scientific Proceedings of the Royal Dublin Society* 24(5): 43–62. [274]

Doyle, J.A. & Donoghue, M.J. 1987 The origin of angiosperms: a cladistic approach. *The Origins of Angiosperms and Their Biological Consequences* (ed. E.M. Friis, W.G. Chaloner & P.R. Crane), pp. 17–49. Cambridge Unversity Press, Cambridge. [372, 373, 374]

Doyle, J.A., Hotton, C.L. & Ward, J.V. 1990a Early Cretaceous tetrads, zonasulcate pollen, and Winteraceae, I. Taxonomy, morphology and ultrastructure. *American Journal of Botany* 77: 1544–1557. [376]

Doyle, J.A., Hotton, C.L. & Ward, J.V. 1990b Early Cretaceous tetrads, zonasulcate pollen, and Winteraceae, II. Cladistic analysis and implications. *American Journal of Botany* 77: 1558–1568. [376, 377]

Drabble, E. & Drabble, H. 1917 The syrphid visitors to certain flowers. *New Phytologist* 16: 105–109. [68]

Drabble, E. & Drabble, H. 1927 Some flowers and their dipteran visitors. *New Phytologist* 26: 115–123. [60, 68, 69, 70, 77]

Dressler, R.L. 1982 Biology of the orchid bees (Euglossini). *Annual Review of Ecology and Systematics* 13: 373–394. [216]

Dressler, R.L. 1993 *Phylogeny and Classification of the Orchid Family*. Cambridge University Press, Cambridge. [188, 213,215]

Drummond, D.C. & Hammond, P.M. 1991 Insects visiting *Arum dioscoridis* Sm. and *A. orientale* M. Bieb. *Entomologist's Monthly Magazine* 127: 151–156. [304]

du Toit, J.T. 1990 Giraffe feeding on *Acacia* flowers: predation or pollination? *African Journal of Ecology* 28: 63–68. [261]

Duckett, J.G. & Pang, W.C. 1984 The origins of heterospory: a comparative study of sexual behaviour in the fern *Platyzoma microphyllum* R.Br. and the horsetail *Equisetum giganteum* L. *Botanical Journal of the Linnean Society* 88: 11–34. [368]

Dudley, P. 1724 Observations on some of the Plants in New-England. *Philosophical Transactions of the Royal Society* 33: 194. [16]

Dukas, R. & Real, L.A. 1993 Learning constraints and floral choice in bumble bees.

Animal Behaviour 46: 637–644. [199, 222]

Duncan, C.D. 1939 A contribution to the biology of North America Vespine wasps. *Stanford University Publications: Biological Science* 8(1): 1–272. [100, 106]

Eastham, L.E.S. & Eassa, Y.E.E. 1955 The feeding mechanism of the butterfly *Pieris brassicae* L. *Philosophical Transactions of the Royal Society B* 239: 1–43. [82]

Edmonds, R.E. 1979 *Aerobiology: the Ecological Systems Approach.* US IBP Synthesis Series 10. Dowden, Hutchinson & Ross, Stroudsburg, Pa. [272]

Edwards, D. 1993 Tansley Review No. 53: Cells and tissues in the vegetative sporophytes of early land plants. *New Phytologist* 125: 225–247. [365]

Ehrlich, P.R. & Raven, P.H. 1965 Butterflies and plants: a study in coevolution. *Evolution* 18: 586–608. [380]

Eijnde, J. van den, Ruijter, A. de & Steen, J. van der 1991 Method for rearing *Bombus terrestris* continuously and the production of bumblebee colonies for pollination purposes. *Sixth International Symposium on Pollination (Acta Horticulturae no. 288)* (ed. C. van Heemert & A. de Ruijter), pp. 154–155. International Soc. Hortic. Sci., Wageningen. [359]

Ellington, C.P., Machin, K.E. & Casey, T.M. 1990 Oxygen consumption of bumblebees in forward flight. *Nature* 347: 472–473. [139]

Ellstrand, N.C. & Marshall, D.L. 1985 Interpopulation gene flow by pollen in wild radish, *Raphanus sativus. American Naturalist* 126: 606–616. [423]

Endress, P.K. 1987a The Chloranthaceae: reproductive structures and phylogenetic position. *Botanische Jahrbucher fur Systematik, Pflanzengeschichte und Pflanzengeographie* 109: 153–226. [375, 376, 377]

Endress, P.K. 1987b The early evolution of the angiosperm flower. *Trends in Ecology and Evolution* 2: 300–304. [377, 378]

Endress, P.K. 1990 Evolution of reproductive structures and functions in primitive angiosperms (Magnoliidae). *Memoirs of the New York Botanical Garden* 55: 5–34. [377, 379]

Endress, P.K. 1994 *Diversity and Evolutionary Biology of Tropical Flowers.* Cambridge University Press, Cambridge. [328, 337, 376, 391, 394]

Erdtman, G. 1969 *Handbook of Palynology.* Munksgaard, Copenhagen. [32]

Erhardt, A. 1990 Pollination of *Dianthus gratianopolitanus. Plant Systematics and Evolution* 170: 125–132. [146]

Erickson, E.H. 1983 Pollination of entomophilous hybrid seed parents. *Handbook of Experimental Pollination Biology* (ed. C.E. Jones & R.J. Little), pp. 493–535. Van Nostrand Reinhold, New York. [353]

Erickson, E.H. & Buchmann, S.L. 1983 Electrostatics and pollination. *Handbook of Experimental Pollination Ecology* (ed. C.E. Jones & R.J. Little), pp. 173–185. Van Nostrand Reinhold, New York. [272]

Esch, H. 1967 The evolution of bee language. *Scientific American* 216: 96–104. [133]

Esch, H., Esch, I. & Kerr, W.E. 1965 Sound: an element common to communication of stingless bees and to dances of the honeybee. *Science* 149: 320–321. [133]

Essig, F.B. 1973 Pollination in some New Guinea palms. *Principes* 17: 75–83. [311]

Evans, M.S. 1895 The fertilisation of *Loranthus kraussianus* and *L. dregei. Nature* 51: 235–236. [236]

Faegri, K. 1986 The solanoid flower. *Transactions of the Botanical Society of Edinburgh, 150th Anniversary Supplement* (ed. I.J. Alexander & N.M. Gregory), pp. 51–59. Botanical Society of Edinburgh, Edinburgh. [179]

Faegri, K., Kaland, P.E. & Krzywinski, K. 1989 *Textbook of Pollen Analysis, ed. 4.* Wiley, Chichester. [32]

Faegri, K. & van der Pijl, L. 1966 *The Principles of Pollination Ecology.* Pergamon, Oxford. [22, 173]

Faegri, K. & van der Pijl, L. 1979 *The Principles of Pollination Ecology, ed. 3.* Pergamon, Oxford, etc. [28, 154, 165, 178, 241, 244, 334]

Fahn, A. 1990 *Plant Anatomy, ed. 4.* Pergamon Press, Oxford. [179]

Fairey, D.T. & Lefkovitch, L.P. 1991 Reproduction of *Megachile rotundata* Fab. foraging on *Trifolium* spp. and *Brassica campestris. Sixth International Symposium on Pollination (Acta Horticulturae no. 288)* (ed. C. van Heemert & A. de Ruijter), pp. 185–189. International Soc. Hortic. Sci., Wageningen. [357]

Faulkner, C.J. 1962 Blow-flies as pollinators of *Brassica* crops. *Commercial Grower* 3457: 807–809. [358]

Faulkner, J.S. 1973 Experimental hybridisation of north-west European species in *Carex* section *Acutae* (Cyperaceae). *Botanical Journal of the Linnean Society* 67: 233–253. [344]

Feinsinger, P. 1976 Organization of a tropical guild of nectarivorous birds. *Ecological Monographs* 46: 257–291. [409, 410]

Feinsinger, P. & Colwell, R.K. 1978 Community organization among neotropical nectar-feeding birds. *American Zoologist* 18: 779–795. [409]

Feinsinger, P. & Swarm, L.A. 1978 How common are ant-repellent nectars? *Biotropica* 10: 238–239. [108]

Feinsinger, P., Swarm, L.A. & Wolfe, J.A. 1985 Nectar-feeding birds on Trinidad and Tobago: comparison of diverse and depauperate guilds. *Ecological Monographs* 55: 1–28. [409]

Fogg, G.E. 1950 Biological Flora of the British Isles. *Sinapis arvensis* L. *Journal of Ecology* 38: 415–429. [172]

Ford, E.B. 1945 *Butterflies.* Collins, London. [80]

Ford, E.B. 1955 *Moths.* Collins, London. [80]

Ford, H.A., Paton, D.C. & Forde, N. 1979 Birds as pollinators of Australian plants. *New Zealand Journal of Botany* 17: 509–519. [236]

Ford, M.A. & Kay, Q.O.N. 1985 The genetics of incompatibility in *Sinapis arvensis* L. *Heredity* 54: 99–102. [325]

Frankel, R. & Galun, E. 1977 *Pollination Mechanisms, Reproduction and Plant Breeding.* Springer, Berlin. [353]

Frankie, G.W. 1975 Tropical forest phenology and pollinator plant coevolution. *Coevolution of Animals and Plants* (ed. L.E. Gilbert & P.H. Raven), pp. 192–209. University of Texas Press, Austin. [394]

Frankie, G.W. 1976 Pollination of widely dispersed trees by animals in Central America, with an emphasis on bee pollination systems. *Variation, Breeding and Conservation of Tropical Forest Trees* (ed. J. Burley & B.T. Stiles), pp. 151–159. Academic Press, London, New York. [134, 416]

Frankie, G.W. & Baker, H.G. 1974 The importance of pollinator behavior in the reproductive biology of tropical trees. *Anales de Instituto de Biología de la universidad nacional de México, ser. Botánica* 45 (1): 1–10. [134, 250, 254]

Frankie, G.W. & Haber, W.A. 1983 Why bees move among mass-flowering neotropical trees. *Handbook of Experimental Pollination Biology* (ed. C.E. Jones & R.J. Little), pp. 360–372. Van Nostrand Reinhold, New York. [416]

Frankie, G.W., Haber, W.A., Opler, P.A. & Bawa, K.S. 1983 Characteristics and organization of the large bee pollination system in the Costa Rican dry forest. *Handbook of Experimental Pollination Biology* (ed. C.E. Jones & R.J. Little), pp. 411–447. Van Nostrand Reinhold, New York. [394]

Frankie, G.W., Opler, P.A. & Bawa, K.S. 1976 Foraging bahaviour of solitary bees: implications for outcrossing of a neotropical forest tree species. *Journal of Ecology* 64: 1049–1057. [134]

Free, J.B. 1962 Studies on the pollination of fruit trees by Honey-bees. *Journal of the Royal Horticultural Society* 87: 302–309. [361]

Free, J.B. 1963 The flower constancy of Honey-bees. *Journal of Animal Ecology* 32: 119–131. [136]

Free, J.B. 1964 Comparison of the importance of insect and wind pollination of apple trees. *Nature* 201: 726–727. [285]

Free, J.B. 1965 The ability of bumblebees and Honey-bees to pollinate Red Clover. *Journal of Applied Ecology* 2: 289–294. [358, 362]

Free, J.B. 1966 The foraging areas of Honey-bees in an orchard of standard apples. *Journal of Applied Ecology* 3: 261–268. [362]

Free, J.B. 1968 Dandelion as a competitor to fruit trees for bee visitors. *Journal of Applied Ecology* 5: 161–178. [362]

Free, J.B. 1970a Effect of flower shapes and nectar guides on the behaviour of foraging honeybees. *Behaviour* 37: 269–285. [127, 129, 139]

Free, J.B. 1970b *Insect Pollination of Crops.* Academic Press. [361]

Free, J.B. 1993 *Insect Pollination of Crops, ed. 2.* Academic Press, London. [352, 361]

Free, J.B. & Butler, C.G. 1959 *Bumblebees.* Collins, London. [141, 357]

Free, J.B. & Spencer-Booth, Y. 1964 The foraging behaviour of Honey-bees in an orchard of dwarf Apple trees. *Journal of Horticultural Science* 39: 78–83. [362]

Freeman, C.E., Worthington, R.D. & Corral, R.D. 1985 Some floral nectar-sugar compositions from Durango and Sinaloa, Mexico. *Biotropica* 17: 309–313. [243, 244]

Freeman, C.E., Worthington, R.D. & Jackson, M.S. 1991 Floral nectar sugar compositions of some South and Southeast Asian species. *Biotropica* 23: 568–574. [245]

Freeman, D.C., Harper, K.T. & Charnov, E.L. 1980 Sex change in plants: old and new observations and new hypotheses. *Oecologia* 47: 222–232. [344]

Frick, K., Potter, H. & Weaver, H. 1960 *Development and maintenance of Alkali Bee nesting sites* (Washington Agricultural Experimental Stations Circular, No. 366). [354]

Friedman, W.E. 1990 Sexual reproduction in *Ephedra nevadensis* (Ephedraceae): further evidence for double fertilisation in a nonflowering seed plant. *American Journal of Botany* 77: 1582–1598. [372]

Friedman, W.E. 1992 Evidence of a pre-angiosperm origin of endosperm: implications for the evolution of flowering plants. *Science* 255: 336–339. [372]

Friese, H. 1923 *Die Europäischen Bienen (Apidae).* De Gruyter, Berlin. [111, 113, 121]

Friis, E.M. 1985 *Actinocalyx* gen. nov., sympetalous angiosperm flowers from the Upper Cretaceous of Southern Sweden. *Review of Palaeobotany and Palynology* 45: 171–183. [379]

Friis, E.M. 1990 *Silvianthemum suecicum* gen. et sp. nov., a new saxifragalean flower from the Late Cretaceous of Sweden. *Biologiske Skrifte* 36: 1–35. [379]

Friis, E.M. & Crepet, W.L. 1987 Time of appearance of floral features. *The Origins of Angiosperms and their Biological Consequences* (ed. E.M. Friis, W.G. Chaloner & P.R. Crane), pp.

145–179. Cambridge University Press, Cambridge. [375, 376]

Friis, E.M. & Endress, P.K. 1990 Origin and evolution of angiosperm flowers. *Advances in Botanical Research* 17: 99–162. [381]

Frisch, K. von 1914 Der Farbsinn und Formensinn der Biene. *Zoologischer Jahrbücher, Abteilung allgemeine Physiologie der Tiere* 35: 1–182. [23]

Frisch, K. von 1950 *Bees, their Vision, Chemical Senses and Language.* Cornell University Press, Ithaca, N.Y. [125, 128, 129, 133]

Frisch, K. von 1954 *The Dancing Bees.* Methuen, London. [126, 129, 134, 136, 351]

Frisch, K. von 1965 *Tanzsprache und Orientierung der Bienen.* Springer, Berlin. [133]

Frisch, K. von 1967 *The Dance Language and Orientation of Bees.* Transl. L.E. Chadwick. Harvard University Press, Cambridge, Mass. [133]

Frisch, K. von 1993 *The Dance Language and Orientation of Bees.* Transl. L.E. Chadwick, with new foreword by T.D. Seeley. Harvard University Press, Cambridge, Mass. [133]

Frost, S.K. & Frost, P.G.H. 1981 Sunbird pollination of *Strelitzia nicolai. Oecologia* 49: 379–384. [240]

Fussell, M. & Corbet, S.A. 1992 Flower usage by bumble-bees: a basis for forage plant management. *Journal of Applied Ecology* 29: 451–465. [364, 403, 405]

Galen, C., Gregory, T. & Galloway, L.F. 1989 Costs of self-pollination in a self-incompatible plant, *Polemonium viscosum. American Journal of Botany* 76: 1675–1680. [336]

Galil, J. 1973 Pollination in dioecious figs: pollination of *Ficus fistulosa* by *Ceratosolen hewitti. Gardens' Bulletin, Singapore* 26: 303–311. [314, 315]

Galil, J., Dulberger, R. & Rosen, D. 1970 The effects of *Sycophaga sycomori* L. on the structure and development of the syconia in *Ficus sycomorus* L. *New Phytologist* 69: 103–111. [316]

Galil, J. & Eisikovitch, D. 1968a On the pollination ecology of *Ficus sycomorus* in East Africa. *Ecology* 49: 259–269. [313, 314, 315, 316]

Galil, J. & Eisikovitch, D. 1968b On the pollination ecology of *Ficus religiosa* in Israel. *Phytomorphology* 18: 356–563. [313, 314]

Galil, J. & Eisikovitch, D. 1969 Further studies on the pollination ecology of *Ficus sycomorus* L. *Tijdschrift voor Entomologie* 112: 1–13. [313, 314, 316]

Galil, J. & Eisikowitch, D. 1974 Further studies on pollination ecology in *Ficus sycomorus,* II. pocket filling and emptying by *Ceratosolen arabicus* Mayr. *New Phytologist* 73: 515–528. [315]

Galil, J., Ramirez, W. & Eisikowitch, D. 1973 Pollination of *Ficus costaricana* and *F.*

hemsleyana by *Blastophaga esterae* and *B. tonduzi* in Costa Rica (Hymenoptera: Chalcidoidea, Agaonidae). *Tijdschrift voor Entomologie* 116: 175–183. [315]

Galil, J. & Snitzer-Pasternak, Y. 1970 Pollination in *Ficus religiosa* L. as connected with the structure and mode of action of the pollen pockets of *Blastophaga quadraticeps* Mayr. *New Phytologist* 69: 775–784. [315]

Galil, J. & Zeroni, M. 1965 Nectar system of *Asclepias curassavica. Botanical Gazette* 126: 144–148. [185]

Galil, J. & Zeroni, M. 1967 On the pollination of *Zizyphus spina-christi* (L.) Willd. in Israel. *Israel Journal of Botany* 16: 71–77. [359]

Galil, J., Zeroni, M. & Bar Shalom, D. (Bogoslavsky) 1973 Carbon dioxide and ethylene effects in co-ordination between the pollinator *Blastophaga quadraticeps* and the syconium in *Ficus religiosa. New Phytologist* 72: 1113–1127. [315]

Gamerro, J.C. 1968 Observaciones sobre la biología floral y morfología de la Potamogetoncáea *Ruppia cirrhosa* (Petag.) Grande (=*R. spiralis* L. ex Dum.). *Darwinia* 14: 575–609. [288]

Ganders, F.R. 1979 The biology of heterostyly. *New Zealand Journal of Botany* 17: 607–635. [327, 328]

Gentry, A.H. 1974 Coevolutionary patterns in Central American Bignoniaceae. *Annals of the Missouri Botanical Garden* 61: 728–759. [401, 408]

Gibbs, P.E. 1986 Do homomorphic and heteromorphic self-incompatibility systems have the same sporophytic mechanism? *Plant Systematics and Evolution* 154: 285–323. [323, 325, 327, 328, 330]

Gibbs, P.E. 1988 Self-incompatibility mechanisms in flowering plants: some complications and clarifications. *Lagascalia* 15: 17–28. [325, 328]

Gibbs, P.E. & Bianchi, M. 1993 Post-pollination events in species of *Chorisia* (Bombacaceae) and *Tabebuia* (Bignoniaceae) with late-acting self-incompatibility. *Botanica Acta* 106: 64–71. [328]

Gifford, E.M. & Foster, A.S. 1989 *Morphology and Evolution of Vascular Plants.* W.H. Freeman and Co., San Fransisco. [368, 373]

Gilbert, F. & Owen, J. 1990 Size, shape, competition, and community structure in hoverflies (Diptera: Syrphidae). *Journal of Animal Ecology* 59: 21–39. [409]

Gilbert, F.S. 1981 The foraging ecology of hoverflies (Diptera, Syrphidae): morphology of the mouthparts in relation to feeding on nectar and pollen in some common urban species. *Ecological Entomology* 6: 245–262. [65, 66, 68, 70]

Gilbert, L.E. 1972 Pollen feeding and

reproductive biology of *Heliconius* butterflies. *Proceedings of the National Academy of Sciences, USA* 69: 1403–1407. [82]

Gill, D.E. 1989 Fruiting failure, pollinator inefficiency, and speciation in orchids. *Speciation and its Consequences* (ed. D. Otte & J.A. Endler), pp. 458–481. Sinauer, Sunderland, Mass. [223]

Gill, F.B. & Wolf, L.L. 1978 Comparative foraging efficiencies of some montane sunbirds in Kenya. *Condor* 80: 391–400. [233]

Givnish, T.J. 1982 Outcrossing versus ecological constraints in the evolution of dioecy. *American Naturalist* 119: 849–865. [340, 342]

Gleeson, S.K. 1982 Heterodichogamy in walnuts: inheritance and stable ratios. *Evolution* 36: 892–902. [344]

Godfery, M.J. 1925 The fertilisation of *Ophrys speculum*, *O. lutea* and *O. fusca*. *Journal of Botany [London]* 63: 33–40. [205, 206]

Godfery, M.J. 1927 The fertilisation of *Ophrys fusca* Link. *Journal of Botany [London]* 65: 350–351. [205, 208]

Godfery, M.J. 1929 Recent observations on the pollination of *Ophrys*. *Journal of Botany [London]* 67: 298–302. [207]

Godfery, M.J. 1930 The fertilisation of *Ophrys fusca* and *O. lutea*. *Journal of Botany [London]* 68: 237–238. [208]

Godfery, M.J. 1933 *Monograph and Iconograph of Native British Orchidaceae.* Cambridge University Press, London. [209]

Godley, E.J. 1979 Flower biology in New Zealand. *New Zealand Journal of Botany* 17: 441–466. [390]

Godley, E.J. & Smith, D.H. 1980 Breeding systems in New Zealand plants. 5. *Pseudowintera colorata* (Winteraceae). *New Zealand Journal of Botany* 19: 151–156. [328]

Godwin, H. 1975 *The History of the British Flora, ed. 2.* Cambridge. [32]

Goldingay, R.L., Carthew, S.M. & Whelan, R.J. 1991 The importance of non-flying mammals in pollination. *Oikos* 61: 79–87. [257, 258, 262, 263]

Goldsmith, T.H. 1980 Hummingbirds see near-ultraviolet light. *Science* 207: 786–788. [242]

Goodman, L.J. & Fisher, R.C. (eds.) 1991 *The Behaviour and Physiology of Bees.* CAB International, Wallingford. [125]

Goodwin, T.W. (ed.) 1976 *The Chemistry and Biochemistry of Plant Pigments, ed. 2.* Academic Press, London. [34, 36]

Goodwin, T.W. (ed.) 1988 *Plant Pigments.* Academic Press, London. [34, 36]

Gottlieb, L.D. 1973 Genetic differentiation, sympatric speciation, and the origin of a diploid species of *Stephanomeria*. *American Journal of Botany* 60: 545–553. [332]

Gottsberger, G. & Amaral, A., jr. 1984

Pollination strategies in Brazilian *Philodendron* species. *Berichte der deutschen botanischen Gesellschaft* 97: 391–410. [310]

Gouin, F. 1949 Recherches sur la morphologie de l'appareil buccal des Diptères. *Mémoires du Museum National d'Histoire Naturelle, Paris, N.S.* 28: 167–269. [62, 65]

Gould, E. 1978 Foraging behaviour of Malaysian nectar-feeding bats. *Biotropica* 10: 184–193. [245, 246, 247, 254]

Gould, J.L. 1976 The dance-language controversy. *Quarterly Review of Biology* 51: 211–244. [133]

Gould, J.L. & Gould, C.G. 1988 *The Honey Bee.* Scientific American Library, New York. [133]

Gould, S.J. 1990 *Wonderful Life.* Hutchinson Radius, London. [377]

Graham-Smith, G.S. 1930 Further observations on the anatomy and function of the proboscis of the Blow-fly, *Calliphora erythrocephala* L. *Parasitology* 22: 47–115. [74]

Grandi, G. 1961 The hymenopterous insects of the superfamily Chalcidoidea developing within the receptacles of figs. *Bolletino dell'Istituto di Entomologia della Università degli Studi di Bologna* 26: 1–3. [314]

Grant, K.A. 1966 A hypothesis concerning the prevalence of red coloration in California hummingbird flowers. *American Naturalist* 100: 85–97. [229]

Grant, K.A. & Grant, V. 1968 *Hummingbirds and Their Flowers.* Columbia University Press, New York. [227, 229]

Grant, V. 1950a The pollination of *Calycanthus occidentalis*. *American Journal of Botany* 37: 294–297. [54, 56]

Grant, V. 1950b The protection of the ovules in flowering plants. *Evolution* 4: 179–201. [56, 172, 239]

Grant, V. 1950c The flower constancy of bees. *Botanical Review* 16: 379–398. [130, 134, 141]

Grant, V. 1994 Historical development of ornithophily in the western North American flora. *Proceedings of the National Academy of Sciences USA* 91: 10407–10411. [388]

Grant, V. 1995 Sexual selection in plants: pros and cons. *Proceedings of the National Academy of Sciences, USA* 92: 1247–1250. [420]

Grant, V. & Grant, K.A. 1965 *Flower Pollination in the Phlox Family.* Columbia University Press, New York. [397, 398]

Greathead, D.J. 1983 The multi-million dollar weevil that pollinates oil palms. *Antenna* 7: 105–107. [361, 398]

Gregory, P.H. 1973 *The Microbiology of the Atmosphere, ed. 2.* Leonard Hill, London. [270, 271, 272]

Grensted, L.W. 1946 An assemblage of Diptera

on Cow-Parsnip. *Entomologist's Monthly Magazine* 82: 180. [58, 73]

Grensted, L.W. 1947 Diptera in the spathes of *Arum maculatum*. *Entomologist's Monthly Magazine* 83: 1–3. [72, 304]

Grew, Nehemiah 1671 The Anatomy of Vegetables Begun. *See 'Account' in Philosophical Transactions of the Royal Society* 78: 3041. [13]

Grew, Nehemiah 1682 *The Anatomy of Plants*. London. [13]

Gribel, R. 1988 Visits of *Caluromys lanatus* (Didelphidae) to flowers of *Pseudobombax tomentosum* (Bombacaceae): a probable case of pollination by marsupials in Central Brazil. *Biotropica* 20: 344–347. [258]

Gribel, R. [1995] *Reproductive biology of two Bombacaceous trees in the Brazilian Central Amazon* (Ph. D. thesis, University of St Andrews). Unpublished. [245]

Griffin, R.P., Macfarlane, R.P. & Ende, H.J. van den 1991 Rearing and domestication of long tongued bumble bees in New Zealand. *Sixth International Symposium on Pollination (Acta Horticulturae no. 288)* (ed. C. van Heemert & A. de Ruijter), pp. 149–153. International Soc. Hortic. Sci., Wageningen. [357]

Griffiths, D.J. 1950 The liability of seed-crops of perennial rye grass (*Lolium perenne*) to contamination by wind-borne pollen. *Journal of Agricultural Science [Cambridge]* 40: 19–38. [271, 422]

Grime, J.P., Hodgson, J.G. & Hunt, R.P. 1988 *Comparative Plant Ecology*. Unwin Hyman, London. [329, 348]

Grime, J.P. & Mowforth, M.A. 1982 Variation in genome size - an ecological interpretation. *Nature* 299: 151–153. [397]

Grubb, P.J. 1977 The maintenance of species-richness in plant communities: the importance of the regeneration niche. *Biological Reviews* 52: 107–145. [415]

Gryj, E., Martinez del Rio, C. & Baker, I. 1990 Avian pollination and nectar use in *Combretum fruticosum* (Loefl.). *Biotropica* 22: 266–271. [242]

Guerrant, E.O. 1989 Early maturity, small flowers and autogamy: a developmental connection? *The Evolutionary Biology of Plants* (ed. J.H. Bock & Y.B. Linhart), pp. 61–84. Westview Press, Boulder, Co. [332]

Guo, Y.H. & Cook, C.D.K. 1989 Pollination efficiency of *Potamogeton pectinatus* L. *Aquatic Botany* 34: 381–384. [287, 288]

Guo, Y.H. & Cook, C.D.K. 1990 The floral biology of *Groenlandia densa* (L.) Fourreau (Potamogetonaceae). *Aquatic Botany* 38: 283–288. [288]

Guo, Y.H., Sperry, R., Cook, C.D.K. & Cox, P.A. 1990 The pollination ecology of

Zannichellia palustris L. (Zannichelliaceae). *Aquatic Botany* 38: 341–356. [291]

Gutowski, J.M. 1990 Pollination of the orchid *Dactylorhiza fuchsii* by longhorn beetles in primeval forests of northeastern Poland. *Biological Conservation* 51: 287–297. [199]

Haber, W.A. & Frankie, G.W. 1989 A tropical hawkmoth community: Costa Rican dry forest Sphingidae. *Biotropica* 21: 155–172. [394, 408]

Hafsten, U. 1960 The Quaternary history of vegetation in the South Atlantic islands. *Proceedings of the Royal Society B* 152: 516–529. [273]

Hagerup, O. 1932 On pollination in the extremely hot air at Timbuctu. *Dansk Botanisk Arkiv* 8(1): 1–20. [386]

Hagerup, O. 1950a Thrips pollination in *Calluna*. *Biologiske Meddelelser* 18(4): 1–16. [50, 51]

Hagerup, O. 1950b Rain pollination. *Biologiske Meddelelser* 18(4): 3–18. [334]

Hagerup, O. 1951 Pollination in the Faroes - in spite of rain and poverty of insects. *Biologiske Meddelelser* 18(15): 1–48. [58, 169, 199, 386]

Hagerup, O. 1952 Bud autogamy in some northern orchids. *Phytomorphology* 2: 51–60. [192]

Hagerup, O. & Hagerup, E. 1953 Thrips pollination of *Erica tetralix*. *New Phytologist* 52: 1–7. [51, 147]

Haldane, J.B.S. 1932 *The Causes of Evolution*. London. [21]

Haldane, J.B.S. 1938 Heterostylism in natural populations of the Primrose, *Primula acaulis*. *Biometrika* 30: 196–198. [224]

Hammond, A. 1874 The mouth of the Crane Fly. *Science Gossip* 1874: 155–160. [59]

Hamrick, J.L. & Godt, M.J. 1990 Allozyme diversity in plant species. *Plant Population Genetics, Breeding and Genetic Resources* (ed. A.H.D. Brown, M.T. Clegg, A.L. Kahler & B.S. Weir), pp. 43–63. Sinauer, Sunderland, Mass. [424]

Hamrick, J.L., Godt, M.J.W. & Sherman-Broyles, S.L. 1995 Gene flow among plant populations: evidence from genetic markers. *Experimental and Molecular Approaches to Plant Biosystematics* (ed. P.C. Hoch & A.G. Stephenson), pp. 215–232. Missouri Botanical Garden, St Louis. [273]

Harberd, D.J. 1961 Observations on population structure and longevity of *Festuca rubra* L. *New Phytologist* 60: 184–206. [347]

Harberd, D.J. 1962 Some observations on natural clones of *Festuca ovina*. *New Phytologist* 61: 85–100. [347]

Harborne, J.B. 1963 Distribution of anthocyanins in higher plants. *Chemical Plant Taxonomy* (ed. T. Swain), pp. 359–388. Academic Press, London. [34, 36]

Harborne, J.B. 1993 *Introduction to Ecological Biochemistry, ed. 4*. Academic Press, London. [40,

43, 221]

Harder, L.D. & Barclay, R.M.R. 1994 The functional significance of poricidal anthers and buzz pollination: controlled pollen removal from *Dodecatheon. Functional Ecology* 8: 509–517. [40, 179, 180]

Harper, J.L. 1957 Biological Flora of the British Isles. *Ranunculus acris, R. repens* and *R. bulbosus. Journal of Ecology* 45: 289–342. [45, 99, 102]

Harper, J.L. 1977 *Population Biology of Plants.* Academic Press, London. [22, 415]

Harper, J.L. & Wood, W.A. 1957 Biological Flora of the British Isles. *Senecio jacobaea* L. *Journal of Ecology* 45: 617–637. [71, 73, 113, 176]

Harrington, J.B. 1979 Principles of deposition of microbiological particles. *Aerobiology: the Ecological Systems Approach.* US IBP Synthesis Series 10 (ed. R.E. Edmonds), pp. 111–137. Dowden, Hutchinson & Ross, Stroudsburg, Pa. [272]

Harris, B.J. & Baker, H.G. 1959 Pollination of flowers by bats in Ghana. *Nigerian Field* 24: 151–159. [246, 247]

Haslerud, H.-D. 1974 Pollination of some Ericaceae in Norway. *Norwegian Journal of Botany* 21: 211–216. [147]

Hatton, R.H.S. 1965 Pollination of mistletoe (*Viscum album* L.). *Proceedings of the Linnean Society of London* 176: 67–76. [285]

Hawkins, R.P. 1965 Factors affecting the yield of seed produced by different varieties of Red Clover. *Journal of Agricultural Science [Cambridge]* 65: 245–253. [356]

Hawthorn, L.R., Bohart, G.E. & Toole, E.H. 1956 Carrot seed yield and germination as affected by different levels of insect pollination. *Proceedings of the American Society for Horticultural Science* 67: 384–389. [173, 358]

Hayes, H.K., Immer, F.R. & Smith, D.C. 1955 *Methods of Plant Breeding, ed. 2.* McGraw-Hill, New York. [351, 353]

Heinrich, B. 1972 Energetics of temperature regulation and foraging in a bumblebee, *Bombus terricola* Kirby. *Journal of Comparative Physiology* 77: 49–64. [139]

Heinrich, B. 1975 Energetics of pollination. *Annual Review of Ecology and Systematics* 6: 139–170. [22, 41, 93, 139, 232, 385]

Heinrich, B. 1976a Resource partitioning among some eusocial insects: bumblebees. *Ecology* 57: 874–889. [388]

Heinrich, B. 1976b The foraging specializations of individual bumblebees. *Ecological Monographs* 46: 105–128. [136, 388]

Heinrich, B. 1976c Bumblebee foraging and the economics of sociality. *American Scientist* 64: 384–395. [136]

Heinrich, B. 1979 *Bumblebee Economics.* Harvard University Press, Cambridge, Mass. [22, 140,

405, 406, 416]

Heinrich, B. 1983 Insect foraging energetics. *Handbook of Experimental Pollination Biology* (ed. C.E. Jones & R.J. Little), pp. 187–214. Van Nostrand Reinhold, New York. [93, 136, 139, 140, 416]

Heinrich, B. & Raven, P.H. 1972 Energetics and pollination ecology. *Science* 176: 597–602. [22, 139]

Heithaus, E.R. 1974 The role of plant-pollinator interactions in determining community structure. *Annals of the Missouri Botanical Garden* 61: 675–691. [404]

Heithaus, E.R., Fleming, T.H. & Opler, P.A. 1975 Foraging patterns and resource utilization in seven species of bats in a seasonal tropical forest. *Ecology* 56: 841–854. [253, 410, 417]

Heithaus, E.R., Opler, P.A. & Baker, H.G. 1974 Bat activity and pollination of *Bauhinia pauletia*: plant-pollinator coevolution. *Ecology* 55: 412–419. [245, 253, 254]

Helversen, O. von 1972 Zur spektralen Unterschiedsempfindlichkeit der Honigbiene. *Journal of Comparative Physiology* 80: 439–472. [131]

Henderson, A. 1986 A review of pollination studies in the Palmae. *Botanical Review* 52: 221–259. [311]

Hepburn, H.R. 1971 Proboscis extension and recoil in Lepidoptera. *Journal of Insect Physiology* 17: 637–656. [82]

Herrera, C.M. 1987 Components of pollinator "quality": a comparative analysis of a diverse insect assemblage. *Oikos* 50: 79–90. [416, 417]

Herrera, J. 1988 Pollination relationships in southern Spanish Mediterranean shrublands. *Journal of Ecology* 76: 274–289. [389]

Hertz, M. 1931 Die Organisation des optischen Feldes bei der Biene. I. *Zeitschrift für vergleichende Physiologie* 8: 693–748. [129]

Hertz, M. 1935 Die Untersuchungen über den Formensinn der Honigbiene. *Naturwissenschaften* 36: 618–624. [128]

Heslop-Harrison, J. 1975a Incompatibility and the pollen-stigma interaction. *Annual Review of Plant Physiology* 26: 403–425. [30]

Heslop-Harrison, J. 1975b The adaptive significance of the exine. *The Evolutionary Significance of the Exine* (ed. I.K. Ferguson & J. Muller), pp. 27–37. Linnean Society, London. [30]

Hesse, M. 1979a Entwicklungsgeschichte und Ultrastruktur von Pollenkitt und Exine bei nahe verwandten entomo- und anemophilen Angiospermen: Salicaceae, Tiliaceae und Ericaceae. *Flora* 168: 540–557. [278, 285]

Hesse, M. 1979b Entwicklungsgeschichte und Ultrastruktur von Pollenkitt und Exine bei nahe verwandten entomo- und anemophilen Angiospermen: Polygonaceae. *Flora* 168: 558–577. [283]

Hesse, M. 1979c Entstehung und Auswirkungen der unterschiedlichen Pollenklebrigkeit von *Sanguisorba officinalis* und *S. minor.* *Pollen et Spores* 21: 399–413. [284]

Hesselman, H. 1919 Iakttagelser över skogsträdspollens spridningsförmåga. *Meddelanden från Statens Skogsförsöksanstalt* 16: 27–53. [272]

Heywood, J.S. 1993 Biparental inbreeding depression in the self-incompatible annual plant *Gaillardia pulchella* (Asteraceae). *American Journal of Botany* 80: 545–550. [422]

Heywood, V.H. (ed.) 1978 *Flowering Plants of the World.* Oxford University Press, Oxford. [247]

Hickman, J.C. 1974 Pollination by ants: a low-energy system. *Science* 184: 1290–1292. [107]

Hildebrand, F. 1867 *Die Geschlechtsverteilung bei den Pflanzen.* Leipzig. [20]

Hill, D.S. 1967 Figs (*Ficus* spp.) and fig-wasps (Chalcidoidea). *Journal of Natural History* 1: 413–434. [316]

Hills, H.G., Williams, N.H. & Dodson, C.H. 1972 Floral fragrances and isolating mechanisms in the genus *Catasetum* (Orchidaceae). *Biotropica* 4: 61–76. [219]

Hobby, B.M. 1933 Diptera and Coleoptera visiting orchids. *Journal of the Entomological Society of Southern England* 1: 105–106. [54]

Hobby, B.M. & Smith, K.G.V. 1961 The bionomics of *Empis tesselata* F. (Dipt. Empididae). *Entomologist's Monthly Magazine* 97: 2–10. [61]

Hocking, B. 1968 Insect-flower associations in the high Arctic with special reference to nectar. *Oikos* 19: 359–388. [388]

Hodges, D. 1952 *Pollen Loads of the Honeybee.* Bee Research Association, London. [113, 124]

Hoekstra, F.A. & Bruinsma, J. 1975 Reduced independence of the male gametophyte in angiosperm evolution. *Annals of Botany* 42: 759–762. [350]

Holloway, B.A. 1976 Pollen-feeding in hover-flies (Diptera: Syrphidae). *New Zealand Journal of Zoology* 3: 339–350. [67, 68, 70]

Holm, E. 1988 *On Pollination and Pollinators in Western Australia.* Eigil Holm, Gedved. [235, 236, 242, 244, 257, 261, 262, 263]

Holsinger, K.E. 1992 Ecological models of plant mating systems and the evolutionary stability of mixed mating systems. *Ecology and Evolution of Plant Reproduction* (ed. R. Wyatt), pp. 169–191. Chapman & Hall, London. [422]

Hopkins, H.C. 1983 The taxonomy, reproductive biology and economic potential of *Parkia* (Leguminosae: Mimosoideae) in Africa and Madagascar. *Botanical Journal of the Linnean Society* 87: 135–167. [246, 247]

Hopkins, H.C. 1984 Floral biology and pollination ecology of the neotropical species of *Parkia. Journal of Ecology* 72: 1–23. [246]

Hopkins, H.C.F. & Hopkins, M.G.J. 1993 Rediscovery of *Mucuna macropoda* (Leguminosae: Papilionoideae), and its pollination by bats in Papua New Guinea. *Kew Bulletin* 48: 297–305. [247]

Hopper, S.D. 1980a Pollen loads on Honeyeaters in a *Grevillea rogersoniana* thicket south of Shark Bay. *Western Australian Naturalist* 14: 186–189. [236]

Hopper, S.D. 1980b Bird and mammal pollen vectors in *Banksia* communities at Cheyne Beach, Western Australia. *Australian Journal of Botany* 28: 61–75. [257, 261, 262]

Howard, R.A. 1970 The ecology of an elfin forest in Puerto Rico. 10. Notes on two species of *Marcgravia. Journal of the Arnold Arboretum* 51: 41–55. [241]

Howell, D.J. 1974 Bats and pollen: physiological aspects of the syndrome of chiropterophily. *Comparative Biochemistry and Physiology* 48A: 263–276. [245]

Howell, D.J. 1979 Flock foraging in nectar-feeding bats: advantages to the bats and to the host plants. *American Naturalist* 114: 23–49. [245, 254]

Hubbard, C.E. 1954 *Grasses.* Penguin, Harmondsworth. [280]

Hulkkonen, O. 1928 Zur Biologie der südfinnischen Hummeln. *Annales Universitatis Aboensis, Ser. A* 3: 1–81. [140]

Hussein, M.Y., Lajis, N.H. & Ali, J.H. 1991 Biological and chemical factors associated with the successful introduction of *Elaeidobius kamerunicus* Faust, the oil palm pollinator in Malaysia. *Sixth International Symposium on Pollination (Acta Horticulturae no. 288)* (ed. C. van Heemert & A. de Ruijter), pp. 81–87. International Soc. Hortic. Sci., Wageningen. [361]

Hutchings, M.J. 1987a The population biology of the early spider orchid, *Ophrys sphegodes* Mill. I. A demographic study from 1975 to 1984. *Journal of Ecology* 75: 711–727. [209]

Hutchings, M.J. 1987b The population biology of the early spider orchid, *Ophrys sphegodes* Mill. II. Temporal patterns in behaviour. *Journal of Ecology* 75: 729–742. [209]

Huxley, J. 1942 *Evolution: the Modern Synthesis.* Allen & Unwin, London. [21]

Hyde, H.A. 1950 Studies in atmospheric pollen. IV. Pollen deposition in Great Britain, 1943. *New Phytologist* 49: 398–420. [266]

Hyde, H.A. & Williams, D.A. 1961 Atmospheric pollen and spores as causes of allergic disease: hay-fever, asthma and the aerospora. *Advancement of Science* 1961: 526–533.

[267]
Iglesias, M.C. & Bell, G. 1989 The small-scale spatial distribution of male and female plants. *Oecologia* 80: 229–235. [342]
Ilse, D. 1928 Über den Farbensinn der Tagfalter. *Zeitschrift für vergleichende Physiologie* 8: 658–692. [93, 94]
Ilse, D. 1932 Zur 'Formwahrnehmung' der Tagfalter. I. Spontane Bevorzugung von Formmerkmalen durch Vanessen. *Zeitschrift für vergleichende Physiologie* 17: 537–556. [94]
Ilse, D. 1941 The colour vision of insects. *Proceedings of the Royal Philosophical Society of Glasgow* 65: 98–112. [94]
Ilse, D. 1949 Colour discrimination in the Drone Fly, *Eristalis tenax. Nature* 163: 255–256. [70]
Imms, A.D. 1947 *Insect Natural History.* Collins, London. [49]
Inoue, T. & Kato, M. 1992 Inter- and intraspecific morphological variation in bumblebee species and competition in flower utilization. *Effects of Resource Distribution on Animal-Plant Interactions* (ed. M.D. Hunter, T. Ohgushi & P.W. Price), pp. 393–427. Academic Press, London. [406]
Inouye, D.W. 1980 The terminology of floral larceny. *Ecology* 61: 1251–1253. [171]
Inouye, D.W. 1983 The ecology of nectar robbing. *The Biology of Nectaries* (ed. B. Bentley & T. Elias), pp. 153–173. Columbia University Press, New York. [171]
Inouye, D.W. & Pyke, G.H. 1988 Pollination biology in the Snowy Mountains of Australia: comparisons with montane Colorado, USA. *Australian Journal of Ecology* 13: 191–210. [390]
Inouye, D.W., Gill, D.E., Dudash, M.R. & Fenster, C.B. 1994 A model and lexicon for pollen fate. *American Journal of Botany* 81: 1517–1530. [184]
Ish-Am, G. & Eisikowitch, D. 1991 Possible routes of Avocado tree pollination by honeybees. *Sixth International Symposium on Pollination (Acta Horticulturae no. 288)* (ed. C. van Heemert & A. de Ruijter), pp. 225–233. International Soc. Hortic. Sci, Wageningen. [359]
Ish-Am, G. & Eisikowitch, D. 1993 The behaviour of honey bees (*Apis mellifera*) visiting avocado (*Persea americana*) flowers and their contribution to its pollination. *Journal of Apicultural Research* 32: 175–186. [360]
Jaeger, P. 1954a Note sur l'anatomie florale, l'anthocinétique et les modes du pollinisation du Fromager (*Ceiba pentandra* Gaertn.). *Bulletin de L'Institut Français de l'Afrique Noire, Sér. A* 16: 370–378. [247]
Jaeger, P. 1954b Les aspects actuels du problème de la chéiroptérogamie. *Bulletin de L'Institut Français de l'Afrique Noire, Sr. A* 16: 786–821. [244, 246, 247]

Jain, S.K. 1976 The evolution of inbreeding in plants. *Annual Review of Ecology and Systematics* 7: 469–495. [334]
James, R.L. 1948 Some hummingbird flowers east of the Mississippi. *Castanea* 13: 97–109. [229]
Janson, C.H., Terborgh, J. & Emmons, L.H. 1981 Non-flying mammals as pollinating agents in the Amazonian forest. *Biotropica: supplement* 13: 1–6. [256, 258]
Janzen, D.H. 1967 Synchronization of sexual reproduction of trees within the dry season in Central America. *Evolution* 21: 620–637. [394]
Janzen, D.H. 1971 Euglossine bees as long-distance pollinators of tropical plants. *Science* 171: 203–205. [134, 416]
Janzen, D.H. 1974 Tropical blackwater rivers, animals, and mast fruiting by the Dipterocarpaceae. *Biotropica* 6: 69–103. [392]
Janzen, D.H. 1979 How to be a fig. *Annual Review of Ecology and Systematics* 10: 13–51. [316, 320]
Jay, S.C. 1986 Spatial management of honey bees on crops. *Annual Review of Entomology* 31: 49–65. [352]
Johnson, S.D. 1992 Plant-animal relationships. *The Ecology of Fynbos* (ed. R.M. Cowling), pp. 175–205. Oxford University Press, Cape Town. [389, 390]
Jones, C.E. & Little, R.J. (eds.) 1983 *Handbook of Experimental Pollination Biology.* Van Nostrand Reinhold, New York. [22]
Jones, D.L. & Gray, B. 1974 The pollination of *Calochilus holtzei* F. Muell. *American Orchid Society Bulletin* 43: 604–606. [103]
Jones, E.C. & Rich, P.V. 1972 Ornithophily and extrafloral color patterns in *Columnea florida* Morton (Gesneriaceae). *Bulletin of the Southern California Academy of Sciences* 71: 113–116 & cover. [242]
Jones, E.W. 1945 Biological Flora of the British Isles. *Acer* L. *Journal of Ecology* 32: 215–252. [99]
Karoly, K. 1994 Inbreeding effects on mating system traits for two species of *Lupinus* (Leguminosae). *American Journal of Botany* 81: 1538–1544. [329]
Karron, J.D. 1989 Breeding systems and levels of inbreeding depression in geographically restricted and widespread species of *Astragalus* (Fabaceae). *American Journal of Botany* 76: 331–340. [335]
Kato, M. & Inoue, T. 1994 Origin of insect pollination. *Nature* 368: 195. [372]
Kauffeld, N.M. & Sorensen, E.L. 1971 Interrelations of honeybee preference of alfalfa clones and flower color, aroma, nectar volume and sugar concentration. *Research Bulletin, Kansas Agricultural Experiment Station* 163: 1–14. [399]
Kaur, A., Ha, C.O., Jong, K., Sands, V.E., Chan, H.T., Soepadmo, E. & Ashton, P.S.

1978 Apomixis may be widespread among trees of the climax rain forest. *Nature* 271: 440–442. [348]

Kausik, S.B. 1939 Pollination and its influences on the behavior of the pistillate flower in *Vallisneria spiralis. American Journal of Botany* 26: 207–211. [286]

Kay, Q.O.N. 1982 Intraspecific discrimination by pollinators and its role in evolution. *Pollination and Evolution* (ed. J.A. Armstrong, J.M. Powell & A.J. Richards), pp. 9–28. Royal Botanic Garden, Sydney. [418]

Kay, Q.O.N. 1985 Nectar from willow catkins as a food source for Blue Tits. *Bird Study* 32: 40–44. [226]

Kay, Q.O.N. 1987 Ultraviolet patterning and ultraviolet-absorbing pigments in flowers of the Leguminosae. *Advances in Legume Systematics 3* (ed. C.H. Stirton), pp. 317–354. Royal Botanic Gardens, Kew, London. [43]

Kay, Q.O.N., Daoud, H.S. & Stirton, C.H. 1981 Pigment distribution, light reflection and cell structure in petals. *Botanical Journal of the Linnean Society* 83: 57–84. [37]

Kay, Q.O.N., Lack, A.J., Bamber, F.C. & Davies, C.R. 1984 Differences between sexes in floral morphology, nectar production and insect visits in a dioecious species, *Silene dioica. New Phytologist* 98: 515–529. [341, 418]

Kay, Q.O.N. & Stevens, D.P. 1986 The frequency, distribution and reproductive biology of dioecious species in the native flora of Britain and Ireland. *Botanical Journal of the Linnean Society* 92: 39–64. [340, 341, 342, 348]

Kearns, C.A. & Inouye, D.W. 1993 *Techniques for Pollination Biologists*. Colorado University Press, Niewot, Col. [22]

Keeley, J.E. 1992 A Californian's view of fynbos. *The Ecology of Fynbos* (ed. R.M. Cowling), pp. 372–388. Oxford University Press, Cape Town. [389, 390]

Keighery, G.J. 1982 Bird-pollinated plants in Western Australia. *Pollination and Evolution* (ed. J.A. Armstrong, J.M. Powell & A.J. Richards), pp. 77–89. Royal Botanic Gardens, Sydney, Sydney. [242]

Kerner von Marilaun, A. 1878 *Flowers and their Unbidden Guests*. Transl. W. Ogle. Kegan Paul, London. [20, 107]

Kerner von Marilaun, A. 1902 *The Natural History of Plants*. Transl. F.W. Oliver. Blackie, London. [20]

Kevan, P. 1972 Insect pollination of high arctic flowers. *Journal of Ecology* 60: 831–867. [388]

Kevan, P.G. 1973 Parasitoid wasps as flower visitors in the Canadian High Arctic. *Anzeiger für Schädlingskunde, Pflanzen- und Umweltschutz* 46: 3–7. [102]

Kevan, P.G. 1978 Floral coloration, its

colorimetric analysis and significance in anthecology. *Pollination of Flowers by Insects* (ed. A.J. Richards), pp. 51–78. Academic Press, London. [133]

Kevan, P.G. 1979 Vegetation and floral colors revealed by ultraviolet light: interpretational difficulties for functional significance. *American Journal of Botany* 66: 749–751. [133]

Kevan, P.G. 1991 Pollination: keystone process in sustainable global productivity. *Sixth International Symposium on Pollination (Acta Horticulturae no. 288)* (ed. C. van Heemert & A. de Ruijter), pp. 103–110. International Soc. Hortic. Sci., Wageningen. [364]

Kevan, P.G. & Lack, A.J. 1985 Pollination in a cryptically dioecious plant *Decaspermum parviflorum. Biological Journal of the Linnean Society* 25: 319–330. [346]

Kheyr-Pour, A. 1981 Wide nucleo-cytoplasmic polymorphism for male-sterility in *Origanum vulgare* L. *Journal of Heredity* 72: 45–51. [345]

Kikuchi, T. 1962 Studies on the coaction among insects visiting flowers. II. Dominance relationships in the so-called drone fly group. *Science Reports of the Tohoku University Series IV (Biology)* 28: 47–51. [409]

Kimmins, D.E. 1939 Empididae (Dipt.) on the flowers of *Orchis elodes* Godf. *Journal of the Society for British Entomology* 2: 37–38. [58]

Kirby, W. & Spence, W. 1815-1826 *Introduction to Entomology, eds. 1 - 7*. London. [19]

Kirk, W.D.J. 1984a Ecological studies on *Thrips imaginis* Bagnall (Thysanoptera) in flowers of *Echium plantagineum* L. in Australia. *Australian Journal of Ecology* 9: 9–18. [51]

Kirk, W.D.J. 1984b Pollen feeding in thrips (Insecta: Thysanoptera). *Journal of Zoology, London* 204: 107–117. [51]

Kirk, W.D.J. 1985 Effect of some floral scents on host finding by thrips (Insecta: Thysanoptera). *Journal of Chemical Ecology* 11: 35–43. [51]

Kloet, G.S. & Hincks, W.D. 1964 - 1978 *A Check List of British Insects. (ed. 2). 1. Minor orders, Hemiptera (1964); 2. Lepidoptera (1972); 3. Coleoptera and Strepsiptera (1977); 4. Hymenoptera (1978); 5. Diptera and Siphonaptera (1975).* Royal Entomological Society, London. [10]

Knight, G.H. 1961 Some observations on pollination. *Biology and Human Affairs* 27: 35–42. [148]

Knight, G.H. 1968 Observations on the behaviour of *Bombylius major* L. and *B. discolor* Mik. (Dipt., Bombyliidae) in the Midlands. *Entomologist's Monthly Magazine* 103: 177–182. [63]

Knight, T. 1799 Experiments on the fecundation of vegetables. *Philosophical Transactions of the Royal Society* 89: 195–204. [19]

Knoll, F. 1921 *Bombylius fuliginosus* und die Farbe

der Blumen. (Insekten und Blumen I). *Abhandlungen der Kaiserlich-Königlichen Zoologisch-botanischen Gesellschaft in Wien* 12: 17–119. [62, 63, 64]

Knoll, F. 1922 Lichtsinn und Blumenbesuch des Falters von *Macroglossum stellatarum*. (Insekten und Blumen III). *Abhandlungen der Kaiserlich-Königlichen Zoologisch-botanischen Gesellschaft in Wien* 12: 121–378. [81, 82, 89, 90, 93]

Knoll, F. 1925 Lichtsinn und Bltenbesuch des Falters von *Deilephila livornica*. *Zeitschrift für vergleichende Physiologie* 2: 329–380. [90, 91, 97]

Knoll, F. 1926 Die *Arum*-Blütenstände und ihre Besucher. (Insekten und Blumen IV). *Abhandlungen der zoologische-botanische Gesellschaft in Wien* 12: 379–481. [21, 56, 77, 302]

Knoll, F. 1927 Über Abendschwärmer und Schwärmerblumen. *Berichte der Deutschen Botanischen Gesellschaft* 45: 510–518. [90, 91, 97]

Knoll, F. 1935 Über den Schwärmflug der Maskenbienen (*Prosopis*). *Biologia Generalis* 11: 115–154. [129]

Knoll, F. 1956 *Die Biologie der Blüte*. Springer, Berlin. [22, 298]

Knudsen, J.T., Tollsten, L. & Bergström, L.G. 1993 Floral scents - a checklist of volatile compounds isolated by head-space techniques. *Phytochemistry* 33: 253–280. [38]

Knuth, P. 1898-1905 *Handbuch der Blütenbiologie* 1 - 3. W. Engelmann, Leipzig. [20]

Knuth, P. 1906-1909 *Handbook of Flower Pollination*. Transl. J.A. Davis 1 - 3. Clarendon Press, Oxford. [20, 50, 58, 60, 61, 63, 71, 72, 74, 81, 85, 87, 99, 102, 103, 109, 114, 169, 304, 384, 386]

Koach, J. & Galil, J. 1986 The breeding system of *Arisarum vulgare* Targ.-Tozz. *Israel Journal of Botany* 35: 79–90. [297]

Kochmer, J.L. & Handel, S.N. 1986 Constraints and competition in the evolution of flowering phenology. *Ecological Monographs* 56: 303–325. [387, 397, 404]

Kohn, J.R. 1988 Why be female? *Nature* 335: 431–433. [335, 346]

Kölreuter, J.G. 1761-1766 *Vorläufiger Nachricht von einigen das Geschlecht der Pflanzen betreffenden Versuchen und Beobachten*. Leipzig. [17]

Krassilov, V.A. 1977 The origin of angiosperms. *Botanical Review* 43: 143–176. [378]

Krassilov, V.A. 1991 The origin of angiosperms: new and old problems. *Trends in Ecology and Evolution* 6: 215–220. [378]

Krebs, S.L. & Hancock, J.F. 1991 Embryonic genetic load in the highbush blueberry, *Vaccinium corymbosum* (Ericaceae). *American Journal of Botany* 78: 1427–1437. [328]

Kress, W.J. 1983 Self-incompatibility in Central American *Heliconia*. *Evolution* 37: 735–744. [231]

Kubitzki, K. & Kurz, H. 1984 Synchronized dichogamy and dioecy in neotropical Lauraceae. *Plant Systematics and Evolution* 147: 253–266. [359]

Kugler, H. 1932a Blütenökologische Untersuchungen mit Hummeln. III. Das Verhalten der Tiere zu Duftstoffen, Duft und Farbe. *Planta* 16: 227–276. [127, 140]

Kugler, H. 1932b Blütenökologische Untersuchungen mit Hummeln. IV. Der Duft als chemischer Nahfaktor bei duftenden und `duftlosen' Blüten. *Planta* 16: 534–53. [127, 140]

Kugler, H. 1938 Sind *Veronica chamaedrys* L. und *Circaea lutetiana* L. Schwebe-fliegenblumen? *Botanisches Archiv [Berlin]* 39: 147–165. [68, 69]

Kugler, H. 1940 Die Bestäubung von Blumen durch Furchenbienen (*Halictus* Latr.). *Planta* 30: 789–799. [110, 127, 134]

Kugler, H. 1943 Hummeln als Blütenbesucher. *Ergebnisse der Biologie* 19: 143–323. [115, 129, 140, 141]

Kugler, H. 1950 Der Blütenbesuch der Schlammfliege (*Eristalomyia tenax*). *Zeitschrift für vergleichende Physiologie* 32: 328–347. [70, 71]

Kugler, H. 1951 Blütenökologische Untersuchungen mit Goldfliegen (Lucilien). *Berichte der Deutschen Botanischen Gesellschaft* 64: 327–341. [77]

Kugler, H. 1955a *Einführung in die Blütenökologie*. Fischer, Stuttgart. [22, 62, 67, 99, 105, 117, 129]

Kugler, H. 1955b Zur Problem der Dipterenblumen. *Österreichische Botanische Zeitschrift* 102: 529–541. [73, 78]

Kugler, H. 1956 Über die optische Wirkung von Fliegenblumen auf Fliegen. *Berichte der Deutschen Botanischen Gesellschaft* 69: 387–398. [74, 77, 78]

Kugler, H. 1963 UV-Musterrungen auf Blüten und ihr Zustandekommen. *Planta* 59: 296–329. [132]

Kugler, H. 1966 UV-Male auf Blüten. *Berichte der Deutschen Botanischen Gesellschaft* 79: 57–70. [132, 242]

Kugler, H. 1984 Die Bestäubung von Blüten durch den Schmalkäfer *Oedemera* (Coleoptera). *Berichte der Deutschen Botanischen Gesellschaft* 97: 383–390. [52, 54, 56]

Kullenberg, B. 1950 Investigations on the pollination of *Ophrys* species. *Oikos* 2: 1–19. [207]

Kullenberg, B. 1956a On the scents and colours of *Ophrys* flowers and their specific pollinators among the Aculeate Hymenoptera. *Svensk Botanisk Tidskrift* 50: 25–46. [103, 206]

Kullenberg, B. 1956b Field experiments with chemical sexual attractants on aculeate Hymenoptera males. I. *Zoologiska Bidrag Uppsala* 31: 253–354. [113]

Kullenberg, B. 1961 Studies in *Ophrys* pollination. *Zoologiska Bidrag Uppsala* 34: 1–340. [206, 207, 209]

Kwak, M. 1977 Pollination ecology of five hemiparasitic, large-flowered Rhinanthoideae with special reference to the pollination behaviour of nectar-thieving, short-tongued bumblebees. *Acta Botanica Neerlandica* 26: 97–107. [168]

Lack, A. 1976 Competition for pollinators and evolution in *Centaurea*. *New Phytologist* 77: 787–792. [402]

Lack, A.J. 1977 Genets feeding on nectar from *Maranthes polyandra* in Northern Ghana. *East African Wildlife Journal* 15: 233–234. [256]

Lack, A.J. 1978 The ecology of the flowers of the savanna tree *Maranthes polyandra* and their visitors, with particular reference to bats. *Journal of Ecology* 66: 287–295. [247, 251, 256]

Lack, A.J. 1982a Competition for pollinators in the ecology of *Centaurea scabiosa* L. and *Centaurea nigra* L. I. Variation in flowering time. *New Phytologist* 91: 297–308. [399]

Lack, A.J. 1982b Competition for pollinators in the ecology of *Centaurea scabiosa* L. and *Centaurea nigra* L. II. Observations on nectar production. *New Phytologist* 91: 309–320. [399]

Lack, A.J. 1982c Competition for pollinators in the ecology of *Centaurea scabiosa* L. and *Centaurea nigra* L. III. Insect visits and the number of successful pollinations. *New Phytologist* 91: 321–339. [403]

Lack, A.J. 1982d The ecology of flowers of chalk grassland and their insect pollinators. *Journal of Ecology* 70: 773–790. [388, 397, 400, 405]

Lack, A.J. & Diaz, A. 1991 The pollination of *Arum maculatum* L. - a historical review and new observations. *Watsonia* 18: 333–342. [304]

Lack, A.J. & Kay, Q.O.N. 1987 Genetic structure, gene flow and reproductive ecology in sand-dune populations of *Polygala vulgaris*. *Journal of Ecology* 75: 259–276. [418]

Lagerberg, T., Holmboe, J. & Nordhagen, R. 1957 *Våre Ville Planter* 6. Tanum, Oslo. [165]

Lawrence, W.J.C. 1939 *Practical Plant Breeding*, ed. 2. Allen & Unwin, London. [351, 353]

le Maitre, D.C. & Midgley, J.J. 1992 Plant reproductive ecology. *The Ecology of Fynbos* (ed. R.M. Cowling), pp. 135–174. Oxford University Press, Cape Town. [389]

Leclercq, J. 1960 Fleurs butinées par les bourdons (Hym. Apidae Bombinae). *Bulletin de l' Institut Agronomique de l'État et des Stations de Recherches de Gembloux* 28: 180–198. [140]

Lederer, G. 1951 Biologie der Nahrungsaufnahme der Imagines von *Apatura* und *Limenitis*, sowie Versuche zur Feststellung der Gustorezeption durch die Mittel- und Hintertarsen dieser Lepidoptera. *Zeitschrift für Tierpsychologie* 18: 41–61. [93]

Lee, T.D. 1988 Patterns of fruit and seed production. *Plant Reproductive Ecology* (ed. J. Lovett Doust & L. Lovett Doust), pp. 179–202. Oxford

University Press, Oxford. [414, 421, 422]

Leius, K. 1960 Attractiveness of different foods and flowers to the adults of some hymenopterous parasites. *Canadian Entomologist* 92: 369–376. [101]

Leppik, E. 1953 The ability of insects to distinguish number. *American Naturalist* 87: 229–236. [140]

Leppik, E. 1956 The form and the function of numeral pattern in flowers. *American Journal of Botany* 43: 445–455. [144]

Leppik, E. 1957 Evolutionary relationships between entomophilous plants and anthophilous insects. *Evolution* 11: 466–481. [144]

Leppik, E.E. 1966 Floral evolution and pollination in the Leguminosae. *Annales Botanici Fennici* 3: 299–308. [354]

Les, D.H. 1991 Genetic diversity in the monoecious hydrophile *Ceratophyllum* (Ceratophyllaceae). *American Journal of Botany* 78: 1070–1082. [290]

Levin, D.A. 1972 Competition for pollinator service: a stimulus for the evolution of autogamy. *Evolution* 26: 668–669. [399]

Levin, D.A. 1989 Inbreeding depression in partially self-fertilising *Phlox*. *Evolution* 43: 1417–1423. [335]

Levin, D.A. & Kerster, H.W. 1967 Natural selection for reproductive isolation in *Phlox*. *Evolution* 21: 679–687. [370]

Levin, D.A. & Kerster, H.W. 1974 Gene flow in seed plants. *Evolutionary Biology* 7: 139–220. [418, 423]

Levin, D.A., Kerster, H.W. & Niedzlek, M. 1971 Pollinator flight directionality and its effect on pollen flow. *Evolution* 25: 113–118. [418]

Lewis, T. 1973 *Thrips, their biology, ecology and economic importance*. Academic Press, London. [51]

Lex, T. 1954 Duftmale an Blüten. *Zeitschrift für vergleichende Physiologie* 36: 212–234. [126]

Lichtenstein, L.M. 1993 Allergy and the immune system. *Scientific American* 269(3): 116–124. [30]

Lidgard, S. & Crane, P.R. 1988 Quantitative analyses of the early angiosperm radiation. *Nature* 331: 344–346. [379]

Liebermann, A. 1925 Korrelation zwischen den antennalen Geruchsorganen und der Biologie der Musciden. *Zeitschrift für Morphologie und Ökologie der Tiere* 5: 1–97. [57, 77]

Lindauer, M. 1961 *Communication Among Social Bees*. Harvard University Press, Cambridge, Mass. [125, 133]

Lindner, E. 1928 *Aristolochia lindneri* Berger und ihre Bestäubung durch Fliegen. *Biologisches Zentralblatt* 48: 93–101. [298]

Linhart, Y.B. 1973 Ecological and behavioral determinants of pollen dispersal in hummingbird-pollinated *Heliconia*. *American*

Naturalist 107: 511–523. [230, 417]

Linhart, Y.B., Feinsinger, P. & 7 others 1987 Disturbance and predictability of flowering patterns in bird-pollinated cloud forest plants. *Ecology* 68: 1696–1710. [393, 394]

Linsley, E.G. & MacSwain, J.W. 1959 Ethology of some *Ranunculus* insects with emphasis on competition for pollen. *University of California Publications in Entomology* 16: 1–46. [136]

Liston, A., Rieseberg, L.H. & Elias, T.S. 1990 Functional androdioecy in the flowering plant *Datisca glomerata*. *Nature* 343: 641–642. [346]

Little, R.J. 1983 A review of floral food deception mimicries with comments on floral mutualism. *Handbook of Experimental Pollination Biology* (ed. C.E. Jones & R.J. Little), pp. 294–309. Van Nostrand Reinhold, New York. [295]

Lloyd, D.G. & Webb, C.J. 1986 The avoidance of interference between the presentation of pollen and stigmas in angiosperms. I. Dichogamy. *New Zealand Journal of Botany* 24: 135–162. [336, 344]

Loew, E. 1895 *Einführung in die Blütenbiologie auf historische Grundlage*. Berlin. [20]

Logan, J. 1739 Experiments concerning the impregnation of the seeds of plants. *Philosophical Transactions of the Royal Society of London* 39: 192. [15]

Loveless, M.D. & Hamrick, J.L. 1984 Ecological determinants of genetic structure in plant populations. *Annual Review of Ecology and Systematics* 15: 65–95. [424]

Lovett Doust, J. 1980 Floral sex ratio in andromonoecious Umbelliferae. *New Phytologist* 85: 265–273. [346]

Lovett Doust, J. & Lovett Doust, L. (eds.) 1988 *Plant Reproductive Ecology: Patterns and Strategies*. Oxford University Press, Oxford. [22]

Lubbock, Sir John (Lord Avebury) 1875 *On British Wild Flowers Considered in Relation to Insects*. Macmillan, London. [23]

Lucas, F.A. 1897 The tongues of birds. *Report of the United States National Museum* 1895: 1001–1020. [228, 233]

Luckwill, L.C., Way, D.W. & Duggan, J.B. 1962 The pollination of fruit crops. II and III. *Scientific Horticulture* 15: 82–122. [361]

Lumer, C. 1980 Rodent pollination of *Blakea* (Melastomataceae) in a Costa Rican cloud forest. *Brittonia* 32: 512–517. [256, 258]

Lumer, C. & Schoer, R.D. 1986 Pollination of *Blakea austin-smithii* and *B. penduliflora* (Melastomataceae) by small rodents in Costa Rica. *Biotropica* 18: 363–364. [256, 258

Lundqvist, A. 1975 Complex self-incompatibility systems in angiosperms. *Proceedings of the Royal Society of London, B* 188: 235–245. [324, 330]

Lundqvist, A. 1990 One-locus sporophytic S-gene system with traces of gametophytic

pollen control in *Cerastium arvense* ssp. *strictum* (Caryophyllaceae). *Hereditas* 113: 203–215. [325]

Mabberley, D.J. 1991 *Tropical Rain Forest Ecology*. Blackie, London. [393]

Macior, L.W. 1964 An experimental study of the pollination of *Dodecatheon meadia*. *American Journal of Botany* 51: 96–108. [179]

Macior, L.W. 1965 Insect adaptation and behaviour in *Asclepias* pollination. *Bulletin of the Torrey Botanical Club* 92: 114–126. [171, 186]

Macior, L.W. 1966 Foraging behavior of *Bombus* (Hymenoptera: Apidae) in relation to *Aquilegia* pollination. *American Journal of Botany* 53: 302–309. [169, 171]

Macior, L.W. 1967 Pollen-foraging behaviour of *Bombus* in relation to pollination of nototribic flowers. *American Journal of Botany* 54: 359–364. [169]

Macior, L.W. 1968a Pollination adaptation in *Pedicularis groenlandica*. *American Journal of Botany* 55: 927–932. [165, 166]

Macior, L.W. 1968b Pollination adaptation in *Pedicularis canadensis*. *American Journal of Botany* 55: 1031–1035. [165, 166]

Macior, L.W. 1970 Pollination ecology of *Dodecatheon amethystinum*. *Bulletin of the Torrey Botanical Club* 97: 150–153. [179]

Macior, L.W. 1973 The pollination ecology of *Pedicularis* on Mount Rainier. *American Journal of Botany* 60: 863–871. [400]

Macior, L.W. 1975 The pollination ecology of *Pedicularis* (Scrophulariaceae) in the Yukon Territory. *American Journal of Botany* 62: 1065–1072. [400]

Macior, L.W. 1982 Plant community and pollinator dynamics in the evolution of pollination mechanisms in *Pedicularis* (Scrophulariaceae). *Pollination and Evolution* (ed. J.A. Armstrong, J.M. Powell & A.J. Richards), pp. 29–45. Royal Botanic Gardens, Sydney, Sydney. [167]

Mackworth-Praed, H.W. 1973 Some observations on insects visiting the martagon lily, *Lilium martagon*. *Annual Report of the Surrey Naturalists' Trust (for 1973)*. [139]

Maheshwari, P. & Kanta, K. 1964 Control of fertilization. *Pollen Physiology and Fertilization* (ed. H.F. Linskens), pp. 187–193. North Holland Publishing Co., Amsterdam. [351]

Manasse, R.S. & Pinney, K. 1991 Limits to reproductive success in a partially self-incompatible herb: fecundity depression at serial life-cycle stages. *Evolution* 45: 712–720. [328]

Manning, A. 1956a The effect of honeyguides. *Behaviour* 9: 114–139. [128, 129]

Manning, A. 1956b Some aspects of the foraging behaviour of bumble-bees. *Behaviour* 9: 164–201. [127, 137, 141]

Manning, A. 1957 Some evolutionary aspects of flower-constancy of bees. *Proceedings of the Royal Physical Society of Edinburgh* 25: 67–71. [127]

Marsden-Jones, E.M. 1935 *Ranunculus ficaria* Linn.: life-history and pollination. *Journal of the Linnean Society (Botany)* 50: 39–55. [56]

Marshall, A.G. 1983 Bats, flowers and fruit: evolutionary relationships in the Old World. *Biological Journal of the Linnean Society* 20: 115–135. [252, 254, 255]

Marshall, D.F. & Abbott, R.J. 1982 Polymorphism for outcrossing frequency at the ray floret locus in *Senecio vulgaris* L. I. Evidence. *Heredity* 48: 227–235. [333]

Marshall, D.F. & Abbott, R.J. 1984 Polymorphism for outcrossing frequency at the ray floret locus in *Senecio vulgaris* L. II. Confirmation. *Heredity* 52: 331–336. [333]

Marshall, D.L. & Ellstrand, N.C. 1988 Effective mate choice in wild radish: evidence for selective seed abortion and its mechanism. *American Naturalist* 131: 739–756. [421, 423]

Marshall, D.L. & Folsom, M.W. 1991 Mate choice in plants: an anatomical to population perspective. *Annual Review of Ecology and Systematics* 22: 37–63. [421]

Mason, C.J. 1979 Principles of atmospheric transport. *Aerobiology: the Ecological Systems Approach.* US IBP Synthesis Series 10 (ed. R.E. Edmonds), pp. 85–95. Dowden, Hutchinson & Ross, Stroudsburg, Pa. [271]

Mauss, V. & Treiber, R. 1994 *Bestimmungsschlüssel für die Faltenwespen (Hymenoptera: Masarinae, Polistinae, Vespinae) der Bundesrepublik Deutschland.* Deutscher Jugendbund für Naturbeobachtung, Hamburg. [105]

Mayer, D.F. 1983 Apple pollination. *American Bee Journal* 123: 272–273. [361]

Maynard Smith, J. 1978 *The Evolution of Sex.* Cambridge University Press, Cambridge. [321, 341]

Mayr, E. 1982 *The Growth of Biological Thought.* Harvard University Press, Cambridge, Mass. [22]

McCall, C. & Primack, R.B. 1992 Influence of flower characteristics, weather, time of day, and season on insect visitation rates in three plant communities. *American Journal of Botany* 79: 434–442. [390]

McCann, C. 1943 'Light-windows' in certain flowers. *Journal of the Bombay Natural History Society* 44: 182–184. [305]

McDade, L.A. 1992 Pollinator relationships, biogeography, and phylogenetics. *BioScience* 42: 21–26. [383]

McGregor, S.E., Alcorn, S.M. & Olin, G. 1962 Pollination and pollinating agents of the saguaro. *Ecology* 43: 259–267. [244]

McKelvey, S.D. 1947 *Yuccas of the Southwestern United States* 2. Arnold Arboretum, Jamaica Plain, Mass. [316]

McLean, R.C. & Cook, W.R. Ivimey- 1956 *Textbook of Theoretical Botany* 2. Longmans, Green & Co., London. [314]

McNaughton, I.H. & Harper, J.L. 1960 The comparative biology of closely related species living in the same area. I. External breeding-barriers between *Papaver* species. *New Phytologist* 59: 15–26. [47]

Meagher, T.R. 1986 Analysis of paternity within a natural population of *Chamaelirium luteum.* I. Identification of most-likely male parents. *American Naturalist* 128: 199–215. [423]

Meagher, T.R. 1988 Sex determination in plants. *Plant Reproductive Ecology* (ed. J. Lovett Doust & L. Lovett Doust), pp. 125–138. Oxford University Press, Oxford. [340, 344]

Meagher, T.R. 1991 Analysis of paternity within a natural population of *Chamaelirium luteum.* II. Patterns of male reproductive success. *American Naturalist* 137: 738–752. [423]

Meagher, T.R. & Thompson, E. 1987 Analysis of parentage for naturally established seedlings of *Chamaelirium luteum* (Liliaceae). *Ecology* 68: 803–812. [423]

Meeuse, B.J.D. 1961 *The Story of Pollination.* Ronald Press, New York. [22]

Meeuse, B.J.D. 1966 The voodoo lily. *Scientific American* 215: 80–88. [304]

Meeuse, B. & Morris, S. 1984 *The Sex Life of Flowers.* Faber & Faber, London. [22, 305]

Menzel, R. 1990 Color vision in *flower visiting insects.* Forschungszentrum Jülich, Jülich. [49, 126, 129]

Menzel, R. & Shmida, A. 1993 The ecology of flower colours and the natural colour vision of insect pollinators: the Israeli flora as a study case. *Biological Reviews* 68: 81–120. [178]

Mesler, M.R., Ackerman, J.D. & Lu, K.L. 1980 The effectiveness of fungus gnats as pollinators. *American Journal of Botany* 67: 564–567. [195]

Michener, C.D. 1979 Biogeography of the bees. *Annals of the Missouri Botanical Garden* 66: 277–347. [382, 383]

Michener, C.D. & Grimaldi, D.A. 1988 The oldest fossil bee: apoid history, evolutionary stasis and antiquity of social behavior. *Proceedings of the National Academy of Sciences, USA* 85: 6424–6426. [382, 383]

Miller, P. 1724 *The Gardener's and Florist's Dictionary.* London. [15]

Miller, P. 1731 *The Gardener's Dictionary.* London. [15]

Miller, P. 1752 *The Gardener's Dictionary,* ed. 6. London. [15]

Mittler, T.E. (ed.) 1962 *Proceedings of the First International Symposium on Pollination* (Swedish Seed Growers' Assoc., Comm. No. 7). [354, 357, 358]

Mogie, M. 1992 *The Evolution of Asexual Reproduction in Plants.* Chapman & Hall, London. [348, 349]

Moldenke, A.R. 1976 California pollination ecology and vegetation types. *Phytologia* 34: 305–361. [389, 404, 406]

Molitor, A. 1937 Zur vergleichenden Psychobiologie der akuleaten Hymenopteren auf experimenteller Grundlage. *Biologia Generalis* 13: 294–333. [126]

Moller, W. 1930 Über die Schnabel- und Zungen-mechanik blüten-besuchender Vögel. I. *Biologia Generalis* 6: 651–726. [227, 228, 233]

Moller, W. 1931a Über die Schnabel- und Zungen-mechanik blüten-besuchender Vögel. II. *Biologia Generalis* 7: 99–154. [227, 232, 235]

Moller, W. 1931b Vorläufige Mitteilung über die Ergebnisse einer Forschungsreise nach Costa Rica zu Studien über die Biologie blütenbesuchender Vögel. *Biologia Generalis* 7: 287–312. [229, 232]

Montalvo, A.M. 1992 Relative success of self and outcross pollen comparing mixed- and single-donor pollinations in *Aquilegia caerulea.* *Evolution* 46: 1181–1198. [335]

Monteith, J.L. & Unsworth, M.H. 1990 *Principles of Environmental Physics, ed. 2.* Edward Arnold, London. [272, 285]

Moore, P.D., Webb, J.A. & Collinson, M.E. 1991 *Pollen analysis.* Blackwell, Oxford. [32]

Mordue, W., Goldsworthy, G.J., Brady, J. & Blaney, W.M. 1980 *Insect Physiology.* Blackwell, Oxford. [49]

Mori, S.A., Prance, G.T. & Bolten, A.B. 1978 Additional notes on the floral biology of neotropical Lecythidaceae. *Brittonia* 30: 113–130. [295]

Morland, S. 1703 Some new observations upon the parts and use of the flower in plants. *Philosophical Transactions of the Royal Society* 287: 1474. [15]

Morse, D.H. 1994 The role of self-pollen in the female reproductive success of common milkweed (*Asclepias syriaca*: Asclepiadaceae). *American Journal of Botany* 81: 322–330. [336]

Morton, A.G. 1981 *History of Botanical Science.* Academic Press, London. [15]

Mosquin, T. 1971 Competition for pollinators as a stimulus for evolution of flowering time. *Oikos* 22: 398–402. [396]

Motten, A.F. 1986 Pollination ecology of the spring wildflower community of a temperate deciduous forest. *Ecological Monographs* 56: 21–42. [387]

Muenchow, G. 1982 A loss-of-alleles model for the evolution of distyly. *Heredity* 49: 81–93. [330]

Muenchow, G. & Grebus, H. 1989 The evolution of dioecy from distyly: reevaluation of the hypothesis of the loss of long-tongued

pollinators. *American Naturalist* 133: 20–41. [342]

Mulcahy, D.L. 1979 The rise of the angiosperms: a genecological factor. *Science* 206: 20–23. [375]

Mulcahy, D.L. & Mulcahy, G.B. 1975 The influence of gametophytic selection on sporophytic quality in *Dianthus chinensis.* *Theoretical and Applied Genetics* 46: 277–280. [421]

Mulcahy, D.L. & Mulcahy, G.B. 1987 The effects of pollen competition. *American Scientist* 75: 44–50. [375]

Mulcahy, D.L., Mulcahy, G.B. & Searcy, K.B. 1992 Evolutionary genetics of pollen competition. *Ecology and Evolution of Plant Reproduction* (ed. R. Wyatt), pp. 25–36. Chapman & Hall, London. [420, 421]

Müller, H. 1873 *Die Befruchtung der Blumen durch Insekten und die gegenseitigen Anpassungen beider.* W. Engelmann, Leipzig. [20]

Müller, H. 1879 Weitere Beobachtungen über Befruchtung der Blumen durch Insekten. II. *Verhandlungen der naturhistorischen Vereines der preussischen Rheinlande* 36: 198–268. [20]

Müller, H. 1881 *Die Alpenblumen, ihre Befruchtung durch Insekten und ihre Anpassungen an dieselben.* Leipzig. [20, 85, 141]

Müller, H. 1883 *The Fertilisation of Flowers.* Transl. D'Arcy W. Thompson. London. [20, 50, 53, 54, 60, 62, 63, 65, 66, 106, 107, 117, 121, 304]

Müller, L. 1926 Zur biologischen Anatomie der Blüte von *Ceropegia woodii* Schlechter. *Biologia Generalis* 2: 799–814. [21, 301]

Murlis, J., Elkinton, J.S. & Card-, R.T. 1992 Odor plumes and how insects use them. *Annual Review of Entomology* 37: 505–532. [271]

Murray, M.G. 1985 Figs (*Ficus* spp.) and fig wasps (Chalcidoidea, Agaonidae): hypotheses for an ancient symbiosis. *Biological Journal of the Linnean Society* 26: 69–81. [320]

Nelson, E.C. 1978 Tropical drift fruits and seeds on coasts in the British Isles and western Europe, I. Irish beaches. *Watsonia* 12: 103–112. [255]

Newton, I. 1995 The contribution of some recent research on birds to ecological understanding. *Journal of Animal Ecology* 64: 675–696. [224]

Nicholls, W.H. 1955, 1958 *Orchids of Australia* 3 - 4. Georgian House, Melbourne. [306]

Niklas, K.J. 1985 The aerodynamics of wind pollination. *Botanical Review* 51: 328–386. [272, 273]

Niklas, K.J. 1987 Aerodynamics of wind pollination. *Scientific American* 257(1): 90–95. [273]

Nilsson, L.A. 1978a Pollination ecology and adaptation in *Platanthera chlorantha* (Orchidaceae). *Botaniska Notiser* 131: 35–51. [201]

Nilsson, L.A. 1978b Pollination ecology of

Epipactis palustris (Orchidaceae). *Botaniska Notiser* 131: 355–368. [105, 190]

Nilsson, L.A. 1979a Anthecological studies on the Lady's Slipper, *Cypripedium calceolus* (Orchidaceae). *Botaniska Notiser* 132: 329–347. [188]

Nilsson, L.A. 1979b The pollination ecology of *Herminium monorchis* (Orchidaceae). *Botaniska Notiser* 132: 537–549. [72, 102, 204]

Nilsson, L.A. 1980 The pollination ecology of *Dactylorhiza sambucina* (Orchidaceae). *Botaniska Notiser* 133: 367–385. [199]

Nilsson, L.A. 1981 The pollination ecology of *Listera ovata* (Orchidaceae). *Nordic Journal of Botany* 1: 461–480. [100, 193, 195, 208]

Nilsson, L.A. 1983a Anthecology of *Orchis mascula* (Orchidaceae). *Nordic Journal of Botany* 3: 157–179. [197, 198, 203]

Nilsson, L.A. 1983b Processes of isolation and introgressive interplay between *Platanthera bifolia* (L.) Rich. and *P. chlorantha* (Custer) Reichb. (Orchidaceae). *Botanical Journal of the Linnean Society* 87: 325–350. [200, 201]

Nilsson, L.A. 1983c Mimesis of bellflower (*Campanula*) by the red helleborine orchid (*Cephalanthera rubra*). *Nature* 305: 799–800. [192]

Nilsson, L.A. 1988 The evolution of flowers with deep corolla tubes. *Nature* 334: 147–149. [202, 382]

Nilsson, L.A. 1992a Animal pollinators adjust plant gender in relation to floral display: evidence from *Orchis morio* (Orchidaceae). *Evolutionary Trends in Plants* 6: 33–40. [415]

Nilsson, L.A. 1992b Orchid pollination biology. *Trends in Ecology and Evolution* 7: 255–259. [222, 414]

Nilsson, L.A., Jonsson, L., Rason, L. & Randrianjohany, E. 1985 Monophily and pollination mechanisms in *Angraecum arachnites* Schltr. (Orchidaceae) in a guild of long-tongued hawk-moths (Sphingidae) in Madagascar. *Biological Journal of the Linnean Society* 26: 1–19. [214, 215, 382, 393, 408]

Nilsson, L.A., Jonsson, L., Ralison, L. & Randrianjohany, E. 1987 Angraecoid orchids and hawkmoths in central Madagascar: specialized pollination systems and generalized foragers. *Biotropica* 19: 310–318. [214, 215, 393, 408]

Nilsson, L.A., Rabakonandrianina, E., Razananivo, R. & Randriamanindry, J.-J. 1992 Long pollinia on eyes: hawk-moth pollination of *Cynorkis uniflora* Lindley (Orchidaceae) in Madagascar. *Botanical Journal of the Linnean Society* 109: 145–160. [220]

Nilsson, L.A., Rabakonandrianina, E., Pettersson, B. & Grünmeir, R. 1993 Lemur pollination in the Malagasy rainforest liana *Strongylodon craveniae* (Leguminosae). *Evolutionary Trends in Plants* 7: 49–56. [259]

Nilsson, S., Praglowski, J. & Nilsson, L. 1977 *Atlas of Airborne Pollen Grains and Spores in Northern Europe.* Natur och Kultur, Stockholm. [268]

Nixon, R.W. 1928 The direct effect of pollen on the fruit of the Date Palm. *Journal of Agricultural Research* 36: 97–128. [360]

Norstog, K. 1987 Cycads and the origin of insect pollination. *American Scientist* 75: 270–279. [369, 370, 372]

O'Donnell, S. & Lawrence, M.J. 1984 The population genetics of the self-incompatibility polymorphism in *Papaver rhoeas*. IV. The estimation of the number of alleles in a population. *Heredity* 53: 495–508. [324]

O'Toole, C. & Raw, A. 1991 *Bees of the World.* Blandford, London. [108, 125]

Obeso, J.R. 1992 Geographic distribution and community structure of bumblebees in the northern Iberian peninsula. *Oecologia* 89: 244–252. [388]

Ockendon, D.J. & Currah, L. 1977 Self-pollen reduces the number of cross-pollen tubes in the styles of *Brassica oleracea* L. *New Phytologist* 78: 675–680. [336]

Oettli, M. 1972 Beobachtungen über den Blütenbesuch von Hummeln auf *Campanula barbata. Saussurea* 4: 55–63. [127, 138]

Ohno, M. 1963 Studies on pollen suspensions for saving labour in the pollination of fruit trees: the effect of alcohol added to the suspension on pollen germination. *Journal of Japanese Horticultural Science* 31: 360–364. [363]

Okubo, A. & Levin, S.A. 1989 A theoretical framework for data analysis of wind dispersal of seeds and pollen. *Ecology* 70: 329–338. [271]

Olesen, J.M. & Warncke, E. 1989 Predation and potential transfer of pollen in a population of *Saxifraga hirculus. Holarctic Ecology* 12: 87–95. [417]

Oliveira, P.E., Gibbs, P.E., Barbosa, A.A. & Talavera, S. 1992 Contrasting breeding systems in two *Eriotheca* (Bombacaceae) species of the Brazilian cerrados. *Plant Systematics and Evolution* 179: 207–219. [348]

Oliver, F.W. 1888 On the sensitive labellum of *Masdevallia muscosa*, Rchb. f. *Annals of Botany* 1: 237–253. [306]

Ollerton, J. & Lack, A.J. 1992 Flowering phenology: an example of relaxation of natural selection? *Trends in Ecology and Evolution* 7: 274–276. [404]

Osborne, J.L., Williams, I.H. & Corbet, S.A. 1991 Bees, pollination and habitat change in the European Community. *Bee World* 72: 99–116. [363]

Østerbye, U. 1975 Self-incompatibility in *Ranunculus acris* L. 1. Genetic interpretation and evolutionary aspects. *Hereditas* 80: 91–112. [324]

Paige, K.N. & Whitham, T.G. 1987 Flexible life history traits: shifts by scarlet gilia in response to pollinator abundance. *Ecology* 68: 1691–1695. [414]

Parkin, J. 1928 The glossy petals of *Ranunculus. Annals of Botany* 42: 739–755. [37]

Parkin, J. 1931 The structure of the starch layer in the glossy petal of *Ranunculus. Annals of Botany* 45: 201–205. [31]

Parkin, J. 1935 The structure of the starch layer in the glossy petal of *Ranunculus.* II. The British species examined. *Annals of Botany* 49: 283–289. [37]

Parmenter, L. 1941 Diptera visiting flowers of Devilsbit Scabious, *Scabiosa succisa* L. *Entomologist's Record and Journal of Variation* 53: 134. [73]

Parmenter, L. 1948 *Rhingia campestris* Mg. (Dipt., Syrphidae), a further note. *Entomologist's Record and Journal of Variation* 60: 119–120. [69]

Parmenter, L. 1952a Flies visiting Greater Stitchwort, *Stellaria holostea* Linn. (Caryophyllaceae). *Journal of the Society for British Entomology* 4: 88–89. [47, 63, 73]

Parmenter, L. 1952b Flies at Ivy-bloom. *Entomologist's Record and Journal of Variation* 64: 90–91. [58, 60, 73]

Parmenter, L. 1956 Beetles visiting the flowers of Dogwood, *Cornus sanguinea* L. *Entomologist's Record and Journal of Variation* 68: 243–244. [54]

Parmenter, L. 1958 Flies (Diptera) and their relations with plants. *London Naturalist* 37: 115–125. [58, 70]

Parrish, J.A.D. & Bazzaz, F.A. 1979 Differences in pollination niche relationships in early and late successional plant communities. *Ecology* 60: 597–610. [403]

Paton, D.C. & Ford, H.A. 1977 Pollination by birds of native plants in South Australia. *Emu* 77: 73–85. [234, 235, 236]

Paton, D.C. & Turner, V. 1985 Pollination of *Banksia ericifolia* Smith: birds, mammals and insects as pollen vectors. *Australian Journal of Botany* 33: 271–286. [258]

Paulus, H.F. & Gack, C. 1981 Neue Beobachtungen zur Bestäubung von *Ophrys* in Südspanien, mit besonderer Berücksichtigung des Formenkreises *Ophrys fusca* agg. *Plant Systematics and Evolution* 137: 241–258. [209]

Paulus, H.F. & Gack, C. 1990a Pollination of *Ophrys* (Orchidaceae) in Cyprus. *Plant Systematics and Evolution* 169: 177–207. [209]

Paulus, H.F. & Gack, C. 1990b Pollinators as prepollinating isolation factors: evolution and speciation in *Ophrys* (Orchidaceae). *Israel Journal of Botany* 39: 43–79. [209]

Peakall, R. 1989 The unique pollination of *Leporella fimbriata* (Orchidaceae): pollination by pseudocopulating male ants (*Myrmecia urens,* Formicidae). *Plant Systematics and Evolution* 167: 137–148. [107]

Peakall, R. 1990 Responses of male *Zaspilothynnus trilobatus* Turner wasps to females and the sexually deceptive orchid it pollinates. *Functional Ecology* 4: 159–167. [211, 212]

Peakall, R. & Beattie, A.J. 1989 Pollination of the orchid *Microtis parviflora* R. Br. by flightless worker ants. *Functional Ecology* 3: 515–522. [221]

Peakall, R., Beattie, A.J. & James, S.H. 1987 Pseudocopulation of an orchid by male ants: a test of two hypotheses accounting for the rarity of ant pollination. *Oecologia* 73: 522–524. [212]

Peakall, R., Handel, S.N. & Beattie, A.J. 1991 The evidence for, and importance of, ant pollination. *Ant - Plant Interactions* (ed. C.R. Huxley & D.F. Cutler), pp. 421–429. Oxford University Press, Oxford. [107]

Pellmyr, O. 1989 The cost of mutualism: interactions between *Trollius europaeus* and its pollinating parasites. *Oecologia* 78: 53–59. [312]

Pellmyr, O. 1992 The phylogeny of a mutualism: evolution and coadaptation. *Biological Journal of the Linnean Society* 47: 337–365. [312]

Pellmyr, O. & Thompson, J.N. 1992 Multiple occurrences of mutualism in the yucca moth lineage. *Proceedings of the National Academy of Sciences, USA* 89: 2927–2929. [319]

Percival, M.S. 1961 Types of nectar in angiosperms. *New Phytologist* 60: 235–281. [40]

Percival, M.S. 1965 *Floral Biology.* Pergamon, Oxford. [22, 177]

Percival, M.S. 1974 Floral ecology of coastal scrub in southeast Jamaica. *Biotropica* 6: 104–129. [394, 408]

Perkins, R.C.L. 1903 Vertebrata. *Fauna Hawaiiensis 1(4)* (ed. D. Sharp), pp. 368–465. [236]

Petanidou, T., den Nijs, J.C.M., Oostermeijer, J.G.B. & Ellis-Adams, A.C. 1995 Pollination ecology and patch-dependent reproductive success of the rare perennial *Gentiana pneumonanthe* L. *New Phytologist* 129: 155–163. [150]

Petanidou, T. & Vokou, D. 1990 Pollination and pollen energetics in Mediterranean ecosystems. *American Journal of Botany* 77: 986–992. [389, 406]

Peterson, A. 1916 The head-capsule and mouth-parts of Diptera. *Illinois Biological Monographs* 3: 173–282. [62, 65]

Pettersson, M.W. 1991 Pollination by a guild of fluctuating moth populations: option for unspecialization in *Silene vulgaris. Journal of Ecology* 79: 591–604. [417, 419]

Pettet, A. 1977 Seasonal changes in

nectar-feeding by birds at Zaria, Nigeria. *Ibis* 119: 291–308. [392]

Philbrick, C.T. 1993 Underwater cross-pollination in *Callitriche hermaphroditica* (Callitrichaceae): evidence from random amplified polymorphic DNA markers. *American Journal of Botany* 80: 391–394. [291]

Philbrick, C.T. & Anderson, G.J. 1987 Implications of pollen/ovule ratios and pollen size for the reproductive biology of *Potamogeton* and autogamy in aquatic angiosperms. *Systematic Botany* 12: 98–105. [287, 290]

Philbrick, C.T. & Anderson, G.J. 1992 Pollination biology in the Callitrichaceae. *Systematic Botany* 17: 282–292. [290]

Philbrick, C.T. & Bernardello, L.M. 1992 Taxonomic and geographic distribution of internal geitonogamy in New World *Callitriche* (Callitrichaceae). *American Journal of Botany* 79: 887–890. [291]

Philipp, M., Böcher, J., Mattsson, O. & Woodell, S.R.J. 1990 A quantitative approach to the sexual reproductive biology and population structure in some arctic flowering plants: *Dryas integrifolia*, *Silene acaulis* and *Ranunculus nivalis*. *Meddelelser om Grønland, Bioscience* 34: 1–60. [388]

Philipp, M. & Schou, O. 1981 An unusual heteromorphic incompatibility system. Distyly, self-incompatibility, pollen load and fecundity in *Anchusa officinalis*. *New Phytologist* 89: 693–703. [329]

Pickens, A.L. 1927 Unique method of pollination by the Ruby-throat. *Auk* 44: 24–27. [237]

Pickens, A.L. 1936 Steps in the development of the bird-flower. *Condor* 38: 150–154. [227]

Pickens, A.L. 1941 A red figwort as the ideal nearctic bird flower. *Condor* 43: 100–102. [326]

Pickens, A.L. 1944 Seasonal territory studies of Ruby-throats. *Auk* 61: 88–92. [229]

Pigott, C.D. 1958 Biological Flora of the British Isles. *Polemonium caeruleum* L. *Journal of Ecology* 46: 507–525. [50]

Piper, J.G., Charlesworth, B. & Charlesworth, D. 1984 A high rate of self-fertilization and increased seed fertility of homostyle primroses. *Nature* 310: 50–51. [329]

Plateau, F. 1885 - 1898 etc. Numerous references listed by Knuth. [23]

Plateau, F. 1899 Nouvelles recherches sur les rapports entre les insects et les fleurs. II. Le choéix des couleurs par les insectes. *Mémoires de la Société Zoologique de France* 12: 336–370. [23]

Pleasants, J.M. 1983 Structure of plant and pollinator communities. *Handbook of Experimental Pollination Biology* (ed. C.E. Jones & R.J. Little), pp. 375–393. Van Nostrand Reinhold, New York. [405]

Pleasants, J.M. 1991 Evidence for short-distance dispersal of pollinia in *Asclepias syriaca*. *Functional Ecology* 5: 75–82. [186]

Pohl, F. 1937 Die Pollenerzeugung der Windblutler. *Beihefte zur botanisches Zentralblatt, A* 56: 365–470. [265]

Pojar, J. 1974 Reproductive dynamics of four plant communities of southwestern British Columbia. *Canadian Journal of Botany* 52: 1819–1834. [388]

Pont, A.C. 1993 Observations on anthophilous Muscidae and other Diptera (Insecta) in Abisko National Park, Sweden. *Journal of Natural History* 27: 631–643. [388]

Popham, E. J. 1961 Earwigs in the British Isles. *Entomologist* 94: 308–310. [50]

Porsch, O. 1924 Vogelblumenstudien. I. *Jahrbücher für wissenschaftliche Botanik* 63: 553–706. [21, 226, 242]

Porsch, O. 1927 Kritische Quellenstudien über Blumenbesuch durch Vögel. II. *Biologia Generalis* 3: 475–548. [242]

Porsch, O. 1930 Kritische Quellenstudien über Blumenbesuch durch Vögel. V. *Biologia Generalis* 6: 135–246. [232]

Porsch, O. 1933 Der Vogel als Blumen Bestäuber. *Biologia Generalis* 9: 239–252. [225]

Porsch, O. 1957 Alte Insektentypen als Blumenausbeuter. *Österreichische Botanische Zeitschrift* 104: 115–164. [50]

Porter, K., Steel, C.A. & Thomas, J.A. 1992 Butterflies and communities. *The Ecology of Butterflies in Britain* (ed. R.L.H. Dennis), pp. 139–177. Oxford University Press, Oxford. [408]

Poulton, E.B. 1932 in Godfery, M. J., Insect carriers of orchid pollinia. *Proceedings of the Entomological Society of London* 6: 70. [99]

Pouyanne, A. 1917 La fécondation des *Ophrys* par les insectes. *Bulletin de la Société d'Histoire Naturelle de l'Afrique du Nord* 8: 6–7. [205]

Powell, J.A. 1992 Interrelationships of yuccas and yucca moths. *Trends in Ecology and Evolution* 7: 10–15. [316, 318, 319]

Powell, J.A. & Mackie, R.A. 1966 Biological interrelationships of moths and *Yucca whipplei* (Lepidoptera: Gelechiidae, Blastobasidae, Prodoxidae). *University of California Publlications in Entomology* 42. [318]

Prance, G.T. & Arias, J.R. 1975 A study of the floral biology of *Victoria amazonica* (Poepp.) Sowerby (Nymphaeaceae). *Acta Amazonica* 5: 109–139. [310]

Primack, R. & Hall, P. 1990 Costs of reproduction in the pink lady's slipper orchid: a four-year experimental study. *American Naturalist* 136: 638–656. [223, 414]

Prime, C.T. 1954 Biological Flora of the British Isles. *Arum neglectum* (Townsend) Ridley. *Journal of*

Prime, C.T. 1960 *Lords and Ladies*. Collins, London. [304]

Proctor, M. & Yeo, P. 1973 *The Pollination of Flowers*. Collins, London. [22, 107]

Prys-Jones, O.E. & Corbet, S.A. 1987 *Bumblebees*. Cambridge University Press, Cambridge. [405]

Ptácek, V. 1991 Trials to rear bumble bees. *Sixth International Symposium on Pollination (Acta Horticulturae no. 288)* (ed. C. van Heemert & A. de Ruijter), pp. 144–148. International Soc. Hortic. Sci., Wageningen. [359]

Purves, W.K, Orians, G.H. & Heller, H.C. 1995 *Life, ed. 4*. Sinauer/W.H. Freeman & Co., Sunderland, Mass. [368]

Pye, J.D. 1983 *Moths and bats: pollen and nectar collection at night*. Central Assoc. of Beekeepers, Ilford. [252, 364]

Pyke, G.H. 1991 What does it cost a plant to produce floral nectar? *Nature* 350: 58–59. [39]

Pyke, G.H. & Cartar, R.V. 1992 The flight directionality of bumblebees: do they remember where they came from? *Oikos* 65: 321–327. [418]

Qiu, Y.-L., Chase, M.W., Les, D.H. & Parks, C.R. 1993 Molecular phylogenetics of the Magnoliidae: cladistic analysis of nucleotide sequences of the plastid gene rbcL. *Annals of the Missouri Botanical Garden* 80: 587–606. [379]

Ramirez, B.W. 1969 Fig wasps: mechanism of pollen transfer. *Science* 163: 580–581. [315]

Ranta, E., Lundberg, H. & Teras, I. 1981 Patterns of resource utilization in two Fennoscandian bumblebee communities. *Oikos* 36: 1–11. [388, 406]

Ranta, E. & Vepsäläinen, K. 1981 Why are there so many species? Spatio-temporal heterogeneity and northern bumblebee communities. *Oikos* 36: 28–34. [406]

Rathcke, B. 1988 Flowering phenologies in a shrub community: competition and constraints. *Journal of Ecology* 76: 975–994. [386, 402]

Rathcke, B. & Lacey, E.P. 1985 Phenological patterns of terrestrial plants. *Annual Review of Ecology and Systematics* 16: 179–214. [396, 400]

Rathcke, B. & Real, L. 1993 Autogamy and inbreeding depression in mountain laurel, *Kalmia latifolia* (Ericaceae). *American Journal of Botany* 80: 143–146. [335, 399]

Raven, J.A. 1993 The evolution of vascular plants in relation to quantitative functioning of dead water-conducting cells and stomata. *Biological Reviews* 68: 337–363. [365]

Raven, P.H. 1973 Why are bird-visited flowers predominantly red? *Evolution* 26: 674. [229]

Raven, P.H. 1977 Erythrina symposium II.; *Erythrina* (Fabaceae: Faboideae): introduction to symposium II. *Lloydia* 40: 401–406. [240]

Ray, J. 1686-1704 *Historia Plantarum* 1 – 3.

London. [13]

Real, L. 1983 *Pollination Biology*. Academic Press, London. [22]

Rebelo, A.G. 1987 Bird pollination in the Cape Flora. *A Preliminary Synthesis of Pollination Biology in the Cape Flora: South African National Science Programmes, report no. 141* (ed. A.G. Rebelo), pp. 83–107. CSIR, Pretoria. [235, 241, 243]

Rebelo, A.G. & Breytenbach, G.J. 1987 Mammal pollination in the Cape flora. *A Preliminary Synthesis of Pollination Biology in the Cape Flora: South African National Science Programmes, report no. 141* (ed. A.G. Rebelo), pp. 109–125. CSIR, Pretoria. [257, 260]

Regal, P.J. 1982 Pollination by wind and animals: ecology of geographic patterns. *Annual Review of Ecology and Systematics* 13: 497–524. [265, 395]

Rempe, H. 1937 Untersuchungen über die Verbreitung des Blütenstaubes durch Luftströmungen. *Planta* 27: 93–147. [272]

Rendle, A.B. 1930 *The Classification of Flowering Plants (ed. 2)* 1. Cambridge University Press, Cambridge. [280]

Ribbands, C.R. 1949 The foraging method of individual honeybees. *Journal of Animal Ecology* 18: 47–66. [135]

Ribbands, C.R. 1953 *The Behaviour and Social Life of Honeybees*. Bee Research Association, London. [125]

Ribbands, C.R. 1955 The scent perception of the Honeybee. *Proceedings of the Royal Society B* 143: 367–379. [126]

Richards, A.J. 1986 *Plant Breeding Systems*. Allen & Unwin, London. [323, 327, 332, 335, 348, 349]

Richards, K.W. 1984 *Alfalfa leafcutter bee management in Western Canada, ed. 2*. Agriculture Canada, pub. 1495/E, Ottawa. [356]

Richards, K.W. 1991 Effectiveness of the alfalfa leafcutter bee as a pollinator of legume forage crops. *Sixth International Symposium on Pollination (Acta Horticulturae no. 288)* (ed. C. van Heemert & A. de Ruijter), pp. 180–184. International Soc. Hortic. Sci., Wageningen. [357]

Richards, O.W. & Davies, R.G. 1977 *Imms' General Textbook of Entomology, ed. 10* 2. Chapman & Hall/John Wiley, London/New York. [80]

Richards, O.W. & Hamm, A.H. 1939 The biology of the British Pompilidae (Hymenoptera). *Transactions of the Society for British Entomology* 6: 51–114. [103]

Rick, C.M. 1950 Pollination relations of *Lycopersicon esculentum* in native and foreign regions. *Evolution* 4: 110–122. [359]

Ridgeway, R. 1891 The hummingbirds. *Report of the United States National Museum* 1890: 253–383. [227, 228]

Ridley, H.N. 1890 On the method of fertilization in *Bulbophyllum macranthum*, and allied orchids. *Annals of Botany* 4: 327–336. [215]

Rieseberg, L.H., Hanson, M.A. & Philbrick, C.T. 1992 Androdioecy is derived from dioecy in Datiscaceae: evidence from restriction site mapping of PCR-amplified chloroplast DNA fragments. *Systematic Botany* 17: 324–336. [346]

Rigney, L.P. 1995 Postfertilization causes of differential success of pollen donors in *Erythronium grandiflorum* (Liliaceae): nonrandom ovule abortion. *American Journal of Botany* 82: 578–584. [421]

Riley, C.V. 1892 The Yucca moth and Yucca pollination. *Report of the Missouri Botanical Garden* 3: 99–158. [316, 317]

Robertson, A.W. 1992 The relationship between floral display size, pollen carryover and geitonogamy in *Myosotis colensoi* (Kirk) Macbride (Boraginaceae). *Biological Journal of the Linnean Society* 46: 333–349. [45]

Robinson, T.R. & Savage, E.M. 1926 *Pollination of the Avocado* (U.S. Department of Agriculture, Circular No. 387). USDA, Washington, D.C. [359]

Robinson, W.S., Nowogrodzki, R. & Morse, R.A. 1989a The value of honey bees as pollinators of U. S. crops, part 1. *American Bee Journal* 129: 411–423. [352, 356, 360, 361, 362, 363]

Robinson, W.S., Nowogrodzki, R. & Morse, R.A. 1989b The value of honey bees as pollinators of U. S. crops, part 2. *American Bee Journal* 129: 477–487. [352, 363]

Ross, M.D. & Abbott, R.J. 1987 Fitness, sexual asymmetry, functional sex and selfing in *Senecio vulgaris* L. *Evolutionary Trends in Plants* 1: 21–28. [347]

Ross-Craig, S. 1956 *Drawings of British Plants* 9(2). G. Bell & Sons, London. [181]

Rothschild, L.W. & Jordan, K. 1903 A revision of the lepidopterous family Sphingidae. *Novitates Zoologicae* 9: 1–972. [214]

Roubik, D.W. 1989 *Ecology and Natural History of Tropical Bees*. Cambridge University Press, Cambridge. [22, 382, 383, 389, 390, 394, 395, 407, 408, 416]

Roubik, D.W. 1992 Loose niches in tropical communities: why are there so few bees and so many trees? *Effects of Resource Distribution on Animal-Plant Interactions* (ed. M.D. Hunter, T. Ohgushi & P.W. Price), pp. 327–354. Academic Press, London. [407]

Rourke, J. & Wiens, D. 1977 Convergent floral evolution in South African and Australian Proteaceae and its possible bearing on pollination by nonflying mammals. *Annals of the Missouri Botanical Garden* 64: 1–17. [259, 261, 262]

Rydell, J. 1986 Feeding territoriality in female northern bats, *Eptesicus nilssoni*. *Ethology* 72: 329–337. [254]

Sacchi, C.F. & Price, P.W. 1988 Pollination of the Arroyo willow, *Salix leptolepis*: role of insects and wind. *American Journal of Botany* 75: 1387–1393. [278]

Sachs, J. 1875 *Geschichte der Botanik vom 16 Jahrhundert bis 1860*. R. Oldenbourg, München. [15]

Sachs, J. 1890 *History of Botany (1530 - 1860)*. Transl. H.E.F. Garnsey & I. Bayley-Balfour. Oxford University Press, Oxford. [15]

Sadamori, S., et al. 1958 Studies in commercial hand pollination methods for apple flowers. I. Examination of pollen diluents, of degree of pollen distribution and pollinating methods. *Bulletin of the Tôhoku National Agricultural Experiment Station* 14: 74–81. [363]

Sampson, D.R. 1967 Frequency and distribution of self-incompatibility alleles in *Raphanus raphanistrum*. *Genetics* 56: 241–251. [325]

Sargent, O.H. 1909 Notes on the life-history of *Pterostylis*. *Annals of Botany* 23: 265–274. [213, 307]

Sargent, O.H. 1918 Fragments on the flower biology of Westralian plants. *Annals of Botany* 32: 215–231. [225]

Sargent, O.H. 1934 Pollination in *Pterostylis*. *Victorian Naturalist* 51: 82–84. [213, 307]

Saunders, E. 1878 Remarks on the hairs of some of our British Hymenoptera. *Transactions of the Entomological Society of London* 1878: 169–172. [115]

Saunders, E. 1890 On the tongues of the British Hymenoptera Anthophila. *Journal of the Linnean Society (Zoology)* 23: 410–432. [118, 122]

Saville, N.M. [1993] *Bumblebee Ecology in Woodlands and Arable Farmland*. PhD thesis, University of Cambridge. Unpublished. [387, 388, 403]

Sazima, I. & Sazima, M. 1977 Solitary and group foraging: two flower-visiting patterns of the lesser spear-nosed bat *Phyllostomus discolor*. *Biotropica* 9: 213–215. [254]

Sazima, M. & Sazima, I. 1975 Quiropterofilia em *Lafoensia pacari* St. Hil. (Lythraceae), na Serra do Cipó, Minas Gerais. *Revista Ciência e Cultura* 27: 405–416. [253]

Sazima, M. & Sazima, I. 1980 Bat visits to *Marcgravia myriostigma* Tr. & Planch. (Marcgraviaceae) in Southeastern Brazil. *Flora* 169: 84–88. [248, 249]

Schaal, B.A. & Levin, D.A. 1976 The demographic genetics of *Liatris cylindracea* Michx. (Compositae). *American Naturalist* 110: 191–206. [424]

Schatz, G.E. 1990 Some aspects of pollination biology in Central American forests. *Reproductive*

Ecology of Tropical Plants (ed. K.S. Bawa & M. Hadley), pp. 69–84. Unesco, Paris. [340]

Schemske, D.W. & Horvitz, C.C. 1984 Variation among floral visitors in pollination ability: a precondition for mutualism specialization. *Science* 225: 519–521. [415]

Schemske, D.W., Willson, M.F. & 5 others 1978 Flowering ecology of some spring woodland herbs. *Ecology* 59: 351–366. [387]

Schiemenz, H. 1957 Vergleichende funktionellanatomische Untersuchungen der Kopfmuskulatur von *Theobaldia* und *Eristalis* (Dipt. Culicid. und Syrphid.). *Deutsche Entomologische Zeitschrift* NF 4: 268–331. [65, 66]

Schlegtendal, A. 1934 Beitrag zum Farbensinn der Arthropoden. *Zeitschrift für vergleichende Physiologie* 20: 545–581. [56]

Schlessman, M.A. 1988 Gender diphasy ("sex choice"). *Plant Reproductive Ecology* (ed. J. Lovett Doust & L. Lovett Doust), pp. 139–153. Oxford University Press, Oxford. [340]

Schmitt, J. & Ehrhardt, D.W. 1990 Enhancement of inbreeding depression by dominance and suppression in *Impatiens capensis*. *Evolution* 44: 269–278. [424]

Schmitt, U. & Bertsch, A. 1990 Do foraging bumblebees scent-mark food sources and does it matter? *Oecologia* 82: 137–144. [139, 140]

Schoen, D.J. 1982 The breeding system of *Gilia achilleifolia*: variation in floral characteristics and outcrossing rate. *Evolution* 36: 352–360. [333, 337]

Schremmer, F. 1941a Sinnesphysiologie und Blumenbesuch des Falters von *Plusia gamma* L. *Zoologischer Jahresbericht. Zoologische Station zu Neapel* 74: 375–434. [88, 97]

Schremmer, F. 1941b Versuche zum Nachweis der Rotblindheit von *Vespa rufa* L. *Zeitschrift für vergleichende Physiologie* 28: 457–466. [126]

Schumann, K. 1890-3 Sterculiaceae. *Die Natürlichen Pflanzenfamilien III.6* (ed. A. Engler), pp. 69–99. Engelmann, Leipzig. [60]

Scorer, A.G. 1913 *The Entomologist's Log-book and Dictionary of the Life-Histories and Food Plants of the British Macro-Lepidoptera*. Routledge, London. [85]

Scott Elliot, G.F. 1890a Note on the fertilisation of *Musa*, *Strelitzia reginae*, and *Ravenala madagascariensis*. *Annals of Botany* 4: 259–263. [240]

Scott Elliot, G.F. 1890b Ornithophilous flowers in South Africa. *Annals of Botany* 4: 265–280. [225]

Scott, A.C. & Taylor, T.N. 1983 Plant/animal interactions during the Upper Carboniferous. *Botanical Reviews* 49: 259–307. [374]

Scott, H. 1953 Discrimination of colours by *Bombylius* (Dipt., Bombyliidae). *Entomologist's Monthly Magazine* 89: 259–260. [63]

Scott, J.A. & Wright, D.M. 1990 Butterfly phylogeny and fossils. *Butterflies of Europe 2* (ed. O. Kudrna), pp. 152–208. Aula-Verlag, Wiesbaden. [382]

Scott, K. 1991 When the bird's away . . . *BBC Wildlife* 9: 683. [264]

Sculthorpe, C.D. 1967 *The Biology of Aquatic Vascular Plants*. Edward Arnold, London. [285]

Seavey, S.R. & Bawa, K.S. 1986 Late-acting self-incompatibility in angiosperms. *Botanical Reviews* 52: 195–219. [323, 328]

Seavey, S.R. & Carter, S.K. 1994 Self-sterility in *Epilobium obcordatum* (Onagraceae). *American Journal of Botany* 81: 331–338. [328, 423]

Shaw, R.J. 1962 The biosystematics of *Scrophularia* in western North America. *Aliso* 5: 147–178. [106]

Shivanna, K.R. & Johri, B.M. 1985 *The Angiosperm Pollen. Structure and Function*. Wiley, New Delhi. [350]

Shoemaker, J.S. 1962 Pollination requirements of Flordagrand Blackberry. *Proceedings of the Florida Station for Horticultural Science* 74: 356–358. [362]

Silberbauer-Gottsberger, I. 1972 Anthese und Bestäubung der Rubiaceen *Tocoyena brasiliensis* und *T. formosa* aus dem cerrado Brasiliens. *Österreichische Botanische Zeitschrift* 120: 1–13. [85]

Silberbauer-Gottsberger, I. 1991 Pollination of two cerrado palms. *Plant Reproductive Ecology, Uppsala 9 - 11 September 1991: Progress and Perspectives, Lecture and Poster Abstracts*: n. p. [311]

Silberglied, R.E. 1984a Visual communication and sexual selection among butterflies. *Symposia of the Royal Entomological Society 11* (ed. R.I. Vane-Wright & P.R. Ackery), pp. 207–223. Royal Entomological Society, London. [87, 96]

Silberglied, R.E. 1984b [collective bibliography for the whole work that includes: Visual communication and sexual selection among butterflies]. *Symposia of the Royal Entomological Society 11* (ed. R.I. Vane-Wright & P.R. Ackery), pp. 355–410. Royal Entomological Society, London. [87,96]

Silén, F. 1906a Blombiologiska iakttagelser i Kittilä Lappmark. *Meddelander af Societas pro Fauna et Flora Fennica* 31: 80–99. [204]

Silén, F. 1906b Blombiologiska iakttagelser i södra Finland. *Meddelander af Societas pro Fauna et Flora Fennica* 32: 120–134. [169]

Silvertown, J.W. & Lovett Doust, J. 1993 *Introduction to Plant Population Biology*. Blackwell Scientific, Oxford. [413]

Simes, J.A. 1946 Behaviour of *Bombylius* (Dipt., Bombyliidae) while feeding. *Entomologist's Monthly Magazine* 89: 234. [62]

Simpson, B.B. 1977 Breeding systems of dominant perennial plants of two disjunct warm desert ecosystems. *Oecologia* 27: 203–226. [389]

Skog, L.E. 1976 A study of the Tribe Gesnerieae, with a revision of *Gesneria* (Gesneriaceae: Gesnerioideae). *Smithsonian Contributions to Botany* 29: 1–182. [244]

Sladen, F.W.L. 1911a How pollen is collected by the social bees, and the part played in the process by the auricle. *British Bee Journal* 39: 491–495. [124]

Sladen, F.W.L. 1911b The pollen-collecting apparatus in the social bees. *British Bee Journal* 39: 506. [124]

Sladen, F.W.L. 1912a How the corbicula is loaded. *British Bee Journal* 40: 138. [124]

Sladen, F.W.L. 1912b (Four papers on loading the corbicula in social bees). *British Bee Journal* 40: 144–145, 164 - 166, 196, 462 - 463. [124]

Sladen, F.W.L. 1912c (Three papers on pollen packing by social bees). *British Bee Journal* 40: 164–166, 196, 462 - 463. [124]

Small, J. 1915 The pollen-presentation mechanism in the Compositae. *Annals of Botany* 29: 457–470. [176]

Small, J. 1917 The origin and development of the Compositae. Introduction and I - III. *New Phytologist* 16: 157–158, 159 - 177, 198 - 221, 253 - 276. [177]

Small, J. 1918 The origin and development of the Compositae. IV - IX. *New Phytologist* 17: 13–40, 69 - 94, 114 - 125, 126 - 132, 133 - 142, 200 - 230. [177]

Smart, J. 1943 *Simulium* feeding on Ivy flowers. *Entomologist* 76: 20–21. [58]

Smith, K.V.G. 1959 The distribution and habits of the British Conopidae (Dipt.). *Transactions of the Society for British Entomology* 13: 113–136. [71]

Smith, K.V.G. 1961 Supplementary records of the distribution and habits of the British Conopidae (Diptera). *Entomologist* 94: 238–239. [71]

Snodgrass, R.E. 1956 *The Anatomy of the Honey Bee.* Comstock/Constable, Ithaca, N.Y./London. [120, 124]

Snow, A.A. 1990 Effects of pollen-load size and number of donors on sporophyte fitness in wild radish (*Raphanus raphanistrum*). *American Naturalist* 136: 742–758. [421]

Snow, A.A. & Spira, T.P. 1991 Differential pollen-tube growth rates and nonrandom fertilization in *Hibiscus moscheutos* (Malvaceae). *American Journal of Botany* 78: 1419–1426. [421]

Snow, A.A. & Spira, T.P. 1993 Individual variation in the vigor of self pollen and selfed progeny in *Hibiscus moscheutos* (Malvaceae). *American Journal of Botany* 80: 160–166. [335, 421]

Snow, B.K. & Snow, D.W. 1972 Feeding niches of hummingbirds in a Trinidad valley. *Journal of Animal Ecology* 41: 471–485. [229]

Snow, B.[K.] & Snow, D.[W.] 1988 *Birds and Berries.* Poyser, Calton. [387]

Sobrevilla, C. & Arroyo, M.T.K. 1982 Breeding systems in a montane tropical cloud forest in Venezuela. *Plant Systematics and Evolution* 140: 19–38. [393]

Soderstrom, R.T. & Calderon, C.E. 1971 Insect pollination in tropical rain forest grasses. *Biotropica* 3: 1–16. [392]

Solomon, A.M. 1979 Pollen. *Aerobiology: the Ecological Systems Approach.* US IBP Synthesis Series 10 (ed. R.E. Edmonds), pp. 41–54. Dowden, Hutchinson & Ross, Stroudsburg, Pa. [267, 385]

Southwick, E.E. 1984 Photosynthate allocation to floral nectar: a neglected energy investment. *Ecology* 65: 1775–1779. [39, 384]

Spooner, G.M. 1930 *The Bees, Wasps and Ants of Cambridgeshire.* Cambridge Natural History Society, Cambridge. [103, 106]

Spooner, G.M. 1941 The characters of the female and distribution in Britain of *Pompilus trivialis* Dahlb., *unguicularis* Thoms. and *wesmaeli* Thomas. (Hymenoptera: Pompilidae). *Transactions of the Society for British Entomology* 7: 85–122. [103]

Sporne, K.R. 1965 *The Morphology of Gymnosperms, ed. 2.* Hutchinson, London. [369, 370, 373, 374]

Sprague, E.F. 1962 Pollination and evolution in *Pedicularis* (Scrophulariaceae). *Aliso* 5: 181–209. [127, 165]

Sprengel, C.K. 1793 *Das entdeckte Geheimniss der Natur im Bau und in der Befruchtung der Blumen.* Friedrich Vieweg, Berlin. [17, 18, 19]

St John, H. 1956 Monograph of the genus *Elodea*: Part 4 and summary. *Rhodora* 67: 1–35. [286]

Stace, C.A. 1961 Some studies in *Calystegia*: compatibility and hybridisation in *C. sepium* and *C. silvatica. Watsonia* 5: 88–105. [150]

Stace, C.A. 1965 Some studies in *Calystegia*. 2. Observations on the floral biology of the British inland taxa. *Proceedings of the Botanical Society of the British Isles* 6: 21–31. [150]

Stanton, M.L. 1994 Male-male competition during pollination in plant populations. *American Naturalist Supplement* 144: 540–568. [421]

Stanton, M.L., Preston, R.E., Snow, A.A. & Handel, S.N. 1986 Floral evolution: attractiveness to pollinators increases male fitness. *Science* 232: 1625–1627. [415]

Stanton, M.L., Snow, A.A., Handel, S.N. & Beresky, J. 1989 The impact of a flower-color polymorphism on mating patterns in experimental populations of wild radish (*Raphanus raphanistrum* L.). *Evolution* 43: 335–346. [418]

Stanton, M.L., Ashman, T.-L., Galloway, L.F. & Young, H.J. 1992 Estimating male fitness of plants in natural populations. *Ecology and Evolution of Plant Reproduction* (ed. R. Wyatt), pp. 62–90. Chapman & Hall, London. [417, 419]

Start, A.N. & Marshall, A.G. 1976 Nectarivorous bats as pollinators of trees in West Malaysia. *Variation, Breeding and Conservation of Tropical Forest Trees* (ed. J. Burley & B.T. Styles), pp. 141–150. Academic Press, London. [246, 247, 251, 253, 417]

Stebbins, G.L. 1950 *Variation and Evolution in Plants.* Columbia University Press, New York. [22]

Stebbins, G.L. 1957 Self-fertilization and population variability in the higher plants. *American Naturalist* 41: 337–354. [333]

Stebbins, G.L. 1974 *Flowering Plants: Evolution Above the Species Level.* Belknap Press, Cambridge, Mass. [372]

Steiner, K.E. 1981 Nectarivory and potential pollination by a neotropical marsupial. *Annals of the Missouri Botanical Garden* 68: 505–513. [256]

Steiner, K.E. 1983 Pollination of *Mabea occidentalis* (Euphorbiaceae) in Panama. *Systematic Botany* 8: 105–117. [249, 251]

Steiner, K.E. 1989 The pollination of *Disperis* (Orchidaceae) by oil-collecting bees in southern Africa. *Lindleyana* 4: 164–183. [215]

Steiner, K.E. & Whitehead, V.B. 1988 The association between oil-producing flowers and oil-collecting bees in the Drakensberg of southern Africa. *Modern Systematic Studies in African Botany, Monographs in Systematic Botany of the Missouri Botanical Garden 25* (ed. P. Goldblatt & P.P. Lowry), pp. 259–277. Missouri Botanical Garden, St Louis. [115]

Steiner, K.E. & Whitehead, V.B. 1990 Pollinator adaptation to oil-secreting flowers - *Rediviva* and *Diascia*. *Evolution* 44: 1701–1707. [282]

Steiner, K.E., Whitehead, V.B. & Johnson, S.D. 1994 Floral and pollinator divergence in two sexually deceptive South African orchids. *American Journal of Botany* 81: 185–194. [205]

Stelleman, P. 1981 Anthecological relations between reputedly anemophilous flowers and syrphid flies. V. Some special aspects of the visiting of *Plantago media* and *P. lanceolata* by insects. *Beiträge zur Biologie der Pflanzen* 55: 157–167. [282]

Stelleman, P. & Meeuse, A.D.J. 1976 Anthecological relations between reputedly anemophilous flowers and syrphid flies. I. The possible role of syrphid flies as pollinators of *Plantago*. *Tijdschrift voor Entomologie* 119: 15–31. [67, 70]

Stephen, W.P. 1961 Artificial nesting sites for the propagation of the Leaf-cutter Bee, *Megachile* (*Eutricharaea*) *rotundata*, for Alfalfa pollination.

Journal of Economic Entomology 54: 989–993. [355]

Stephen, W.P. 1965 *Artificial beds for Alkali Bee propagation* (Agricultural Experiment Station, Oregon State University, Cornwallis, Bulletin No. 598). [354, 355]

Stephenson, A.G. & Bertin, R.I. 1983 Male competition, female choice, and sexual selection in plants. *Pollination Biology* (ed. L. Real), pp. 109–149. Academic Press, London. [330, 334, 414]

Stephenson, A.G., Lau, T.-C., Quesada, M. & Winsor, J.A. 1992 Factors that affect pollen performance. *Ecology and Evolution of Plant Reproduction* (ed. R. Wyatt), pp. 119–136. Chapman & Hall, London. [420, 421]

Stephenson, A.G. & Winsor, J.A. 1986 *Lotus corniculatus* regulates offspring quality through selective fruit abortion. *Evolution* 40: 453–458. [421]

Stevens, D.P. 1988 On the gynodioecious polymorphism in *Saxifraga granulata*. *Biological Journal of the Linnean Society* 35: 15–28. [345, 346]

Stevens, D.P. & Kay, Q.O.N. [1991] *The distribution and reproductive biology of gynodioecy in the native flora of Britain and Ireland.* Unpublished manuscript. [345, 346]

Stevens, D.P. & van Damme, J.M.M. 1988 The evolution and maintenance of gynodioecy in sexually and vegetatively reproducing plants. *Heredity* 61: 329–337. [346]

Stevens, J.P. & Bougourd, S.M. 1988 Inbreeding depression and the outcrossing rate in natural populations of *Allium schoenoprasum* L. (wild chives). *Heredity* 60: 257–261. [335]

Stevens, J.P. & Kay, Q.O.N. 1988 The number of loci controlling the sporophytic self-incompatibility system in *Sinapis arvensis* L. *Heredity* 61: 411–418. [325]

Stiles, F.G. 1975 Ecology, flowering phenology, and hummingbird pollination of some Costa Rican *Heliconia* species. *Ecology* 56: 285–301. [230, 409]

Stiles, F.G. 1981 Geographical aspects of bird-flower coevolution, with particular reference to Central America. *Annals of the Missouri Botanical Garden* 68: 323–351. [226, 227, 229, 230, 231, 232, 233, 236, 241, 243, 255, 388, 392]

Stix, E. & Grossebrauckmann, G. 1970 Der Pollen- und Sporengehalt der Luft und seine tages- und jahreszeitlich Schwankungen unter mitteleropäischen Verhältnisse. *Flora* 159: 1–37. [267]

Stoutamire, W.P. 1974 Australian terrestrial orchids, Thynnid wasps, and pseudocopulation. *American Orchid Society Bulletin* 43: 13–18. [103, 212]

Stoutamire, W.P. 1975 Pseudocopulation in Australian terrestrial orchids. *American Orchid*

Society Bulletin 44: 226–233. [212]

Sukopp, H. 1987 On the history of plant geography and plant ecology in Berlin. *Englera* 7: 85–103. [16]

Sussman, R.W. & Raven, P.H. 1978 Pollination by lemurs and marsupials: an archaic coevolutionary system. *Science* 200: 731–736. [259, 393]

Sutherland, S. 1986 Floral sex ratios, fruit set, and resource allocation in plants. *Ecology* 67: 991–1001. [341, 342]

Swihart, C.A. 1971 Colour discrimination in the butterfly, *Heliconius charitonius* Linn. *Animal Behaviour* 19: 156–164. [96]

Swihart, C.A. & Swihart, S.L. 1970 Colour selection and learned feeding preferences in the butterfly, *Heliconius charitonius* Linn. *Animal Behaviour* 18: 60–64. [95]

Swihart, S.L. 1969 Colour vision and the physiology of the superposition eye of a butterfly (Hesperidae). *Journal of Insect Physiology* 15: 1347–1365. [96]

Swihart, S.L. 1970 The neural basis of colour vision in the butterfly, *Papilio troilus. Journal of Insect Physiology* 16: 1623–1636. [96]

Swihart, S.L. 1972 The neural basis of colour vision in the butterfly, *Heliconius erato. Journal of Insect Physiology* 18: 1015–1025. [96]

Swynnerton, C.F.M. 1916a Short cuts by birds to nectaries. *Journal of the Linnean Society (Botany)* 43: 381–416. [225, 243]

Swynnerton, C.F.M. 1916b Short cuts to nectaries by Blue Tits. *Journal of the Linnean Society (Botany)* 43: 417–422. [226]

Syed, R.A. 1979 Studies on oil palm pollination by insects. *Bulletin of Entomological Research* 69: 213–224. [360]

Talavera, S., Gibbs, P.E. & Herrera, J. 1993 Reproductive biology of *Cistus ladanifer* (Cistaceae). *Plant Systematics and Evolution* 186: 123–134. [179]

Tanaka, H. 1982 Relationship between ultraviolet and visual spectral guidemarks of 93 species of flowers and their pollinators. *Journal of Japanese Botany* 57: 146–159. [132]

Taylor, D.W. & Hickey, L.J. 1992 Phylogenetic evidence for the herbaceous origin of angiosperms. *Plant Systematics and Evolution* 180: 137–156. [379]

Taylor, L.R. & Taylor, R.A.J. 1977 Aggregation, migration and population mechanics. *Nature* 265: 415–421. [330, 395]

Thien, L.B. 1969a Mosquito pollination of *Habenaria obtusata* (Orchidaceae). *American Journal of Botany* 56: 232–237. [60]

Thien, L.B. 1969b Mosquitoes can pollinate orchids. *Morris Arboretum Bulletin* 20(2): 19–23. [60]

Thien, L.B. 1974 Floral biology of *Magnolia*.

American Journal of Botany 61: 1037–1045. [54]

Thien, L.B., Bernhardt, P., et mult. al. 1985 The pollination of *Zygogonum* (Winteraceae) by a moth, *Sabatinca* (Micropterigidae): an ancient association? *Science* 227: 540–543. [87]

Thompson, M.M. 1979 Genetics of incompatibility in *Corylus avellana* L. *Theoretical and Applied Genetics* 54: 113–116. [344]

Thomson, J.D. & Brunet, J. 1990 Hypotheses for the evolution of dioecy in seed plants. *Trends in Ecology and Evolution* 5: 11–16. [342]

Thomson, J.D. & Plowright, R.C. 1980 Pollen carryover, nectar rewards, and pollinator behavior with special reference to *Diervilla lonicera. Oecologia* 46: 68–74. [419]

Thomson, J.D. & Thomson, B.A. 1992 Pollen presentation and viability schedules in animal-pollinated plants: consequences for reproductive success. *Ecology and Evolution of Plant Reproduction* (ed. R. Wyatt), pp. 1–24. Chapman & Hall, London. [417]

Tillyard, R.J. 1923 On the mouthparts of the Micropterygoidea (Order Lepidoptera). *Transactions of the Entomological Society of London* 1923: 181–206. [87]

Tollsten, L. & Øvstedal, D.O. 1994 Differentiation in floral scent chemistry among populations of *Conopodium majus* (Apiaceae). *Nordic Journal of Botany* 14: 361–367. [173]

Tonsor, S.J. 1985 Leptokurtic pollen-flow, non-leptokurtic gene-flow in a wind-pollinated herb, *Plantago lanceolata* L. *Oecologia* 67: 442–446. [271, 418]

Torchio, P.F. 1991 Bees as crop pollinators and the role of solitary species in changing environments. *Sixth International Symposium on Pollination (Acta Horticulturae no. 288)* (ed. C. van Heemert & A. de Ruijter), pp. 49–61. International Soc. Hortic. Sci, Wageningen. [358, 363, 364]

Totland, Ø. 1993 Pollination in alpine Norway: flowering phenology, insect visitors and visitation rates in two plant communities. *Canadian Journal of Botany* 71: 1072–1079. [45, 75]

Totland, Ø. 1994 Intraseasonal variation in pollination intensity and seed set in an alpine population of *Ranunculus acris* in southwestern Norway. *Ecography* 17: 159–165. [45]

Traveset, A. 1994 Reproductive biology of *Phillyrea angustifolia* L. (Oleaceae) and effect of galling-insects on its reproductive output. *Botanical Journal of the Linnean Society* 114: 153–166. [346]

Trelease, W. 1881 The fertilisation of *Scrophularia. Bulletin of the Torrey Botanical Club* 8: 133–140. [106, 181]

Trelease, W. 1893 Further studies of yuccas and their pollination. *Annual Report of the Missouri Botanical Garden* 4: 181–226. [316]

Turner, V. 1982 Marsupials as pollinators in Australia. *Pollination and Evolution* (ed. J.A. Armstrong, J.M. Powell & A.J. Richards), pp. 55–66. Royal Botanic Gardens, Sydney, Sydney. [262, 263]

Turner, V. 1984 *Banksia* pollen as a source of protein in the diet of two Australian marsupials, *Cercartetus nanus* and *Tarsipes rostratus*. *Oikos* 43: 53–61. [263]

Tyre, A.J. & Addicott, J.F. 1993 Facultative non-mutualistc behaviour by an "obligate" mutualist: "cheating" by yucca moths. *Oecologia (Berlin)* 94: 173–175. [318, 320]

Upchurch, G.R. & Wolfe, J.A. 1987 Mid-Cretaceous to Early Tertiary vegetation and climate: evidence from fossil leaves and woods. *The Origins of Angiosperms and their Biological Consequences* (ed. E.M. Friis, W.G. Chaloner & P.R. Crane), pp. 75–105. Cambridge University Press, Cambridge. [379]

Vaillant, S. 1718 *Discours sur la Structure des Fleurs.* Leiden. [16]

van Damme, J.M.M. & van Delden, W. 1982 Gynodioecy in *Plantago lanceolata* L. I. Polymorphism for plasmon type. *Heredity* 49: 303–318. [345]

van Damme, J.M.M. & van Delden, W. 1984 Gynodioecy in *Plantago lanceolata* L. IV. Fitness components of sex types in different life cycle stages. *Evolution* 38: 1326–1336. [345, 346]

van der Meijden, E. 1989 Mechanisms in plant population control. *Toward a More Exact Ecology* (ed. P.J. Grubb & J.B. Whittaker), pp. 163–181. Blackwell Scientific, Oxford. [424]

van der Pijl, L. 1936 Fledermäuse und Blumen. *Flora, Jena* 131: 1–40. [244, 245, 247, 251, 255]

van der Pijl, L. 1937 Disharmony between Asiatic flower-birds and American bird-flowers. *Annales du Jardin Botanique de Buitenzorg* 48: 17–26. [233, 240]

van der Pijl, L. 1953 On the flower biology of some plants from Java, with general remarks on fly-traps. *Annales Bogorienses* 1: 77–99. [304, 308, 311]

van der Pijl, L. 1954 *Xylocopa* and flowers in the Tropics. *Proceedings Koninklijke Nederlandse Akademie van Wetenschappen, Ser. C* 57: 413–423, 541 - 562. [108, 139, 294]

van der Pijl, L. 1956 Remarks on pollination by bats in the genera *Freycinetia* etc. and on chiropterophily in general. *Acta Botanica Neerlandica* 5: 135–144. [251]

van der Pijl, L. 1961 Ecological aspects of flower evolution. II. Zoophilous flower classes. *Evolution* 15: 403–416. [143, 178]

van der Pijl, L. 1969 *Principles of Dispersal in Higher Plants.* Springer-Verlag, Berlin. [423]

van der Pijl, L. & Dodson, C.H. 1966 *Orchid*

Flowers: their Pollination and Evolution. University of Miami Press, Coral Gables. [213]

van Emden, H.F. 1963 Observations on the effect of flowers on the activity of parasitic Hymenoptera. *Entomologist's Monthly Magazine* 98: 265–270. [101]

van Leeuwen, B.H. 1981 The role of pollination in the population biology of the monocarpic species *Cirsium palustre* and *Cirsium vulgare*. *Oecologia* 51: 28–32. [422]

Varapoulos, A. 1979 Breeding systems in *Myosotis scorpioides* L. (Boraginaceae) I. Self-incompatibility. *Heredity* 42: 149–157. [329]

Varley, G.C., Gradwell, G.R. & Hassell, M.P. 1973 *Insect Population Ecology.* Blackwell, Oxford. [330]

Vaughton, G. 1991 Variation between years in pollen and nutrient limitation of fruit set in *Banksia spinulosa*. *Journal of Ecology* 79: 389–400. [413]

Verdcourt, B. 1948 Scarcity of *Rhingia campestris*, Mg. (Dipt. Syrphidae). *Entomologist's Record and Journal of Variation* 60: 108. [69]

Verlaine, L. 1932a L'instinct et l'intelligence chez les Hymenoptères. XVIII. L'odorat et la généralisation, le relatif et l'absolu chez les guêpes. *Bulletin Annuel de la Société Entomologique Belge* 72: 311–322. [125]

Verlaine, L. 1932b L'instinct et l'intelligence chez les Hymenoptères. XV. Les guêpes ont-elles un langage? *Mémoires de la Société Royale des Sciences de Liège, Sér. 3* 17(13): 1–16. [106, 126]

Verrall, G.H. 1909 *British Flies* 5. Gurney & Jackson, London. [63]

Vickery, M.L. & Vickery, B. 1981 *Secondary Plant metabolism.* Macmillan, London. [38]

Vink, W. 1970 The Winteraceae of the Old World. I. *Pseudowintera* and *Drimys* - morphology and taxonomy. *Blumea* 18: 225–354. [377, 378]

Vogel, S. 1954 *Blütenbiologische Typen als Elemente der Sippengliederung.* Fischer, Jena. [62, 92, 307]

Vogel, S. 1958 Fledermausblumen in Sudamerika. *Österreichische Botanische Zeitschrift* 104: 491–530. [245, 247, 248, 249, 253, 254]

Vogel, S. 1961 Die Bestäubung der Kesselfallen-Blüten von *Ceropegia*. *Beiträge zur Biologie der Pflanzen* 36: 159–237. [300, 301]

Vogel, S. 1963 Duftdrüsen im Dienste der Bestäubung: über Bau und Funktion der Osmophoren. *Abhandlungen. Mathematisch-natur-wissenschaftliche Klasse, Akademie der Wissenschaften und der Literatur, Mainz* 1962(10): 599–763. [305, 308]

Vogel, S. 1966 Scent organs of orchid flowers and their relation to insect pollination. *Proceedings of the Fifth World Orchid Conference* (ed. L. R. de Garmo), pp. 253–259. Fifth World Orchid Conference, Long Beach, California. [115]

Vogel, S. 1968-9 Chiropterophilie in der

neotropischen Flora, neue Mitteilungen I. *Flora, Abt. B* 157: 562–602; II. *ibid.* 158: 185–222; III. *ibid.* 158: 289–323. [244, 245, 248, 249, 254, 255]

Vogel, S. 1969 Flowers offering fatty oil instead of nectar. *Proceedings of XI International Botanical Congress, Seattle, Abstract*: 229. [43]

Vogel, S. 1972 Pollination von *Orchis papilionacea* L. in der Schwarmbahnen von *Eucera tuberculata* F. *Jahresberichte der Naturwissenschaftlichen Vereins in Wuppertal* 25: 67–74. [199]

Vogel, S. 1974 Ölblumen und ölsammelnde Bienen. *Tropische und Subtropische Pflanzenwelt* 7: 1 (281)–267 (545). [115]

Vogel, S. 1976a Zur *Ophrys*-Bestäubung auf Kreta. *Jahresberichte der Naturwissenschaftlichen Vereins in Wuppertal* 29: 131–139. [208]

Vogel, S. 1976b *Lysimachia*: Ölblumen der Holarktis. *Naturwissenschaften* 63: 44. [112, 115]

Vogel, S. 1978a Evolutionary shifts from reward to deception in pollen flowers. *The Pollination of Flowers by Insects* (ed. A.J. Richards), pp. 89–96. Academic Press, London. [295]

Vogel, S. 1978b Pilzmückenblumen als Pilzmimeten, erster Teil. *Flora* 167: 329–366. [295, 300, 305, 307]

Vogel, S. 1978c Pilzmückenblumen als Pilzmimeten, Fortsetzung und Schluss. *Flora* 167: 367–398. [295, 296, 300, 305, 307]

Vogel, S. 1984 The *Diascia* flower and its bee - an oil-based symbiosis in southern Africa. *Acta Botanica Neerlandica* 33: 509–518. [115]

Vogel, S. 1986 Ölblümen und ölsammelnde Bienen, zweite Folge, *Lysimachia* und *Macropis*. *Tropische und Subtropische Pflanzenwelt* 54: 1 (145)–168 (312). [112]

Vogel, S. 1990 Ölblümen und ölsammelnde Bienen, dritte Folge, *Momordica*, *Thladiantha* und die Ctenoplectridae. *Tropische und Subtropische Pflanzenwelt* 73: 1–186. [115]

Vöth, W. 1982 Die "ausgeborgten" Bestäuber von *Orchis pallens* L. *Orchidee [Hamburg]* 33: 196–203. [222]

Vöth, W. 1984 *Echinomyia magnicornis* Zett. Bestäuber von *Orchis ustulata* L. *Orchidee (Hamburg)* 35: 189–192. [73]

Vroege, P.W. & Stelleman, P. 1990 Insect and wind pollination in *Salix repens* L. and *Salix caprea* L. *Israel Journal of Botany* 39: 125–132. [278]

Wagner, H.O. 1946 Food and feeding habits of Mexican hummingbirds. *Wilson Bulletin* 58: 69–93. [239, 241]

Waite, S., Hopkins, N. & Hitchings, S. 1991 Levels of pollinia export, import and fruit set among plants of *Anacamptis pyramidalis*, *Dactylorhiza fuchsii*, and *Epipactis helleborine*. *Population Ecology of Terrestrial Orchids* (ed. T.C.E. Wells & J.H. Willems), pp. 103–110. SPB Academic Publishing, The Hague. [192, 199]

Waite, S. & Hutchings, M.J. 1991 The effects of different management regimes on the population dynamics of *Ophrys sphegodes*: analysis and description using matrix models. *Population Ecology of Terrestrial Orchids* (ed. T.C.E. Wells & J.H. Willems), pp. 161–175. SPB Academic Publishing, The Hague. [209]

Walker, J.W., Brenner, G.J. & Walker, A.G. 1983 Winteraceous pollen in the Lower Cretaceous of Israel: early evidence of a magnolialean angiosperm family. *Science* 220: 1273–1275. [376]

Walker, J.W. & Walker, A.G. 1984 Ultrastructure of Lower Cretaceous angiosperm pollen and the origin and early evolution of flowering plants. *Annals of the Missouri Botanical Garden* 71: 464–521. [375, 376]

Waller, D.M. 1984 Differences in fitness between seedlings derived from cleistogamous and chasmogamous flowers in *Impatiens capensis*. *Evolution* 38: 427–440. [335]

Wang, C.W., Perry, T.O. & Johnson, A.G. 1960 Pollen dispersal of Slash Pine (*Pinus elliottii* Engelm.) with special reference to seed orchard management. *Silvae Genetica* 9: 78–86. [270, 271]

Waser, N.M. 1983 Competition for pollination and floral character differences among sympatric plant species: a review of evidence. *Handbook of Experimental Pollination Biology* (ed. C.E. Jones & R.J. Little), pp. 277–293. Van Nostrand Reinhold, New York. [400, 417]

Waser, N.M. 1988 Comparative pollen and dye transfer by pollinators of *Delphinium nelsonii*. *Functional Ecology* 2: 41–48. [419]

Waser, N.M., Chittka, L., Price, M.V., Williams, N. & Ollerton, J. [1995] Generalisation in pollination systems, and why it matters. *Ecology* (submitted). [381]

Waser, N.M. & Price, M.V. 1983 Optimal and actual outcrossing in plants and the nature of plant-pollinator interaction. *Handbook of Experimental Pollination Biology* (ed. C.E. Jones & R.J. Little), pp. 341–359. Van Nostrand Reinhold, New York. [423]

Waser, N.M. & Price, M.V. 1989 Optimal outcrossing in *Ipomopsis aggregata*: seed set and offspring fitness. *Evolution* 43: 1097–1109. [423]

Waser, N.M. & Price, M.V. 1991 Reproductive costs of self-pollination in *Ipomopsis aggregata* (Polemoniaceae): are ovules usurped? *American Journal of Botany* 78: 1036–1043. [328]

Waser, N.M. & Price, M.V. 1994 Crossing-distance effects in *Delphinium nelsonii*: outbreeding and inbreeding depression in progeny fitness. *Evolution* 48: 842–852. [423]

Waser, N.M. & Real, L.A. 1979 Effective mutualism between sequentially flowering plant species. *Nature* 281: 670–672. [388, 403]

Watson, J.D. 1968 *The Double Helix*. Weidenfeld

& Nicolson, London. [21]

Watson, J.D. 1970 *Molecular Biology of the Gene,* ed. 2. Benjamin, New York. [21]

Watson, J.D. & Crick, F.H.C. 1953 Molecular structure of nucleic acids. A structure for deoxyribose nucleic acid. *Nature* 171: 737–738. [21]

Watts, L.E. 1958 Natural cross-pollination in lettuce, *Lactuca sativa* L. *Nature* 181: 1084. [177]

Webb, C.J. 1981 Test of a model predicting equilibrium frequencies of females in populations of gynodioecious angiosperms. *Heredity* 46: 397–405. [346]

Webb, C.J. & Lloyd, D.G. 1986 The avoidance of interference between the presentation of pollen and stigmas in angiosperms. II. Herkogamy. *New Zealand Journal of Botany* 24: 163–178. [338, 339]

Weiss, M.R. 1995 Floral colour change: a widespread functional convergence. *American Journal of Botany* 82: 167–185. [44]

Weller, S.G. & Denton, M.F. 1976 Cytogeographic evidence for the evolution of distyly from tristyly in the North American species of *Oxalis* section *Ionoxalis*. *American Journal of Botany* 63: 120–125. [329]

Weller, S.G. & Ornduff, R. 1977 Cryptic self-incompatibility in *Amsinckia grandiflora*. *Evolution* 31: 47–51. [329]

Wells, T.C.E. & Cox, R. 1991 Demographic and biological studies on *Ophrys apifera*: some results from a 10 year study. *Population Ecology of Terrestrial Orchids* (ed. T.C.E. Wells & J.H. Willems), pp. 47–61. SPB Academic Publishing, The Hague. [210]

Werfft, R. 1951 Über die Lebensdauer der Pollenkörner in der freien Atmosphäre. *Biologische Zentralblatt* 70: 354–367. [272]

Werth, E. 1956 *Bau und Leben der Blumen*. Enke, Stuttgart. [243]

Westerkamp, Chr. 1990 Bird flowers: hovering versus perching exploitation. *Botanica Acta* 103: 366–371. [227, 234]

Westerkamp, Chr. 1991 Honeybees are poor pollinators - why? *Plant Systematics and Evolution* 177: 71–75. [364]

Westrich, P. 1990 *Die Wildbienen Baden-Württembergs, ed. 2* 1 and 2. Ulmer, Stuttgart-Hohenheim. [110, 352]

Weymouth, R.D., Lasiewski, R.C. & Berger, A.J. 1964 The tongue apparatus of hummingbirds. *Acta Anatomica* 58: 252–270. [228]

White, A. & Sloane, B.L. 1937 *The Stapelieae* 3. Pasadena, California. [296]

Whitehead, D.R. 1969 Wind pollination in the angiosperms; evolutionary and environmental considerations. *Evolution* 23: 28–35. [265, 379]

Whitehead, D.R. 1983 Wind pollination: some

ecological and evolutionary perspectives. *Pollination Biology* (ed. L. Real), pp. 97–108. Academic Press, Orlando, Florida. [265]

Whitehead, V.B., Giliomee, J.H. & Rebelo, A.G. 1987 Insect pollination in the Cape flora. *A Preliminary Synthesis of Pollination Biology in the Cape Flora: South African National Science Programmes, report no. 141* (ed. A.G. Rebelo), pp. 52–82. CSIR, Pretoria. [52, 61, 62]

Whitehouse, H.L.K. 1950 Multiple-allelo-morph incompatibility of pollen and style in the evolution of the angiosperms. *Annals of Botany, New Series* 14: 198–216. [329, 375]

Whittaker, R.H. 1970 *Communities and Ecosystems*, ed. 2. Macmillan, New York. [384]

Widen, B. [1982] *Reproductive Biology in the* Helianthemum oelandicum *(Cistaceae) complex on Öland, Sweden*. PhD thesis, University of Lund. [386]

Wiebes, J.T. 1963 Taxonomy and host preferences of Indo-Australian figwasps of the genus *Ceratosolen* (Agaonidae). *Tijdschrift voor Entomologie* 106: 1–112. [314, 316]

Wiens, D. 1985 Secrets of a cryptic flower. *Natural History* 94: 70–77. [259, 260]

Wiens, D. & Rourke, J.P. 1978 Rodent pollination in southern African *Protea* spp. *Nature* 276: 71–73. [259]

Wiens, D., Rourke, J.P., Casper, B.B., Rickert, E.A., Lapine, T.R., Peterson, C.J. & Channing, A. 1983 Nonflying mammal pollination of southern African Proteas: a non-coevolved system. *Annals of the Missouri Botanical Garden* 70: 1–31. [259]

Wiklund, C., Eriksson, T. & Lundberg, H. 1979 The wood white butterfly, *Leptidea sinapis*, and its nectar plants: a case of mutualism or parasitism? *Oikos* 33: 358–362. [416]

Willemstein, S.C. 1987 *An Evolutionary Basis for Pollination Ecology*. Brill/Leiden University Press, Leiden. [54, 370, 378, 381]

Williams, I.H., Corbet, S.A. & Osborne, J.L. 1991 Beekeeping, wild bees and pollination in the European Community. *Bee World* 72: 170–180. [363]

Williams, N.H. 1982 The biology of orchids and euglossine bees. *Orchid Biology: Reviews and Perspectives* 2: 120–171. [216]

Williams, P.H. 1988 Habitat use by bumble bees (*Bombus* spp.). *Ecological Entomology* 13: 223–237. [387]

Willis, J.C. & Burkill, I.H. 1895 - 1908 Flowers and insects in Great Britain. I. *Annals of Botany* 9: 227–273 (1895); II. *ibid.* 17: 313–349 (1903a); III. *ibid.* 17: 539–570 (1903b); IV. *ibid.* 22: 603–649 (1908). [50, 58, 60, 61, 69, 71, 72, 73, 74, 75, 77, 84, 85, 86, 99, 102, 104, 106, 107]

Willis, J.C. & Burkill, I.H. 1903a Flowers and insects in Great Britain. II. *Annals of Botany* 17:

313–349. [169]

Willson, M.F. 1983 *Plant Reproductive Ecology*. J. Wiley, Chichester. [341, 347]

Willson, M.F. 1994 Sexual selection in plants: perspective and overview. *American Naturalist (supplement)* 144: S13–S39. [221]

Willson, M.F. & Ågren, J. 1989 Differential floral rewards and pollination by deceit in unisexual flowers. *Oikos* 55: 23–29. [341]

Wilson, J.W., III 1974 Analytical zoogeography of North American mammals. *Evolution* 28: 124–140. [254]

Wolf, L.L. & Hainsworth, F.R. 1971 Time and energy budgets of territiorial hummingbirds. *Ecology* 52: 980–988. [232]

Wolf, L.L., Hainsworth, F.R. & Stiles, F.G. 1972 Energetics of foraging: rate and efficiency of nectar extraction by hummingbirds. *Science* 176: 1351–1352. [231, 233]

Wolf, L.L., Hainsworth, F.R. & Gill, F.B. 1975 Foraging efficiencies and time budgets in nectar-feeding birds. *Ecology* 56: 117–128. [227, 232, 233]

Wolfe, L.M. 1993 Inbreeding depression in *Hydrophyllum appendiculatum*: role of maternal effects, crowding and parental mating history. *Evolution* 47: 374–386. [335]

Wolfe, L.M. & Barrett, S.C.H. 1989 Patterns of pollen removal and deposition in tristylous *Pontederia cordata* L. (Pontederiaceae). *Biological Journal of the Linnean Society* 36: 317–329. [326]

Wolff, T. 1961 Pollination of the Fly Ophrys, *Ophrys insectifera* L. in Allendelille fredskov, Denmark. *Oikos* 2: 20–59. [207]

Wollin, H. 1975 Kring nornans biologi. *Fauna och Flora* 3: 89–98. [222]

Wright, J.W. 1953 Pollen-dispersion studies: some practical applications. *Journal of Forestry* 51: 114–118. [270, 271]

Wyatt, R. 1981 Ant-pollination of the granite outcrop endemic *Diamorpha smallii* (Crassulaceae). *American Journal of Botany* 68: 1212–1217. [107]

Wyatt, R. 1982 Inflorescence architecture: how flower number, arrangement, and phenology affect pollination and fruit-set. *American Journal of Botany* 69: 585–594. [28]

Wyatt, R. 1983 Pollinator-plant interactions and the evolution of breeding systems. *Pollination Biology* (ed. L.Real), pp. 51–95. Academic Press, London. [399]

Wyatt, R. 1986 Ecology and evolution of self-pollination in *Arenaria uniflora* (Caryophyllaceae). *Journal of Ecology* 74: 403–418. [399]

Wyatt, R. 1988 Phylogenetic aspects of the evolution of self-pollination. *Plant Evolutionary*

Biology (ed. L.D. Gottlieb & S.K. Jain), pp. 109–131. Chapman & Hall, New York. [332, 333, 334] **Wyatt, R. (ed.)** 1992 *Ecology and Evolution of Plant Reproduction*. Chapman & Hall, New York. [221]

Yarrow, I.H.H. 1945 Collecting bees and wasps, in The Hymenopterist's Handbook. *Amateur Entomologist* 7: 55–81. [106]

Yeoboah Gyan, K. & Woodell, S.R.J. 1987 Flowering phenology, flower colour and mode of reproduction of *Prunus spinosa* L. (blackthorn); *Crataegus monogyna* Jacq. (hawthorn); *Rosa canina* L. (dog rose); and *Rubus fruticosus* L. (bramble) in Oxfordshire, England. *Functional Ecology* 1: 261–268. [386]

Yeo, P.F. 1966 The breeding relationships of some European *Euphrasiae*. *Watsonia* 6: 216–245. [333]

Yeo, P.F. 1972 Miscellaneous notes on pollination and pollinators. *Journal of Natural History* 6: 667–686. [139]

Yeo, P.F. 1993 *Secondary Pollen Presentation: Form, Function and Evolution* (Plant Systematics and Evolution Supplementum 6). Springer, Wien. [151, 176]

Young, A.M., Schaller, M. & Strand, M. 1984 Floral nectaries and trichomes in relation to pollination in some species of *Theobroma* and *Herrania* (Sterculiaceae). *American Journal of Botany* 71: 466–480. [60]

Young, H.J. & Stanton, M.L. 1990 Influence of environmental quality on pollen competitive ability. *Science* 248: 1631–1633. [420]

Zahavi, A., Eisikowitch, D., Zahavi, A.K. & Cohen, A. 1983 A new approach to flower constancy in honey bees. *Veme Symposium International sur la Pollinisation, Versailles, 27 - 30 septembre 1983* (ed. INRA), pp. 89–95. [137]

Zavada, M.S. & Taylor, T.N. 1986 The role of self-incompatibility and sexual selection in the gymnosperm-angiosperm transition: a hypothesis. *American Naturalist* 128: 538–550. [330]

Zimmerman, J.K. & Aide, T.M. 1989 Patterns of fruit production in a neotropical orchid: pollinator vs. resource limitation. *American Journal of Botany* 76: 67–73. [414]

Zimmerman, M. 1988 Nectar production, flowering phenology, and strategies for pollination. *Plant Reproductive Ecology* (ed. J. Lovett Doust & L. Lovett Doust), pp. 157–178. Oxford University Press, Oxford. [414]

Zucchi, H. 1989 Nektarnutzung durch Blaumeisen *Parus caeruleus*. *Vogelwelt* 110: 236–237. [226]

Index

Numbers in **bold** refer to main entries; those in *italics* refer to illustrations.

Carduus, 174, 388, 405
Carex, 86, 281, 322, 343
Carex binervis, 70
Carex demissa, *281*
Carex dioica, 343
Caribbean, 293
carnation, 83
Carnegiea, 249
Carollia, 251
Carollia perspicillata, 250, 253
carotenoids, 34-37
carpel, 24-6, 368, 374-5, *377*
carpet moth, common marbled, 85
carpet moth, striped twin-spot, 85
Carpinus, 266, *267*
Carpinus betulus, 343
carrion flies, 37, 41
carrot, 36, 173
carrot, wild, sea, *49*, *104*, 358
carrot family, see Apiaceae
Caryocaraceae, 249
Caryophyllaceae, 27, 34, 67-8, 85,
    146, 322, 325, 337, 379
Caryophyllales, 34
Castanea, 266, *267*
Castanea sativa, 276
Castilleja, 388
castor-oil plant, 13
Catasetum, *218*, 221
Catasetum macrocarpum, 219
Catasetum tridentatum, 219
catchfly, 85
catchfly, night-scented, 339
catchfly, nottingham, 182, *183*, 339
catchfly, Spanish, 58, 60, 339
catkin, 265, 275-9, 344, 378-9
catsear, common, 51, 68, 70
caudicles, 197, 200, 220
cauliflory, 245
Cayaponia, *248*
Caytoniales, 373-4
Cebus, 256
cedar, 274
cedar, Atlas, 270
Cedrus, 274
Cedrus atlantica, 270
Ceiba, 254
Ceiba acuminata, 403
Ceiba pentandra, 247, 255-6, 259
Celaena haworthii, 85
celandine, greater, 338
celandine, lesser, 25, 387
Celerio livornica, see *Hyles livornica*
Celonites abbreviatus, 105
Centaurea (knapweed), 61, 67, 85,
    112, 174, 178, 388, *399*, 402
Centaurea cyanus, 34
Centaurea montana, 177
Centaurea nigra, 95, *399*, *402*, 403,
    Plate 2e
Centaurea scabiosa, 95, *177*, 402-3
Centaurium erythraea, 146
centaury, 146
Centranthus ruber, 85, 182, *Plate 3d*
Centris, 215, 416
Centrospermae, 34
Cephalanthera, 190, 192, 222
Cephalanthera damasonium, 192, 223
Cephalanthera longifolia, 99, 192, 223
Cephalanthera rubra, 192, 222
Cephidae, 100
Cerambycidae, 54, *55*

Cerapteryx graminis, 85
Cerastium arvense, 325
Cerastium litigiosum, 63
Ceratina, 189
Ceratophyllaceae, 376, 379
Ceratophyllum, 285, 289, 344, 379
Ceratophyllum demersum, *290*
Ceratopogon, 301
Ceratopogonidae, 304
Ceratosolen arabicus, *313*, 314-5
Ceratosolen hewitti, 314
Ceratothrips ericae, 51
Cercartetus caudatus, 262
Cercartetus concinnus, 262-3
Cercartetus nanus, 257, 262
cereals, 34, 352
Ceropegia, 300, *301*, 307-8
Ceropegia ampliata, 300
Ceropegia elegans, 300
Ceropegia euracme, 300
Ceropegia haygarthii, 300
Ceropegia nilotica, 300
Ceropegia radicans, 300
Ceropegia robynsiana, 300
Ceropegia sandersonii, 300
Ceropegia stapeliiformis, 300
Ceropegia woodii, 300-2
Cetonia aurea, 54
Chaenomeles speciosa, 25
chafer, garden, *52*, 54
chafer, rose, 54
chalaza, 32, 34
chalcid wasps, 102
chalk grassland, 400, 405
Chamaelirium luteum, 423
Chamaerops humilis, 16
Charaxes, 97
Charaxes jasius, 93
charlock, 172, 325, 336
Cheiranthus cheiri, 36, 83, *112*, *143*,
    145
Cheirostemon platanoides, *248*
Chelidonium majus, 338
Chelone glabra, 139
Chelostoma, 112, 192
Chenopodiaceae, 36, 265-6, *267*,
    283
cherry, *26*, 145
cherry, wild, 132
chestnut, horse, 266, *267*
chestnut, sweet, 266, *267*, 276, 285
Chiastochaeta, 312
Chile, 385, 388-90
Chiloglottis, 212
China, 164, 370
chiropterophily, see bat-pollination
chives, 335
Chloranthaceae, 375-8, *376*, 379
Chloranthus, 375
Chloroclysta citrata, 85
Chloromyia formosa, *61*
Chloropidae, 72
Chlorops, 72
chromosomes, 21, 29, 323, in
    gynodioecy, 345
Chrysididae, 102
Chrysobalanaceae, 247, 249, 256
Chrysomelidae, 54-6, *55*
Chrysosplenium, 58
Cichoriae, 266
Cicuta virosa, 77
Cimbex, 100

Cimbicidae, 99
Cinnyris, 226
Cinnyris niassae, 226
cinquefoil, 348
cinquefoil, creeping, 131-2
cinquefoil, marsh, 60
cinquefoil, shrubby, 27, *72*
Circaea lutetiana, *68*, 69
Cirrhopetalum, 305
Cirrhopetalum ornatissimum, 307
Cirsium, 61, 67, 85, 174, 388, 405
Cirsium acaule, 177
Cirsium arvense, 103, 340
Cirsium palustre, 422
Cirsium vulgare, 422
Cistus, 40, 179
Cistus albidus, 389
Citrus, 37-8, 322, 348
clary, 164
clary, meadow, 163, *164*
Clavatipollenites, 375-6
cleistogamy, 44, 280
climax forest, 266
clinandrium, 190
clonal growth, 342, 347-9, 395, 413
cloudberry, 341
clouds, effect on pollen, 272
clover, 27, 28, 155-7, 172
clover, alpine, *Plate 1a*
clover, red, 67, 139, *156*, 357
clover, white, 42, *113*, 114
cloves, 37, 38
clubmosses, 365-8, 392
Clusia, 43
Clusiaceae, 43
coal, 369
cob and papillate stigmas, *327*
Cobaea scandens, *248*, 250
cockchafer, 54
cockroach, *49*, 50
cocksfoot, 70, 270
cocoa, see *Theobroma cacao*
coconut, 413
Coelioxys, 122
Coeloglossum viride, *204*
Coelonia solani, 214
Coenonympha pamphilus, 81
Coenotephria salicata, 85
Colchicum, 152
Coleoptera, see beetles
Collembola, 50
Colletes, 111-2, 118
Colletes daviesanus, 114, *Plate 2b*
Colletidae,111, 118, 390
Colocasia antiquorum, 305
colonisation, 335, 346, 348-9, 395
colour preference, hummingbirds,
    228-9
colour vision, in insects, *49*, 56-8,
    70, 79, 87-97, **130-3**, *131*, 141
coltsfoot, 63, *76*, 177
columbine (*Aquilegia* spp.), 152,
    171, 229, 239, 335, 418
Columnea florida, see *Dalbergaria florida*
Columnea microphylla, *237*, 238
Combretaceae, 256
Combretum, 242
Combretum fruticosum, 256
comfrey, 146, *172*, *Plate 2d*
Commelinaceae, 294, 324
communication, among bees and
    wasps, 126, 133, 136, 141